HBJ
Fundamentals of Mathematics

Linda Dritsas

Wilmer L. Jones

Marian L. Rasmussen

HBJ Harcourt Brace Jovanovich, Publishers

Orlando   San Diego   Chicago   Dallas

About the Authors

Linda Dritsas
Coordinator/Secondary Mathematics
Fresno Unified School District
Fresno, California

Marian L. Rasmussen
Secondary Mathematics Teacher
Sonoma Valley Unified School District
Sonoma, California

Dr. Wilmer L. Jones
Headmaster/Teacher
Baltimore Lutheran Middle School
Formerly Coordinator of Mathematics
Baltimore City Public Schools
Baltimore, Maryland

Editorial Advisors

Cheryl McCormick-Baker
Mathematics Teacher
Akron Public Schools
Akron, Ohio

Dr. Mary Ann DuPont
Formerly Secondary Mathematics
  Curriculum Specialist
Palm Beach County
West Palm Beach, Florida

Hilde Howden
District Mathematics Coordinator
Albuquerque Public Schools
Albuquerque, New Mexico

Mary E. Shepherd Lester
Director of Mathematics
Dallas Independent School District
Dallas, Texas

Maria Parker, Ed.D.
Chairperson, Department of Mathematics
Vailsburg High School
Newark, New Jersey

ISBN 0-15-353001-4

CONTENTS

Using Whole Numbers

In 1903, the Wright brothers' first airplane flight covered 120 feet and lasted 12 seconds.

- Would you multiply 120 and 12 or divide 120 by 12 to find how many feet the plane traveled in one second?

- At that speed, how far could the plane travel in one minute (60 seconds)?

- At that speed, how far could the plane travel in one hour (60 minutes)?

- At that speed, <u>about</u> how many miles (5280 feet = 1 mile) would the plane travel in one hour?

Estimating with Large Numbers

Your social studies class is reading about the United States. You read that Arizona, New Mexico, Oklahoma, and Texas are the southwestern states. This map shows the population of these states in the 1980 census.

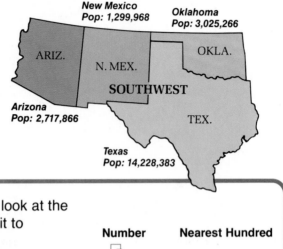

New Mexico
Pop: 1,299,968

Oklahoma
Pop: 3,025,266

ARIZ.

N. MEX.

OKLA.

SOUTHWEST

Arizona
Pop: 2,717,866

TEX.

Texas
Pop: 14,228,383

1. How would you estimate the total population of the southwestern states?

To round whole numbers, look at the digit to the right of the digit to be rounded.

If it is 5 or greater, round up.

If it is less than 5, round down.

Number		Nearest Hundred
875	⟶	900
719	⟶	700

EXAMPLE 1   Estimate the total population of the southwestern states.

a. To the nearest thousand   **b.** To the nearest million

Think: To estimate, first round to the nearest specified place. Then add.

		a. Nearest thousand	b. Nearest million
Arizona:	2,717,866 ⟶	2,718,000 ⟶	3,000,000
New Mexico:	1,299,968 ⟶	1,300,000 ⟶	1,000,000
Oklahoma:	3,025,266 ⟶	3,025,000 ⟶	3,000,000
Texas:	14,228,383 ⟶	+ 14,228,000 ⟶	+ 14,000,000
		21,271,000	**21,000,000**

Add mentally.

2. Which estimate, **a** or **b,** is closer to the actual sum? Why?

3. Which sum, **a** or **b,** is easier to compute? Why?

When performing computations, estimating first will help you to know whether an answer is reasonable. When estimating, round to a place that makes it convenient to use mental computation.

EXAMPLE 2

According to the last census, how many more people were living in Texas than in Oklahoma?

1. Estimate the difference.
2. Compute the actual difference.

$$\begin{array}{r} 14{,}000{,}000 \\ -\ \ 3{,}000{,}000 \\ \hline 11{,}000{,}000 \end{array}$$
◀ **Subtract mentally.**

$$\begin{array}{r} 14{,}228{,}383 \\ -\ \ 3{,}025{,}266 \\ \hline 11{,}203{,}117 \end{array}$$

3. Compare. Since **11,203,117** is close to the estimate, 11,000,000, the answer is reasonable.

EXERCISES

Round each number to the nearest ten and to the nearest hundred.

1. 541 **2.** 806 **3.** 1452 **4.** 20,896 **5.** 97,900

Round to the nearest thousand and to the nearest ten thousand.

6. 57,860 **7.** 89,294 **8.** 100,098 **9.** 28,293 **10.** 206,922

Estimate. Find the actual sum or difference. Compare.

11. 3969 + 8592 **12.** 7806 − 1922 **13.** 816 − 593

14. 63,472 + 80,002 **15.** 61,005 − 22,876 **16.** 9560 + 8111

This table shows the area in square miles of each of the southwestern states. Refer to this table for Exercises 17–21.

State	Area (square miles)
Arizona	113,909
New Mexico	121,666
Oklahoma	69,919
Texas	267,339

·18. Which state has an area that is about 145,000 square miles less than that of Texas?

19. How much greater than the area of Oklahoma is the area of New Mexico?

20. Estimate (nearest hundred thousand) the total area of the southwestern states.

21. Estimate (nearest hundred thousand) the combined area of Texas and Arizona.

Tell whether you would use exact numbers or estimates. Give a reason for each answer.

22. You are writing a social studies report about your state's population growth.

23. You are reporting on the number of students who voted in a student council election.

Average

Clare, Bob, and Neil are carrying a stack of American history books to the book room. Clare has 3 books, Bob has 7 books, and Neil has 8.

1. How can the three students "even off" the number of books each is carrying?

2. How many books will each student be carrying when the numbers are "evened off?"

The number 6 is the arithmetic **average** (or **mean**) of 3, 7, and 8. To compute the average number of books for the three students, you add the number of books. Then you divide by the number of persons carrying the books.

This table shows the largest cities in the United States in 1830 and in 1980.

City	*Population in 1830	City	*Population in 1980
New York	215	New York	7,072
Philadelphia	161	Chicago	3,005
Boston	86	Los Angeles	2,967
Baltimore	81	Philadelphia	1,688
New Orleans	46	Houston	1,595

*In thousands

3. In 1980, was the population of Los Angeles 2,967 or 2,967,000?

EXAMPLE   What was the average population of the three largest cities in 1830?

1 Find the sum of the populations of New York, Philadelphia, and Boston in 1830.

$$\begin{array}{r} 215 \\ 161 \\ +\ 86 \\ \hline 462 \end{array}$$

◄ **The sum is 462,000.**

2 Divide by the number of cities.

$$\begin{array}{r} 154 \\ 3\overline{)462} \end{array}$$ ◄─── **Average: 154,000**

The average population of the three largest cities was **154,000**.

EXERCISES

Find the average.

1.
Number of Pages in Ten Books				
335	465	303	289	399
385	561	447	467	479

2.
Price for Six Cars		
$8445	$7665	$12,580
$13,888	$15,650	$20,480

3.
Passengers on Seven Buses			
42	46	44	40
38	48	43	

4.
Grades on a Math Quiz				
80	74	82	77	90
92	85	86	78	86

Complete.

5. The average of 0 and 100 is __?__ .

6. The average of 49 and 51 is __?__ .

7. The average of 3 and 97 is __?__ .

8. The average of 46, 72, 28, 50, and 54 is __?__ .

For Exercises 9–10, refer to your answers for Exercises 5–8.

9. Is the average of a set of numbers always equal to one of the numbers? Explain.

10. Is the average of a set of numbers always close to all of the numbers? Explain.

Nine students in a social studies class were part of an experiment. Each student was asked to name the thirteen original American colonies.

Student	Time (Seconds)
Emily Coalter	39
Dave Allan	45
Tim Lance	46
Teresa Stuart	43
Meg De Carlo	39
Kim Sun	48
Jose Rivera	44
Tony Arico	42
Lucy Ames	41

11. What was the average time for the nine students?

12. How many students took longer than the average time?

13. How many students took less than the average time?

14. How many students took the same amount of time as the average?

Although there is no proof, legend has it that the Philadelphia seamstress Betsy Ross made the nation's first flag.

For Exercises 15–16, refer to the table on page 4.

15. Estimate to the nearest ten thousand the average population of Philadelphia, Boston, Baltimore, and New Orleans in 1830.

16. Find the actual average population of the cities in Exercise 15. How close is the actual average to the estimate?

Math and Travel

Travel agents assist with travel plans, make transportation and hotel reservations, and arrange for tours. **Time zones** are involved in calculating the number of hours and minutes of flights as well as hours of arrival and departure. This map shows four of the standard time zones in the United States.

Pacific Time 4:00 PM   **Mountain Time** 5:00 PM   **Central Time** 6:00 PM   **Eastern Time** 7:00 PM

1. When it is 9 P.M. in Chicago, what time is it in Cleveland?

2. When it is 7 A.M. in Denver, what time is it in Seattle?

> **A.** Traveling from east to west, **subtract** one hour for each time zone.
>
> **B.** Traveling from west to east, **add** one hour for each time zone.

EXAMPLE

A plane leaves Denver at 2:00 P.M. (Mountain time). It arrives in Atlanta three hours later. What time (Eastern time) does it arrive?

1. In what direction does the plane fly? ⟶ West to east

2. How many time zones does it cross? ⟶ MT CT ET → 1 2 **Two zones**

3. What is the hour of departure in Eastern time?

   2:00 P.M. + 2 hours = **4:00 P.M.**

4. What is the hour of arrival in Eastern time?

   4:00 P.M. + 3 hours (flight time) = **7:00 P.M.**

EXERCISES

List the number of time zones through which you would travel for each flight.

1. From Denver to Chicago

2. From Miami to Denver

3. From Cleveland to Seattle

4. From Houston to San Francisco

Give the Mountain time for each given standard time.

5. 3:00 A.M. Pacific time

6. 9 P.M. Eastern time

7. 12 noon Central time

8. Midnight Central time

Solve each problem. Refer to the map of time zones on page 6.

9. A plane leaves Chicago at 10:00 A.M. (Central time). The flight to Newark takes 2 hours. What is the arrival time (Eastern time)?

10. Melissa's flight left Los Angeles at 11:00 A.M. The flight to Boston took 6 hours. At what time (Eastern time) did the plane arrive in Boston?

11. A flight leaves Atlanta for Los Angeles at 10:50 A.M. (Eastern time). The flight takes 4 hours 20 minutes. What time will it be in Los Angeles (Pacific time) when the plane arrives?

12. Gary's flight left Houston at 2:00 P.M. (Central time). The flight took 2 hours 30 minutes. What time will it be in Chicago when the plane lands?

13. A plane leaves Seattle at 10:45 A.M. (Pacific time). It arrives in Boston at 6:30 P.M. (Eastern time). How long did the trip take?

14. Sandra's flight left Wichita at 9:30 A.M. (Central time). The plane arrived in Cleveland at 12:35 P.M. (Eastern time). How long did the trip take?

15. Philip flew out of Minneapolis at 9:45 P.M. (Central time) and arrived in Portland, Oregon 2 hours 30 minutes later. At what time (Pacific time) did he arrive?

The Wright brothers' first flight covered 120 feet and took 12 seconds.

Order of Operations

The coastline of Oregon is about twice as long as the Washington coastline. The coastline of Washington is about 150 miles long.

Here's how Nancy and Joel found the total length of the two coastlines. Whose answer is correct?

WASHINGTON

OREGON

Joel's Computation

Nancy's Computation

Length
150 + 2 × 150
152 × 150
~~22,800 miles~~

Length
150 + 2 × 150
150 + 300
450 miles

The answers are not the same.

1. Did Joel add or multiply first?

2. Did Nancy add or multiply first?

Only one answer can be correct. When computations involve more than one operation you need to know which operation to do first.

> **Rule 1:** First do the operations within parentheses or the computations above or below a division bar.
>
> **Rule 2:** Then do multiplication and division from left to right. Multiply or divide <u>before</u> you add or subtract.
>
> **Rule 3:** Finally, do addition and subtraction from left to right.

Examples	Order	Computations
a. $36 \times 3 \div 2$	☐1 Multiply first. ☐2 Divide.	$36 \times 3 \div 2 = 108 \div 2$ $= 54$
b. $19 + 8 \div 4$	☐1 Divide first. ☐2 Add.	$19 + 8 \div 4 = 19 + 2$ $= 21$
c. $\frac{21 - 9}{4}$	☐1 First find the difference above the fraction bar. ☐2 Divide.	$\frac{21 - 9}{4} = \frac{12}{4}$ $= 3$
d. $6 \times (9 - 2)$	☐1 Do the work in parentheses. ☐2 Multiply.	$6 \times (9 - 2) = 6 \times 7$ $= 42$

EXERCISES

Which operation would you perform first? Choose your answers from the box at the right.

1. $36 \div 9 \times 2$

2. $(7 + 3) \times 2$

3. $9 - 4 + 3$

4. $17 - 12 \div 4$

5. $\frac{2 \times 8}{4}$

6. $16 \div (9 - 5)$

7. Solve Exercise 1.

8. Solve Exercise 2.

9. Solve Exercise 3.

10. Solve Exercise 4.

11. Solve Exercise 5.

12. Solve Exercise 6.

Find each answer.

13. $9 - (4 - 3)$

14. $12 + 8 \div 2$

15. $12 - 8 \div 4$

16. $20 \div 4 + 1$

17. $18 \times (2 + 3)$

18. $(3 + 8) \times 9$

19. $24 - 3 \times 6$

20. $15 \times 2 + 5 \times 6$

21. $3 \times 4 + 5 \times 2$

22. $16 - (5 + 2)$

23. $16 - 5 + 2$

24. $10 + 53 - 60$

25. $33 - \frac{7 + 5}{2}$

26. $21 + \frac{6 \times 9}{3}$

27. $\frac{48}{8 + 4} + 12$

28. $(12 - 2) \div 5$

29. $32 \div 4 \times 7$

30. $14 + 3 \times 12$

31. $42 - 3 \times 4$

32. $(5 + 3) \times 2$

33. $(50 - 4) + 2 \times 9$

34. $25 - 7 \times 3 + 10$

35. $4 \times 12 - 5 \times 2$

36. $32 - 4 \times 2 + 10$

37. Which is greater?

  a. $2 \times 8 - 5$

  b. $2 \times (8 - 5)$

38. Which is less?

  a. $24 \div 8 - 2$

  b. $24 \div (8 - 2)$

Insert parentheses to make each statement true.

39. $3 + 17 - 8 \times 2 = 21$

40. $3 + 7 \times 2 - 8 = 12$

41. $7 \times 8 - 3 + 2 = 51$

42. $54 \div 6 \times 14 \div 7 = 18$

43. Joan has $28. She bought two T-shirts at $8 each. Which describes the amount of money she has left?

  a. $28 + $2 + $8

  b. $28 + (2 \times $8)$

  c. $28 - (2 \times $8)$

  d. $28 - $2 - $8

44. Paul bought three kites at $7 each and four puzzles at $3 each. Which describes how much he spent?

  a. $(3 \times $7) - (4 \times $3)$

  b. $(3 \times $7) + (4 \times $3)$

  c. $3 + $7 + $4 + $3

  d. $(3 \times $7) + $4 + 3

Strategy: GUESS AND CHECK

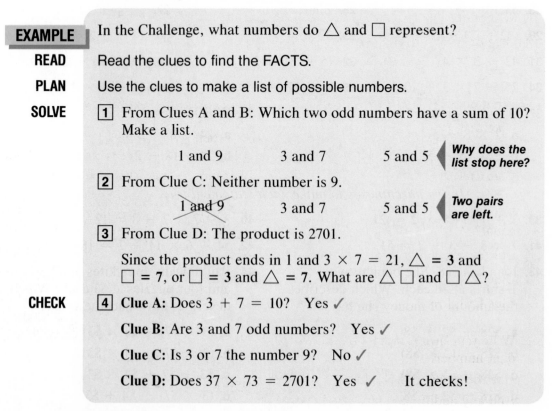

CHALLENGE

What are $\triangle \square$ and $\square \triangle$?

Clues

A. $\square + \triangle = 10$

B. $\square$ and $\triangle$ represent odd numbers.

C. Neither $\square$ nor $\triangle$ is 9.

$$\begin{array}{r} \triangle \square \\ \times\, \square \triangle \\ \hline 2\ 7\ 0\ 1 \end{array}$$ ◀ Clue D

Even whole numbers end in 0, 2, 4, 6, or 8.
All others are **odd numbers.**

---

1. Name the odd numbers from 30 to 40.

2. How many two-digit numbers can you write using both 5 and 6? What are they?

EXAMPLE   In the Challenge, what numbers do $\triangle$ and $\square$ represent?

READ   Read the clues to find the FACTS.

PLAN   Use the clues to make a list of possible numbers.

SOLVE   **1** From Clues A and B: Which two odd numbers have a sum of 10? Make a list.

       1 and 9   3 and 7   5 and 5 ◀ *Why does the list stop here?*

     2 From Clue C: Neither number is 9.

       ~~1 and 9~~   3 and 7   5 and 5 ◀ *Two pairs are left.*

     3 From Clue D: The product is 2701.

       Since the product ends in 1 and $3 \times 7 = 21$, $\triangle = \mathbf{3}$ and $\square = \mathbf{7}$, or $\square = \mathbf{3}$ and $\triangle = \mathbf{7}$. What are $\triangle \square$ and $\square \triangle$?

CHECK   **4** **Clue A:** Does $3 + 7 = 10$?   Yes ✓

     Clue B: Are 3 and 7 odd numbers?   Yes ✓

     Clue C: Is 3 or 7 the number 9?   No ✓

     Clue D: Does $37 \times 73 = 2701$?   Yes ✓   It checks!

EXERCISES

For Exercises 1–2, find $\square \triangle$ *and* $\triangle \square$.

1. **Clue A:** $\square + \triangle = 9$

   Clue B: $\square$ is an even number; $\triangle$ is an odd number.

   Clue C: $\square$ is greater than $\triangle$.

   Clue D

2. **Clue A:** $\triangle + \square = 9$

   Clue B: $\triangle$ is less than $\square$.

   Clue C: $\triangle$ is an even number; $\square$ is an odd number.

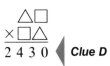

   Clue D

3. What does $\square \times \triangle \times \bigcirc$ equal?

   Clue A: $\square + \triangle + \bigcirc = 10$

   Clue B: $\square + \triangle = 5$

   Clue C: $\triangle \times \bigcirc = 15$

   HINT: **1** Start with Clue B. Possible values of $\square$: 0 and _?_
   1 and _?_

   2 Use Clues B and A to find $\bigcirc$.

4. What does $\square \times \square \times \square$ equal?

   Clue A: $\square + \triangle + \bigcirc = 10$

   Clue B: $\square = 2 \times \triangle$

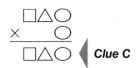

   Clue C

5. What does $\square \times \bigcirc \times \triangle$ equal?

   Clue A: $\square + \bigcirc + \triangle = 15$

   Clue B: $\bigcirc + 1 = \triangle$

   Clue C: $\bigcirc - 1 = \square$

   HINT: Make a guess for $\bigcirc$.
   1 Let $\bigcirc = 4$.
   Then $\triangle = 5$ and $\square = 3$.
   2 Check: Does $3 + 4 + 5 = 15$? No
   3 Guess again. Check each guess.

6. What does $\square \times \square + \bigcirc - \triangle$ equal?

   Clue A: $\triangle + \square + \bigcirc = 21$

   Clue B: $\square + 2 = \bigcirc$

   Clue C: $\square - 2 = \triangle$

7. **Write your own problem** about a two-digit number.

   Use the clues at the right.

   Add any additional clues you need.

   Clue A: $\square$ is less than 10.

   Clue B: $\square = 2 \times \triangle$

Mid-Chapter Review

Estimate. Find the actual sum or difference. Compare.   (Pages 2–3)

1. 915 + 489

2. 9106 − 2876

3. 3510 + 7214

4. 28,795 + 50,018

5. 45,111 − 39,909

6. 175,678 − 48,872

Find the average.   (Pages 4–5)

7.

Grades on a Science Test				
71	85	92	95	88
80	96	79	84	90

8.

Miles per Gallon for Twelve Cars					
15	21	24	18	20	30
32	28	26	35	25	14

Find each answer.   (Pages 8–9)

9. 8 + (5 − 2)

10. 9 ÷ 3 × 6

11. (4 + 7) × 5

12. 13 + $\frac{18 - 4}{2}$

13. (25 − 16) + 3 × 7

14. 16 + 5 × 2 − 15

Find □△ and △□.   (Pages 10–11)

15. Clue A: □ + △ = 13

Clue B: □ is an odd number; △ is an even number.

Clue C: □ is less than △.

Clue D

16. In North Carolina, the area of Cape Hatteras is 30,319 acres and the area of Cape Lookout is 28,415 acres. Estimate their combined area. (Pages 2–3)

17. Toby bought a pair of tennis shorts for $14 and two shirts for $15 each. Which describes how much he spent?   (Pages 8–9)
a. ($14 × 2) − $15
b. $14 + (2 × $15)

MAINTENANCE

The table at the right shows the heights of several waterfalls in the United States.

18. How much higher than Minnehaha Falls is Twin Falls?

19. About how many times as high as Lower Yosemite Falls is Middle Yosemite Falls?

Waterfall	Height (in feet)
Minnehaha	53
Twin	120
Lower Yosemite	320
Middle Yosemite	675

20. Is the combined heights of Minnehaha Falls, Twin Falls, and Lower Yosemite Falls greater than or less than the height of Middle Yosemite Falls?

Math In Social Studies

Morgan's social studies class was studying the migration rates of people in the United States. He made a table showing the gains and losses in population of some states during a recent year. In the table, $^+62{,}000$ means a population **gain** of 62,000 people; $^-262{,}000$ means a population **loss** of 262,000 people.

Migration Rates			
Alaska	$^+62{,}000$	Michigan	$^-453{,}000$
Arizona	$^+209{,}000$	Nevada	$^+79{,}000$
California	$^+940{,}000$	New York	$^-147{,}000$
Florida	$^+1{,}090{,}000$	Rhode Island	$^+1{,}000$
Illinois	$^-262{,}000$	Texas	$^+1{,}009{,}000$
Kentucky	$^-38{,}000$	Virginia	$^+135{,}000$

1. What does the word "migration" mean?

2. Did Michigan have a loss or gain of 453,000 people?

EXERCISES *Use the information in the table to answer the questions.*

1. Which state had the greatest number of people move into it?

2. Which state lost the most people?

3. Which state showed the least amount of change?

4. What is the combined migration rate for Illinois and Michigan?

Conestoga wagons, the *camels of the prairies*, carried the American pioneers westward.

5. How many people moved into Florida and Texas combined?

6. How many more people moved to Arizona than to Alaska?

7. How many fewer people moved to Nevada than to Virginia?

8. What is the combined migration rate for Illinois and New York?

PROJECT Do research to find the number of students who have moved into your school district and the number of students who have moved out of your district in the last three months. Make a table showing the loss or gain of students for each month.

Exponents

The United States bought Alaska from Russia for $7,200,000. This was considered a huge amount to pay for a "Polar Bear Garden."

You can use *powers* to write large numbers. For example,

$$1000 = 10 \times 10 \times 10$$
$$= 10^3 \xleftarrow{\text{Exponent}}_{\text{Base}}$$

The State Flag

The State Seal

The number 1000 is a **power** of 10. In 10^3, the exponent 3 tells how many times the base 10 is used as a factor. For example, in the expression $10 \times 10 \times 10$, 10 is used as a **factor** 3 times.

NOTE: 10^3 is read as "10 cubed" or "10 to the third power."

1. *Complete:* $10^2 = 10 \times \underline{\ ?\ } = 100$

2. *Complete:* $2^5 = 2 \times 2 \times 2 \times 2 \times 2 = \underline{\ ?\ }$

Since $2^5 = 32$, the number 32 is the **standard form** of 2^5.

EXAMPLE 1   Write the standard form for each number or product.

a. 3^4

b. 6×2^3

a. Think: $3^4 = 3 \times 3 \times 3 \times 3$

So $3^4 =\ 9\ \times\ 9$
$3^4 = \mathbf{81}$

b. Think: First find 2^3.

$2^3 = 2 \times 2 \times 2 = 8$
So $6 \times 2^3 = 6 \times 8$
$6 \times 2^3 = \mathbf{48}$

3. Which is greater, 3^4 or 4^3?

EXAMPLE 2   Complete: $7,200,000 = 72 \times 10^?$

Think: Since $7,200,000 = 72 \times 100,000$
$7,200,000 = 72 \times 10 \times 10 \times 10 \times 10 \times 10$ ◀ *Five 10's*
$7,200,000 = \mathbf{72 \times 10^5}$

4. *Complete:* $1500 = 15 \times 10^?$

EXERCISES

Complete. Choose your answers from the box at the right.

1. In 6^2, the number 6 is the __?__ and the number 2 is the __?__ .

2. The number 36 is called a __?__ of 6.

3. In 6^2, the 2 tells how many times 6 is used as a __?__ .

4. The standard form of 6^2 is __?__ .

> 2
> base
> 36
> exponent
> factor
> power

Write the standard form of each number or product.

5. 2^3	6. 6^3	7. 4^3	8. 4^4	9. 8^2	10. 10^4
11. 10^5	12. 9^2	13. 4^1	14. 9^3	15. 2^6	16. 2^8

17. 9×10^2 18. 7×3^2 19. 17×10^3

20. 28×2^2 21. 25×10^4 22. 90×10^2

Complete.

23. $10^? = 1000$ 24. $10^? = 10,000$ 25. $10^? = 100,000$

26. $\underline{?}^2 = 81$ 27. $\underline{?}^3 = 27$ 28. $\underline{?}^4 = 16$

29. $3 \times 10^? = 300$ 30. $7 \times 10^? = 7000$ 31. $9 \times 10^? = 90,000$

Write the standard form of each number.

32. 1^2 33. 1^3 34. 1^4 35. 1^5 36. 1^6 37. 1^7

38. *Complete:* The number 1 raised to any power equals __?__ .

For Exercises 39–50, choose which is the greater number in each pair.

39. 1^3 or 3^1 40. 5^2 or 5×2 41. 2^4 or 2×4 42. 1^4 or 1^8

43. 3^2 or 2^3 44. 4^3 or 3^4 45. 3^4 or 3×4 46. 6^2 or 6×2

47. 2^4 or 4^2 48. 3^3 or 3×3 49. 5^3 or 5×3 50. 2^5 or 5^2

For Exercises 51–56, find each answer.
Look for a pattern in the answers.

51. $1^3 + 2^3$

52. $1^3 + 2^3 + 3^3$

53. $1^3 + 2^3 + 3^3 + 4^3$

54. $1^3 + 2^3 + 3^3 + 4^3 + 5^3$

55. $1^3 + 2^3 + 3^3 + 4^3 + 5^3 + 6^3$

56. $1^3 + 2^3 + 3^3 + 4^3 + 5^3 + 6^3 + 7^3$

57. What is the pattern in the answers to Exercises 51–56?

Reading/Writing/Comparing Numbers

Karen bought two books on life in colonial America. When she checked the cash register tape against the sales slip, she noticed a difference.

TAPE
Thank you Call again
TOTAL $2.75

SALES SLIP

Subtotal	26.19
Tax	1.31
TOTAL	$27.50

Check one:
Cash ☑   Check ☐
Credit card ☐

1. How are the numbers different?

2. Which do you think is the correct amount?

   Numbers such as 2.75 and 27.50 are **decimals.**

> Each place in a number has a value.
>
> **a.** Ten times the value of the place to its right.
>
> **b.** One-tenth the value of the place to its left.

EXAMPLE 1

Read "and" for the decimal point.

Millions | Hundred thousands | Ten thousands | Thousands | Hundreds | Tens | Ones | . | Tenths | Hundredths | Thousandths | Ten-Thousandths

Standard Numeral	In Words
a. 1, 0 0 5, 6 2 7 . 5	One million, five thousand, six hundred twenty-seven and five tenths
b. 2 . 7 5	Two and seventy-five hundredths
c. 0 . 0 0 9	Nine thousandths

To compare numbers, compare digits that are in the same place. Start at the left. Use the symbols < **(is less than)** and > **(is greater than).**

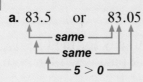

EXAMPLE 2 Which number is greater?

a. 83.5   or   83.05
same
same
5 > 0

b. 7.021   or   7.022
same
same
same
1 < 2

So **7.021 < 7.022.**

3. Is 0.3 greater than, equal to, or less than 0.30. Why?

NOTE: It is customary to write 0.3 for .3.

EXERCISES

In what place is the underlined digit?

1. 42.87<u>3</u>   **2.** 5<u>7</u>6,207   **3.** 0.3<u>09</u>   **4.** 5.<u>9</u>18   **5.** <u>9</u>00,000,000

6. 1.999<u>9</u>   **7.** 10.<u>5</u>406   **8.** 9,916,<u>00</u>0   **9.** 23<u>0</u>   **10.** 41.00000<u>6</u>

Write in words.

11. 9.6   **12.** 15.68   **13.** 0.625   **14.** 54,540,540   **15.** 486.5

16. 5.129   **17.** 51.29   **18.** 512.9   **19.** 5129   **20.** 0.005129

Write the standard numeral.

21. 2 thousandths   **22.** 2 thousand   **23.** 2 tenths

24. 172 and 64 hundredths   **25.** 819 million   **26.** 819 ten-thousandths

27. Fifteen hundred-thousandths   **28.** Forty and three ten-thousandths

29. Six hundred and five hundredths   **30.** Thirty-one and eight thousandths

Replace the ● with <, >, or = .

31. 7.01 ● 7.89   **32.** 0.450 ● 0.45   **33.** 14,682 ● 14,689

34. 1.09 ● 1.9   **35.** 8.421 ● 8.422   **36.** 569,998 ● 569,909

37. 0.1005 ● 1.0036   **38.** 18.72 ● 18.7   **39.** 48.07 ● 48.7

40. 0.941 ● 92   **41.** 1.1 ● 1.01   **42.** 896.3 ● 897

43. Write the greatest possible number having four different digits and a 5 in the hundredths place.

44. Write the smallest possible number having four different digits and a 7 in the tenths place.

Writing Checks

Karen decided to write a check to pay for the two books on colonial America. On a check, the amount paid is written both in <u>decimals</u> and in <u>words</u>.

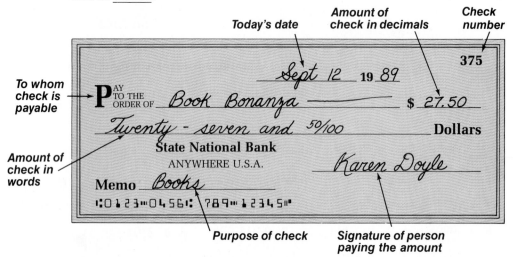

Today's date

Amount of check in decimals

Check number

To whom check is payable

Amount of check in words

Purpose of check

Signature of person paying the amount

1. What are some advantages of paying by check?

EXERCISES

For Exercises 1–10, write each amount in words as it would appear on a check.

1. $25.12   **2.** $36.72   **3.** $14.00   **4.** $118.00   **5.** $325.75

6. $481.29   **7.** $19.08   **8.** $206.05   **9.** $1725.00   **10.** $3005.00

What is missing on each check? Correct any errors.

11.

12.

ROBERT PEREZ   138

January 15, 19 *89*

PAY TO THE ORDER OF *Stereo Barn* $

Four hundred and ⁷⁸/₁₀₀ ——————— Dollars

State National Bank
ANYWHERE U.S.A.

Robert Perez

Memo *Stereo*

⑈0⃦23⃧04⃦56⃧ 789⃧12345⃧⑊

13.

LYDIA FULLER   489

December 12 19 *89*

PAY TO THE ORDER OF *Bill's Bicycle Shop* $ *162.75*

One hundred sixty-two and ⁷⁵/₁₀₀ —— Dollars

State National Bank
ANYWHERE U.S.A.

Memo *Bicycle*

⑈0⃦23⃧04⃦56⃧ 789⃧12345⃧⑊

14.

ROSEMARY KERUKI   603

June 28, 19 *89*

PAY TO THE ORDER OF $ *1605.00*

One thousand six hundred five and ⁰⁰/₁₀₀ Dollars

State National Bank
ANYWHERE U.S.A.

Rosemary Keruki

Memo *Tuition*

⑈0⃦23⃧04⃦56⃧ 789⃧12345⃧⑊

*For each of Exercises 15–16, first copy
a check form. Then write out a check as
indicated.*

15. On August 30, Janet Keefe wrote
check number 305 to Midway
Dental Clinic for $68.75.

16. On October 8, Anthony Bi Blasi
wrote check number 117 to Uptown
Electric Company for $61.00.

Patterns and Codes

When Neal purchased an atlas of the United States, he noticed the universal product code on the atlas. Codes that appear on items that you buy are used to give information.

Study this code.

0   16000 75440

Code	1	2	3
	4	5	6
	7	8	9

$\longrightarrow$

1 = ⌐   4 = ⌐   7 = _?_

2 = ⊔   5 = □   8 = _?_

3 = ⌐   6 = ⊏   9 = _?_

1. What are the codes for 7, 8, and 9?

   You can use the code to write numbers.

   2406 $\longrightarrow$ ⊔ ⌐ 0 ⊏

2. What is the code for 3917?

   Sometimes codes are used to show the price a merchant pays for an item (**dealer's cost**). If you can read the code and you know the selling price, you can find the **markup,** or how much more you pay than the dealer.

A computer coded ⌐ □ ⊔ 0 is to be sold for $5650. What is the markup?

1️⃣ Find the dealer's cost.
   ⌐ □ ⊔ 0 $\longrightarrow$ $4520

2️⃣ Find the markup.
   $5650 − $4520 = $1130

$$\text{Markup} = \frac{\text{Selling}}{\text{Price}} - \frac{\text{Dealer's}}{\text{Cost}}$$

EXERCISES

Write the code for each amount.

1. $9308   2. $5862   3. $1976   4. $8009   5. $8764   6. $3015

7. A car coded ⌐ ⊏ 00 is to be sold for $12,800. What is the markup?

8. A television set coded ⌐ ⌐ ⌐ is to be sold for $615. Find the markup.

Chapter Summary

1. To round whole numbers:

    1 Look at the digit to the right of the place to which you are rounding.

    2 **a.** If the digit is 5 or more, round up.
    b. If the digit is less than 5, round down.

2. To find the average:

    1 Add the measures.

    2 Divide the sum by the number of measures.

3. When computations involve more than one operation:

    1 First do the operations within parentheses or the computations above or below a division bar.

    2 Then do multiplication and division from left to right.

    3 Finally, do addition and subtraction from left to right.

4. To write a power in standard form:

    1 Read the exponent.

    2 Use the exponent to determine how many times the base is multiplied by itself.

5. To compare decimals:

    1 Starting from the left, compare the first digits that are <u>not</u> alike. Annex final zeros when necessary.

    2 Compare the decimals in the same order.

---

1. | Number | Nearest Ten |
   |--------|-------------|
   | 27 | 30 |

Number	Nearest Thousand
15,511	16,000

2. Find the average.

    10, 15, 18, 21

    $10 + 15 + 18 + 21 = 64$

    $64 \div 4 = \mathbf{16}$ **◀ Average: 16**

3. $(4 + 7) \times 3 = \underline{}$

    $(4 + 7) \times 3 = 11 \times 3$
    $= \mathbf{33}$

    $2 \div 2 \times 6 = \underline{}$

    $2 \div 2 \times 6 = 1 \times 6$
    $= \mathbf{6}$

    $9 - 3 \times 2 = \underline{}$

    $9 - 3 \times 2 = 9 - 6$
    $= \mathbf{3}$

4. $10^4 = 10 \times 10 \times 10 \times 10$
    $= \mathbf{10,000}$

5. Which number is greater?

    24.63 or 24.56
    $6 > 5$

    So **24.63** > **24.56**.

Chapter Review

Part 1: VOCABULARY

For Exercises 1–8, choose from the box at the right the word(s) that complete(s) each statement.

	multiplication
	average
	value
	estimated
	addition
	hundredth
	standard form
	exponent

1. An answer obtained by using rounded numbers is an __?__ answer.   (Page 2)

2. To find the __?__ of three scores, divide the sum of the scores by the number of scores.   (Page 4)

3. In $(9 + 5) \div 2$, perform the __?__ first.   (Page 8)

4. In $3 \times 5 - 4 \div 2$, perform the __?__ first.   (Page 8)

5. In 5^3, the number 3 is an __?__ .   (Page 14)

6. The number 27 is the __?__ for 3^3.   (Page 14)

7. The 2 in 253 has a __?__ of 200.   (Page 16)

8. The place name of the 5 in 22.35 is __?__ .   (Page 16)

Part 2: SKILLS

Round each number to the nearest ten and to the nearest hundred. (Pages 2–3)

9. 64   10. 229   11. 1655   12. 17,105   13. 63,998

Round each number to the nearest thousand and to the nearest ten thousand.   (Pages 2–3)

14. 8710   15. 50,255   16. 68,500   17. 145,456   18. 218,781

Estimate. Find the actual sum or difference. Compare.   (Pages 2–3)

19. $4278 + 3052$ 20. $9641 - 5863$ 21. $7001 - 3599$

22. $10,615 - 4044$ 23. $20,972 + 14,891$ 24. $16,382 + 47,006$

25. $131,812 - 69,850$ 26. $138,602 + 510,120$ 27. $249,678 - 109,659$

Find the average.   (Pages 4–5)

28.	Cost of Lettuce at Six Stores		
	78¢	68¢	88¢
	59¢	63¢	76¢

29.	Prices for Four Boats	
	$4726	$3860
	$5682	$5220

30. **Points Scored in Five Basketball Games**

82		90		68
	88		77	

31. **Passengers on Seven Airline Flights**

104		88		38		110
	92		68		102	

Find each answer. (Pages 8–9)

32. $12 \times 2 + 6$

33. $11 + 24 - 8$

34. $(8 - 6) \times 10$

35. $14 - \dfrac{5 \times 8}{4}$

36. $24 \div 4 + 6 \times 3$

37. $82 - 50 + 10 \div 5$

Complete. (Pages 14–15)

38. $10^? = 100,000$

39. $10^? = 1,000,000$

40. $5^? = 3125$

41. $\underline{\ ?\ }^2 = 36$

42. $\underline{\ ?\ }^4 = 81$

43. $6 \times 10^? = 6000$

Write each number in words.
(Pages 16–19)

44. 8.4   **45.** 12.25   **46.** 1024   **47.** 57.35   **48.** 38.14   **49.** 182.4

Write the standard numeral.
(Pages 16–17)

50. Three hundred   **51.** Two hundredths   **52.** Five tenths

Replace the ● *with* <, >, *or* =. (Pages 16–17)

53. 6.15 ● 6.12

54. 0.356 ● 0.351

55. 1564 ● 1574

56. 25.06 ● 24.99

57. 13.17 ● 13.71

58. 41,010 ● 41,100

Part 3: APPLICATIONS

For Exercises 59–60, use the table on page 3. (Pages 2–3)

59. Which state has an area that is about 200,000 square miles less than the area of Texas?

60. How much greater than the area of Arizona is the area of Texas?

61. What does $\square \times \triangle + \bigcirc$ equal?

Clue A: $\square + \triangle = 7$

Clue B: $\square + 4 = \bigcirc$

Clue C: $\bigcirc - 5 = \triangle$

(Pages 10–11)

62. What does $\square \times \triangle \times \bigcirc$ equal?

Clue A: $\square + \triangle + \bigcirc = 15$

Clue B: $\square + \bigcirc = 10$

Clue C: $\square \times \triangle = 20$

(Pages 10–11)

63. Write the greatest possible number having three different digits and a 6 in the tenths place. (Pages 16–17)

64. Write the smallest possible number greater than one having four different digits and a 3 in the thousandths place. (Pages 16–17)

Chapter Test

Round each number as indicated.

1. 7; to the nearest ten

2. 6732; to the nearest hundred

3. 950; to the nearest hundred

4. 15,876; to the nearest thousand

5. 5500; to the nearest thousand

6. 42; to the nearest ten

Estimate. Find the actual sum or difference. Compare.

7. $988 + 2663$

8. $5810 + 1099$

9. $12,455 - 6,109$

10. $81,071 - 62,010$

11. $134,498 + 98,915$

12. $152,873 - 49,904$

Find the average.

13.

Points Scored in Eight Football Games			
21	13	3	9
42	30	14	20

14.

Number of Books Sold During Six Days		
80	25	47
62	54	56

Find each answer.

15. $18 + 2 \times 4$

16. $9 \times (12 - 8)$

17. $20 \div 2 + 3 \times 4$

Complete.

18. $10^? = 1000$

19. $\underline{\ ?\ }^4 = 256$

20. $5 \times 10^? = 50,000$

Replace the ● with <, >, or =.

21. 5014 ● 5104

22. 0.318 ● 0.138

23. 46.25 ● 46.24

24. Refer to the table at the right to determine how many more people were living in Colorado than in Montana in 1980.

Population	1980 Census
Montana	786,690
Idaho	943,935
Colorado	2,889,964

25. What does $\square \times \triangle \times \bigcirc$ equal?

Clue A: $\square + \triangle = 14$

Clue B: $\triangle \times \bigcirc = 30$

Clue C: $\square + \triangle + \bigcirc = 19$

Whole Numbers and Decimals

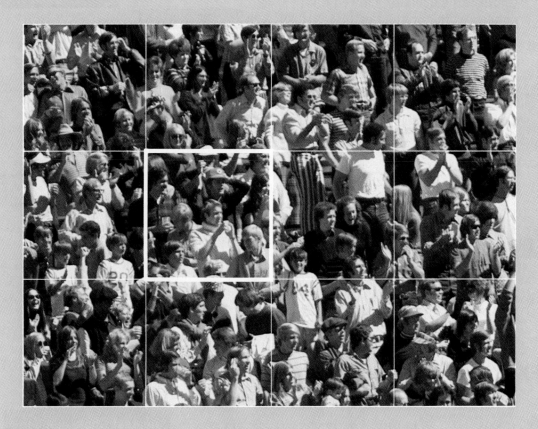

CHAPTER 2

- How can you determine how many people are in the square outlined in white?

- How can you use the number of people in one square to <u>estimate</u> the number of people in five squares?

- Estimate the total number of people in the photograph above.

- Describe a procedure that could be used to estimate how many trees there are in a photograph of a forest.

Rounding Decimals

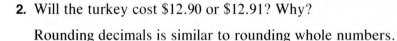

FRESH TURKEY

89¢/pound

TODAY ONLY!!!!!

`12.905`

Clarence works after school at a supermarket. A customer asked him how much a 14.5-pound turkey would cost on sale.

1. How do you write a decimal for 89¢?

Clarence's calculator showed this number as the product of 14.5 × 0.89.

2. Will the turkey cost $12.90 or $12.91? Why?

Rounding decimals is similar to rounding whole numbers.

> To round a decimal, look at the digit to the right of the place to which you are rounding.
>
> If the digit is less than 5, round down.
>
> If the digit is 5 or more, round up.

EXAMPLES

1. 6.8358 rounded to the nearest whole number is **7.**

2. 6.8358 rounded to the nearest tenth is **6.8.**

3. 6.8358 rounded to the nearest hundredth is **6.84.**

4. 6.8358 rounded to the nearest thousandth is **6.836.**

EXERCISES

To which decimal place was each number rounded? Choose your answers from the box at the right.

1. 12.905 ⟶ 12.91   **2.** 12.905 ⟶ 13

3. 0.00693 ⟶ 0.0069   **4.** 0.00693 ⟶ 0.007

5. 5.447 ⟶ 5.4   **6.** 10.555 ⟶ 11

> nearest whole number
> nearest tenth
> nearest hundredth
> nearest thousandth
> nearest ten-thousandth

Round to the nearest whole number.

7. 2.76   **8.** 3.82   **9.** 10.375   **10.** 0.75   **11.** 10.83

Round to the nearest tenth.

12. 4.642   **13.** 3.731   **14.** 7.025   **15.** 0.65   **16.** 9.482

Round to the nearest hundredth.

17. 0.978 **18.** 4.025 **19.** 60.074 **20.** 7.154 **21.** 0.029

Round to the nearest cent.

22. $0.951 **23.** $6.534 **24.** $157.605 **25.** $9.848 **26.** $0.0096

Round to the nearest thousandth.

27. 1.5196 **28.** 0.12745 **29.** 0.0036 **30.** 0.13095 **31.** 6.0104

Round to the nearest ten–thousandth.

32. 0.37054 **33.** 6.00009 **34.** 1.739428 **35.** 9.00815 **36.** 0.05055

Round to the nearest ten dollars.

37. $18.20 **38.** $52.80 **39.** $65.20 **40.** $87.50 **41.** $125.98

Round to the nearest hundred dollars.

42. $101.25 **43.** $193.80 **44.** $1590.00 **45.** $3650.00 **46.** $1205.25

This table shows the population (in millions) of the five largest metropolitan areas in the United States in 1980. It also shows the projected population in 1990.

Refer to the table for Exercises 47–52. Round each decimal to the nearest tenth.

47. What was the population of the Los Angeles-Long Beach area in 1980?

48. What is the projected population for the Chicago area in 1990?

49. What is the projected population for the New York area in 1990?

50. What was the population of the Detroit area in 1980?

51. What was the population of the Philadelphia area in 1980?

52. Which cities expect an increase in population from 1980 to 1990?

Metropolitan Areas Population in Millions		
	1980	**1990**
New York	8.30	8.27
Los Angeles-Long Beach	7.49	8.05
Chicago	6.07	6.32
Philadelphia	4.72	4.85
Detroit	4.49	4.71

Decimals: ADDITION/SUBTRACTION

This table shows the weekly spending habits of girls in the 13–15 age group.

Girls ages 13 to 15	
Clothes	$7.10
Food	$4.95
Entertainment	$3.00
Cosmetics	$3.10
Records and tapes	$1.20
Hobbies	70¢
School supplies	50¢
Books	45¢
Video games	10¢

1. On which item do these teenagers spend the most per week?

2. On which items do these teenagers spend less than $1.00 per week?

3. Write a decimal for 60¢.

EXAMPLE 1

What is the total amount these teenagers spend per week on clothes, food, entertainment, and cosmetics?

Estimate: Round to the nearest whole dollar.

$$
\begin{array}{rcl}
\$7.10 & \longrightarrow & \$\ 7 \\
4.95 & \longrightarrow & 5 \\
3 & \longrightarrow & 3 \\
3.10 & \longrightarrow & +\ 3 \\
\hline
& & \$18
\end{array}
$$

Computation: Line up the decimal points and add.

$$
\begin{array}{r}
\$\ 7.10 \\
4.95 \\
3.00 \\
+3.10 \\
\hline
\$18.15
\end{array}
$$

Insert a decimal point and annex two zeros.

— Compare —

The teenagers spend **$18.15** per week on these four items.

---

4. How can you check the answer to a subtraction problem?

EXAMPLE 2

Subtract. Check your answers.

    a. 27.58 − 18.627   **b.** 20 − 12.94

Think: Line up the decimal points.

a.
$$
\begin{array}{r}
{\scriptstyle 1\ 16\ 15\ 7\ 10} \\
2\ 7.5\ 8\ 0 \\
-1\ 8.6\ 2\ 7 \\
\hline
8.9\ 5\ 3
\end{array}
$$
Annex one zero.

b.
$$
\begin{array}{r}
{\scriptstyle 1\ 9\ 9\ 10} \\
2\ 0.0\ 0 \\
-1\ 2.9\ 4 \\
\hline
7.0\ 6
\end{array}
$$
Insert the decimal point. Annex two zeros.

Check: a.
$$
\begin{array}{r}
8.953 \\
+18.627 \\
\hline
27.580
\end{array}
$$

b.
$$
\begin{array}{r}
7.06 \\
+\ 12.94 \\
\hline
20.00
\end{array}
$$

EXERCISES

Add or subtract as indicated.

1. 6.72
    4.36
    +0.28

2. 8.72
    3.6
    +0.63

3. 8.91
    −2.65

4. 41.53
    −33.48

5. 9.6
    −1.582

6. 2.7
    −0.986

7. 35.84
    9.21
    + 0.79

8. 0.005
    1.68
    +27.6

9. 2.7
    −0.935

10. 68
    7.59
    +108.8

11. 35.84
    79.6
    + 3.615

12. 41.53
    −39.4

13. 0.294
    −0.2507

14. 10
    − 0.86

15. 7.35
    10.1
    + 8.006

16. 6.27 + _?_ = 10.19

17. 12 + 0.075 + 9.2

18. 9.3 − 7.6

19. 16.41 − 6.172

20. 12 − 4.72

21. 42.5 + 97.8 + 1.09

22. 48.65 + _?_ + 15.9 = 67.82

23. 0.075 + 91.2 + 316

24. _?_ − 0.407 = 8.993

25. 0.9 − _?_ = 0.18

26. 0.075 + 9.2 + 11.5

27. 3.6 − 1.728

For Exercises 28–31, refer to the table on page 28.

28. What is the total amount spent per week on the items in the table?

29. How much more do the teenagers spend per week for entertainment than for books?

30. How much more is spent for clothes and food than for entertainment and cosmetics?

31. How much more is spent for cosmetics than for hobbies?

32. Jessica has a $20–bill. She spends $6.95 on cosmetics and $4.95 on magazines. How much change should she receive?

33. Christi paid for books and school supplies with a $50–bill. She received $14.85 in change. How much did the books and school supplies cost?

Patterns and Powers of 10

Numbers such as 10, 100, and 1000 are **powers of 10.**

$10 = 10^1$ $100 = 10 \times 10$, or 10^2 $1000 = 10 \times 10 \times 10$, or 10^3

Look for the pattern in each column.

Number $\times$ 10	**Number $\times$ 100**	**Number $\times$ 1000**
$8 \times 10 = 80$	$8 \times 100 = 800$	$8 \times 1000 = 8000$
$89 \times 10 = 890$	$89 \times 100 = 8900$	$89 \times 1000 = 89,000$
$893 \times 10 = 8930$	$893 \times 100 = 89,300$	$893 \times 1000 = 893,000$
$8930 \times 10 = 89,300$	$8930 \times 100 = 893,000$	$8930 \times 1000 = 8,930,000$

Complete.

1. To multiply a number by 10, first write the number. Then annex _?_ zero(s).

2. To multiply a number by 100, first write the number. Then annex _?_ zero(s).

3. To multiply a number by 1000, first write the number. Then annex _?_ zero(s).

EXAMPLE   Tricia's heart beats 76 times per minute.

a. How many times does it beat in 100 minutes?

2 zeros

$76 \times 100 = 7600$ ◀ *Beats in 100 min*

b. How many times does it beat in 10,000 minutes?

$76 \times 10,000 = 760,000$ ◀ *Beats in 10,000 min*

4 zeros

4. Write a rule you could use to multiply a number by ten thousand.

5. Write a rule you could use to multiply a number by one hundred thousand.

6. Write a rule you could use to multiply a number by one million.

EXERCISES

How many zeros should you annex to the product? Write 1, 2, 3, 4, or 5.

1. $46 \times 10 = 46$▇   **2.** $983 \times 100 = 983$▇   **3.** $32 \times 1000 = 32$▇

4. $9 \times 100,000 = 9$▇   **5.** $1600 \times 10 = 1600$▇   **6.** $290 \times 100 = 290$▇

Complete.

7. $2005 \times \underline{\ ?\ } = 20,050$ **8.** $58 \times \underline{\ ?\ } = 58,000$ **9.** $109 \times \underline{\ ?\ } = 10,900$

10. $10 \times \underline{\ ?\ } = 75,000$ **11.** $1000 \times \underline{\ ?\ } = 2,700,000$ **12.** $\underline{\ ?\ } \times 100 = 36,000$

13. $5893 \times 1000 = \underline{\ ?\ }$ **14.** $465 \times \underline{\ ?\ } = 4,650,000$ **15.** $398,000 = 3980 \times \underline{\ ?\ }$

16. $2700 = 270 \times \underline{\ ?\ }$ **17.** $250 \times 10,000 = \underline{\ ?\ }$ **18.** $90 \times \underline{\ ?\ } = 90,000$

19. 15×2000
     $15 \times 1000 \times \underline{\ ?\ }$
     $15,000 \times \underline{\ ?\ }$
     $\underline{\ ?\ }$

20. 780×300
     $780 \times \underline{\ ?\ } \times 3$
     $\underline{\ ?\ } \times 3$
     $\underline{\ ?\ }$

21. $625 \times 40,000$
     $625 \times 4 \times \underline{\ ?\ }$
     $\underline{\ ?\ } \times 4$
     $\underline{\ ?\ }$

Which is more?

22. Fifty $100-bills or thirty-five $20-bills

23. Two hundred $50-bills or six hundred fifty $20-bills

24. One thousand $5-bills or one hundred sixty $20-bills

25. One hundred eighty $100-bills or seven thousand $10-bills

26. Six hundred $5-bills or three hundred $10-bills

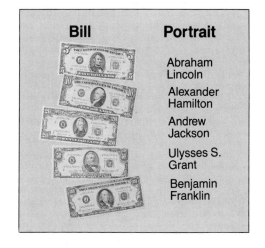

Bill	Portrait
	Abraham Lincoln
	Alexander Hamilton
	Andrew Jackson
	Ulysses S. Grant
	Benjamin Franklin

You are a bank teller. How would you count out each amount? Use only the bills shown above. Use as few bills as possible.

27. $650   **28.** $255   **29.** $260   **30.** $490   **31.** $725   **32.** $345

Find the error.

33. You exchange seven $50-bills for eighty $5-bills.

34. You exchange three $100-bills for twelve $20-bills and one $5-bill.

35. You exchange nineteen $20-bills for seven $50-bills and four $10-bills.

Multiplying Whole Numbers

Steve took a cross-country motorcycle trip. He traveled an average of 281 miles each day during the first week.

1. Did Steve travel exactly 281 miles each day?

2. Did Steve travel more than, or fewer than, 281 miles each day?

EXAMPLE 1

If Steve maintains the same rate, how far will he travel in 29 days?

Estimate: Round to the nearest ten.

$281 \times 29 \longrightarrow 280 \times 30$

Think: $280 \times 30 = 280 \times 10 \times 3$

So $280 \times 10 = 2800$, and

$2800 \times 3 = 8400.$ ◄ *Estimate*

Computation

$$
\begin{array}{r}
281 \\
\times \quad 29 \\
\hline
2529 \\
562\,0 \\
\hline
8149
\end{array}
$$

◄ *You can omit this zero.*

Steve will travel **8149 miles.**

---

3. In Example 1, compare the answer with the estimate. Is the answer reasonable?

Sometimes one or both of the factors in a multiplication problem contain zeros. The **factors** are the numbers that are multiplied. The **product** is the answer.

EXAMPLE 2

Find the product: 507×603

Estimate

Round to the nearest hundred.

$507 \longrightarrow 500 \qquad 603 \longrightarrow 600$

$500 \times 600 = \mathbf{300,000}$

Compare the estimate with the answer. Is the answer reasonable?

Computation

$$
\begin{array}{r}
507 \\
\times 603 \\
\hline
1\,521 \\
304\,20 \\
\hline
305{,}721
\end{array}
$$

◄ *Write a zero in the tens place. Then multiply 507 by 6.*

---

4. Will 402×301 be closer to 120,000 or 130,000? Explain.

5. Will 690×710 be closer to 420,000 or 490,000? Explain.

EXERCISES

Identify each underlined number as a factor, F, or as a product, P.

1. $29 \times \underline{16} = 464$ **2.** $\underline{2184} = 52 \times 42$ **3.** $3999 \times 0 = \underline{0}$

For Exercises 4–9:
a. *Estimate the product.*
b. *Replace* ● *with* < *(less than) or* > *(greater than).*

4. 69×71 ● 5000 **5.** 82×52 ● 4000 **6.** 188×27 ● 6000

7. $\$4.98 \times 29$ ● $\$140$ **8.** $\$10.44 \times 48$ ● $\$500$ **9.** $11 \times \$128$ ● $\$1200$

For Exercises 10–39, multiply.

10. $\begin{array}{r} 49 \\ \times 95 \\ \hline \end{array}$ **11.** $\begin{array}{r} 82 \\ \times 74 \\ \hline \end{array}$ **12.** $\begin{array}{r} 41 \\ \times 95 \\ \hline \end{array}$ **13.** $\begin{array}{r} 86 \\ \times 70 \\ \hline \end{array}$ **14.** $\begin{array}{r} 348 \\ \times 116 \\ \hline \end{array}$

15. $\begin{array}{r} 612 \\ \times 529 \\ \hline \end{array}$ **16.** $\begin{array}{r} 283 \\ \times 296 \\ \hline \end{array}$ **17.** $\begin{array}{r} 356 \\ \times 767 \\ \hline \end{array}$ **18.** $\begin{array}{r} 43 \\ \times 72 \\ \hline \end{array}$ **19.** $\begin{array}{r} 63 \\ \times 58 \\ \hline \end{array}$

20. $\begin{array}{r} 81 \\ \times 54 \\ \hline \end{array}$ **21.** $\begin{array}{r} 89 \\ \times 98 \\ \hline \end{array}$ **22.** $\begin{array}{r} 654 \\ \times 703 \\ \hline \end{array}$ **23.** $\begin{array}{r} 309 \\ \times 265 \\ \hline \end{array}$ **24.** $\begin{array}{r} 820 \\ \times 506 \\ \hline \end{array}$

25. $\begin{array}{r} 940 \\ \times 209 \\ \hline \end{array}$ **26.** $\begin{array}{r} 702 \\ \times 207 \\ \hline \end{array}$ **27.** $\begin{array}{r} 802 \\ \times 405 \\ \hline \end{array}$ **28.** $\begin{array}{r} 709 \\ \times 830 \\ \hline \end{array}$ **29.** $\begin{array}{r} 598 \\ \times 607 \\ \hline \end{array}$

30. 46×75 **31.** 89×91 **32.** 826×736 **33.** 823×111 **34.** 460×507

35. 930×202 **36.** 310×810 **37.** 440×330 **38.** 622×407 **39.** 919×808

40. When 7 is subtracted from a number, you get 20. What do you get when you multiply the number by 147?

41. When 24 is added to a number, you get 108. What do you get when you multiply the number by 315?

42. When 31 is subtracted from a number, you get 76. What do you get when you multiply the number by itself?

43. Replacing each □ with the same digit will give the product. □ = ?

$\begin{array}{r} \square\,\square \\ \times\ \square\,\square \\ \hline 4\ 3\ 5\ 6 \end{array}$

Find the missing digits.

44. $\begin{array}{r} 3\,\square \\ \times\ 4\ 8 \\ \hline 2\ 8\,\square \\ 1\ 4\,\square \\ \hline 1\ 6\,\square\ 0 \end{array}$ **45.** $\begin{array}{r} 2\ 9 \\ \times\ 3\,\square \\ \hline \square\ 9 \\ 8\,\square \\ \hline 8\ 9\,\square \end{array}$ **46.** $\begin{array}{r} \square\,\square \\ \times\ 5\ 9 \\ \hline \square\,\square\ 0 \\ 1\ 0\,\square \\ \hline 1\,\square\,\square\ 0 \end{array}$ **47.** $\begin{array}{r} \square\ 7 \\ \times\,\square\ 2 \\ \hline 7\ 4 \\ \square\,\square\ 8 \\ \hline \square\,\square\ 5\ 4 \end{array}$ **48.** $\begin{array}{r} \square\ 0\,\square \\ \times\ \ \ 3\ 5 \\ \hline \square\ 0\,\square\ 0 \\ 1\ 8\ 1\ 8 \\ \hline 2\,\square\ 2\,\square\ 0 \end{array}$

WHOLE NUMBERS AND DECIMALS   **33**

Mid-Chapter Review

Round each decimal to the nearest indicated place. (Pages 26–27)

1. 5.43; whole number   **2.** 0.35; tenth   **3.** 50.061; hundredth

4. 2.1537; thousandth   **5.** 5.01354; ten-thousandth   **6.** $7.541; cent

Add or subtract as indicated. (Pages 28–29)

7. 5.35
2.78
+ 0.56

8. 7.43
− 3.31

9. 15.07
4.18
+ 12.9

10. 5.64
− 2.58

11. 12
− 0.53

12. 1.172 + 0.56 + 3.9   **13.** 18.8 − _?_ = 10.885   **14.** 14.5 + _?_ + 9.17 = 33.92

Complete. (Pages 30–31)

15. 54 × _?_ = 540   **16.** _?_ × 1000 = 24,000   **17.** 312 × _?_ = 31,200

Multiply. (Pages 32–33)

18. 52
× 25

19. 91
× 37

20. 248
× 756

21. 208
× 307

22. 580
× 290

23. 43 × 56   **24.** 219 × 175   **25.** 308 × 105   **26.** 518 × 240   **27.** 602 × 460

28. The population of Detroit in 1990 is projected to be 4.71 million. Round this decimal to the nearest tenth. (Pages 26–27)

29. Selena has a $20-bill. She spends $2.50 on a teen magazine and $6.95 on a tape. How much change should she receive? (Pages 28–29)

MAINTENANCE

Replace each ● with <, >, or = (Pages 14–15)

30. 2^3 ● 3^2 **31.** 3^3 ● 3×3 **32.** 1^6 ● 6^1 **33.** 9^1 ● 3^2

34. 6^2 ● 2×6 **35.** 2^3 ● 4^2 **36.** 2^4 ● 4×4 **37.** 6^3 ● 9^2

Use the table at the right. (Pages 2–3)

38. Estimate how much longer the Golden Gate Bridge is than the Bayonne Bridge.

39. Which bridge is 45 feet longer than the Sciotoville Bridge?

Modern Bridges	
Name	**Length (feet)**
Brooklyn Bridge	1595
Golden Gate Bridge	4200
Sciotoville Bridge	1550
Bayonne Bridge	1152

Math and Making Change

Lauren is a cashier at Cinema Center. She reads the amount of change on the cash register. Then Lauren uses this rule to count out the change.

Use as few bills and coins as possible.

EXERCISES

*Choose the best way to make change. Choose **a, b,** or **c**.*

1.

| TOTAL | CHANGE | SUBTOTAL |
4.93

a. Four $1-bills, 3 quarters, one dime, one nickel

b. Four $1-bills, 9 dimes, 3 pennies

c. Four $1-bills, 3 quarters, one dime, one nickel, 3 pennies

2.

| TOTAL | CHANGE | SUBTOTAL |
3.52

a. Three $1-bills, 5 dimes, 2 pennies

b. Three $1-bills, 2 quarters, two pennies

c. Three $1-bills, 1 quarter, 2 dimes, 1 nickel, and 2 pennies

3.

| TOTAL | CHANGE | SUBTOTAL |
2.35

a. Eight quarters, 3 dimes, 1 nickel

b. Nine quarters, 1 dime

c. Two $1-bills, 1 quarter, 1 dime

4.

| TOTAL | CHANGE | SUBTOTAL |
9.45

a. One $5-bill, four $1-bills, one quarter, 2 dimes

b. Nine $1-bills, 4 dimes, 1 nickel

c. One $5-bill, 17 quarters, 2 dimes

For Exercises 5–8, make a chart like the one below. First, find the change due. Then write the number of bills and coins in the boxes to show the best way to make change.

	Amount of Sale	Money Received	Change Due	Change: Number of						
				$10-bills	$5-bills	$1-bills	Quarters	Dimes	Nickels	Pennies
5.	$ 4.55	$10	?	?	?	?	?	?	?	?
6.	$ 9.81	$20	?	?	?	?	?	?	?	?
7.	$24.63	$40	?	?	?	?	?	?	?	?
8.	$18.19	$20	?	?	?	?	?	?	?	?

Deposits

You want to open a checking account with Jupiter Bank. To open the account, you must make a deposit. A **deposit** is an amount of money you put into the account. You use a **deposit slip** to record the amount of cash (bills and coins) and the checks you deposit.

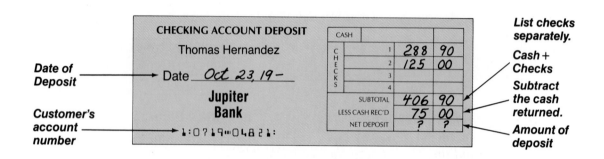

List checks separately.

Cash + Checks

Subtract the cash returned.

Amount of deposit

Date of Deposit

Customer's account number

CHECKING ACCOUNT DEPOSIT

Thomas Hernandez

Date _Oct 23, 19–_

Jupiter Bank

1:0719m0482 1:

CASH			
CHECKS 1		288	90
2		125	00
3			
4			
SUBTOTAL		406	90
LESS CASH REC'D		75	00
NET DEPOSIT		?	?

1. On what date was the deposit made?
2. Was any cash deposited?
3. What was the total of cash and checks?
4. How much cash did the bank return to Thomas Hernandez?
5. How much was actually deposited (net deposit)?

$$ 4\;0\;6\;\cdot\;9\;0\;-\;7\;5\;=\;????????? $$

When Tom deposits a check, he knows that he must **endorse,** or sign, it on the back. There are three ways to do this.

Thomas Hernandez

For deposit only
Thomas Hernandez

Pay to the order of
Julia Foote
Thomas Hernandez

Tom endorses the check on the reverse side <u>after</u> he arrives at the bank.

This check can be deposited only in Tom's checking or savings account.

This check is signed over to Julia Foote. Both Tom and Julia must endorse the check.

6. Why is it important to endorse a check only after you arrive at the bank?

EXERCISES

Copy the figure at the right. Write in the boxes the letters of the words that correspond to each description in Exercises 1–5.

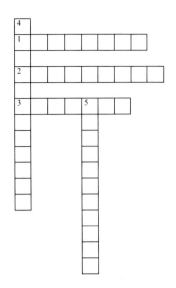

1. The operation you use to determine the total of cash and checks on a deposit slip (1 across)

2. What signing a check is called (2 across)

3. An amount of money you put into a bank account (3 across)

4. One thing you have to do in order to open a checking account (4 down)

5. The operation you use to compute the amount deposited when you receive some cash back (5 down)

Find the TOTAL and the NET DEPOSIT.

6.

		Dollars	Cents
CASH		47	12
CHECKS	1	15	80
List	2	246	15
Each	3	38	50
Check	4		
TOTAL		?	
▶ Less Cash Rec'd		0	
NET DEPOSIT		?	

7.

		Dollars	Cents
CASH			
CHECKS	1	241	25
List	2	208	13
Each	3		
Check	4		
TOTAL		?	
▶ Less Cash Rec'd		125	00
NET DEPOSIT		?	

8.

		Dollars	Cents
CASH			
CHECKS	1	51	15
List	2	135	27
Each	3		
Check	4		
TOTAL		?	
▶ Less Cash Rec'd		25	00
NET DEPOSIT		?	

Correct any errors in these deposit slips.

9.

		Dollars	Cents
CASH		25	19
CHECKS	1	423	50
List	2		
Each	3		
Check	4		
TOTAL		459	69
▶ Less Cash Rec'd		0	
NET DEPOSIT		459	59

10.

		Dollars	Cents
CASH			
CHECKS	1	128	14
List	2	17	81
Each	3		
Check	4		
TOTAL		145	95
▶ Less Cash Rec'd		50	00
NET DEPOSIT		195	95

11.

		Dollars	Cents
CASH		33	63
CHECKS	1	415	04
List	2	12	95
Each	3		
Check	4		
TOTAL		461	62
▶ Less Cash Rec'd		0	
NET DEPOSIT		461	00

12. Karen deposits checks for $135.40 and $62.10. She receives 2 twenty-dollar bills, 1 ten-dollar bill, 3 quarters and two dimes. What is Karen's net deposit?

13. Molly deposited 17 fifty-dollar bills, 12 five-dollar bills, 17 quarters, and 112 dimes into her checking account. What was the total deposit?

Decimals and Powers of Ten

Look for the pattern in each column.

Number × 10	Number × 100	Number × 1000
0.1243 × 10 = 1.243	0.6382 × 100 = 63.82	0.76814 × 1000 = 768.14
1.243 × 10 = 12.43	6.382 × 100 = 638.2	7.6814 × 1000 = 7681.4
12.43 × 10 = 124.3	63.82 × 100 = 6382	76.814 × 1000 = 76814
124.3 × 10 = 1243	638.2 × 100 = 63820	768.14 × 1000 = 768140

Annex one zero.

1. *Complete:* To multiply a number by 10, move the decimal point __?__ place to the right.

2. *Complete:* To multiply a number by 100, move the decimal point __?__ places to the right.

3. *Complete:* To multiply a number by 1000, move the decimal point __?__ places to the __?__ .

4. When you multiply 638.2 by 100, why do you have to annex a zero?

Look for the pattern in each column.

Number × 0.1	Number × 0.01	Number × 0.001
986.3 × 0.1 = 98.63	3265 × 0.01 = 32.65	97,613 × 0.001 = 97.613
98.63 × 0.1 = 9.863	326.5 × 0.01 = 3.265	9761.3 × 0.001 = 9.7613
9.863 × 0.1 = 0.9863	32.65 × 0.01 = 0.3265	976.13 × 0.001 = 0.97613
0.9863 × 0.1 = 0.09863	3.265 × 0.01 = 0.03265	97.613 × 0.001 = 0.097613

Insert one zero.

Complete.

5. To multiply a number by 0.1, move the decimal point __?__ place to the left.

6. To multiply a number by 0.01, move the decimal point __?__ places to the left.

7. To multiply a number by 0.001, move the decimal point __?__ places to the left.

8. Why do you move the decimal point to the <u>right</u> when you multiply by 10, 100, or 1000?

9. Why do you move the decimal point to the <u>left</u> when you multiply by 0.1, 0.01, or 0.001?

EXERCISES

Complete. Choose your answers from the box at the right.

1. When you multiply by 0.001, move the decimal point to the <u>?</u> .

2. When you multiply by 10,000, move the decimal point to the <u>?</u> .

3. When you multiply 5.6 by 10, the product will be <u>?</u> 5.6.

4. When you multiply 5.6 by 0.01, the product will be <u>?</u> 5.6.

> greater than
> less than
> right
> left
> equal to

Multiply. Use mental computation. Write only the answer.

5. 2.7×10
6. 10×0.27
7. 0.062×100
8. 6.2×100

9. 5.36×1000
10. 1000×0.072
11. 0.018×100
12. 0.18×100

13. 4.8×0.01
14. 32×0.01
15. 0.136×0.001
16. 1.36×0.001

17. 0.1×97.6
18. 9×0.1
19. 0.01×36
20. 48.5×0.01

21. 0.1×0.1
22. 32×0.001
23. 25×0.1
24. 0.1×10

25. 100×0.01
26. 0.001×1000
27. 0.4×100
28. 0.07×1000

Complete.

29. $10 \times \underline{\ ?\ } = 34.2$
30. $6.72 \times \underline{\ ?\ } = 672$
31. $1000 \times \underline{\ ?\ } = 4590$

32. $3.2 \times \underline{\ ?\ } = 0.032$
33. $6.3 \times \underline{\ ?\ } = 630$
34. $7.5 \times \underline{\ ?\ } = 0.075$

35. A quality control inspector estimates that about 0.001 of the light bulbs produced each day in a factory are defective. About how many defective bulbs are there in a day in which 50,000 are produced?

36. In a shipment of batteries, about 0.01 are defective. Predict the number of defective batteries in a shipment of 120,000.

Placing the Decimal Point

Estimate to know where to place the decimal point in this product.

$$39 \times 2.136 = 83304$$

Estimate: Round 39 to _?_ .

Round 2.136 to _?_ . $\left.\rule{0pt}{18pt}\right\}$ ⟶ $40 \times 2 = 80$

Place the decimal point: $39 \times 2.136 = \mathbf{83.304}$
▲

1. **a.** In $39 \times 2.136 = 83.304$, what is the first factor?
   b. What is the second factor?
   c. What is the product?

   Now look for a pattern for placing the decimal point in $39 \times 2.136 = 83304$.

2. **a.** How many decimal places are there in the first factor?
   b. How many decimal places are there in the second factor?
   c. What is the **sum** of your answers to **a** and **b**?
   d. How many decimal places are there in the product?

EXERCISES

*Complete this table. Look for a pattern for placing the decimal point. For each problem, answer questions **a–d** in Exercise 2 above.*

	Problem	Estimate	Answers to a.	b.	c.	d.	Place the decimal point.
1.	$8.1 \times 9.2 = 7452$	$8 \times 9 =$ _?_	?	?	?	?	7452
2.	$12.005 \times 19 = 228095$	$12 \times 20 =$ _?_	?	?	?	?	228095
3.	$62.01 \times 2.03 = 1258803$	_?_	?	?	?	?	1258803
4.	$6.201 \times 2.03 = 1258803$	_?_	?	?	?	?	1258803
5.	$62.01 \times 20.3 = 1258803$	_?_	?	?	?	?	1258803
6.	$3.001 \times 1.001 = 3004001$	_?_	?	?	?	?	3004001

Complete.

7. **a.** To multiply with decimals, first multiply as with _?_ numbers.
   b. Then count the number of decimal places in the _?_ .
   c. Write the decimal point so that the number of decimal places in the product equals the _?_ of the decimal places in the factors.

Decimals: MULTIPLICATION

You sell pennants at your school's sports activities.

1. How much will you charge for four pennants?

2. How did you know where to place the decimal point in your answer to Exercise 1?

Remember: Estimating the answer will help you to know where to place the decimal point when you multiply with decimals.

EXAMPLE 1

You sell 71 pennants at a soccer match. What are your total sales?

Estimate	Computation
Round to the nearest dollar and to a convenient number.	3.9 8
	$\times$ 7 1
	3 9 8
$3.98 \longrightarrow \$4 \qquad 71 \longrightarrow 70$	2 7 8 6
$\$4 \times 70 = \$280$ ◀ **Estimate**	2 8 2.5 8 ◀ **Total Sales: $282.58**

Compare the answer with the estimate. Is the answer reasonable?

---

3. How does the estimate help you to place the decimal point?

4. How close is the actual answer to the estimate?

This rule can help you when multiplying with decimals.

> To multiply with decimals, multiply as with whole numbers. Then place the decimal point so that the product has the same number of decimal places as the sum of the decimal places in the factors.

EXAMPLE 2

	Decimal Places			Decimal Places
a.	11.2 ◀———— 1	**b.**	11.2 ◀———— 1	
	$\times$ 0.04 ◀———— 2		$\times$ 0.004 ◀———— 3	
	0.448 ◀———— 1 + 2 = 3		**0.0448** ◀———— 1 + 3 = 4	

Insert one zero to make 4 decimal places

EXERCISES

Complete. Choose your answers from the box at the right.

1. To multiply 7.95 × 2.03, first multiply as with ? .

2. Then count the number of decimal places in both ? .

3. The product of 7.95 and 2.03 will have ? decimal places.

4. The product of 7.95 and 2.3 will have ? decimal places.

Copy the product. Place the decimal point.

5. 18 × 0.16 = 288

6. 1.8 × 1.6 = 288

7. 0.18 × 1.6 = 288

8. 0.018 × 16 = 288

9. 0.18 × 0.16 = 288

10. 0.0018 × 0.16 = 288

Multiply.

11. $4.75
× 12

12. $8.50
× 18

13. 7.65
× 6.8

14. 3.5
×9.2

15. 2.08
× 4.3

16. 10.9
×0.03

17. 0.03
×0.001

18. 0.12
×0.27

19. 6.05
×0.13

20. 0.0029
× 1.51

21. 7.842
× 1.23

22. 91.31
×0.015

23. 0.188
×0.022

24. 3.05
×1.06

25. 8.001
×0.309

26. $8.12 × 25

27. $23.65 × 20

28. 2.65 × 3.8

29. 0.412 × 0.003

30. 4.37 × 5.1

31. 0.278 × 0.13

Place the decimal point correctly in the second factor.

32. 6.24 × 31 = 19.344

33. 0.11 × 28 = 0.308

34. 0.072 × 48 = 0.03456

35. Karl ordered 6 dozen pennants from Furly Flag Company. He will pay $35.40 per dozen for the flags. How much does he owe?

36. At the next football game, Karl sold 5 dozen of the pennants he ordered. He charged $4.10 each for them. Did he make enough to pay the amount he owes?

37. The students at Boone High purchase 58 pennants at $3.85 each to decorate the gym for a dance. They spend $225.80 for refreshments and $110 for a band.
   a. Have they spent more, or less, than $500?
   b. How much more or less?

38. Find the missing digits.

```
   9.□ 3 4 □
 ×      0.9 8
 7 2 2 7 6 □
 8 1 □□ 1 4
 8.8 5 3 9 0 8
```

Check Registers

You use a **check register** to keep a record of deposits made and checks written. The **balance** is the amount of money in the account.

Check
Register

CHECK NO.	DATE	CHECK ISSUED TO	AMOUNT OF CHECK		AMOUNT OF DEPOSIT		BALANCE	
			BALANCE BROUGHT FORWARD →				425	10
125	9/5	C&P Telephone Co.	48	65			?	?
	9/13	Deposit			500	00	?	?

← Balance from previous page

1. When you enter a check amount, do you add or subtract? Why?

2. When you enter a deposit amount, do you add or subtract? Why?

EXAMPLE

Find the balance for the check register after both transactions.

Line 1: Check written ⟶ Subtract.

$425.10
− 48.65
$376.45 ◀ **New Balance**

Line 2: Deposit ⟶ Add.

$376.45
+ 500.00
$876.45 ◀ **New Balance**

The balance after both transactions is **$876.45.**

---

3. Why is it important not to delay entering new deposits and checks and computing the new balance in your check register?

EXERCISES

Find the new balance after each check written or each deposit.

1.

AMOUNT OF CHECK		AMOUNT OF DEPOSIT		BALANCE	
BALANCE BROUGHT FORWARD				181	35
18	92			?	?
		137	60	?	?
152	90			?	?

2.

AMOUNT OF CHECK		AMOUNT OF DEPOSIT		BALANCE	
BALANCE BROUGHT FORWARD				563	80
		20	25	?	?
		196	00	?	?
218	20			?	?
310	76			?	?

3.

AMOUNT OF CHECK		AMOUNT OF DEPOSIT		BALANCE	
BALANCE BROUGHT FORWARD				216	38
95	70			?	?
43	80			?	?
		151	60	?	?
25	17			?	?

Correct any errors in these check registers.

4.

CHECK NO.	DATE	CHECK ISSUED TO	AMOUNT OF CHECK		AMOUNT OF DEPOSIT		BALANCE	
		BALANCE BROUGHT FORWARD →					618	35
412	3/8	Band T. Products	45	90			572	45
413	3/9	Martin's Florist	25	00			557	45
414	3/10	Line's Department Store	300	00			257	45
	3/11	Deposit			400	00	257	45
415	3/11	G. & R. Books	29	95			227	50

5.

CHECK NO.	DATE	CHECK ISSUED TO	AMOUNT OF CHECK		AMOUNT OF DEPOSIT		BALANCE	
		BALANCE BROUGHT FORWARD →					196	27
	3/19	Deposit			582	31	878	58
286	3/21	Mana Dupont	300	00			578	58
287	3/21	American Telephone	27	85			551	73

For Exercises 6–9, copy the check register shown on page 43.
Use the information given to complete the check register.

6. On September 15, Mario had a balance of $229 in his checking account. He made a deposit of $100 the next day. On September 17, he wrote check number 354 for $27.38 to Smith Hardware.

7. On November 2, Jennifer had a balance of $179.60 in her checking account. She wrote check number 228 for $49.50 to Klein Shoes for a pair of shoes and check number 229 for $15.86 to Chancer Books, Inc.

8. On January 28, Maria had a balance of $987.15 in her checking account. On February 1, she made a deposit of $67.25 and wrote check number 421 for $223.40 to A-Z Hardware for tools.

9. On March 20, Wilmer wrote check number 501 for $245.81 to Johnson's Pharmacy and check number 502 for $345.80 to Garrison's Furniture Store. His new balance was $182.50. How much was in the account before he wrote the checks on March 20?

Strategy: USING PATTERNS

Number patterns have fascinated people since the beginning of recorded history. For example, these numbers are called **triangular numbers** because of the pattern of dots associated with each number.

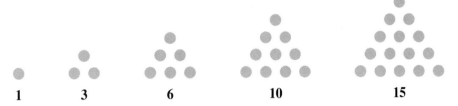

1   3   6   10   15

1. What is the next triangular number? Draw it.

A list of numbers such as 1, 3, 6, 10, 15, . . . is called a **sequence**. When you know the pattern in a sequence, you can write other numbers, or **terms** of the sequence.

EXAMPLE Find the pattern in each sequence. Then write two more terms.

a. 2, 4, 6, 8, 10, _?_ , _?_

$$2 \quad\quad 4 \quad\quad 6 \quad\quad 8 \quad\quad 10$$

Think: 2 + 2 = 4   4 + 2 = 6   6 + 2 = 8   8 + 2 = 10

Pattern: Each term is **2 more than the one before.**

Next two terms: 10 + 2 = **12**   12 + 2 = **14**

b. 50, 45, 40, 35, _?_ , _?_

$$50 \quad\quad 45 \quad\quad 40 \quad\quad 35$$

Think: 50 − 5 = 45   45 − 5 = 40   40 − 5 = 35

Pattern: Each term is **5 less than the one before.**

Next two terms: 35 − 5 = **30**   30 − 5 = **25**

c. 15, 30, 60, 120, _?_ , _?_

$$15 \quad\quad 30 \quad\quad 60 \quad\quad 120$$

Think: 15 × 2 = 30   30 × 2 = 60   60 × 2 = 120

Pattern: Each term is **twice the one before.**

Next two terms: 120 × 2 = **240**   240 × 2 = **480**

EXERCISES

Find the missing terms.

1. 3, 6, 9, __?__, 15, __?__, 21

2. 4, 8, 12, __?__, 20, __?__, __?__

3. 100, 95, 90, __?__, 80, __?__, __?__

4. 24, 20, 16, 12, __?__, __?__

5. 3, 6, 12, __?__, 48, __?__, 192

6. 4, 12, 36, __?__, __?__, 972

7. 15, 21, 27, 33, __?__, __?__

8. 93, 86, 79, __?__, __?__, 58

9. 1, 5, 25, 125, __?__, __?__

10. 12, 23, 34, __?__, __?__, 67

11. 3, 5, 8, 12, __?__, 23, __?__

12. 53, 51, 47, 41, __?__, __?__,

13. What is the sum of the six terms in Exercise 4?

14. What is the sum of the six terms in Exercise 6?

15. a. Write a sequence having three missing terms.
b. Explain how to find the missing terms.

16. a. Write a sequence that uses a different rule than Exercise 15. Have three missing terms.
b. Explain how to find them.

Examine this arrangement of numbers.

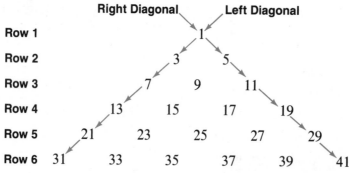

17. Look for a pattern in the rows of numbers. What numbers will be in Row 7?

18. Look for a pattern in the right diagonal. Start at 1. What two numbers will come after 41?

19. Examine the diagonal that starts with 3, 9, and so on. Is the pattern the same as the pattern in Exercise 17?

20. Examine the left diagonal.
a. How is the pattern the same as that in the right diagonal?
b. How is it different?

PROJECT Work in small groups or as your teacher directs. Have a contest to see which group or person can discover the most patterns in the arrangement of numbers.

Planning A Vacation

Michael is planning a vacation that will last 7 days and 6 nights. He collected the following information from tour brochures, advertisements, travel agencies, and friends. Then he made charts to help him plan the trip and organize the information.

1. Round trip air fare: $468
   This includes lunch and dinner each way.

2. Round trip train fare: $367
   This includes one night's lodging each way. Meals cost an additional $25 each way.

3. Round trip driving costs: $159
   It will take one and a half days each way. The driving costs include one night's lodging each way.

HOTEL/MOTEL

1. Budget Inn: $23.00 per night
   Small room and no television

2. Inn at the Lake: $54.00 per night
   Large room with television

3. Pheasant Lodge: $127.00 per night
   Luxurious room. Television, heated pool, spa, health club.

	Cost/Night	Tax	Total
Sun.			
Mon.			
Tues.			
Wed.			
Thurs.			
Fri.			

Total Hotel Cost _____

FOOD

1. Fast food restaurants
   Breakfast: $2   Lunch: $4
   Dinner: $5

2. Mom's Diner
   Breakfast: $3   Lunch: $5
   Dinner: $7

3. The Gracious Gourmet
   Breakfast: $9   Lunch: $11
   Dinner: $19

	B	L	D	Total
Sun.				
Mon.				
Tues.				
Wed.				
Thurs.				
Fri.				
Sat.				

Total Food Cost _____

ENTERTAINMENT

(No more than 2 events per day)

1. Movies
   Afternoon: $3   Evening: $5

2. Live Theater (plays, musicals)
   Afternoon: $17   Evening: $22

3. Amusement Park
   One-day: $27   Three-day pass: $75

	Event 1 Cost	Event 2 Cost	Total
Sun.	_____	_____	_____
Mon.	_____	_____	_____
Tues.	_____	_____	_____
Wed.	_____	_____	_____
Thurs.	_____	_____	_____
Fri.	_____	_____	_____
Sat.	_____	_____	_____

Total Entertainment Cost _____

TOTAL EXPENSES: Transportation, Lodging, Food, Entertainment _____

EXERCISES

For Exercises 1–2, suppose that you are Michael. Use the information he collected to find the cost of each vacation. Make charts similar to those on pages 47–48 to organize the information.
NOTE: Travel time will affect the costs for food and hotel/motel.

1. Travel by train
   Stay at Pheasant Lodge
   Eat at the Gracious Gourmet
   Evening movies: 3
   Amusement park: 3-day pass

2. Travel by car
   Stay at the Inn on the Lake
   Eat at fast food restaurants
   Amusement park: One day
   Afternoon movies: 2

3. Suppose you are Michael. Plan the vacation if you have $900 to spend.

PROJECT

Suppose that you have graduated from junior college and are looking for a full-time job.

a. Search through newspapers to find advertisements of jobs you would like to have.

b. What is the yearly salary for one of the jobs you found?

c. What is the salary for two weeks?

d. Plan a vacation of 7 days and 6 nights to a place of your choosing in the United States. The amount you can spend will equal two weeks salary for the job you chose in a, b, and c.

e. Your completed project should provide an outline of your vacation and include the following.

   (1) Your destination   (2) The method and cost of transportation
   (3) Lodging, Meal, and Entertainment costs   (4) Total costs

Chapter Summary

1. To round a decimal, look at the digit to the right of the place to which you are rounding.
   a. If the digit is less than 5, round down.
   b. If the digit is 5 or more, round up.

2. To add and subtract with decimals:
   1. Line up the decimal points one under the other.
   2. Annex final zeros when necessary.
   3. Add or subtract as with whole numbers.

3. To find the net deposit:
   1. Add the amounts of the cash and the checks deposited to find the total.
   2. Subtract the cash returned from the total. This is the net deposit.

4. To multiply a number by 10, by 100, or by 1000, annex one, two, or three zeros respectively.

5. To multiply with whole numbers:
   1. Multiply by the ones, by the tens, and so on.
   2. Add the products obtained in Step 1.

6. To multiply a decimal by 10, by 100, or by 1000:
   1. Move the decimal point one, two, or three places to the right respectively.
   2. When necessary, annex zeros in order to have the correct number of decimal places.

7. To multiply with decimals:
   1. Multiply as with whole numbers.
   2. Count the number of decimal places in the factors. Starting at the right and moving to the left, count the same number of decimal places in the product. Insert the decimal point.

8. To find the new balance in a check register, subtract the amount of the check from the balance, or add the amount of the deposit to the balance.

---

1. 5.67 rounded to the nearest whole number is **6.**

   0.455 rounded to the nearest hundredth is **0.46.**

2.
   a.
   $$\begin{array}{r} \overset{1}{1}\overset{1}{0} \\ 16.070 \\ 5.203 \\ +\ 91.532 \\ \hline \mathbf{112.805} \end{array}$$
   b.
   $$\begin{array}{r} \overset{2}{}\overset{9}{}\overset{9}{4}\overset{10}{4} \\ 3.\cancel{4}\cancel{4}\cancel{4} \\ -\ 1.999 \\ \hline \mathbf{1.001} \end{array}$$

3.

CASH		Dollars	Cents
CHECKS	1	4 2 1	9 0
List	2	1 2 5	0 0
Each	3		
Check	4		
TOTAL		5 4 6	9 0
▶ Less Cash Rec'd		7 5	0 0
NET DEPOSIT		4 7 1	9 0

4. $471 \times \mathbf{10} = 47\mathbf{10}$
   $6 \times \mathbf{100} = 6\mathbf{00}$
   $135 \times \mathbf{1000} = 135{,}\mathbf{000}$

5.
$$\begin{array}{r} 172 \\ \times\ 349 \\ \hline 1\ 548 \\ 6\ 88 \\ 51\ 6 \\ \hline \mathbf{60{,}028} \end{array}$$

6. $5.84 \times 10 = \mathbf{58.4}$
   $0.9 \times 100 = \mathbf{90}$
   $21.3 \times 1000 = \mathbf{21{,}300}$

7.
$$\begin{array}{rl} & \text{Decimal Places} \\ 11.08 & \longleftarrow\ 2 \\ \times\ \ 1.05 & \longleftarrow\ 2 \\ \hline 55\ 40 & \\ 0\ 00\ 0 & \\ 11\ 08 & \\ \hline \mathbf{11.63\ 40} & \longleftarrow\ 2 + 2 = 4 \end{array}$$

8. $\begin{array}{r} \$331.15 \\ -\ 25.00 \\ \hline \mathbf{\$306.15} \end{array}$ ← Check
   ← New balance

   $\begin{array}{r} \$306.15 \\ +\ 85.50 \\ \hline \mathbf{\$391.65} \end{array}$ ← Deposit
   ← New balance

Chapter Review

Part 1: VOCABULARY

For Exercises 1–8, choose from the box at the right the word(s) that complete(s) each statement.

1. When adding or subtracting with decimals, first line up the __?__ .   (Page 28)

2. In a multiplication problem, the __?__ are the numbers that are multiplied.   (Page 32)

3. In a multiplication problem, the __?__ is the answer. (Page 32)

4. The amount of money you put into an account is a __?__ . (Page 36)

5. When multiplying by 10, by 100, or by 1000, move the decimal point to the __?__ .   (Page 38)

6. When multiplying by 0.1, by 0.01, or by 0.001, move the decimal point to the __?__ .   (Page 38)

7. To place the decimal point in a product, you find the sum of the number of __?__ places in the factors.   (Page 41)

8. The amount of money in a checking account is the __?__ .   (Page 43)

| balance |
| product |
| decimal |
| decimal points |
| left |
| factors |
| right |
| deposit |
| check register |

Part 2: SKILLS

Round to the nearest whole number.   (Pages 26–27)

9. 36.84   **10.** 9.63   **11.** 0.723   **12.** 900.3   **13.** 6.578   **14.** 187.44

Round to the nearest tenth.   (Pages 26–27)

15. 2.856   **16.** 41.145   **17.** 5.968   **18.** 0.631   **19.** 83.453   **20.** 91.72

Round to the nearest thousandth.   (Pages 26–27)

21. 4.6809   **22.** 24.1283   **23.** 12.3594   **24.** 0.9316   **25.** 38.4601   **26.** 1.9997

Add or subtract as indicated.   (Pages 28–29)

27.	28.	29.	30.	31.
36.86	1.38	23.728	2.516	6.27
9.25	16.24	2.597	6.24	7.4
+ 15.38	+ 8.39	+ 6.362	+4.518	+21.836

32.   7.64
      $-$ 3.59

33.   18.5
       $-$ 9.8

34.   28.03
       $-$ 6.95

35.   2.36
       $-$ 1.034

36.   28.6
       $-$ 19.732

37. 3.17 + 14.38 + 6.53 + 2.45   **38.** 700 $-$ 35.64   **39.** 3000 $-$ 981.42

Complete.   (Pages 30–31)

40. 21 $\times$ _?_ = 210   **41.** 55 $\times$ _?_ = 55,000   **42.** _?_ $\times$ 100 = 3400

Multiply.   (Pages 32–33)

43. 45 $\times$ 18   **44.** 62 $\times$ 31   **45.** 625 $\times$ 73   **46.** 470 $\times$ 224   **47.** 308 $\times$ 409

Multiply.   (Page 38–39)

48. 6.21 $\times$ 10   **49.** 6.28 $\times$ 100   **50.** 6.28 $\times$ 1000   **51.** 23.8 $\times$ 10

52. 23.8 $\times$ 0.01   **53.** 1.62 $\times$ 0.1   **54.** 4.8 $\times$ 0.1   **55.** 5.62 $\times$ 0.001

Multiply.   (Pages 41–42)

56.   47
      $\times$ 0.8

57.   92
      $\times$ 0.6

58.   321
       $\times$ 0.9

59.   462
       $\times$ 2.1

60.   3.24
       $\times$ 9.06

61.   6.8
      $\times$ 0.002

62.   3.5
      $\times$ 4.7

63.   7.07
       $\times$ 0.23

64.   0.198
       $\times$ 0.035

65.   7.015
       $\times$ 0.039

Part 3: APPLICATIONS

66. June paid for some clothing purchases with a $50-bill. She received $15.28 in change. How much did the clothes cost? (Pages 28–29)

67. Ward deposits checks for $235.50 and $72.60. He receives 4 twenty-dollar bills, 3 ten-dollar bills, 2 quarters and 6 pennies. What is Ward's net deposit?   (Pages 36–37)

68. In a shipment of auto parts, about 0.01 are defective. Predict the number of defective parts in a shipment of 500.   (Pages 38–39)

69. Eduardo ordered 8 dozen school T-shirts to sell at football games. He paid $42.24 per dozen for the shirts. How much did he pay for 8 dozen? (Pages 41–42)

70. On October 3, Mrs. Rutherford had a balance of $436 in her checking account. The next day she wrote check number 24 for $88.63 to a grocery store. On October 8, she made a deposit of $120.00. Find the balance after each transaction. (Pages 43–44)

71. Find the pattern in the sequence below. Then write the next two terms.   (Pages 45–46)

2, 6, 18, 54, _?_ , _?_

Chapter Test

Round each number as indicated.

1. 7.83; to the nearest whole number

2. 81.839; to the nearest tenth

3. 29.645; to the nearest hundredth

4. 4.73416; to the nearest thousandth

Perform the indicated operations.

5.
$$\begin{array}{r} 32.6 \\ 143.72 \\ +\ \ 8.61 \\ \hline \end{array}$$

6.
$$\begin{array}{r} 26.4 \\ -19.8 \\ \hline \end{array}$$

7.
$$\begin{array}{r} 32.5 \\ -\ 1.62 \\ \hline \end{array}$$

8. $632.1 + 93.62 + 8.461 + 0.7$

9. $300 - 19.96$

10. $250 + 9.7 + 25.036$

11.
$$\begin{array}{r} 32 \\ \times 47 \\ \hline \end{array}$$

12.
$$\begin{array}{r} 81 \\ \times 79 \\ \hline \end{array}$$

13.
$$\begin{array}{r} 312 \\ \times\ \ 45 \\ \hline \end{array}$$

14.
$$\begin{array}{r} 340 \\ \times 210 \\ \hline \end{array}$$

15.
$$\begin{array}{r} 806 \\ \times 204 \\ \hline \end{array}$$

16.
$$\begin{array}{r} 52 \\ \times 0.9 \\ \hline \end{array}$$

17.
$$\begin{array}{r} 3.8 \\ \times 1.2 \\ \hline \end{array}$$

18.
$$\begin{array}{r} 6.21 \\ \times\ \ 0.3 \\ \hline \end{array}$$

19.
$$\begin{array}{r} 8.8 \\ \times 0.004 \\ \hline \end{array}$$

20.
$$\begin{array}{r} 21.5 \\ \times 0.012 \\ \hline \end{array}$$

21. 62×10

22. 8.6×0.1

23. 32.8×1000

24. 57×100

25. 1.4×0.01

26. 3.9×0.001

27. 15×1000

28. 0.5×100

29. Rosita paid for some school supplies with a \$20-bill. She received \$6.47 in change. How much did the supplies cost?

30. Mr. Sulu deposits checks for \$75.50 and \$53.45. He receives 2 twenty-dollar bills, 5 five-dollar bills, and 7 dimes. What is Mr. Sulu's net deposit?

31. Find the pattern in the sequence below. Then write the next two terms.

    82, 70, 58, 46, __?__ , __?__

32. On May 10, Valerie had a balance of \$641.80 in her checking account. On May 12, she made a deposit of \$110.50 and wrote check number 18 for \$375.55 to a credit card company. Find the balance after each transaction.

33. A quality control inspector estimates that about 0.001 of the pocket calculators produced each day are defective. About how many defective calculators are there in a day in which 4000 are produced?

Cumulative Maintenance   Chapters 1–2

*Choose the correct answer. Choose **a, b, c,** or **d.***

1. Add.

$$4323$$
$$215$$
$$+4315$$

   a. 8753   **b.** 8853
   c. 8843   **d.** 8743

2. Round 136,835 to the nearest thousand.

   a. 136,000   **b.** 140,000
   c. 137,000   **d.** 100,000

3. Kim earned $124 in tips during April, $185 in May, and $210 in June. Find the average monthly tips.

   a. $150   **b.** $167
   c. $165   **d.** $173

4. Subtract.
$$5684 - 2696$$

   a. 3088   **b.** 2988
   c. 2978   **d.** 2982

5. Multiply.

$$38$$
$$\times\,10000$$

   a. 38,000   **b.** 380,000
   c. 3800   **d.** 380

6. Round 8.09 to the nearest tenth.

   a. 8.9   **b.** 8.1
   c. 9.0   **d.** 8.0

7. Multiply.
$$30.7 \times 1000$$

   a. 30,700   **b.** 307,000
   c. 3070   **d.** 37,000

8. Jennifer bought three sweaters at $27 each and six pairs of socks at $4 each. Which best describes how much money she spent?

   a. $3 \times (\$27 + 6) \times \$4$
   b. $(3 + \$27) \times (6 + \$4)$
   c. $10 \times (\$27 + \$4)$
   d. $(3 \times \$27) + (4 \times \$6)$

9. Add.
$$12 + 6 + 83 + 9$$

   a. 108   **b.** 110
   c. 98   **d.** 107

10. Multiply.

$$279$$
$$\times\ 52$$

   a. 14,788   **b.** 15,788
   c. 14,508   **d.** 15,408

11. Multiply.

$$306 \times 350$$

a. 107,000   b. 10,700
c. 10,710   d. 107,100

12. Subtract.

$$8003 - 276$$

a. 7627   b. 7827
c. 7727   d. 2737

13. Multiply.

$$\begin{array}{r} 4.25 \\ \times \ 3.7 \\ \hline \end{array}$$

a. 15.725   b. 4.250
c. 12.75   d. 157.25

14. Find the answer.

$2 \times 5 + 4 \div 2$

a. 7   b. 9
c. 12   d. 14

15. Which shows the numbers listed in order from least to greatest?

a. 0.99,   0.9,   1.9,   1.09
b. 1.9,   1.09,   0.99,   0.9
c. 0.99,   0.9,   1.09,   1.9
d. 0.9,   0.99,   1.09,   1.9

16. Multiply: 31.7×0.24

a. 7.608   b. 760.8
c. 76.08   d. 7608

17. Add.

$$\begin{array}{r} 3.48 \\ 2.5 \\ + \ 14.608 \\ \hline \end{array}$$

a. 20.66   b. 20.588
c. 14.981   d. 20.016

18. Subtract.

$$\begin{array}{r} 400 \\ - \ 13.68 \\ \hline \end{array}$$

a. 387.68   b. 413.68
c. 386.32   d. 386.42

19. What does $\square + \triangle + \bigcirc$ equal?

Clue A: $\square + \triangle = 13$
Clue B: $4 \times \square = 32$
Clue C: $\square + \bigcirc = 17$

a. 22   b. 31
c. 28   d. 21

20. Write 4^5 in standard form.

a. 625   b. 20
c. 256   d. 1024

21. Multiply.

$$46 \times 28$$

a. 1087   b. 1287
c. 1188   d. 1288

22. What is the total value of eighteen $5-bills, five $10-bills, and twelve $20-bills?

a. $380   b. $350
c. $370   d. $320

Division: Whole Numbers/Decimals

3 CHAPTER

- How do you find the unit price of an item?

- Which would be the more efficient way to find the unit price of a product, use paper and pencil or use a calculator?

- Will the larger size of a product always have the lower unit price?

- When comparing two items to find the better buy, what else should be considered besides unit price?

Divisibility

1. Is 29 an even number? Why or why not?

During a gasoline shortage, a state had an "even-odd" system for buying gasoline. Car owners with license plate numbers that could be evenly divided by 2 could buy gasoline on even-numbered calender days; all other car owners could buy gas only on odd-numbered calendar days.

2. Your car had this license plate number. Could you buy gasoline on October 22?

3. Can 3782 be divided by 2 without having a remainder?

One number is **divisible** by another if the remainder is 0. In the division problem at the right, 630 is divisible by 9.

$$\begin{array}{r} 70 \\ 9\overline{)630} \\ \underline{630} \\ 0 \\ \underline{0} \\ 0 \end{array}$$

Divisor ⟶ 9) 630 ⟵ Dividend

70 ⟵ Quotient

0 ⟵ Remainder

There are a few simple tests for determining whether a number is divisible by 2, 3, 4, 5, 9, or 10.

Rules for Divisibility

A number is divisible by:
- **2** if it ends in 0, 2, 4, 6, or 8.
- **5** if it ends in 0 or 5.
- **10** if it ends in 0.
- **3** if the sum of its digits is divisible by 3.
- **4** if its last two digits are divisible by 4.
- **9** if the sum of its digits is divisible by 9.

---

4. What is the sum of the digits of the number 726?

5. Is the sum of the digits of 726 divisible by 3? Is 726 divisible by 3?

6. Is 352 divisible by 4? Why or why not?

EXERCISES

Find the sum of the digits of each number.

1. 29
$2 + 9 = \underline{\ ?\ }$

2. 74
$7 + 4 = \underline{\ ?\ }$

3. 217
$2 + 1 + 7 = \underline{\ ?\ }$

4. 345
$3 + 4 + 5 = \underline{\ ?\ }$

5. 2215
$2 + 2 + 1 + 5 = \underline{\ ?\ }$

For Exercises 6–11, choose numbers from the box at the right.

117	938	792
52	811	834
507	920	333

6. Which numbers are divisible by 2?

7. Which numbers are divisible by 3?

8. Which numbers are divisible by 4?

9. Which numbers are divisible by 5?

10. Which numbers are divisible by 9?

11. Which numbers are divisible by 10?

Copy the chart shown below. Fill in each row of the chart with Yes or No. The first row is done for you.

	Number	Is the number divisible by:					
		2?	**3?**	**4?**	**5?**	**9?**	**10?**
12.	195	No	Yes	No	Yes	No	No
13.	240	?	?	?	?	?	?
14.	216	?	?	?	?	?	?
15.	184	?	?	?	?	?	?
16.	330	?	?	?	?	?	?
17.	160	?	?	?	?	?	?
18.	435	?	?	?	?	?	?
19.	983	?	?	?	?	?	?
20.	217	?	?	?	?	?	?

21. The number 456 is divisible by 3. Using only the digits 4, 5, and 6, what other 3-digit numbers also divisible by 3 can be formed?

22. Replace the missing digit in 73,8 $\underline{\ ?\ }$ 2 so that the resulting number is divisible by 9.

23. A garden store sells shrubs for $3 each. Can you buy exactly $136 worth of shrubs?

24. Rose bushes are on sale in groups of 4 only. Can you buy exactly 168 rose bushes?

Division: WHOLE NUMBERS

Evelyn is saving to buy a stereo that costs $648. She thinks that she can save $18 a week.

1. *Complete:* To find the number of weeks it will take to save $648, you __?__ 648 by 18.

EXAMPLE 1

Divide 648 by 18.

1 **Estimate.**

Round the divisor and the dividend to **compatible** (convenient) numbers. Then estimate the quotient. This will help you to place the first digit in the quotient.

$$18\overline{)649} \longrightarrow 20\overline{)600}^{\,30}$$ ◀ **The first digit is in the tens place.**

2 **Divide.**

a. Try 3 in the tens place.

$$18\overline{)648}^{\,3}$$
$$\underline{54}$$
$$10$$

b. Bring down the 8.

Estimate: $20\overline{)108}^{\,5}$
Try 5.

$$18\overline{)648}^{\,35}$$
$$\underline{54}\!\!\downarrow$$
$$108$$
$$\underline{90}$$
$$18$$ ◀ **18 is too much.**

c. Since 5 was too small, try 6.

$$18\overline{)648}^{\,36}$$ ◀ **Quotient**
$$\underline{54}$$
$$108$$
$$\underline{108}$$
$$0$$ ◀ **Remainder**

2. In step 2b of Example 1, why is 18 "too much"?

3. How many weeks will it take Evelyn to pay for the stereo?

4. To estimate $78\overline{)262}$, which numbers would be more compatible, $80\overline{)300}$ or $80\overline{)240}$? Explain.

In some problems, the divisor does not divide <u>evenly</u> into the dividend.

EXAMPLE 2

Divide and check: $42\overline{)3858}$

1 **Estimate:** $42\overline{)3858} \longrightarrow 40\overline{)3600}^{\,90}$ ◀ **The first digit is in the tens place.**

2 **Divide.**

a. Try 9 in the tens place.

$$
\begin{array}{r}
9 \\
42 \overline{\smash{)}3858} \\
378 \\
\hline
7
\end{array}
$$

b. Bring down the 8. **Estimate:** $40\overline{\smash{)}78}$ Try 1.

$$
\begin{array}{r}
1 \\
40 \overline{\smash{)}78}
\end{array}
$$

$$
\begin{array}{r}
91 \\
42 \overline{\smash{)}3858} \\
378\downarrow \\
\hline
78 \\
42 \\
\hline
36 \quad\longleftarrow \text{ Remainder}
\end{array}
$$

Quotient: **91 r36**

3 **Check.** Use a calculator.

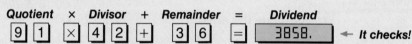

Quotient	×	Divisor	+	Remainder	=	Dividend	
9 1	×	4 2	+	3 6	=	3858.	← It checks!

5. Does "divide evenly" mean the same as "is divisible by"? Explain.

EXERCISES

For Exercises 1–10, use compatible numbers to estimate each quotient.

1. $58\overline{\smash{)}236}$ **2.** $89\overline{\smash{)}278}$ **3.** $22\overline{\smash{)}616}$ **4.** $79\overline{\smash{)}453}$ **5.** $32\overline{\smash{)}650}$

6. $41\overline{\smash{)}8663}$ **7.** $61\overline{\smash{)}1862}$ **8.** $18\overline{\smash{)}7156}$ **9.** $41\overline{\smash{)}1657}$ **10.** $72\overline{\smash{)}3566}$

Estimate to determine how many digits there will be in each quotient. Write one, two, or three.

11. $36\overline{\smash{)}432}$ **12.** $91\overline{\smash{)}809}$ **13.** $45\overline{\smash{)}8995}$ **14.** $72\overline{\smash{)}3509}$ **15.** $59\overline{\smash{)}3599}$

Divide. Check each answer.

16. $19\overline{\smash{)}84}$ **17.** $28\overline{\smash{)}93}$ **18.** $19\overline{\smash{)}95}$ **19.** $21\overline{\smash{)}63}$ **20.** $25\overline{\smash{)}525}$

21. $75\overline{\smash{)}450}$ **22.** $63\overline{\smash{)}495}$ **23.** $44\overline{\smash{)}283}$ **24.** $52\overline{\smash{)}994}$ **25.** $37\overline{\smash{)}262}$

26. $55\overline{\smash{)}392}$ **27.** $65\overline{\smash{)}7750}$ **28.** $84\overline{\smash{)}3924}$ **29.** $54\overline{\smash{)}3672}$ **30.** $43\overline{\smash{)}2659}$

31. $82\overline{\smash{)}3582}$ **32.** $42\overline{\smash{)}6367}$ **33.** $43\overline{\smash{)}5701}$ **34.** $37\overline{\smash{)}8838}$ **35.** $47\overline{\smash{)}9365}$

36. $56\overline{\smash{)}6339}$ **37.** $89\overline{\smash{)}9226}$ **38.** $97\overline{\smash{)}9634}$ **39.** $84\overline{\smash{)}415}$ **40.** $62\overline{\smash{)}753}$

41. $28\overline{\smash{)}934}$ **42.** $34\overline{\smash{)}506}$ **43.** $43\overline{\smash{)}4921}$ **44.** $38\overline{\smash{)}8230}$ **45.** $79\overline{\smash{)}1663}$

46. Use the numbers 1, 2, 3, and 4 to complete this problem.

$$23\overline{\smash{)}\blacksquare 8\blacksquare}$$

47. Use the numbers 5, 6, and 7 to complete the problem.

$$12\overline{\smash{)}6\blacksquare 2}$$

Zeros in the Quotient

Brian is thinking about what he might like to do when he finishes school. He read a newspaper article that gave the average yearly salary for a mail carrier as $24,240.

1. How would you find the average monthly salary for a mail carrier?

EXAMPLE 1

Divide 24,240 by 12.

1 **Estimate** the answer. ⟶
$$\begin{array}{r} 2000 \\ 12 \overline{)24000} \end{array}$$ ◀ *Estimate*

2 **Divide.**

  a. Try 2 in the thousands place.

$$\begin{array}{r} 20 \\ 12 \overline{)24240} \\ \underline{24}\!\downarrow \\ 02 \end{array}$$ ◀ **2 is less than 12. Write a "0" in the quotient.**

The **check** is left for you.

  b. Bring down the 4.

Estimate: $\begin{array}{r} 2 \\ 12 \overline{)24} \end{array}$

$$\begin{array}{r} 2020 \\ 12 \overline{)24240} \\ \underline{24}\!\downarrow \\ 024 \\ \underline{24}\!\downarrow \\ 00 \\ \underline{00} \end{array}$$

◀ *Quotient*

◀ **0 is less than 12. Write a "0" in the ones place.**

2. What is the average monthly salary for a mail carrier?

EXAMPLE 2

Divide 23,873 by 78.

1 **Estimate.** ⟶
$$\begin{array}{r} 300 \\ 80 \overline{)24000} \end{array}$$

2 **a. Divide.** Try 3 in the hundreds place.

$$\begin{array}{r} 30 \\ 78 \overline{)23873} \\ \underline{234} \\ 47 \end{array}$$ ◀ **47 is less than 78. Write a "0" in the quotient.**

The **check** is left for you.

  b. Bring down the 3.

Estimate: $\begin{array}{r} 6 \\ 80 \overline{)480} \end{array}$

$$\begin{array}{r} 306 \text{ r5} \\ 78 \overline{)23873} \\ \underline{234} \\ 473 \\ \underline{468} \\ 5 \end{array}$$

Quotient: **306 r5**

3. *Complete:* To check your answer in Example 2, first multiply 306 and __?__. Then add __?__ to the product.

EXERCISES

For Exercises 1–5, estimate each quotient. Use compatible numbers.

1. $30 \overline{) 656}$ **2.** $39 \overline{) 899}$ **3.** $43 \overline{) 267}$ **4.** $58 \overline{) 5916}$ **5.** $28 \overline{) 1260}$

6. Is $560 \div 62$ closer to 9 or to 90? Why?

7. Is $986 \div 78$ closer to 10 or to 100? Why?

8. Is $2208 \div 53$ closer to 40 or to 400? Why?

9. Is $6719 \div 32$ closer to 20 or to 200? Why?

10. Is $4321 \div 24$ closer to 200 or to 300? Why?

11. Is $4316 \div 83$ closer to 50 or to 500? Why?

Find the missing quotient.

12.
$$58 \overline{) 5916}$$
$$\underline{58}$$
$$116$$
$$\underline{116}$$
$$0$$

13. r 3
$$63 \overline{) 12918}$$
$$\underline{126}$$
$$318$$
$$\underline{315}$$
$$3$$

14. r 6
$$39 \overline{) 90018}$$
$$\underline{78}$$
$$120$$
$$\underline{117}$$
$$318$$
$$\underline{312}$$
$$6$$

15. r 18
$$24 \overline{) 38610}$$
$$\underline{24}$$
$$146$$
$$\underline{144}$$
$$210$$
$$\underline{192}$$
$$18$$

Divide.

16. $68 \overline{) 715}$ **17.** $57 \overline{) 609}$ **18.** $43 \overline{) 4491}$ **19.** $54 \overline{) 5715}$

20. $67 \overline{) 17432}$ **21.** $57 \overline{) 20538}$ **22.** $41 \overline{) 11085}$ **23.** $51 \overline{) 14291}$

24. $43 \overline{) 872}$ **25.** $21 \overline{) 849}$ **26.** $23 \overline{) 927}$ **27.** $14 \overline{) 985}$

28. $37 \overline{) 7696}$ **29.** $43 \overline{) 4429}$ **30.** $42 \overline{) 4410}$ **31.** $75 \overline{) 30525}$

32. $39 \overline{) 80124}$ **33.** $72 \overline{) 15108}$ **34.** $48 \overline{) 9718}$ **35.** $97 \overline{) 9809}$

36. $12 \overline{) 2488}$ **37.** $24 \overline{) 9120}$ **38.** $69 \overline{) 5541}$ **39.** $34 \overline{) 17218}$

40. $42 \overline{) 25563}$ **41.** $29 \overline{) 879}$ **42.** $85 \overline{) 1747}$ **43.** $72 \overline{) 7513}$

44. $64 \overline{) 4480}$ **45.** $56 \overline{) 8427}$ **46.** $72 \overline{) 94075}$ **47.** $48 \overline{) 52065}$

48. The newspaper article Brian read said that accountants earned an average yearly salary of $31,296. What is the average monthly salary?

49. Brian also read that plumbers earn an average yearly salary of $22,880. What is their average weekly salary?

Mid-Chapter Review

Determine whether each number is divisible by 2, by 3, by 4, by 5, by 9, or by 10.   (Pages 56–57)

1. 28   2. 252   3. 85   4. 506   5. 550   6. 633   7. 450

8. 48   9. 245   10. 920   11. 248   12. 117   13. 1197   14. 2196

Divide. Check each answer.   (Pages 58–59)

15. 23) 46   16. 29) 464   17. 34) 728   18. 52) 884   19. 27) 642

20. 75) 450   21. 63) 495   22. 43) 2659   23. 55) 3905   24. 82) 3582

Divide. Check each answer.   (Pages 60–61)

25. 31) 938   26. 23) 927   27. 74) 7795   28. 56) 5779

29. 75) 3055   30. 72) 15108   31. 57) 20538   32. 65) 26615

33. A sporting goods store sells cans of tennis balls for $2 each. Can you buy exactly $75 worth of cans? (Pages 56–57)

34. A total of 56 campers signed up for a hike. Can they be divided into groups of 4 hikers each to share equipment? (Pages 56–57)

35. Patti read that police officers earn an average yearly salary of $20,508. What is their average monthly salary? (Pages 60–61)

36. Ron read that automobile mechanics earn an average yearly salary of $15,704. What is their average weekly salary? (Pages 60–61)

MAINTENANCE

Multiply.   (Pages 41–42)

37.
$$\begin{array}{r} 1.25 \\ \times\ \ \ 4 \\ \hline \end{array}$$

38.
$$\begin{array}{r} 61.4 \\ \times\ \ \ 3 \\ \hline \end{array}$$

39.
$$\begin{array}{r} 89.7 \\ \times\ 1.5 \\ \hline \end{array}$$

40.
$$\begin{array}{r} 153 \\ \times\ 4.2 \\ \hline \end{array}$$

41.
$$\begin{array}{r} 2.06 \\ \times\ 3.8 \\ \hline \end{array}$$

42.
$$\begin{array}{r} 64 \\ \times 0.75 \\ \hline \end{array}$$

43.
$$\begin{array}{r} 47.2 \\ \times\ 0.3 \\ \hline \end{array}$$

44.
$$\begin{array}{r} 1.85 \\ \times 0.25 \\ \hline \end{array}$$

45. Mrs. Peacock deposits checks for $45 and $160. She receives $30. What is her net deposit? (Pages 36–37)

46. Mr. Fawlty deposits checks for $282 and $79. He receives 3 twenty-dollar bills. What is his net deposit? (Pages 36–37)

Math and Finance Charges

Loan officers assist customers who borrow money to determine the interest or **finance charge,** on a loan. Usually, the amount borrowed plus the finance charge is paid back in equal monthly payments over a specified number of months.

1. *Complete:* Amount owed = Amount of loan + _?_

2. *Complete:* Monthly payment = _?_ ÷ Number of months

EXAMPLE   Consolidated Bank lends a customer $4000 for 12 months. Finance charges are $356.

a. Find the total amount owed.
b. Find the amount of the monthly payment.

a. Amount owed:
$4000 + $356 = **$4356** 〈 Amount Borrowed + Finance Charges = Amount Owed

b. Divide the amount owed by the number of months.

$4356 ÷ 12 = **$363**   Monthly payment: **$363**

EXERCISES   *Complete the table.*

	Amount Borrowed	Finance Charges	Total Amount Owed	Number of Months to Repay	Monthly Payment
1.	$5000	$256	?	12	?
2.	$3860	$352	?	18	?
3.	$4500	$516	?	24	?
4.	$2800	$248	?	24	?

5. Paul agreed to pay back $1000 in 18 equal monthly payments. Finance charges are $62. Find the amount of each monthly payment.

6. Joan agreed to repay a loan of $3000 in equal monthly installments over 2 years. Finance charges are $288. Find the monthly payment.

7. A loan was repaid in 12 equal installments of $107 each. An interest charge of $96 was included in the payments. What was the original amount of the loan?

Dividing a Decimal by a Whole Number

A loaf of bread 43.2 centimeters long is cut into 36 slices of the same size.

1. Will the thickness of each slice be closer to 1 centimeter or 2 centimeters? Why?

2. When you divide 43.2 by 36, over what number in the dividend will you place the first digit in the quotient? Why?

EXAMPLE 1

Divide 43.2 by 36. Check your answer.

$$36 \overline{)\, 43.2}$$
◀ *Place the decimal point directly above the decimal point in the dividend.*

◀ *Divide as with whole numbers.*

$$
\begin{array}{r}
1.2 \\
36 \overline{)\, 43.2} \\
\underline{36} \\
72 \\
\underline{72} \\
0
\end{array}
$$

These should be the same. Why?

Check:
$$
\begin{array}{r}
1.2 \\
\times\ 36 \\
\hline
7\,2 \\
36 \\
\hline
43.2
\end{array}
$$

Each slice will be **1.2 centimeters** thick.

Sometimes it is useful to round a quotient to a given decimal place.

3. What is 23.685 rounded to the nearest tenth?

4. What is 23.685 rounded to the nearest hundredth?

EXAMPLE 2

Divide 23.65 by 41. Round the quotient to the nearest hundredth.

$$
\begin{array}{r}
0.576 \\
41 \overline{)\, 23.650} \\
\underline{205} \\
315 \\
\underline{287} \\
280 \\
\underline{246} \\
34
\end{array}
$$

◀ *Annex one zero.*

The quotient is **0.58** (rounded to the nearest hundredth).

EXERCISES

For Exercises 1–4, replace each __?__ with a word or words chosen from the box at the right.

1. To divide a decimal by a whole number, first place the __?__ in the quotient directly above the __?__ in the dividend.

2. To round the quotient of 74.1 ÷ 23 to the nearest tenth, carry the division to the __?__ place.

3. To round the quotient of 93.9 ÷ 18 to the nearest hundredth, carry the division to the __?__ place.

4. To round the quotient of 42.2 ÷ 56 to the nearest thousandth, carry the division to the __?__ place.

> decimal point
> hundredths
> tenths
> thousandths
> ten-thousandths

For Exercises 5–9, select the correct quotient for each problem.
   Quotients: 17.40   0.174   1.74   0.0174   0.00174

5. 3) 5.22   6. 3) 0.0522   7. 3) 52.20   8. 3) 0.00522   9. 3) 0.522

10. Will 19.8 ÷ 6 be closer to 3 or to 4? Why?

11. Will 7.03 ÷ 3 be closer to 2 or to 3? Why?

12. Will 49.56 ÷ 42 be closer to 1 or to 2? Why?

13. Will 191.58 ÷ 93 be closer to 2 or to 3? Why?

14. Will 0.45 ÷ 8 be closer to 0.5 or to 0.05? Why?

15. Will 3.782 ÷ 37 be closer to 1 or to 0.1? Why?

Divide. Check your answers.

16. 8) 9.28   17. 7) 43.96   18. 11) 1.21   19. 25) 1.325   20. 42) 49.56

21. 26) 35.88   22. 82) 33.62   23. 23) 8.28   24. 93) 191.58   25. 86) 474.72

Divide. Round each quotient to the nearest hundredth.

26. 94) 334.2   27. 76) 6.2   28. 24) 8.18   29. 63) 35.3   30. 92) 5.49

31. 54) 0.48   32. 44) 105.6   33. 97) 347.3   34. 28) 824.1   35. 87) 521.6

Divide. Round each quotient to the nearest tenth.

36. 3) 1.3   37. 8) 46.7   38. 83) 52.7   39. 66) 59.1   40. 35) 28.5

41. 29) 88.4   42. 71) 43.8   43. 17) 1.09   44. 58) 34.1   45. 78) 287.6

Complete.

46. 6) ■.6 → 1.■   47. 6) 0.9■ → 0.■6   48. 6) ■■.6 → 16.■   49. 6) ■.09■ → 0.■■6   50. 6) ■0.03■ → 1■.0■6

Comparison Shopping

Shoppers use unit price to compare the cost of different sizes of the same product.

Unit Price = Price of Item ÷ Number of Units

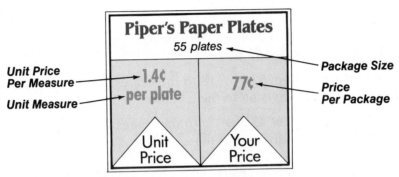

1. What is the cost of one package of Piper's paper plates?

2. How many plates are there in each package?

3. *Complete:* To find the unit price of Piper's paper plates, divide $0.77 by __?__ .

> Shoppers compare the unit prices of two or more sizes of the same product to determine the **better buy.**

When only the price of two or more items is considered, the item with the lowest unit price is the better buy.

EXAMPLE — A box of 20 trash bags sells for $2.29. A box of 40 trash bags of the same brand sells for $4.35.

a. Find the unit price of each box to the nearest tenth of a cent.

Box of 20: $2.29

$$\begin{array}{r} 0.1145 \\ 20 \overline{)2.2900} \end{array}$$

Unit price: 20) 2.2900 ⟶ 11.45¢, or **11.5¢**   *Rounded to the nearest tenth of a cent*

Box of 40: $4.35

$$\begin{array}{r} 0.1087 \\ 40 \overline{)4.3500} \end{array}$$

Unit price: 40) 4.3500 ⟶ 10.87¢, or **10.9¢**   *Rounded to the nearest tenth of a cent*

b. Which is the better buy?
The **40-bag package** has the lower unit price. Therefore, it is the better buy.

EXERCISES

Round each amount to the nearest tenth of a cent.

1. 7.583¢   **2.** 5.908¢   **3.** 3.915¢   **4.** 21.006¢   **5.** 8.054¢

6. $0.7297   **7.** $0.4645   **8.** $0.1772   **9.** $0.5050   **10.** $0.0104

Find the unit price to the nearest tenth of a cent.

Item	Size	Price	Unit Price
11. Muffins	8 per box	$1.89	_?_
12. Toothpaste	5 ounces	$0.92	_?_
13. Lemons	6	$0.68	_?_
14. Bread	16-slice loaf	$1.35	_?_
15. Grass seed	5 kilograms	$2.69	_?_

16. A brand of soap sells at 3 bars for $1.10 and at 5 bars for $1.62. Which is the better buy?

17. A 3-kilogram bag of Doggie Goodies costs $1.25. A 5-kilogram bag costs $2.18. Which is the better buy?

18. The cost of a 5-pound bag of potatoes is $1.03. The potatoes can also be bought for 23¢ per pound. Which would be the cheaper way to buy 10 pounds of potatoes?

19. In Exercise 18, suppose you need exactly 3 pounds of potatoes. Which would be the cheaper way to buy what you need?

20. Treadtop Tires advertises a sale of 4 tires for $135.50. Tireless Rims sells the same brand at 3 tires for $93.35. Which is the better buy?

21. King Corn Flakes sells in 7-ounce packages for 72¢ or in 12-ounce packages for $1.13. The store brand corn flakes sells in 10-ounce packages for 91¢. Which has the lowest unit price?

For Exercises 22–23, discuss the answers in small groups or as your teacher directs.

22. Give two examples to illustrate how the larger size of a product may have the lower unit price and not be the better buy.

23. Give at least one reason why the smaller size of a product may have a lower unit price than a larger size.

Dividing by a Decimal

1. Find the quotient: $8 \overline{)24}^{\,?}$

2. In $24 \div 8$, multiply both the dividend and the divisor by 100. Divide the new dividend by the new divisor. What is the quotient?

3. Compare the quotients in **1** and **2** above. Are they the same?

4. *Complete:* Multiplying a divisor and a dividend by the same nonzero number does not change the __?__ .

EXAMPLE 1

Divide 3.0875 by 0.25. Check your answer.

1. Since the final decimal place in the divisor is hundredths, multiply both the divisor and the dividend by 100.

$$0.25 \overline{)3.0875}$$ ◄ **Move the decimal points two places to the right.**

2. Now the divisor is a whole number. Divide. ──►

$$\begin{array}{r} 12.35 \\ 25 \overline{)308.75} \\ \underline{25} \\ 58 \\ \underline{50} \\ 87 \\ \underline{75} \\ 125 \\ \underline{125} \\ 00 \end{array}$$

3. Check your answer. Use a calculator.

Quotient × Divisor = Dividend

12.35 ☒ **.25** ═ 【 **3.0875** 】

Quotient: **12.35**

EXAMPLE 2

Divide: **a.** $18 \div 0.72$ **b.** $1.7 \div 6.8$

a. 1. $0.72 \overline{)1800.}$ ◄ **Annex two zeros**

$$\begin{array}{r} 25 \\ 72 \overline{)1800} \\ \underline{144} \\ 360 \\ \underline{360} \\ 00 \end{array}$$

Quotient: **25**

b. 1. $6.8 \overline{)1.7}$

2. $\begin{array}{r} 0.25 \\ 68 \overline{)17.00} \\ \underline{13\ 6} \\ 3\ 40 \\ \underline{3\ 40} \\ 00 \end{array}$ ◄ **Annex two zeros.**

Quotient: **0.25**

The checks are left for you.

EXERCISES

Complete.

1. $0.91 \times \underline{} = 91$ **2.** $0.6 \times \underline{} = 6$ **3.** $0.009 \times \underline{} = 9$

Multiply the divisor and the dividend by 10, by 100, or by 1000 to make the divisor a whole number.

4. $0.04 \overline{)1.96}$ **5.** $0.55 \overline{)2.75}$ **6.** $0.05 \overline{)9}$ **7.** $0.6 \overline{)36}$

8. $0.007 \overline{)1.4}$ **9.** $0.015 \overline{)0.72}$ **10.** $0.25 \overline{)1.045}$ **11.** $0.16 \overline{)1.6544}$

Divide.

12. $0.5 \overline{)5.25}$ **13.** $0.4 \overline{)3.24}$ **14.** $2.8 \overline{)20.16}$ **15.** $7.2 \overline{)338.4}$

16. $0.38 \overline{)273.6}$ **17.** $0.18 \overline{)576}$ **18.** $2.1 \overline{)714}$ **19.** $0.023 \overline{)133.4}$

20. $0.026 \overline{)67.6}$ **21.** $0.12 \overline{)4.2}$ **22.** $0.017 \overline{)62.9}$ **23.** $0.082 \overline{)32.8}$

24. $0.68 \overline{)61.2}$ **25.** $0.08 \overline{)20.4}$ **26.** $0.7 \overline{)238}$ **27.** $8.2 \overline{)328}$

28. $0.004 \overline{)14.68}$ **29.** $0.76 \overline{)3.876}$ **30.** $0.02 \overline{)0.072}$ **31.** $2.9 \overline{)26.68}$

Divide. Round each quotient to the nearest tenth.

32. $0.17 \overline{)8.274}$ **33.** $0.52 \overline{)27.1}$ **34.** $1.8 \overline{)36.7}$ **35.** $5.3 \overline{)27.4}$

36. $5.3 \overline{)27.8}$ **37.** $0.016 \overline{)10.9}$ **38.** $3.9 \overline{)19}$ **39.** $2.1 \overline{)35.1}$

Divide. Round each quotient to the nearest hundredth.

40. $2.4 \overline{)7.63}$ **41.** $0.7 \overline{)19.2}$ **42.** $0.51 \overline{)86.3}$ **43.** $0.76 \overline{)294}$

44. $0.58 \overline{)312}$ **45.** $0.26 \overline{)19.3}$ **46.** $0.6 \overline{)16.4}$ **47.** $3.8 \overline{)2.53}$

48. In a speed test on a small island in the Indian Ocean, a giant tortoise covered about 2.958 feet in 8.7 seconds. How many feet per second was this?

49. Pamela Clapper drove 156.3 miles in 3.1 hours. To the nearest tenth, what was her average speed in miles per hour?

50. A hardware dealer has 20 pounds of nails. If each nail weighs 0.01 pound, how many nails does the dealer have?

Dividing by 10, by 100, by 1000

Look for a pattern for dividing by 10.

1. a. $35 \div 10 = \underline{\ ?\ }$ **b.** $3.5 \div 10 = \underline{\ ?\ }$ **c.** $0.35 \div 10 = \underline{\ ?\ }$

2. *Complete:* To divide a number by 10, first write the number. Then move the decimal point $\underline{\ ?\ }$ place(s) to the left.

Look for a pattern for dividing by 100.

3. a. $67 \div 100 = \underline{\ ?\ }$ **b.** $6.7 \div 100 = \underline{\ ?\ }$ **c.** $0.67 \div 100 = \underline{\ ?\ }$

4. *Complete:* To divide a number by 100, first write the number. Then move the decimal point $\underline{\ ?\ }$ place(s) to the left.

Look for a pattern for dividing by 1000.

5. a. $976 \div 1000 = \underline{\ ?\ }$ **b.** $97.6 \div 1000 = \underline{\ ?\ }$ **c.** $9.76 \div 1000\ \underline{\ ?\ }$

6. *Complete:* To divide a number by 1000, first write the number. Then move the decimal point $\underline{\ ?\ }$ place(s) to the left.

EXAMPLE	Divide without using paper and pencil.

a. $3.98 \div 10 = 0.398$ **b.** $1.8 \div 100 = 0.018$ ◀ **Insert one zero.**

c. $100 \overline{)\,96} = 0.96$ **d.** $7.1 \div 1000 = 0.0071$ ◀ **Insert two zeros.**

e. $0.5 \div 10 = 0.05$ **f.** $0.5 \div 1000 = 0.0005$

EXERCISES

For Exercises 1–9, choose the correct answer from the box at the right.

0.000763
0.00763
0.0763
0.763
7.63
76.3

1. $763 \div 1000$ **2.** $7.63 \div 10$ **3.** $76.3 \div 100$

4. $0.763 \div 10$ **5.** $7.63 \div 100$ **6.** $7.63 \div 1000$

7. $76.3 \div 10$ **8.** $0.763 \div 1000$ **9.** $0.0763 \div 10$

Complete without using paper and pencil.

10. $9.3 \div 10 = \underline{\ ?\ }$ **11.** $7 \div 10 = \underline{\ ?\ }$ **12.** $8.92 \div \underline{\ ?\ } = 0.0892$

13. $0.9 \div 10 = \underline{\ ?\ }$ **14.** $1000 \overline{)\,0.725}$ **15.** $4.35 \div \underline{\ ?\ } = 0.00435$

16. $\underline{\ ?\ } \div 10 = 0.734$ **17.** $100 \overline{)\,8892}$ **18.** $255.7 \div 1000 = \underline{\ ?\ }$

19. $\underline{\ ?\ } \div 1000 = 0.00398$ **20.** $\underline{\ ?\ } \div 100 = 0.493$ **21.** $100 \overline{)\,13}$

22. $\underline{\ ?\ } \div 10 = 193.6$ **23.** $17.21 \div \underline{\ ?\ } = 1.721$ **24.** $0.39 \div \underline{\ ?\ } = 0.0039$

25. $\underline{\ ?\ } \div 100 = 0.11$

26. $0.6 \div \underline{\ ?\ } = 0.06$

27. $18 \div \underline{\ ?\ } = 0.018$

28. $\underline{\ ?\ } \div 1000 = 365$

29. $\underline{\ ?\ } \div 100 = 0.5$

30. $9 \div \underline{\ ?\ } = 0.9$

31. a. Divide 900 by 30.

   b. Divide 900 by 10. Then divide the result by 3.

   c. Compare your answers in a and b.

   d. *Complete:* Dividing a number by 10 and then dividing the result by 3 gives the same quotient as dividing by $\underline{\ ?\ }$.

Use the results of Exercise 31 to complete Exercises 32 and 33.

32. To divide a number by 70, first divide by $\underline{\ ?\ }$. Then divide the result by $\underline{\ ?\ }$.

33. To divide a number by 400, first divide by $\underline{\ ?\ }$. Then divide the result by $\underline{\ ?\ }$.

Use the results of Exercises 32 and 33 to divide. Write the answer only.

34. $70\overline{)\,280}$

35. $4800 \div 60$

36. $6300 \div 90$

37. $30\overline{)\,3600}$

38. $4400 \div 40$

39. $120\overline{)\,360}$

40. $150\overline{)\,4500}$

41. $48{,}000 \div 160$

42. $400\overline{)\,1200}$

43. $900\overline{)\,2700}$

44. $400{,}000 \div 500$

45. $320{,}000 \div 800$

46. $400\overline{)\,44000}$

47. $3000 \div 200$

48. $24{,}000 \div 1200$

49. $1600\overline{)\,480000}$

50. $80{,}000 \div 2000$

51. $18{,}000 \div 3000$

52. $5000\overline{)\,75000}$

53. $4000\overline{)\,480000}$

54. A weaver has 64,000 yards of yarn. He wants to weave 20 rugs of equal size. How much yarn will be used for each rug?

55. A worker has 5700 feet of twine to be wound equally on 30 spools. How many feet of twine will be on each spool?

56. Expenses of $2565 are to be shared equally by 100 club members. How much does each member pay?

57. In Averagetown, USA, 20,000 homeowners paid a total of $1,200,000 for water taxes. What was the average tax for each homeowner?

58. A builder paid $273,000 for 30 lots. What was the average cost for one lot?

59. A 10-pound bag of potatoes costs $2.01. Find the cost per pound. Round your answer to the nearest cent.

60. A 200-gram box of cereal costs $2.25. Find the cost per gram. Round your answer to the nearest cent.

Strategy: CHOOSING COMPUTATION METHODS

In everyday situations, you often have to consider whether you need an **exact answer** to a problem or whether an **estimate** will give an answer that is close enough. Then you have to decide whether to use **mental computation,** a **calculator, paper and pencil,** or some combination of these to solve the problem efficiently.

PROBLEM   Ricardo went to Record Outlet to buy the record albums shown in the advertisement at the right below. On his way to the check-out counter, he discovered that he had exactly $26.75 in his wallet. Ricardo thought: "I'll have to find the total cost to be sure I have enough money to pay for everything."

1. Does Ricardo need to find the exact total or will an estimate be close enough? Explain.

2. Which would be the more efficient way to estimate the sum, using a calculator or using mental computation? Explain.

The Bunkers in Concert $4.99	The Best of the Surfers $6.99
New Day $5.10	Moonfire $7.15
Tax: $1.50	

EXAMPLE 1

Ricardo used **front-end estimation** in this way.

1 Add the dollars first. $4 + $6 + $5 + $7 + $1 = **$23**

2 **Estimate** the cents. $0.99 ◄——— **About $1**
                                        0.99 ◄——— **About $1**
                                        0.10 ⎫
                                        0.15 ⎬ ◄——— **75¢**
                                        0.50 ⎭

   Estimate: $23 + $2 + 75¢ = **$25.75**

---

3. How far was Ricardo's estimate from the actual cost?

4. Did Ricardo have enough money to buy all the records?

EXERCISES

Each of five students computed his or her earnings for one week at The Burger Palace. This table shows the number of hours worked and the hourly pay rate for each employee.

Employee	Hours	Rate Per Hour	Earnings
Rosa	18.5	$4.20	?
Stan	20.25	$3.36	?
Julio	15.75	$3.60	?
Wendy	22.25	$3.48	?
Juanita	24.5	$3.80	?

1. Should students estimate their earnings, or is an exact answer needed? Why?

2. What method or methods of computation would you use to compute the earnings? Give reasons for your answer.

3. Find the earnings for each student.

The owner of Videorama listed the top ten movie video rentals for one month. The movies are to be ranked in order, starting with the greatest number of rentals.

Movie	Rentals	Rank
Dawn of the Lead	81	?
Kung Fu Kid	51	?
Pontoon	76	?
Mars Base I	53	?
Easy Driver	70	?
Tanker	66	?
Giant Jellyfish	74	?
Escape from Kran	59	?

4. Is an exact rental listing or an approximate rental listing appropriate? Why?

5. Do you think the owner will use a calculator or paper and pencil to do the rankings? Explain your answer.

6. Rank the movies in order, starting with the greatest number of rentals.

The scorekeeper of the qualifying race for the Texas 500 times each driver during each of the three laps around the track. The greatest time is dropped and the two remaining times are added to determine the qualifying time.

Car	Lap 1 (min)	Lap 2 (min)	Lap 3 (min)	Qualifying Time
9	1.257	1.506	1.340	?
22	1.734	1.629	1.453	?
88	1.463	1.322	1.381	?
43	1.408	1.384	1.298	?
71	1.671	1.637	1.569	?

7. Should the scorekeeper compute an exact qualifying time or use an estimate? Why?

8. To find the qualifying time, would it be more efficient to use a calculator or to use paper and pencil? Why?

9. Find the qualifying time for each driver.

10. Rank the drivers according to qualifying time, from lowest to highest.

11. What computation method did you use to arrange the rankings?

Making a Model

A class treasurer reported the cost of a class party as $35, rounded to the nearest dollar. Dave wanted to know the greatest and least amounts the party could have cost.

Dave drew a number line to model the problem.

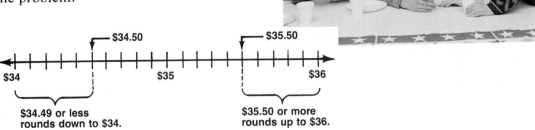

The drawing shows that the party cost **at least $34.50 but no more than $35.49.**

EXERCISES

For Exercises 1–6, each statement contains a number that has been rounded to the nearest dollar. Write the least and greatest possible amounts that could have produced the number. Draw a number line to model each problem.

1. Jane bought a computer for $100.

2. Brian spent $30 to have his bicycle repaired.

3. Marina spent $25 for groceries.

4. Jennifer spent $69 for a new uniform.

5. Eric paid $9500 for a new car.

6. Ella bought a box of cards for $2.

7. Tom estimated that repairs to the family lawnmower would cost $70, rounded to the nearest five dollars. What is the least and greatest amounts the repairs could cost?

8. Maureen estimates that she spends $35 a week for groceries, rounded to the nearest dollar. If this is a good estimate, what is the greatest amount she can expect to spend for groceries in a year?

Chapter Summary

1. A number is divisible by:
   2 if it ends in 0, 2, 4, 6, or 8.
   5 if it ends in 0 or 5.
   10 if it ends in 0.
   3 if the sum of its digits is divisible by 3.
   4 if its last two digits are divisible by 4.
   9 if the sum of its digits is divisible by 9.

2. To divide whole numbers:
   1 Use compatible numbers to estimate where to place the first digit in the quotient.
   2 Use the first digit in the estimate as a trial divisor.
   3 Divide.

3. To divide a decimal by a whole number (not zero):
   1 Place the decimal point in the quotient directly above the decimal point in the dividend.
   2 Divide as with whole numbers.

4. To determine which item is the better buy:
   1 Find the unit price of each item to the nearest tenth of a cent.
   2 Compare the unit prices. For items of the same quality, the item with the lower unit price is the better buy.

5. To divide by a decimal:
   1 Multiply both the divisor and the dividend by 10, or by 100, or by 1000, and so on, to obtain a whole number divisor.
   2 Divide.

6. To divide a decimal by 10, by 100, or by 1000:
   1 Move the decimal point in the dividend one, two, or three places to the left respectively.
   2 When necessary, insert zeros in order to have the correct number of places.

1.
240 is divisible by **2**, by **3**, by **4**, by **5**, and by **10**.
135 is divisible by **3**, by **5**, and by **9**.
297 is divisible by **3** and by **9**.

2.
$$12\overline{)288} \longrightarrow 10\overline{)290}^{\,29}$$

$$\begin{array}{r} 24 \\ 12\overline{)288} \\ 24 \\ \hline 48 \\ 48 \\ \hline 0 \end{array}$$

3.
$$21\overline{)48.3}^{\,.}$$

$$21\overline{)48.3}^{\,2.3}$$

4. Choose the better buy.
Pears: $0.76/16 oz or $1.08/24 oz

$$16\overline{)0.7600}^{\,0.0475} \longrightarrow \textbf{4.8¢}$$

$$24\overline{)1.080}^{\,0.045} \longrightarrow \textbf{4.5¢}$$

Since 4.5¢ < 4.8¢, the **24-ounce can** is the better buy.

5.
$$0.25\overline{)62.50}^{\,.}$$ **Annex one zero.**

$$25\overline{)6250}^{\,250}$$

6.
2.54 ÷ 10 = **0.254**
57.6 ÷ 100 = **0.576**
4.2 ÷ 1000 = **0.0042**

Chapter Review

Part 1: VOCABULARY

For Exercises 1–4, choose from the box at the right the word(s) that complete(s) each statement.

<table>
<tr><td>

1. You can estimate a quotient by rounding the divisor and the dividend to __?__ numbers. (Page 58)

2. The item with the __?__ unit price is usually the better buy. (Page 66)

3. When dividing by a decimal, sometimes you have to annex __?__ to the dividend. (Page 68)

4. To divide by 100, move the decimal point 2 places to the __?__ . (Page 70)

</td><td>

zeros
lower
compatible
right
left

</td></tr>
</table>

Part 2: SKILLS

Determine whether each number is divisible by 2, by 3, by 4, by 5, by 9, or by 10. (Pages 56–57)

5. 35   **6.** 214   **7.** 51   **8.** 792   **9.** 750   **10.** 52   **11.** 1197

12. 865   **13.** 333   **14.** 336   **15.** 344   **16.** 507   **17.** 920   **18.** 834

Divide. Write the remainder with the quotient. (Pages 58–59)

19. $18\overline{)54}$ **20.** $24\overline{)72}$ **21.** $29\overline{)463}$ **22.** $52\overline{)597}$ **23.** $31\overline{)4561}$

24. $36\overline{)4858}$ **25.** $37\overline{)2368}$ **26.** $95\overline{)6555}$ **27.** $78\overline{)2560}$ **28.** $81\overline{)4212}$

Divide. Write the remainder with the quotient. (Pages 60–61)

29. $21\overline{)648}$ **30.** $42\overline{)6325}$ **31.** $38\overline{)7850}$ **32.** $48\overline{)9624}$ **33.** $43\overline{)5615}$

34. $97\overline{)9809}$ **35.** $74\overline{)7795}$ **36.** $73\overline{)3668}$ **37.** $58\overline{)2355}$ **38.** $82\overline{)17042}$

Divide. Round each quotient to the nearest hundredth. (Pages 64–65)

39. $6\overline{)2.5}$ **40.** $7\overline{)35.3}$ **41.** $32\overline{)9.1}$ **42.** $65\overline{)3.185}$ **43.** $21\overline{)0.462}$

44. $29\overline{)42.68}$ **45.** $53\overline{)29.12}$ **46.** $66\overline{)2.442}$ **47.** $45\overline{)256}$ **48.** $56\overline{)32.8}$

Complete. (Pages 64–65)

49. $6\overline{)\blacksquare.2}$ quotient $1.\blacksquare$ **50.** $4\overline{)\blacksquare.40}$ quotient $0.8\blacksquare$ **51.** $6\overline{)\blacksquare\blacksquare.0}$ quotient $13.\blacksquare$ **52.** $7\overline{)\blacksquare.59\blacksquare}$ quotient $0.\blacksquare\blacksquare8$ **53.** $7\overline{)\blacksquare4.24\blacksquare}$ quotient $1\blacksquare.0\blacksquare5$

Find the unit price to the nearest tenth of a cent.   (Pages 66–67)

Item	Size	Price	Unit Price
54. Bread	20-slice loaf	$1.15	?
55. Bottled water	16 ounces	$0.79	?
56. Paper towels	50 sheets	$0.61	?
57. Soap	2 bars	$0.81	?

Divide.   (Pages 68–69)

58. $0.4\overline{)3.76}$ **59.** $0.8\overline{)0.496}$ **60.** $5.6\overline{)43.68}$ **61.** $0.08\overline{)20.4}$ **62.** $0.12\overline{)4.2}$

63. $0.38\overline{)2682.8}$ **64.** $0.7\overline{)20.02}$ **65.** $4.3\overline{)2.365}$ **66.** $8.2\overline{)328}$ **67.** $0.9\overline{)657}$

Complete without using paper and pencil.   (Pages 70–71)

68. $8.7 \div 10 = \underline{?}$ **69.** $5.41 \div \underline{?}\ 0.541$ **70.** $\underline{?} \div 100 = 0.311$ **71.** $100\overline{)15}$

72. $317.2 \div 1000 = \underline{?}$ **73.** $\underline{?} \div 1000 = 13.9$ **74.** $0.26 \div \underline{?} = 0.0026$ **75.** $100\overline{)\,?}^{\ 0.4}$

Part 3: APPLICATIONS

76. A store sells pairs of socks for $3 a pair. Can you buy exactly $15 worth of socks?   (Pages 56–57)

77. Geraldo read that food counter workers earn an average yearly wage of $7308. What is their average monthly salary?   (Pages 60–61)

78. A 10-pound bag of Super Dog dog food costs $4.69. A 20-pound bag costs $9.05. Which is the better buy? (Pages 66–67)

79. A new car dealer paid $3,200,000 for 400 new cars. What was the average cost for each car?   (Pages 70–71)

80. Danielle sold boxes of light bulbs to raise money for the junior class. She listed the people who ordered bulbs. The light bulbs sold for $4.75 per box.   (Pages 72–73)

  a. Should Danielle estimate the amount that each person owes, or is an exact amount needed? Why?

Person	Boxes of Light Bulbs Ordered
Lisa Cheatham	5
Sarah Smith	2
Doug Stumbaugh	8
Jeremy Coleman	3
Kara Ainsley	4
Jim Paine	3

  b. What method or methods of computation will Danielle use to compute the amount each person owes? Give reasons for your answer.

  c. What is the total amount of money she needs to collect?

Chapter Test

Determine whether each number is divisible by 2, by 3, by 4, by 5, by 9, or by 10.

1. 16 **2.** 50 **3.** 84 **4.** 65 **5.** 144 **6.** 618 **7.** 1350

Divide.

8. $34 \overline{)952}$ **9.** $17 \overline{)959}$ **10.** $47 \overline{)506}$ **11.** $53 \overline{)16043}$ **12.** $21 \overline{)0.42}$

13. $30 \overline{)17.1}$ **14.** $26 \overline{)90.22}$ **15.** $0.6 \overline{)21.06}$ **16.** $0.15 \overline{)4.5}$ **17.** $0.008 \overline{)0.12}$

Complete without using paper and pencil.

18. $5.4 \div 10 = \underline{?}$ **19.** $16.5 \div \underline{?} = 1.65$ **20.** $100 \overline{)20}^{?}$ **21.** $\underline{?} \div 1000 = 0.0152$

Solve.

22. Fred read that psychologists earn an average yearly salary of $20,904. What is their average weekly salary?

23. A 32-ounce bottle of apple juice costs $1.31. A 64-ounce bottle costs $2.68. Which is the better buy?

24. While at the supermarket, you want to purchase the following items: milk for $2.49, bread for $0.92, meat for $4.10, paper towels for $1.21, paper plates for $1.45, detergent for $5.89, and apple juice for $1.52. On your way to the check-out counter, you discover that you have exactly $16.75 in your wallet. You want to be sure you have enough money to pay for the items.

a. Should you estimate the total cost or is an exact amount needed? Why?

b. What method or methods of computation will you use to compute the total price? Give reasons for your answer.

c. Will you have enough money in your wallet to pay for all of the items? Explain.

Cumulative Maintenance Chapters 1–3

Choose the correct answer. Choose a, b, c, or d.

1. Round 465 to the nearest ten.

    a. 460   **b.** 400
    c. 470   **d.** 500

2. Subtract.

$$\begin{array}{r} 15 \\ -\ 4.75 \end{array}$$

    a. 11.75   **b.** 11.25
    c. 19.75   **d.** 10.25

3. Multiply.

$$\begin{array}{r} 236 \\ \times\ 153 \end{array}$$

    a. 2124   **b.** 36,108
    c. 14,868   **d.** 12,844

4. Round 3.174 to the nearest tenth.

    a. 3   **b.** 3.17
    c. 3.1   **d.** 3.2

5. Add.

$$\begin{array}{r} 4587 \\ +6914 \end{array}$$

    a. 11,491   **b.** 11,501
    c. 11,401   **d.** 10,501

6. Multiply.

$$3006 \times 0.07$$

    a. 21.042   **b.** 2104.2
    c. 210.42   **d.** 2.1042

7. Marcel had grades of 80, 63, 88, 92, and 82 on his math quizzes. What was his average grade for the quizzes?

    a. 68   **b.** 82
    c. 81   **d.** 90

8. Divide.

$$5.2\,\overline{)\,1568.32}$$

    a. 301.6   **b.** 310.6
    c. 30.16   **d.** 31.6

9. Multiply.

$$\begin{array}{r} 3.071 \\ \times\ 5.2 \end{array}$$

    a. 1596.21   **b.** 1.59692
    c. 159.692   **d.** 15.9692

10. Rhoda deposits checks for $120. She receives 2 twenty-dollar bills and 3 ten-dollar bills. What is Rhoda's net deposit?

    a. $100   **b.** $120
    c. $50   **d.** $70

11. Divide.

$$75\,\overline{)\,30225}$$

    a. 43   **b.** 400
    c. 403   **d.** 4003

12. Subtract.
$$5603 - 3852$$

    a. 1751   **b.** 9455
    c. 2851   **d.** 1851

13. Written in words, 7.05 is

    a. Seven and five tenths
    b. Seven and five hundredths
    c. Seven and five thousandths
    d. Seven hundred and five

14. Find the missing terms.

    4, 12, 36, __?__, 324, __?__, 2916

    a. 108, 972   **b.** 48, 648
    c. 72, 648   **d.** 108, 648

15. Complete: $305 \times$ __?__ $= 305,000$

    a. 10   **b.** 100
    c. 1000   **d.** 10,000

16. Multiply.
$$\begin{array}{r} 6.4 \\ \times 100 \\ \hline \end{array}$$

    a. 6.4   **b.** 6400
    c. 64   **d.** 640

17. Divide.
$$37 \overline{)1665}$$

    a. 45   **b.** 405
    c. 44   **d.** 450

18. $30 - 4 \times 2 + 5 =$ __?__

    a. 57   **b.** 182
    c. 33   **d.** 27

19. Add.
$$\begin{array}{r} 6.35 \\ 12.7 \\ 6.853 \\ +687.6 \\ \hline \end{array}$$

    a. 714.603   **b.** 714.503
    c. 713.603   **d.** 713.503

20. Divide.
$$0.7 \overline{)232.4}$$

    a. 232.4   **b.** 332
    c. 33.2   **d.** 330

21. Ben spends $12.98 on cassette tapes. He pays for them with a $20-bill. How much change will he receive?

    a. $7.02   **b.** $8.02
    c. $6.98   **d.** $8.98

22. At which price will one banana cost the least?

    a. 3 for 92¢   **b.** 4 for $1.16
    c. 5 for $1.50   **d.** 6 for $1.68

23. What is the total value of four $10-bills, three $5-bills, and six $1-bills?

    a. $16   **b.** $51
    c. $81   **d.** $61

24. Which decimal has the smallest value?

    a. 0.508   **b.** 0.00058
    c. 5.8   **d.** 0.50008

Graphs and Applications

CHAPTER 4

Favorite Sports

Softball Basketball Tennis Swimming Track

Key: *Boys*   *Girls*

- What kinds of information can be shown on bar graphs?

- When would you use line graphs to show information?

- What kind of graph would you use to show related information?

- How do the horizontal and vertical scales on graphs help you to interpret information correctly?

Pictographs and Applications

In a **pictograph,** a picture or symbol is used to represent a number. The **key** tells how many are represented by each symbol.

The pictograph at the right shows how much you can expect to earn each week at certain jobs.

Average Weekly Earnings for Certain Jobs

Engineers	⑤ ⑤ ⑤ ⑤ ⑤ ◖
Architects	⑤ ⑤ ⑤ ⑤ ◣
Dental hygienists	⑤ ⑤ ⑤ ◖
Secretaries	⑤ ⑤ ◢
Food attendants	⑤ ◖

Key: *Each* ⑤ *represents $100.*

Use the key for the pictograph above to answer the following questions.

1. What does the key tell you?

2. What part of the circle do you think ◖ represents?

3. How many dollars does ◖ represent?

4. What part of a circle do you think ◣ represents?

5. How many dollars does ◣ represent?

EXERCISES

Use the pictograph above for Exercises 1–10.

1. Which job pays the greatest amount per week?

2. Which job pays the least amount per week?

3. How much do engineers earn per week?

4. How much do food attendants earn per week?

5. How much do dental hygienists earn per week?

6. How much do secretaries earn per week?

7. How much less per week do food attendants earn than engineers?

8. Which job pays $75 less per week than an architect is paid?

9. Which job pays $200 more per week than a secretary is paid?

10. Which job pays $200 less per week than a dental hygienist is paid?

The pictograph below shows the approximate number of aluminum cans collected by six schools as part of a service project.

Number of Aluminum Cans Collected

Allen School	🥫🥫🥫🥫🥫🥫🥫🥫🥫🥫
Glencove School	🥫🥫🥫🥫🥫🥫🥫(
Justin High School	🥫🥫🥫🥫🥫
Bradley School	🥫🥫🥫🥫🥫🥫
Madison School	🥫🥫🥫🥫🥫🥫(
Brooks School	🥫🥫🥫🥫🥫🥫🥫🥫

Key: *Each* 🥫 *represents 200 cans.*

11. How many cans are represented by one symbol?

12. How many cans are represented by one-half of the symbol?

13. Which schools collected more than 1400 cans?

14. Which schools collected fewer than 1300 cans?

15. Which schools collected about the same number of cans?

16. Which school collected 200 more cans than Madison School?

For Exercises 17–18:
   a. *Use the information given to construct a pictograph.*
   b. *Write two problems that relate to your graph.*

17.

Number of Freshman Enrolled in Local High Schools					
School	**Tehipite**	**Carver**	**Central**	**Union**	**Roosevelt**
Number of Students	425	125	350	450	275

Key: ⚣ represents 50 students.

18.

Number of Space Pirate Software Games Sold					
Year	**1983**	**1984**	**1985**	**1986**	**1987**
Number of Consoles	700,000	650,000	300,000	250,000	200,000

Key: Each 🚀 represents 100,000 software games.

Bar Graphs and Applications

A **bar graph** has two axes, a **horizontal axis** and a **vertical axis.** On a bar graph, the length of each bar represents a number. The scale on one of the axes tells you how to find the number.

This bar graph below shows the total number of hurricanes that occurred in the world during each month of the hurricane season.

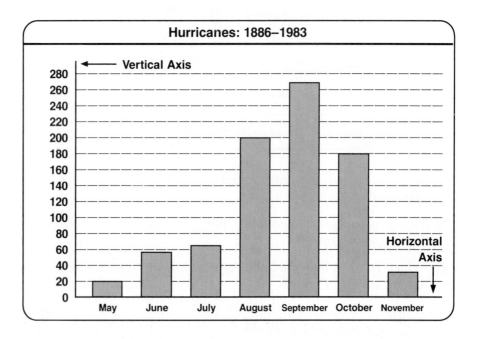

Use the bar graph above to answer these questions.

1. Which month has the fewest number of hurricanes?

2. What are the two most "dangerous" months?

3. Which month has between 40 and 60 hurricanes?

4. November has between ___?___ and ___?___ hurricanes.

EXERCISES

Use the bar graph above for Exercises 1–4.

1. On which axis, the horizontal or vertical, is the scale located?

2. Why does the scale go to 280?

3. Did more hurricanes occur during August or October?

4. About how many hurricanes occurred in September?

The horizontal bar graph at the right shows the number of items of food collected by six school groups for a charity drive. Use this graph for Exercises 5–11.

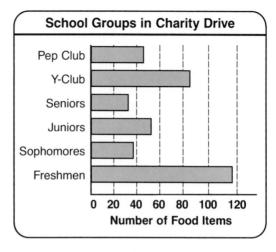

School Groups in Charity Drive

5. Which group collected the most food items?

6. Which group collected almost as many food items as the sophomores?

7. Which group collected about half as many food items as the freshmen?

8. Which estimate, 47 items or 55 items, is closest to the number of food items collected by the Pep Club?

9. Which group collected about 12 more food items than the Pep Club?

The vertical bar graph at the right shows the areas of the seven continents. Use this graph for Exercises 10–14.

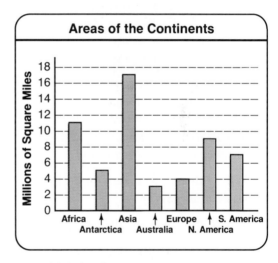

Areas of the Continents

10. List the continents in order from the largest to the smallest area.

11. South America is slightly more than twice the size of what continent?

12. What is the area of Antarctica?

13. What continent is closest in area to the combined areas of Europe and Antarctica?

14. Which two continents have a combined area which is about the same as the area of South America?

15. **a.** Use the given information to construct a bar graph.
    b. Write two problems that relate to this graph.

Seating Capacity for Some Football Stadiums					
Stadium	**Rose Bowl**	**Gator Bowl**	**Cotton Bowl**	**Sugar Bowl**	**Liberty Bowl**
Capacity	106,721	70,000	72,000	80,982	50,180

Scale: Let 1 centimeter represent 10,000 seats.

Line Graphs and Applications

Line graphs show the amount of change over a certain period of time. A line graph has a <u>horizontal</u> axis and a <u>vertical</u> axis. One axis shows the <u>period of time</u>. The other shows the <u>amount of change</u>.

The line graph below shows the average amount of rainfall in Tampa, Florida for each month of the year.

The line graph below shows the average temperature in Tampa, Florida for each month of the year.

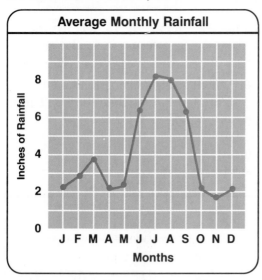

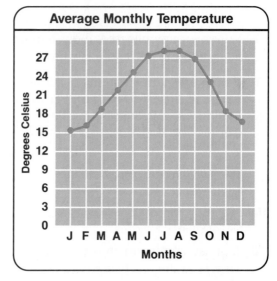

1. Which axis indicates the average number of inches of rainfall?

2. What information is given on the horizontal axis?

3. Which month is the driest?

4. *Complete:* In July, there is more than __?__ inches but less than __?__ inches of rainfall.

5. What information is given on the horizontal axis?

6. What temperature scale is given on the vertical axis?

7. What are the two warmest months?

8. *Complete:* In June, the temperature is more than __?__ degrees Celsius but less than __?__ degrees Celsius.

EXERCISES

You are planning to spend one month in Tampa next year. You want to be there when the temperature is mild <u>and</u> when the amount of rainfall is small. Use the two graphs above for Exercises 1–6.

1. Between which two months does the amount of rainfall increase the most?

2. Between which two months does the amount of rainfall decrease the most?

3. Which months have less than 3 inches of rainfall?

4. What are the four warmest months?

5. Which months have temperatures between 20° and 26°?

6. Which months are the best choices for staying in Tampa?

This line graph shows the number of visitors to the White House for each month of a recent year. Use this graph for Exercises 7–12.
NOTE: *The numbers on the vertical axis stand for thousands. Thus, 20 means 20,000.*

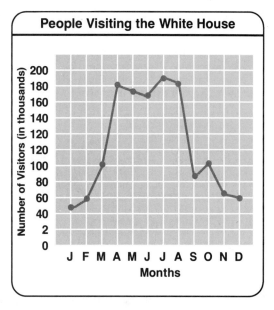

People Visiting the White House

7. During which month are there the fewest visitors?

8. What is the approximate number of visitors in January?

9. Which two months had the most visitors?

10. Why do you think July and August had the most visitors?

11. Notice the large decrease in attendance from August to September. How do you account for this?

12. Why do you think there was a large increase in attendance from March to April?

Use graph paper to construct line graphs to show the data.

13. **Average Temperature (Degrees Fahrenheit) in Albany, N.Y.**

Month	J	F	M	A	M	J	J	A	S	O	N	D
Degrees	23	24	33	46	58	67	71	70	62	51	39	26

14. **Alice's Growth in Inches**

Age	2	4	6	8	10	12	14	16	18
Height	34	40	45	48	51	57	62	63	65

15. **Cars in the Town's Parking Lot**

Time	6	8	10	Noon	2	4	6	8	10
Number	4	17	46	40	51	34	9	14	7

Mid-Chapter Review

The graph at the right shows the total area of each of the Great Lakes. Use this graph for Exercises 1–3. (Pages 84–85)

1. Which lake is the largest?

2. Which lake is the smallest?

3. Which lakes have areas between 60,000 square miles and 80,000 square miles?

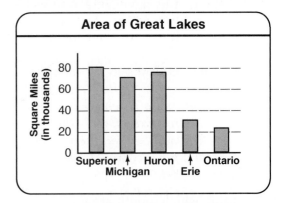

The graph at the right shows the number of cars sold at a dealership during an 8-month period. Use this graph for Exercises 4–7. (Pages 86–87)

4. In which months were the same number of cars sold?

5. Between which two months did car sales increase the most?

6. During which month was the least number of cars sold?

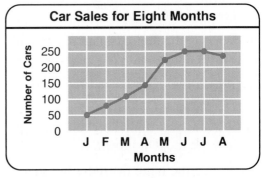

7. Between which two months did car sales decrease?

MAINTENANCE

Multiply or divide as indicated. (Pages 41–42, 64–65, and 68–69)

8. 5.2×1.7 **9.** 4.9×3.4 **10.** 1.75×6.5 **11.** 8.15×8.2

12. $14 \overline{)\, 3.78}$ **13.** $3.6 \overline{)\, 19.08}$ **14.** $0.16 \overline{)\, 27.2}$ **15.** $0.24 \overline{)\, 29.28}$

16. The area of Arkansas is 53,187 square miles. The area of Missouri is 69,697 square miles. Estimate the combined area of the two states. (Pages 2–3)

17. Margaret deposits checks for $72 and $136 in her checking account. She receives three $20-bills back. What is her net deposit? (Pages 36–37)

18. The average yearly salary at Chapman Supply is $16,380. What is the average weekly salary? (Pages 58–59)

19. Which is the better buy, a 12-ounce box of cereal for $1.49, or an 18-ounce box of cereal for $2.15? (Pages 66–67)

Math and Scattergrams

Linda Curran is a manager of an auto supply parts company. From her records, she made the table below to show the number of days it took to send shipments by train.

Distance (miles)	200	300	350	500	550	750	900	1000
Days	3	5	5	8	7	10	13	11

Linda used the data in the table to make the scattergram at the right. She could not draw a straight line to connect all the points so she drew a **line of best fit** through the points. She used the line of best fit to estimate the number of days needed to send shipments.

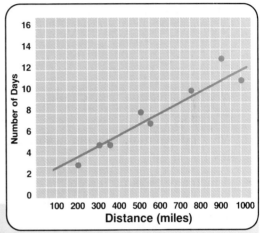

EXAMPLE

Estimate the number of days it will take to send a shipment 700 miles.

1. Locate 700 miles on the horizontal axis.

2. Move straight up to the line and find the corresponding point on the vertical axis.

It will take about **9 days** to send a shipment 700 miles.

EXERCISES

Use the line of best fit in the scattergram above to estimate the number of days needed for each distance.

1. 400 miles

2. 800 miles

3. 900 miles

4. 500 miles

5. 590 miles

6. 208 miles

7. Linda is sending a shipment of parts 650 miles. Should she estimate the number of days needed as 8 or 9? Explain.

8. It took 12 days for a shipment of parts to travel 800 miles. Is the amount of time greater than or less than the estimate Linda would have made from the scattergram?

Other Graphs and Applications

You can compare related information by placing two graphs on the same axes.

This **multiple bar graph** shows the separate responses of boys and girls to a survey on favorite sports. Use the graph to answer these questions.

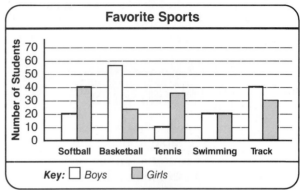

1. What scale is given on the vertical axis?

2. What information is given on the horizontal axis?

3. What information is given in the key?

4. How many boys preferred softball?

5. How many girls preferred softball?

EXERCISES

Use the graph above to answer Exercises 1–12.

1. What was the most popular sport among girls?

2. What was the least popular sport among boys?

3. Which sport was preferred by an equal number of boys and girls?

4. Did more girls prefer basketball or track?

5. Which sports did more than 30 boys prefer?

6. Which sports did fewer than 30 girls prefer?

7. How many more boys preferred basketball than track?

8. How many more girls preferred softball than tennis?

9. How many more boys than girls preferred track?

10. How many more girls than boys preferred softball?

11. How many girls responded to the survey?

12. How many boys responded to the survey?

This graph shows the number of certain registered breeds of dogs in 1983 and in 1985. Use this graph for Exercises 13–21.

13. What information is given on the horizontal axis?

14. What information is given on the vertical axis?

15. Which breed had the greatest number registered in 1985?

16. Which breed had the fewest number registered in 1985?

17. Which breed had the greatest number registered in both 1983 and 1985?

18. Which breeds had about the same number registered in 1983?

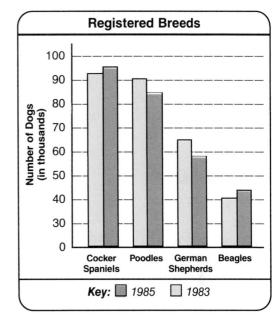

19. Which breed had about the same number registered in both 1983 and in 1985?

20. For which breeds did the number of registrations decrease from 1983 to 1985?

21. Which breeds had fewer than 70,000 registered in both 1983 and 1985?

This graph shows the average monthly temperatures in Seattle, Washington (solid line) and Spokane, Washington (dashed line). Use this graph for Exercises 22–26.

22. In Seattle, which month is the warmest?

23. During which months is the average temperature the same in both Seattle and Spokane?

24. In July, how much warmer is it in Spokane than in Seattle?

25. During which three months is the temperature difference between the two cities the greatest?

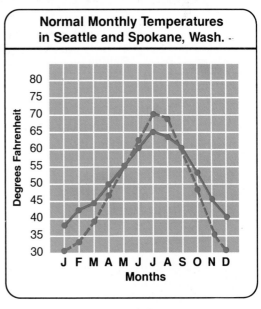

26. Which city has the greater **range** (the difference between the highest and lowest) of temperatures?

Strategy: INTERPRETING GRAPHS

Sometimes the information shown in bar graphs can be misinterpreted because of the way in which the information is displayed. Special attention should be paid to the horizontal and vertical scales used in the graphs.

EXERCISES

The two bar graphs below show the number of books sold by two book stores during a 4-month period. Look at these graphs to answer Exercises 1–6.

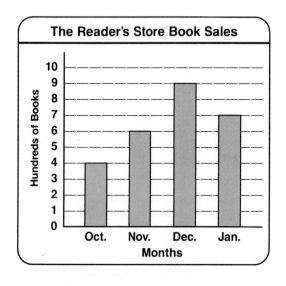

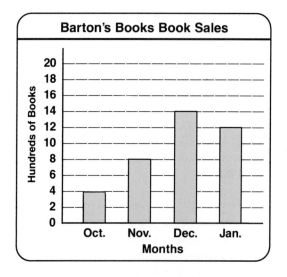

1. Look at the bars on the two graphs. Which store appears to have sold more books during October? during December?

2. Looking at the bars on the two graphs, which store appears to have sold more books during the 4-month period?

3. How many books did The Reader's Store actually sell during the 4 months?

4. How many books did Barton's Books actually sell during the 4 months?

5. Does placing the graphs side by side make it appear that The Reader's Store sold more books than Barton's Books during the 4-month period? Explain.

6. Suppose the two graphs were not placed side by side. Would it still appear that The Reader's Store sold more books than Barton's Books during the 4-month period? Explain.

The graph at the right shows approximate driving mileages from Dallas, Texas to two other cities in Texas. Use this graph for Exercises 7–11.

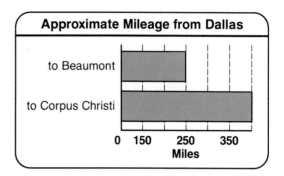

Approximate Mileage from Dallas

to Beaumont

to Corpus Christi

0   150   250   350
Miles

7. How many times as long as the bar for the mileage to Beaumont is the bar for the mileage to Corpus Christi?

8. The mileage to Beaumont is 250 miles. By the appearance of the bars, what would be the mileage to Corpus Christi?

9. What is the actual mileage to Corpus Christi?

10. What is the actual difference in the mileage to each city?

11. Is this graph misleading? Explain.

A team of salespersons made graphs A and B to show the number of airplanes they sold in one year. Use these graphs for Exercises 12–17.

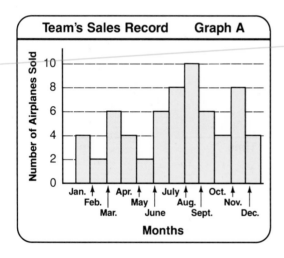

Team's Sales Record   Graph A

Number of Airplanes Sold

Jan. Feb. Mar. Apr. May June July Aug. Sept. Oct. Nov. Dec.

Months

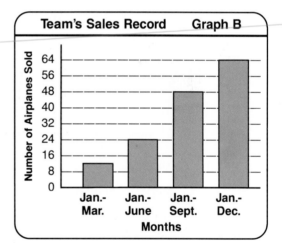

Team's Sales Record   Graph B

Number of Airplanes Sold

Jan.-Mar.   Jan.-June   Jan.-Sept.   Jan.-Dec.

Months

12. Use Graph A to determine how many planes were sold from January through June.

13. Use Graph B to determine how many planes were sold from January through June.

14. How do the graphs differ in the way in which they show sales from January through June?

15. How would you use Graph B to determine the total sales from April through June?

16. Which graph seems to indicate that sales are continually going up? Explain.

17. How does Graph A show that the impression given by Graph B is not true?

Using Graphs

A car or a train is always doing one of the following.

a. At a stop (doing nothing)   **b.** Increasing in speed
c. Decreasing in speed   **d.** Moving at a steady (constant) speed

These line graphs show you a picture of **a, b, c,** and **d.**

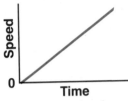

A car starting from
stop and increasing
in speed

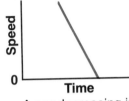

A car decreasing in
speed and coming
to a stop

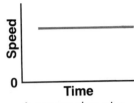

A car moving at a
steady speed

EXAMPLE

Which graph fits this situation?
The traffic light changes to green. The car moves forward and then travels at the speed limit.

A.

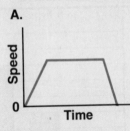

B.

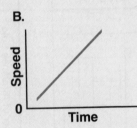

C.

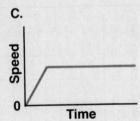

READ   What are the facts?
Starts from a stop. Increases in speed. Moves at a constant speed.

PLAN   List the facts about each graph. Compare with the facts from the situation.

SOLVE   **Graph A:** Starts from a stop. Increases in speed. Moves at a constant speed. Then decreases in speed and comes to a stop.
Graph B: This graph does <u>not</u> show the car starting from a stop. It is in motion and its speed is increasing.
Graph C: Starts from a stop. Increases in speed. Moves at a constant speed.
Answer: **Graph C** fits the facts.

CHECK   Why doesn't Graph A fit the situation? Why doesn't Graph B?

EXERCISES   *Choose the graph that fits the situation.*

1. You are waiting for the light to turn green. The light turns green and you accelerate the car.

   A. 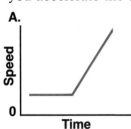   B. 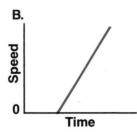   C.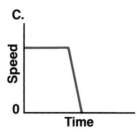

2. You are driving up a hill at a constant speed. You reach the top and drive down the hill at a constant speed.

   A.    B. 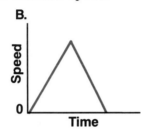   C.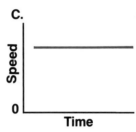

3. You are driving at the speed limit. You park the car to have lunch. You then drive on and drive at the speed limit.

   A. 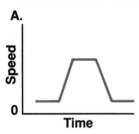   B. 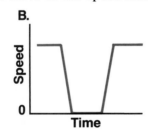   C.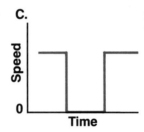

4. You drop a ping pong ball and watch it bounce.

   A. 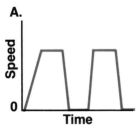   B. 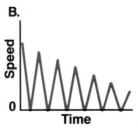   C.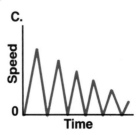

5. Sketch a graph to show the following situation.

   You are waiting for the light to turn green. It turns green and you accelerate the car. You then drive at the speed limit for a few minutes. You then pull into a gas station.

Math in Art

Mr. Jackson's art class is studying different styles of painting.

On one field trip, the class viewed paintings at the Price Gallery of Fine Arts. Each student made a bar graph to show the number of paintings of each art style in the gallery's collection.

EXERCISES

1. Estimate the total number of paintings in the collection.

2. About how many pop art paintings does the gallery have?

3. How many more impressionistic paintings than rococo paintings are there?

4. Which type of painting has twice as much representation as rococo paintings?

5. Suppose that the average worth of each painting in the gallery is $175,000. What is the approximate total value of the paintings?

6. Suppose that five new paintings, each valued at $600,000 are added to the collection. What will the average value of the collection be then?

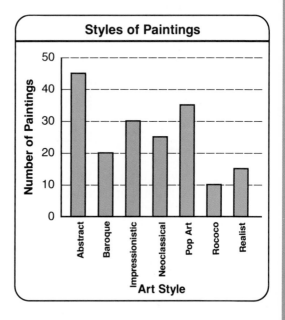

PROJECT

a. Use the library to find examples of at least four of the painting styles mentioned in the graph.

b. Survey at least 20 persons by showing them the examples and asking them to choose the style of painting they like best.

c. Construct a bar graph to show the results of your survey.

Chapter Summary

1. To read a pictograph:
   1️⃣ Look for the key.
   2️⃣ Use the key to interpret the graph.
   Example How many boxes of oranges were produced in 1970?
   Solution 50,000,000 × 4 = 200,000,000
   200 million boxes of oranges

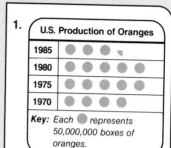

1.
U.S. Production of Oranges

1985	
1980	
1975	
1970	

Key: *Each ● represents 50,000,000 boxes of oranges.*

2. To read a bar graph:
   1️⃣ Look for the number scale on one of the axes.
   2️⃣ Use the scale to interpret the graph.
   Example What is the normal July temperature for Duluth?
   Solution Duluth: About 18°C

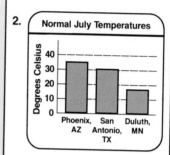

2.
Normal July Temperatures

3. To read a line graph:
   1️⃣ Find the given number or time on one axis.
   2️⃣ Use the second axis to find the unknown number or time.
   Example When was the racer's pulse rate the fastest?
   Solution At 12 noon the pulse rate was fastest.

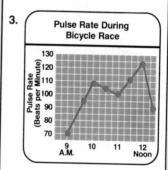

3.
Pulse Rate During Bicycle Race

4. To read a multiple bar graph:
   1️⃣ Read the information on each axis.
   2️⃣ Look for the key.
   3️⃣ Use both of these to interpret the graph.
   Example In what year was the attendance the greatest? For what school?
   Solution 1987; Martin High School

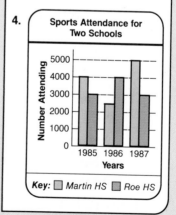

4.
Sports Attendance for Two Schools

Key: ☐ *Martin HS* ☐ *Roe HS*

Chapter Review

Part 1: VOCABULARY

For Exercises 1–6, choose from the box at the right the word(s) that complete(s) each statement.

1. In a pictograph, a picture or symbol is used to represent __?__ . (Page 82)

2. In a pictograph, the __?__ tells how many are represented by each symbol. (Page 82)

3. On a bar graph, the __?__ of the bar represents a number. (Page 84)

4. On a bar graph, the __?__ tells you how to find the number represented by each bar. (Page 84)

5. On a bar graph, the bars can be either vertical or __?__ . (Page 85)

6. Line graphs show the amount of __?__ over a certain period of time. (Page 86)

> bar
> key
> width
> horizontal
> scale
> multiple
> numbers
> length
> change

Part 2: SKILLS AND APPLICATIONS

This pictograph shows the number of cloudy days in certain cities of the United States during 1985. Use this graph for Exercises 7–12. (Pages 82–83)

Number of Cloudy Days in Selected Cities in 1985	
Albany, NY	☁ ☁ ☁ ☁ ☁ ☁ ☁ ☁ ☁ ◖
Cleveland, OH	☁ ☁ ☁ ☁ ☁ ☁ ☁ ☁ ☁ ☁ ☁
Kansas City, MO	☁ ☁ ☁ ☁ ☁ ☁ ☁ ☁
Norfolk, VA	☁ ☁ ☁ ☁ ☁ ☁ ☁ ◖
San Antonio, TX	☁ ☁ ☁ ☁ ☁ ☁ ☁

Key: *Each ☁ represents 20 days.*

7. How many cloudy days does one symbol represent?

8. How many cloudy days are represented by one-half of the symbol?

9. Which city had 140 cloudy days during 1985?

10. How many cloudy days were there in Albany during 1985?

11. Which city had 10 more cloudy days than San Antonio during 1985?

12. What is the difference in the number of cloudy days between the cities with the greatest and the fewest number of cloudy day?

The horizontal bar graph below shows the top speeds of some animals. Use this graph for Exercises 13–20. (Pages 84–85)

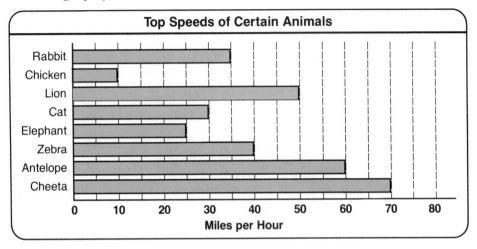

Top Speeds of Certain Animals

Miles per Hour

13. What scale is given on the horizontal axis?

14. What information is given on the vertical axis?

15. Which is the slowest of the animals shown?

16. Which is the fastest of the animals shown?

17. How fast can a zebra run?

18. How fast can an antelope run?

19. Which animal can run twice as fast as a rabbit?

20. Which animals have a top speed greater than 35 miles per hour?

The line graph at the right shows the average amount of rainfall in New Orleans for each month of the year. (Pages 86–87)

21. Which month has the most rainfall?

22. Which month has the least amount of rainfall?

23. Which two months have the same amount of rainfall?

24. Between which two months does the amount of rainfall decrease the most?

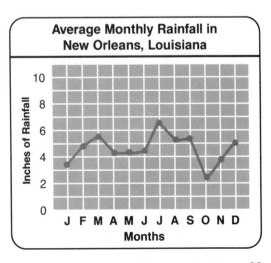

Average Monthly Rainfall in New Orleans, Louisiana

Inches of Rainfall

J F M A M J J A S O N D

Months

Chapter Test

This pictograph shows the number of speakers of the six most commonly spoken languages.

1. How many speakers does 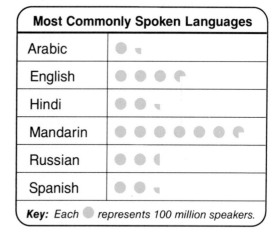 represent?

2. How many speakers does (represent?

3. About how many speakers of Spanish are there?

4. About how many speakers of English are there?

Most Commonly Spoken Languages	
Arabic	● ❜
English	● ● ● ❜
Hindi	● ● ❜
Mandarin	● ● ● ● ● ● ❜
Russian	● ● (
Spanish	● ● ❜

Key: *Each* ● *represents 100 million speakers.*

5. About how many more speakers of Mandarin than English are there?

6. The number of speakers of one language multiplied by 5.4 equals the number of speakers of Mandarin. Which language is this?

This bar graph shows the heights of five buildings in various Texas cities.

7. Which building is the tallest?

8. Which building is the shortest?

9. Which building is 2.5 times as tall as the Tower Life building?

10. Which building is 200 feet taller than One Tandy Center?

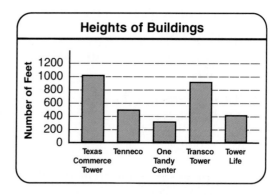

This line graph shows average temperatures in Galveston, Texas for each month of the year.

11. Which are the two warmest months?

12. Which is the coldest month?

13. What is the average temperaure during May?

14. For how many months are the temperatures between 15 degrees and 25 degrees Celsius?

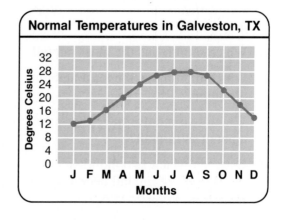

Cumulative Maintenance

Choose the correct answer. Choose a, b, c, or d.

1. Add.

```
   729
  3762
+   53
```

a. 4544   **b.** 4543
c. 4534   **d.** 4454

2. Divide.

15) 4290

a. 275   **b.** 28
c. 286   **d.** 280

3. Subtract.

```
  5003
−  736
```

a. 4267   **b.** 4263
c. 4257   **d.** 4157

4. This bar graph shows the amount of time Allan watched television during the week. On which day did he watch between 60 and 90 minutes?

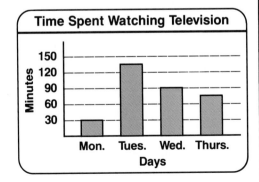

a. Monday   **b.** Tuesday
c. Wednesday   **d.** Thursday

5. Which decimal has the smallest value?

a. 0.731   **b.** 0.079
c. 0.371   **d.** 7.31

6. Multiply.

```
   306
×   42
```

a. 12,962   **b.** 12,752
c. 12,762   **d.** 12,852

7. Add: 73 + 546 + 3592

a. 4011   **b.** 4111
c. 4201   **d.** 4211

8. This table gives the long-distance telephone rates from Central City to several towns.

Long-distance Rates From Central City	Day Rate 8 A.M. to 5 P.M. Mon.–Fri.	
Towns	Initial Two Min	Each Addtl Min
Franklin	$ .90	$ .40
Millbridge	.52	.22
Pleasant Hill	.80	.35
Grifton	.90	.40

What is the cost of a 3-minute call to Franklin at 3:00 P.M. on Tuesday?

a. $1.30   **b.** $0.90
c. $1.70   **d.** $0.50

9. Add.

$$4.35$$
$$12.7$$
$$6.854$$
$$+327.6$$

a. 351.504   b. 351.604
c. 350.504   d. 350.604

10. Multiply 0.73 by 0.06.

a. 438   b. 0.438
c. 0.00438   d. 0.0438

11. Divide: $7 \overline{)2.52}$

a. 0.36   b. 3.6
c. 0.036   d. 36

12. Which of these represents fifteen and three hundredths?

a. 15.3   b. 0.153
c. 15.03   d. 15.003

13. Divide: $4.2 \overline{)12.81}$

a. 30.5   b. 0.305
c. 3.05   d. 305

14. Choose the set of numbers that are correctly ordered from least to greatest.

a. 5.93, 59.7, 0.59, 0.059

b. 65, 6.5, 0.65, 0.065

c. 3.42, 0.342, 3.4, 0.34

d. 0.90, 0.906, 0.91, 1.05

15. Multiply: 8.73×1000

a. 8730   b. 87.3
c. 8.73   d. 87300

16. Subtract: $37.008 - 9.39$

a. 36.069   b. 27.718
c. 27.618   d. 28.628

17. At which price will one pound of hamburger cost the least?

a. 2 lbs for $3.50
b. 3 lbs for $4.50
c. 4 lbs for $6.20
d. 5 lbs for $7.75

18. The cash register in Joe's Diner contains four $20-bills, eight $10-bills, twelve $5-bills, and nine $1-bills. How much money is in the register?

a. $249   b. $229
c. $44   d. $219

19. Bill bought two shirts for $9.00 each and a pair of shorts for $15.00. How much did he pay in all?

a. $39.00   b. $24.00
c. $6.00   d. $33.00

20. How many pizzas are represented by ○ ○ ○ ○ ○?

Key: Each ○ represents 8 pizzas.

a. 50   b. 45
c. 40   d. 32

Statistics

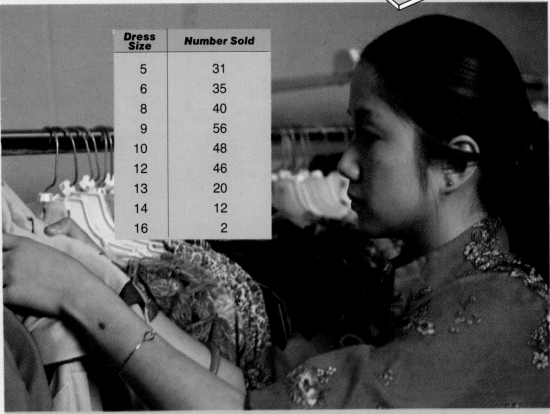

Dress Size	Number Sold
5	31
6	35
8	40
9	56
10	48
12	46
13	20
14	12
16	2

- How do you determine the mean, median, and mode of a list of data?

- How can you use the mean, median, and mode to interpret information and to help you make consumer decisions?

- How can you use frequency tables to organize information?

- What information about data can be read directly from a histogram?

Meaning of Average

ACTIVITY

The students in your math class are making a bar graph to show their **average** scores on 8 quizzes. Each quiz is worth ten points. Suppose that the bar graph at the right (see Figure 1) shows your scores.

1. Copy the bar graph. Use strips of paper to make the bars.

   To find your average quiz score, follow this procedure.

2. Cut unit squares from the larger bars and add them to the tops of the shorter bars. Try to make all the bars as close to the same height as possible (see Figure 2).

3. In Figure 2, do most of the bars have the same height?

4. What score is shown by the seven equal bars? This score, 7, is the approximate average score.

5. How can you use Figure 2 to tell whether the exact average score is greater than 7 or less than 7? Explain.

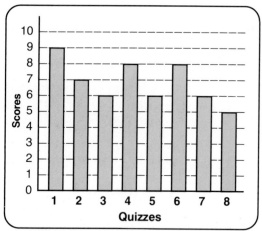

Figure 1

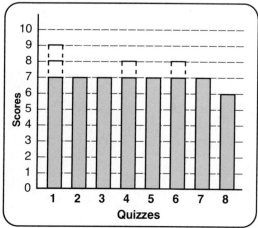

Figure 2

EXERCISES

1. Copy this bar graph.
   Use strips of paper to make the bars.

2. Cut unit squares from the bars and rearrange them to find the approximate average score.

3. Is the exact average greater than or less than the approximate average?

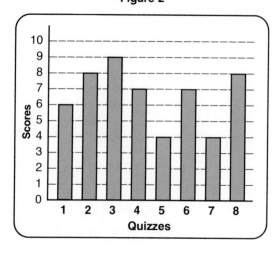

The Mean and the Mode

Ms. Tompson's fourth-period math class listed the number of days each of the thirty students had been absent so far this year. Now they want to find the **mean** number of absences.

Number of Days Absent Math 106: Fourth Period									
7	8	3	9	4	2	3	0	4	5
3	2	4	2	3	5	3	8	1	6
6	5	0	3	4	6	1	4	1	5

1. How do you compute the mean (or average) number of absences?

EXAMPLE What is the mean number of absences for Ms. Tompson's fourth-period math class?

$$\text{Mean} = \frac{\text{Total Number of Absences}}{\text{Number of Students}}$$ *Use a calculator.*

$$= \tfrac{117}{30} \longrightarrow \boxed{1}\,\boxed{1}\,\boxed{7}\,\boxed{\div}\,\boxed{3}\,\boxed{0}\,\boxed{=}\; \boxed{\quad 3.9 \quad}$$

The mean number of absences is **3.9 days.**

2. Was every student in Ms. Tompson's class absent for about four days?

3. How many students were absent four or more days?

4. How many students were absent fewer than four days?

5. Is the average of a list of whole numbers always a whole number? Give an example to illustrate your answer.

> In a list of numbers, the **range** is the difference between the greatest and smallest number in the list.

6. *Complete:* The range for the number of absences in Ms. Tompson's class is 9 − _?_, or _?_.

> The **mode** in a list of measures or items is the one that appears most often. There can be **more than one mode,** or **no mode.**

7. Which number of days appears most often in the list of absences?

8. What is the mode for the number of absences?

EXERCISES

Complete. Choose your answers from the box at the right.

1. The sum of a group of measures divided by the number of measures is called the __?__ .

2. In a listing of data, the measure that occurs most often is the __?__ .

3. For the numbers 7, 6, 6, 3, 3, 5, the mean is __?__ .

4. For the numbers in Exercise 3, the modes are __?__ and __?__ .

| mean |
| mode |
| 3 |
| 5 |
| 6 |

Find the mean, the range, and the mode for each list of numbers.

5. 5, 7, 9, 11, 13

6. 102, 103, 104, 105

7. 10, 10, 10, 10, 15

8. 176, 176, 176, 176, 176, 176

9. 12, 12, 12, 9, 0, 0

10. 7.1, 7.1, 7.2, 7.3, 7.3, 7.4

11.
Earnings Per Hour for Eight Workers

| $6.50 | $5.75 | $5.25 | $4.95 |
| $6.25 | $5.00 | $8.15 | $7.35 |

12.
Number of Hours of Sleep for Ten Adults

| 8 | 6 | 6.5 | 8.5 | 7 |
| 7.5 | 7 | 8 | 5.5 | 8 |

Ten students received these scores on a test.
75, 80, 85, 80, 70, 90, 60, 70, 85, 85

13. What is the range of the scores?

14. What is the mode of the scores?

15. What is the mean score?

16. Suppose that the teacher adds 6 points to each score. What is the new mean score?

17. What is the mode of the scores in Exercise 16?

18. The mean of five numbers is 11.6. What is the sum of the five numbers?

19. Alice has a mean score of 7 for 6 games. She plays again and scores 0. What is her new mean score?

20. At Compuquick, Inc., two employees earn $20,000 per year, six earn $24,000 per year, and two earn $40,000 per year. What is the mean salary?

21. In Exercise 20, how close is the mode to the mean salary?

22. In Exercise 20, find the mean salary for the eight lowest paid employees.

Math and Managing a Store

Lydia Carson is the manager of a small shoe store. She uses **dot diagrams** to help her analyze information that will make her business more efficient and profitable.

EXAMPLE 1

Make a dot diagram to show the sizes of women's shoes sold in one day.

Shoe Size	5	$5\frac{1}{2}$	6	$6\frac{1}{2}$	7	$7\frac{1}{2}$	8
Number Sold	1	2	8	4	4	5	6

EXERCISES

For Exercises 1–4, refer to the dot diagram in the Example.

1. What is the mode of the sizes bought?

2. How many pairs of shoes larger than size 6 were sold?

3. How many pairs of sizes $6\frac{1}{2}$, 7, and $7\frac{1}{2}$ were sold?

4. Which sizes would you expect Lydia to keep in stock?

The table below shows the number of customers entering Lydia's store from noon until 5:00 P.M. on a typical day.

Hours	12:00–1:00	1:00–2:00	2:00–3:00	3:00–4:00	4:00–5:00
Number of Customers	25	18	10	8	12

5. Make a dot diagram to show the number of customers entering Lydia's store from noon to 5:00 P.M.

6. During what hour did the least number of customers enter Lydia's store?

7. During what hour does the mode occur?

8. During which two hours will Lydia need to have the most employees on duty? Why?

9. How could you use the dot diagram to find the mean number of customers (see the Activity on page 104)?

10. Name one advantage of using dot diagrams to show data.

The Median

High Jump Scores

Allan	40 inches
Meg	28 inches
Donald	43 inches
Carol	30 inches
Pierre	60 inches
Melanie	35 inches
Susie	42 inches
Greg	36 inches
Christopher	41 inches
Natalie	34 inches
Tomaso	54 inches

Eleven students in gym class recorded their high jump scores. You want to find the middle score.

1. How can your organize the listing to make the middle score easier to find?

2. Whose score is the middle score?

3. How many scores will be higher than the middle score?

4. How many scores will be lower than the middle score?

In a list or data organized from least to greatest (or from greatest to least), the **median** is the number in the middle.
When a list of data has two middle numbers, the median is the <u>mean</u> of the two numbers.

Suppose that Noam came late to gym class. His high jump score was 38 inches.

5. Which two numbers will now be the middle numbers?

6. What is the mean of these two numbers?

7. What is the new median high jump score?

8. How many scores will be greater than or equal to the new median score?

9. How many scores will be less than or equal to the new median score?

EXERCISES

Complete. Choose your answers from the box at the right.

1. The median of a list of scores arranged in order can be described as a _?_ score.

2. In a list of eight numbers arranged in order, the median is the _?_ of the two _?_ numbers.

3. In a list of nineteen numbers arranged in order, the median is the _?_ number.

4. In a list of eight numbers arranged in order, exactly _?_ are greater than or equal to the median and exactly _?_ are less than or equal to the median.

For Exercises 5–10, find the median.

5.
Heights in Inches
59   71   64   61   60   68   64

6.
Low Temperatures for 8 Days
65° 60° 57° 61° 63° 59° 68° 70°

7.
Shoe Size
$8\frac{1}{2}$ 6 $7\frac{1}{2}$ 8 $9\frac{1}{2}$ $5\frac{1}{2}$ $8\frac{1}{2}$

8.
Employees' Weekly Salaries
$504 $498 $365 $450 $610 $390

9. **Number of Characters in Five Plays**

Play	Number of Characters
Julius Caeser	34
Hamlet	26
Romeo and Juliet	28
The Tempest	16
Twelfth Night	12

10. **Years in Office: Eight Presidents**

President	Years in Office
F. Roosevelt	14
H. Truman	7
D. Eisenhower	8
J. Kennedy	3
L. Johnson	5
R. Nixon	6
G. Ford	2
J. Carter	4

11. **a.** Find the median of these scores.
    92, 96, 85, 89, 109

    b. Replace 85 with 60. Find the median.

    c. Are the medians in **a** and **b** the same? Explain why.

12. **a.** Find the median of these scores.
    28, 30, 30, 32, 23, 29, 28

    b. Omit the largest and the smallest scores. Find the new median.

    c. Are the medians in **a** and **b** the same? Explain why.

13. Insert one number in this list so that the median is not changed.
    21   25   27   33

14. **a.** Add 5 to each number in Exercise 13. Then find the new median.

    b. Compare the new median to the median in Exercise 13.

Strategy: INTERPRETING INFORMATION

Consumers have to make many decisions each day. To make these decisions, they often have to ask questions such as the following.

"Which of these, the mean, median, or mode, will best help me in making the decision?"

The following exercises will help you to understand how this problem-solving process works.

EXERCISES

Cathy Locke owns a small boutique. Because her shop has limited storage space, Cathy has to determine how many dresses of each size to keep in stock. After checking sales receipts for 10 days, she made a list of the number of dresses sold.

Size	Number of Dresses Sold
6	5
8	12
10	19
12	17
14	15
16	7
18	6
20	3

1. What is the mean size sold?

2. What is the median size sold?

3. What is the mode of the sizes sold?

4. Explain why information about the mean and median sizes is not useful to Cathy.

5. What sizes do you think Cathy should keep in stock? Why?

Jeff is a college student who lives at home and drives to classes. At the beginning of his first semester, he timed the number of minutes it took to travel each of two possible routes.

First week: Route A	35	40	31	43	36
Second week: Route B	23	29	31	46	51

6. What is the mean time for each route?

7. What is the range of times for each route?

8. What route would you take if you were Jeff? Explain.

Careful Car Rentals has 10 employees. They are paid the following weekly salaries.

$385	$385	$385	$390	$ 390
$400	$410	$500	$700	$1050

9. What is the mean weekly salary for the ten employees?

10. How many employees make more than this "average" weekly salary?

11. How many employees make less?

12. Find the median weekly salary.

13. What is the mode of the weekly salaries?

14. Which gives the best description of weekly salaries at Careful Car Rentals, the mean, median, or mode? Why?

An elevator in an apartment building is designed to carry 2000 pounds.

15. Which would be the more effective sign to post in the elevator? Why?

   a. Minimum capacity: 2000 pounds

   b. Maximum capacity: 12 persons

16. If the second sign is posted, what is assumed about the "average" weight of each passenger?

Mid-Chapter Review

Find the mean, the range, and the mode. (Pages 105–106)

1.

Miles Per Gallon for Ten Cars				
22	30	12	26	18
32	22	28	18	22

2.

Number of Pages in Eight Textbooks			
480	512	560	528
544	560	464	528

Find the median. (Pages 108–109)

3.

Scores on a Math Test				
68	72	55	83	90
98	76	80	81	87

4.

Speeds of 5 Race Cars				
135	160	143	150	138

The table at the right below shows yearly salaries for six persons who work at Cliff Insurance Company. Use this table for Exercises 5–8. (Pages 110–111)

5. What is the mean yearly salary?

6. What is the median yearly salary?

7. What is the mode of the yearly salaries?

8. Which gives the best description of yearly salaries at the company, the mean, median, or mode? Why?

Yearly Salaries	
Manager	$45,000
Agent	$25,000
Secretary	$14,500
Clerk	$10,000
Clerk	$ 9,500
Clerk	$ 9,000

MAINTENANCE

The line graph at the right shows normal monthly temperatures for a certain city. Use the graph to answer Exercises 9–12. (Pages 86–87)

9. What is the lowest monthly temperature?

10. What is the highest monthly temperature?

11. Between which two months is the increase in temperature greatest?

12. Between which two months is the decrease in temperature greatest?

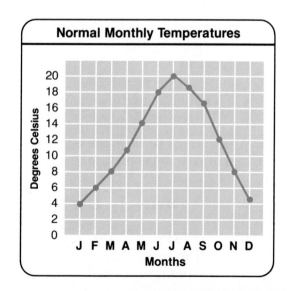

Normal Monthly Temperatures

Math in Science

Eduardo is doing research in science class on the various kinds of trees that grow in the United States. He made a table to organize his information.

Name of Tree	Height in feet	Location
Bald Cypress	83	Louisiana
Chinaberry	75	Hawaii
Coconut	93	Hawaii
Mangrove	75	Florida
Oak	134	Kentucky
Pecan	130	Mississippi
Ponderosa Pine	223	California
Redwood	362	California
Sequoia	275	California
Spruce	216	Oregon

EXERCISES

For Exercises 1–6, refer to the information in the table.

1. Which tree is closest in height to the mean height of the trees?

2. What is the median height of the trees?

3. What is the mode of the data in the table?

4. What is the range of the heights of the trees?

5. What is the mean height of the trees located in California?

6. How much greater is the mean height in Exercise 5 than the mean height of the trees located in Hawaii?

7. Which best represents the heights of the ten trees, the mean, median, or mode? Explain.

8. Would you expect the mean and the median of the heights of the three California trees to be closer together than the mean and the median of the ten trees in the table? Explain.

PROJECT Do research to find the average heights of five of the tallest trees that grow in your state. Find the mean, median, mode, and range of the heights of these trees.

Organizing Data: FREQUENCY TABLES

For a lesson on organizing data, Ms. Tompson divided her fourth-period math class into small groups. Then she wrote this assignment on the chalkboard.

A. Collect the full names (first and last) of 15 students, friends or relatives. Do not use nicknames.

B. List the names in a chart like the one at the right. Count the number of letters in each first, last, and complete name.

	Number of letters in		
Student	First	Last	Complete
1. Karl Swifteagle	4	10	14
2. Susan Clark	5	5	10
3. Antonio Pecci	7	5	12

C. Use **equal intervals** and a **frequency table** to show the number of letters in the last names.

    1. What are "equal intervals"? Give an example.

    2. What does the word "frequency" mean?

    If your teacher wishes, work together in small groups to collect the data, make the chart and frequency table in the Example, and do the Exercises.

EXAMPLE Use the sample data at the right (or use your own data) to make a frequency table. For the sample data, use the intervals 2–4, 5–7, and 8–10.

Number of Letters	Tally	Frequency
2–4	HH IIII	9
5–7	III	3
8–10	III	3

Student	Number of Letters in Last Name
1. Karl Swifteagle	10
2. Susan Clark	5
3. Antonio Pecci	5
4. John Matheny	7
5. Roger Toma	4
6. William Foy	3
7. Mae Wu	2
8. Jorge Rodriguez	9
9. Karen Hue	3
10. Caroline Shim	4
11. Ida Cantiuna	9
12. David Wall	4
13. Emily West	4
14. Alicia Ziff	4
15. William Byrd	4

3. What is the range of the number of letters in the last names?

4. Does the overall range of the intervals in the frequency table correspond to the range in Exercise 3?

EXERCISES

For Exercises 1–2, use the sample data in the Example or use your own data as your teacher directs.

1. Make a frequency table to show the number of letters in the first names of the 15 students. For the sample data, use the intervals 2–3, 4–5, 6–7, and 8–9.

2. Make a frequency table to show the number of letters in the complete names (first and last) of the 15 students. For the sample data, use the intervals 3–6, 7–10, and 11–14.

For Exercises 3–8, refer to the table below.

3. Complete the "frequency" column in the table showing pulse rates.

4. How many people were in the survey?

5. How many people had pulse rates of 80 or higher?

6. How many people had pulse rates of at least 70? (Hint: "At least 70" means 70, 71, or 72, and so on.)

Pulse Rates, Beats per Minute	Tally	Frequency
65–69	⊬⊬⊬	?
70–74	⊬⊬⊬ ⊬⊬⊬ ⊬⊬⊬ ⊬⊬⊬ //	?
75–79	⊬⊬⊬ ⊬⊬⊬ ⊬⊬⊬ ⊬⊬⊬ ///	?
80–84	⊬⊬⊬ ////	?
85–89	/	?

7. What is the most common number of beats per minute?

8. What is the mode of the number of beats per minute?

An airline asked passengers on a flight to rate the quality of service. The table shows the ratings of 21 passengers.

9. How many ratings were "Good" or better?

10. What was the "average" rating?

11. What conclusions can the airline draw about how the typical passenger feels about the service? Use a frequency table to explain your answer.

Service Ratings						
1	5	4	4	3	3	3
4	5	1	2	2	4	4
5	4	4	3	3	4	5

5: Superior   2: Fair
4: Excellent   1: Poor
3: Good

Histograms

A **histogram** is a special kind of bar graph.

EXAMPLE

Draw a histogram to show the number of letters in the last names of the 15 students shown in the table.

Student	No. of Letters in Last Name
1. K. SWIFTEAGLE	10
2. S. CLARK	5
3. A. PECCI	5
4. J. MATHENY	7
5. R. TOMA	4
6. W. FOY	3
7. M. WU	2
8. J. RODRIGUEZ	9
9. K. HUE	3
10. C. SHIM	4
11. I. CANTILINA	9
12. D. WALL	4
13. E. WEST	4
14. A. ZIFF	4
15. W. BYRD	4

1 Use intervals to make a frequency table.

Number of Letters	Frequency
2–4	9
5–7	3
8–10	3

2 Draw a bar for each interval.

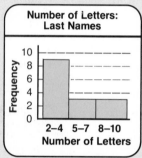

1. What does the height of the bars of the histogram show?

2. What is the range of the number of letters?

3. Which interval is the mode of the number of letters?

4. In which interval does the median occur?
(Remember: The median is the middle number.)

EXERCISES

This histogram shows how many nations joined the United Nations (U.N.) from 1945 to 1984.

1. In what ten-year interval did most nations join?

2. How many members joined the U.N. between 1975-1984?

3. Can you tell how many nations joined in 1968? Explain.

4. Between 1945 and 1964, 115 members joined the U.N. How many members joined the U.N. between 1955 and 1964?

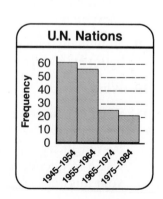

This graph shows the test scores of students in a school-wide exam.

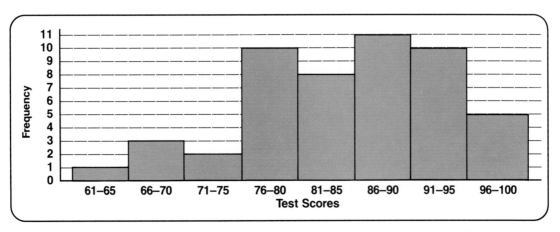

5. How many students received test scores between 81 and 85?

6. How many students received test scores of 61 through 65?

7. How many students took the test?

8. What is the range of the test scores?

9. What is the mode of the test scores?

10. In which score intervals are there the same number of students?

11. Which has the greater number of students, the highest or the lowest score interval?

12. How many students scored 83 on the test? Explain your answer.

13. How many students received scores higher than 80?

14. How many students received scores of 75 or lower?

For Exercises 15–18, use the frequency table to draw a histogram.

15. **Earnings of 12 Women Golfers**

Earnings	Frequency
$6,000 – $10,000	6
$11,000 – $15,000	3
$16,000 – $20,000	3

16. **Bowling Scores**

Scores	Frequency
121–140	2
141–160	14
161–180	10
181–200	1

17. **Lunch Costs of 25 Students**

Cost	Frequency
50¢ – 89¢	8
90¢ – $1.29	9
$1.30 – $1.69	4
$1.70 – $2.09	3
$2.10 – $2.49	1

18. **Diameters of 50 Trees**

Centimeters	Frequency
17–23	9
24–30	23
31–37	14
38–44	4

Sampling

Work in small groups or as your teacher directs.

GOAL   You want to take a survey to determine the color preferences of students in these age groups.

Age Groups: 7–9   10–12   13–15   16–18

You will want to discuss considerations such as these <u>before</u> taking the survey.

Discussion   1. How will you and your class-mates organize yourselves to survey each group?

2. How many students in each age group will you survey?

3. What questions(s) will you ask?

4. How will you classify answers such as "chartreuse," "magenta," "aqua," and so on?

5. Will you assign a length of time in which the survey is to be completed?

6. What information about each person surveyed will you need?

<u>After</u> completing the survey, you will need to consider these questions and make decisions.

Discussion   7. How will you summarize the data? (Will you use tables, graphs, a histogram?)

8. What conclusions can you draw from the data? For example:

   a. What are the differences (if any) in color preferences according to age group?

   b. What are the differences (if any) in color preferences between boys and girls in each age group?

   c. Do the differences (if any) in color preferences between boys and girls appear to decrease with age?

PROJECT   Make a bulletin board showing how you planned the survey, the data you collected (tables, graphs, frequency tables, histograms, and so on), and the conclusions resulting from your work.

Chapter Summary

1. To find the mean:
- ☐1 Find the sum of the measures.
- ☐2 Divide the sum by the number of measures.

2. To find the range:
- ☐1 Identify the greatest and the smallest measure.
- ☐2 Subtract the measures.

3. To find the mode, identify the number(s) that appear most often.
- **a.** When one number appears most often, that number is the mode.
- **b.** When two or more different numbers appear most often, each of the numbers is the mode.
- **c.** When no number appears more than once, there is no mode.

4. To find the median:
- ☐1 Arrange the data in order.
- ☐2 **a.** For an **odd** number of items, the median is the middle measure listed.
- **b.** For an **even** number of items, the median is the mean of the two middle measures.

5. To draw a histogram:
- ☐1 Count how many times each measure occurs.
- ☐2 List the data in equal intervals in a frequency table.
- ☐3 Draw a bar for each interval.

---

1. Test Scores: 88, 70, 91
Find the mean.
$88 + 70 + 91 = 249$
Mean: $249 \div 3 = \textbf{83}$

2. Find the range of the scores above.
Greatest measure: 91
Smallest measure: 70
Range: $91 - 70 = \textbf{21}$

3.
Mileages: 24, 32, 15, 12, 32
Mode: 32

Prices: $10, $8, $5, $8, $5
Modes: $8, $5

Scores: 87, 92, 94, 80, 68
Mode: None

4.
Heights: 62, 71, 58, 65, 68, 71, 68, 65, 62, 58
Median: 65

Temperatures: 57°, 65°, 43°, 72°, 72°, 65°, 57°, 43°
Median: 61°

5.

Hours	Tally	Frequency
1-2	II	2
3-4	ⅢⅠ	5
5-6	III	3

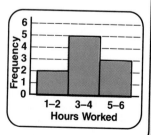

Chapter Review

Part 1: VOCABULARY

For Exercises 1–6, choose from the box at the right the word(s) that complete(s) each statement.

1. The sum of a group of measures divided by the number of measures is called the __?__. (Page 105)

2. The difference between the greatest and smallest number in a list of data is called the __?__. (Page 105)

3. The measure that occurs most often in a set of data is the __?__. (Page 105)

4. In a listing of data, the middle measure is the __?__. (Page 108)

5. Data in a frequency table is listed in equal __?__. (Page 114)

6. A special bar graph that shows data in equal intervals is called a __?__. (Page 116)

| intervals |
| mean |
| histogram |
| range |
| median |
| mode |

Part 2: SKILLS AND APPLICATIONS

For Exercises 7–8, find the mean. (Pages 105–106)

7.

Cost of Cereal at Six Stores		
$2.41	$2.65	$2.37
$2.47	$2.45	$2.53

8.

Miles Driven Per Day on a Vacation Trip			
348	256	310	430
500	383	365	180

9. Find the range of the data in Exercise 7. (Pages 105–106)

10. Find the range of the data in Exercise 8. (Pages 105–106)

For Exercises 11–12, find the mode. (Pages 105–106)

11.

Car Output Per Day For 10 Factories				
175	215	192	180	200
180	207	195	210	183

12.

Distance in Meters for Shot-Put			
12.5	12.7	13.2	16.8
12.7	15.4	12.5	14.9

For Exercises 13–14, find the median.   (Pages 108–109)

13.   **Age on Taking Oath of Office**
of Some Recent Presidents

Presidents	LBJ	RN	GF	JC	RR
Age	55	56	61	52	69

14.   **Number of Points Scored by**
a Football Team

Game	1	2	3	4
Points	22	16	35	21

A salesperson traveled the following number of miles each week.
Use the given information for Exercises 15–18.   (Pages 110–111)

50   435   515   420   382   515   470   319

15. What is the mean number of miles traveled?

16. What is the median number of miles traveled?

17. What is the mode of the number of miles traveled?

18. Which gives the best description of the number of miles traveled, the mean, median, or mode? Why?

For Exercises 19–20, use the given information to complete the table.   (Pages 114–115)

19.   ***Number of Commuters on 20 Trains***

375	387	579	575	369
510	460	300	424	347
490	528	615	495	481
503	604	650	333	580

Commuters	Tally	Frequency
300–399	?	?
400–499	?	?
500–599	?	?
600–699	?	?

20.   ***Prices of 1 Tennis Racquet***

$68.71	$72.45	$69.99	$70.35
$65.25	$50.42	$65.67	$71.20
$73.50	$70.60	$67.55	$68.88

Prices	Tally	Frequency
$50–$55.99	?	?
$56–$61.99	?	?
$62–$67.99	?	?
$68–$73.99	?	?

For Exercises 21–22, use the frequency table to draw a histogram.
(Pages 116–117)

21.   **Number of Books in 50 Libraries**

Number of Books	Frequency
0–300	8
301–600	10
601–900	22
901–1200	10

22.   **Scores for 18 Holes of Golf**

Scores	Frequency
66–70	1
71–75	5
76–80	3
81–85	4

Chapter Test

For Exercises 1–2, find the mean and the range.

1.
Number of Weeks on the Chart for 10 Top Ten Records				
10	5	8	16	12
2	9	8	8	7

2.
Number of Cars Per Hour Passing Through a Toll Gate			
85	124	110	255
208	166	198	230

For Exercises 3–4, find the mode.

3.
Times in Seconds for a 100-Meter Dash				
10.9	11.4	11.7	11.8	12.1
12.1	11.5	12.4	13.2	12.0

4.
Heights in Centimeters of 12 Students					
152	174	163	158	160	162
158	170	160	159	167	171

For Exercises 5–6, find the median.

5.
Number of T-shirts Sold					
12	25	16	32	8	15
15	18	10	14	20	22

6.
Weekly Salaries for Five Clerks				
$150	$162	$160	$185	$175

7. Mr. Prime's math students scored the following number of points on a quiz.

85	94	77	76	92	94
77	98	76	94	83	77

   Which gives the best description of the quiz scores, the mean, median, or mode?

8. Seven students play a game. They list their score in order. The mean and the median score is 5. The modes of the scores are 2 and 7. List all possible scores in order.

9. Use the given information to complete the table.

Heights in Meters of 16 Buildings

235	214	207	184
225	244	229	188
181	219	204	206
209	225	223	239

Heights	Tally	Frequency
180–199	?	?
200–219	?	?
220–239	?	?
240–259	?	?

10. The frequency table at the right shows the ages of young people employed by a company during one summer. Use the given information to draw a histogram.

Age	Tally	Frequency
14–16	ЖН IIII	9
17–19	ЖН ЖН ЖН	15
20–22	ЖН ЖН	10

Cumulative Maintenance   Chapters 1–5

Choose the correct answer. Choose a, b, c, or d.

1. Add: 5287
          2425
      + 318

  a. 8030   **b.** 8020
  c. 7930   **d.** 7010

2. Subtract: 6104
            − 238

  a. 5966   **b.** 5937
  c. 6134   **d.** 5866

3. Round 585 to the nearest ten.

  a. 600   **b.** 580
  c. 590   **d.** 500

4. Add: 6.25 + 1.8 + 14.7 + 6.91

  a. 29.6   **b.** 14.81
  c. 16.43   **d.** 29.66

5. Find the missing terms.

  3, 7, 11, __?__ , __?__ , 23, 27

  a. 15, 19   **b.** 13, 15
  c. 12, 14   **d.** 17, 19

6. Multiply: 24.5 × 100

  a. 245   **b.** 24,500
  c. 2450   **d.** 0.245

7. Jane had scores of 92, 88, 90, and 84 on her quizzes. What was her mean score?

  a. 88.5   **b.** 89
  c. 88   **d.** 90

8. Which of these represents three and seven thousandths?

  a. 3700   **b.** 3.007
  c. 3007   **d.** 3.07

9. This bar graph shows the number of employees absent from work Monday through Thursday. On which day was the greatest number of employees absent?

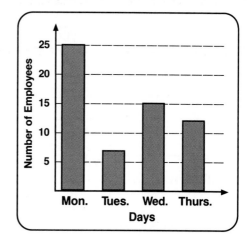

  a. Monday   **b.** Tuesday
  c. Wednesday   **d.** Thursday

10. Write 3^4 in standard form.

    a. 12   **b.** 34
    c. 81   **d.** 64

11. Find the answer.

    $(15 + 4) \times 2 + 7$

    a. 171   **b.** 45
    c. 30   **d.** 28

12. John deposits checks for $83 and
$72 in his checking account. He
receives $60 in cash. What was
John's net deposit?

    a. $115   **b.** $215
    c. $95   **d.** $69

13. At which price will one can of
dog food cost the least?

    a. 3 for $1.00   **b.** 2 for $0.70
    c. 4 for $1.36   **d.** 5 for $1.50

14. Estimate: 59×81

    a. 4000   **b.** 5400
    c. 4800   **d.** 4500

15. Multiply.

$$\begin{array}{r} 3.14 \\ \times\, 0.25 \\ \hline \end{array}$$

    a. 2.198   **b.** 78.5
    c. 21.98   **d.** 0.785

16. Divide.

$$12\,\overline{)\,43.2}$$

    a. 3.6   **b.** 36
    c. 306   **d.** 360

17. Subtract: $46.01 - 18.397$

    a. 27.627   **b.** 27.613
    c. 28.723   **d.** 28.627

18. Mark has a mean score of 9 for
4 games. He plays again and
scores 1. What is his new mean
score?

    a. 9   **b.** 7
    c. 7.4   **d.** 8

19. Estimate: $38\,\overline{)\,282}$

    a. 9   **b.** 8
    c. 6   **d.** 7

20. Multiply:

$$\begin{array}{r} 408 \\ \times\, 57 \\ \hline \end{array}$$

    a. 4896   **b.** 206,856
    c. 23,256   **d.** 2736

21. Sarah has a $20-bill. She spends
$4.75 on a movie and $2.50 on a
magazine. How much change
should she receive?

    a. $12.75   **b.** $32.75
    c. $15.25   **d.** $17.50

Three pans of lasagna are exactly the same size. The lasagna in one pan is cut into thirds, in another pan it is cut into fourths, and in the third pan it is cut into twelfths.

- Which pan has the largest pieces?

- Which pan has one-fourth as many pieces as another pan?

- Make a drawing to show how the lasagna that is divided into thirds can be cut so that there will be 12 pieces in all.

- How many pieces in the "twelfths-pan" are needed to make one piece in the "fourths-pan?"

Meaning of Fractions

ACTIVITY

Draw these four rectangles on a sheet of graph paper.

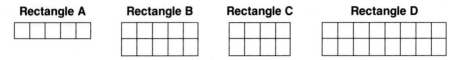

Rectangle A   Rectangle B   Rectangle C   Rectangle D

1. Shade the number of squares in each rectangle as indicated.
   a. Rectangle A: 1   b. Rectangle B: 2
   c. Rectangle C: 4   d. Rectangle D: 8

2. Write a fraction that tells what part of each rectangle is shaded.

   For example, $\frac{1}{5}$ of Rectangle A is shaded.

   $\frac{1}{5}$ ← Shaded parts
   ← Total parts

3. Write a fraction that tells what part of each rectangle is <u>not</u> shaded.

   For Exercises 4–7, refer to the groups at the right below.

4. Which figure shows that $\frac{3}{4}$ of the group is shaded?

5. Which figure shows that $\frac{3}{3}$ of the group is shaded?

6. Which figure shows that $\frac{0}{2}$ of the group is shaded?

7. Which figure shows that $\frac{5}{6}$ of the group is shaded?

EXERCISES

On graph paper, draw a rectangle that is 18 units long and 9 units wide.

1. Divide the rectangle into 6 equal parts. Shade $\frac{4}{6}$ of the rectangle.

2. Draw another rectangle of the same size as the first. Repeat the procedure of Exercise 1.

3. Is there more than one way to choose 4 out of 6 parts? Draw at least 4 ways to show $\frac{4}{6}$.

4. Find as many ways as you can of dividing the rectangle into six equal parts.

Copy the figure. Complete the whole. The first one is done for you.

5. ← $\frac{1}{2}$ 6. ← $\frac{1}{4}$ 7.  ← $\frac{1}{3}$ 8.  ← $\frac{3}{4}$

Fractions

Claire made a pan of lasagna. She divided it into 12 equal portions.

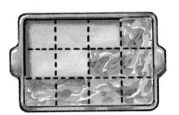

1. What part of the lasagna has been eaten?

2. What part of the lasagna has not been eaten?

Horace collects old coins. Of this group of coins, 7 are quarters.

3. What part of the coins are quarters?

A **fraction** can be used to represent a part of a whole or part of a group. The numbers 7 and 12 are called the **terms** of the fraction.

$\dfrac{7}{12}$ ← **Numerator**
← **Denominator**

---

4. In a fraction such as $\frac{7}{12}$, what does the denominator identify?

5. In a fraction such as $\frac{7}{12}$, what does the numerator identify?

On this number line, the distance from 0 to 1 is divided into 5 equal parts.

$$0 \qquad X \qquad 1$$

$$\frac{0}{5} \quad \frac{1}{5} \quad \frac{2}{5} \quad ? \quad \frac{4}{5} \quad \frac{5}{5} \quad \frac{6}{5}$$

6. What fraction can be used to label the point at X?

A fraction can be used to represent the quotient of two whole numbers.

$$5 \div 5 \longrightarrow \frac{5}{5} = 1 \quad 4 \div 5 \longrightarrow \frac{4}{5} \quad 9 \div 3 \longrightarrow \frac{9}{3} = 3 \quad 6 \div 5 \longrightarrow \frac{6}{5}$$

7. Will $\frac{6}{5}$ be greater than 1 or less than 1? Explain.

8. Write three fractions that equal 1.

A whole number can be written as a fraction with a denominator of 1.

$$6 = \frac{6}{1} \qquad 18 = \frac{18}{1} \qquad 200 = \frac{200}{1}$$

EXERCISES

Complete. Choose your answers from the box at the right.

1. A fraction can be used to represent a __?__ of a whole.

2. In $\frac{2}{3}$, the numbers 2 and 3 are called the __?__ .

3. The term used to represent the part is called the __?__ .

4. The term used to represent the whole is called the __?__ .

5. The fraction $\frac{0}{4}$ equals __?__ .

6. A whole number can be written as a fraction with a denominator of __?__ .

> denominator
> zero
> part
> numerator
> quotient
> one
> terms

Write a fraction that tells what part of each figure is:

a. shaded   *b. not shaded*

7.    8.    9.

For Exercises 10–13, write the fraction that represents the point at X.

10.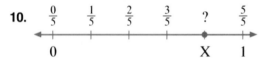

11.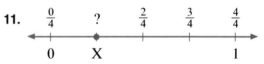

12.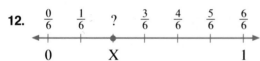

13.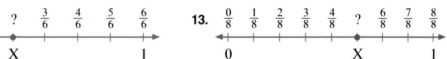

Write each quotient as a fraction or as a whole number.

14. $1 \div 8$ 15. $3 \div 3$ 16. $7 \div 10$ 17. $12 \div 3$ 18. $7 \div 5$

19. Which of the fractions $\frac{1}{8}$, $\frac{5}{8}$, or $\frac{7}{8}$ is closest to 1?

20. Which of the fractions $\frac{2}{6}$, $\frac{3}{6}$, or $\frac{5}{6}$ is closest to 0?

21. Which of the fractions $\frac{5}{6}$, $\frac{1}{2}$, $\frac{9}{9}$, or $\frac{0}{3}$ is equal to 1?

22. Which of the fractions $\frac{5}{12}$, $\frac{3}{12}$, or $\frac{13}{12}$ is greater than 1?

23. Juan carries six textbooks. Two are science textbooks. What fraction are science books?

24. A pie is cut into eight slices. Three slices have been eaten. What fraction has been eaten?

25. A quilt patch has 16 squares of cloth in all. Three of the squares are blue. What fraction are not blue?

26. There are 9 boys in a class of 24 students. What fraction of the class are girls?

Equivalent Fractions

ACTIVITY

Fold a sheet of paper in half. Shade one part.

Fold the paper into four equal parts.

Now fold the paper in half the other way.

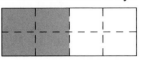

1. What fraction does the shaded part represent?

2. What fraction does the shaded part represent?

3. What fraction does the shaded part represent?

4. Does $\frac{1}{2}$ of the paper equal $\frac{2}{4}$ of the paper? Why?

5. Does $\frac{2}{4}$ of the paper equal $\frac{4}{8}$ of the paper? Why?

6. Does $\frac{1}{2} = \frac{2}{4} = \frac{4}{8}$? Why?

The fractions $\frac{1}{2}$, $\frac{2}{4}$, and $\frac{4}{8}$ are *equivalent*.

> Fractions that name the same number are **equivalent** fractions.

EXERCISES

1. Fold another sheet of paper to show that $\frac{1}{3} = \frac{2}{6}$.

2. Fold another sheet of paper to show that $\frac{1}{4} = \frac{2}{8} = \frac{4}{16}$.

For Exercises 3–6, draw each figure.
   a. *Use the unshaded bars to draw and shade equivalent fractions.*
   b. *Write the equivalent fraction represented by each of your drawings. The first one is begun for you.*

3. $\frac{1}{4}$ ⟶
   $\frac{?}{8}$ ⟶
   $\frac{?}{16}$ ⟶

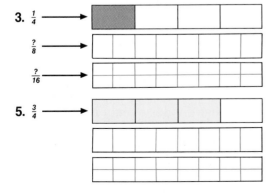

4. $\frac{1}{3}$ ⟶
   $\frac{?}{6}$ ⟶
   $\frac{?}{12}$ ⟶

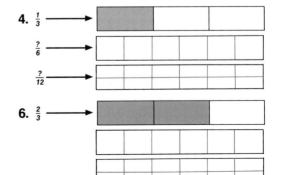

5. $\frac{3}{4}$ ⟶

6. $\frac{2}{3}$ ⟶

Lowest Terms

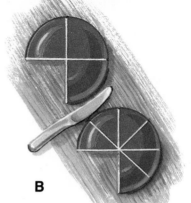

Paul bought two cheese wheels of the same size. The pictures show how much of each wheel remains.

A

1. Into how many pieces was Wheel A cut?

2. What fraction of Wheel A remains?

3. Into how many pieces was Wheel B cut?

B

4. What fraction of Wheel B remains?

The pictures show that the <u>same amount</u> of both wheels remains.

So $\frac{3}{4} = \frac{6}{8}$. Since $\frac{3}{4}$ and $\frac{6}{8}$ name the same number, they are equivalent fractions. The fraction $\frac{3}{4}$ is in *lowest terms*.

5. What number other than 1 divides evenly into both 6 and 8?

> A fraction is in **lowest terms** when the only whole number that will divide evenly into <u>both</u> its numerator and denominator is 1.

6. *Complete:* Since $1 = \frac{2}{2}$, $\frac{6}{8} \div 1 = \frac{6}{8} \div \frac{\Box}{2}$.

7. *Complete:* $\frac{6}{8} \div \frac{2}{2} = \frac{6 \div 2}{8 \div 2} = \frac{\Box}{4}$

EXAMPLE

Write $\frac{18}{24}$ in lowest terms.

Method 1

1 **Think:** 18 and 24 are <u>both</u> divisible by 2.

$$\frac{18}{24} = \frac{18 \div 2}{24 \div 2}$$

$$\frac{18}{24} = \frac{9}{12} \quad \blacktriangleleft \textbf{ Not in lowest terms}$$

2 **Think:** 9 and 12 are <u>both</u> divisible by 3.

$$\frac{9}{12} = \frac{9 \div 3}{12 \div 3}$$

$$\frac{9}{12} = \frac{3}{4} \quad \blacktriangleleft \textbf{ Lowest terms} \qquad \text{So } \frac{18}{24} = \frac{3}{4}.$$

Method 2

Think: What is the <u>greatest number</u> that divides evenly into 18 and 24?

$$\frac{18}{24} = \frac{18 \div 6}{24 \div 6} = \frac{3}{4}$$

EXERCISES

Complete. Choose your answers from the box at the right.

1. Fractions that name the same number are __?__.

2. The fraction $\frac{2}{3}$ is in __?__ since 1 is the only number that will divide evenly into both 2 and 3.

3. The fraction $\frac{2}{4}$ __?__ in lowest terms.

4. To write $\frac{8}{18}$ in lowest terms, __?__ the numerator and denominator by 2.

> **lowest terms**
> **is**
> **divide**
> **equivalent fractions**
> **is not**

Write the greatest number that will divide evenly into both the numerator and denominator of each fraction.

5. $\frac{2}{4}$ 6. $\frac{3}{6}$ 7. $\frac{3}{12}$ 8. $\frac{10}{15}$ 9. $\frac{10}{4}$ 10. $\frac{6}{15}$ 11. $\frac{4}{18}$ 12. $\frac{50}{12}$

13. $\frac{14}{32}$ 14. $\frac{8}{36}$ 15. $\frac{18}{10}$ 16. $\frac{14}{16}$ 17. $\frac{12}{18}$ 18. $\frac{6}{4}$ 19. $\frac{12}{9}$ 20. $\frac{32}{64}$

Complete.

21. $\frac{4}{10} = \frac{\square}{5}$ 22. $\frac{\square}{9} = \frac{2}{3}$ 23. $\frac{21}{36} = \frac{\square}{12}$ 24. $\frac{12}{\square} = \frac{3}{16}$ 25. $\frac{45}{80} = \frac{9}{\square}$

Write in lowest terms.

26. $\frac{25}{90}$ 27. $\frac{13}{39}$ 28. $\frac{6}{12}$ 29. $\frac{12}{60}$ 30. $\frac{14}{35}$ 31. $\frac{13}{15}$ 32. $\frac{15}{45}$ 33. $\frac{8}{12}$

34. $\frac{10}{80}$ 35. $\frac{7}{12}$ 36. $\frac{16}{18}$ 37. $\frac{5}{25}$ 38. $\frac{9}{36}$ 39. $\frac{16}{21}$ 40. $\frac{11}{22}$ 41. $\frac{1}{8}$

42. $\frac{34}{6}$ 43. $\frac{14}{8}$ 44. $\frac{60}{16}$ 45. $\frac{28}{16}$ 46. $\frac{45}{6}$ 47. $\frac{45}{20}$ 48. $\frac{38}{24}$ 49. $\frac{36}{24}$

50. $\frac{55}{25}$ 51. $\frac{75}{50}$ 52. $\frac{60}{58}$ 53. $\frac{21}{18}$ 54. $\frac{16}{10}$ 55. $\frac{45}{10}$ 56. $\frac{24}{21}$ 57. $\frac{36}{8}$

58. $\frac{15}{6}$ 59. $\frac{55}{15}$ 60. $\frac{32}{14}$ 61. $\frac{60}{42}$ 62. $\frac{100}{9}$ 63. $\frac{54}{16}$ 64. $\frac{85}{60}$ 65. $\frac{84}{90}$

66. Paul bought two pizzas of the same size. He cut pizza A into fourths and pizza B into twelfths. Paul gave Anita 1 slice of pizza A. He gave Mike 4 slices of pizza B. Did he give Anita and Mike the same amount of pizza? Explain.

67. Aimee has just completed 12 miles of a 26-mile run. What fraction of the run has she completed so far? Write the answer in lowest terms.

68. The numerator of a fraction is a whole number and the denominator is 1. Does the value of the fraction equal the numerator? Explain.

69. The numerator and the denominator of a fraction are even numbers. Is that fraction in lowest terms? Explain.

Mixed Numbers

Tim practices the piano for 60 minutes every morning and for another 30 minutes every afternoon.

1 minute (min) = 60 seconds (s)
1 hour (h) = 60 min
1 day (d) = 24 hr
1 week (w) = 7 d
1 year (yr) = 52 w
1 year = 12 months (mo)
1 year = 365 d

1. Does Tim practice more than, or less than, 2 hours each day? Explain.

2. What part of an hour is 30 minutes?

EXAMPLE 1 How many hours does Tim practice each day?

Think: Since 60 min = 1 h,   30 min = $\frac{30}{60}$ = $\frac{1}{2}$ h. ◀ **Lowest terms**

So Tim practices 1 h + $\frac{1}{2}$ h, or **1$\frac{1}{2}$ h** each day.

3. *Complete:* $\frac{30}{60} = \frac{30 \div ?}{60 \div ?} = \frac{1}{2}$

In Example 1, the number $1\frac{1}{2}$ is a **mixed number.**

4. *Complete:* A mixed number is the sum of a whole number and a __?__ .

5. Are mixed numbers greater than 1 or less than 1? Explain.

> When the numerator of a fraction is greater than the denominator, the fraction is greater than 1.

You can write a mixed number for a fraction greater than 1.

6. Is $\frac{19}{5}$ greater than 3 or less than 3?

EXAMPLE 2 Write a mixed number for $\frac{19}{5}$.

Method 1: ① $\frac{19}{5} \longrightarrow 19 \div 5 \longrightarrow 5\overline{)19}$ $\begin{array}{r} 3 \\ \underline{15} \\ 4 \end{array}$

② Quotient: 3 r 4,   or $3\frac{4}{5}$ ◀━━ **Remainder**
━━ **Divisor**

So $\frac{19}{15} = 3\frac{4}{5}$ ◀ **Mixed number**

Method 2: $\frac{19}{5} = \underbrace{\frac{5}{5} + \frac{5}{5} + \frac{5}{5}} + \frac{4}{5}$

$= \quad 3 \quad + \frac{4}{5}$, or $3\frac{4}{5}$ ◀ **Mixed number**

EXERCISES

For Exercises 1–3, replace the __?__ with one of the following.
    equal to   *less than*   *greater than*

1. In $\frac{1}{2}$, the numerator is __?__ the denominator. Therefore $\frac{1}{2}$ is __?__ 1.

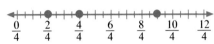

2. In $\frac{9}{4}$, the numerator is __?__ the denominator. Therefore $\frac{9}{4}$ is __?__ 1.

3. In $\frac{8}{8}$, the numerator is __?__ the demonitor. Therefore $\frac{8}{8}$ is __?__ 1.

4. When the numerator of a fraction equals the denominator, will the fraction be less than 1, greater than 1, or equal to 1? Give an example to illustrate your answer.

5. When the numerator of a fraction is less than the denominator, will the fraction be less than 1, greater than 1, or equal to 1? Give an example to illustrate your answer.

Replace each ● *with* <, =, *or* >. *Remember! The "arrow" points to the smaller number.*

6. $\frac{8}{9}$ ● 1   7. $\frac{9}{8}$ ● 1   8. $\frac{8}{8}$ ● 1   9. $\frac{5}{4}$ ● 1   10. $\frac{4}{5}$ ● 1

11. $\frac{5}{5}$ ● 1   12. $\frac{10}{5}$ ● 1   13. $\frac{19}{20}$ ● 1   14. $\frac{8}{5}$ ● 1   15. $\frac{1}{1}$ ● 1

Write a mixed number in lowest terms for each fraction. Try to do these without paper and pencil.

16. $\frac{8}{5}$ 17. $\frac{9}{7}$ 18. $\frac{9}{6}$ 19. $\frac{6}{4}$ 20. $\frac{7}{4}$ 21. $\frac{8}{3}$ 22. $\frac{10}{3}$

23. $\frac{10}{4}$ 24. $\frac{11}{2}$ 25. $\frac{17}{6}$ 26. $\frac{15}{2}$ 27. $\frac{13}{4}$ 28. $\frac{15}{4}$ 29. $\frac{18}{5}$

30. Will $\frac{19}{5}$ be closer to 3 or to 4?

31. Will $\frac{10}{3}$ be closer to 3 or to 4?

32. Will $\frac{21}{4}$ be closer to 5 or to 6?

33. Will $\frac{23}{6}$ be closer to 3 or to 4?

34. Will $\frac{17}{4}$ be closer to 4 or to 5?

35. Will $\frac{35}{3}$ be closer to 11 or to 12?

Use the table on page 132 to complete. Write a mixed number in lowest terms for each fraction.

36. 18 mo = __?__ yr

37. 36 h = __?__ d

38. 120 s = __?__ min

39. 20 d = __?__ wk

40. 150 min = __?__ h

41. 80 h = __?__ d

For Exercises 42–47, write the fractions in order from greatest to smallest.

42. $\frac{6}{6}, \frac{2}{5}, \frac{7}{6}, \frac{0}{3}$

43. $\frac{0}{9}, \frac{7}{8}, \frac{8}{2}, \frac{4}{4}$

44. $\frac{4}{5}, \frac{8}{4}, \frac{0}{3}, \frac{6}{5}$

45. $\frac{8}{8}, \frac{10}{2}, \frac{5}{6}, \frac{7}{4}$

46. $\frac{5}{5}, \frac{1}{2}, \frac{2}{1}, \frac{8}{3}$

47. $\frac{4}{2}, \frac{10}{10}, \frac{3}{1}, \frac{4}{3}$

Addition and Subtraction: LIKE FRACTIONS

Fractions such as $\frac{3}{10}$ and $\frac{9}{10}$ are **like fractions** because they have a common denominator, 10.

1. Will $\frac{3}{10} + \frac{9}{10}$ be greater than 1 or less than 1? Explain.

2. Will $\frac{11}{12} - \frac{5}{12}$ be greater than 1 or less than 1? Explain.

EXAMPLE 1 | Add or subtract as indicated.

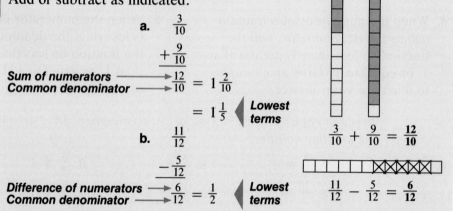

a.
$$\begin{array}{r} \frac{3}{10} \\ + \frac{9}{10} \\ \hline \end{array}$$

Sum of numerators ──→ $\frac{12}{10} = 1\frac{2}{10}$
Common denominator ──→

$$= 1\frac{1}{5}$$ ◄ **Lowest terms**

b.
$$\begin{array}{r} \frac{11}{12} \\ - \frac{5}{12} \\ \hline \end{array}$$

Difference of numerators ──→ $\frac{6}{12} = \frac{1}{2}$ ◄ **Lowest terms**
Common denominator ──→

$$\frac{3}{10} + \frac{9}{10} = \frac{12}{10}$$

$$\frac{11}{12} - \frac{5}{12} = \frac{6}{12}$$

To add or subtract with mixed numbers having <u>like</u> denominators, <u>first</u> add or subtract the fractions. Then add or subtract the whole numbers.

3. Will $2\frac{1}{8} + 1\frac{5}{8}$ be closer to 3 or to 4? Explain.

4. Will $3\frac{3}{4} - 1\frac{1}{4}$ be less than 2 or greater than 2? Explain.

EXAMPLE 2 | Add or subtract as indicated.

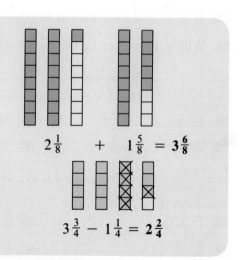

a.
$$\begin{array}{r} 2\frac{1}{8} \\ + 1\frac{5}{8} \\ \hline \end{array}$$ **Add the fractions first.**

$$3\frac{6}{8} = 3\frac{3}{4}$$ ◄ **Lowest terms**

$$2\frac{1}{8} + 1\frac{5}{8} = 3\frac{6}{8}$$

b.
$$\begin{array}{r} 3\frac{3}{4} \\ - 1\frac{1}{4} \\ \hline \end{array}$$ **Subtract the fractions first.**

$$2\frac{2}{4} = 2\frac{1}{2}$$ ◄ **Lowest Terms**

$$3\frac{3}{4} - 1\frac{1}{4} = 2\frac{2}{4}$$

EXERCISES

1. Will $\frac{2}{5} + \frac{2}{5}$ be greater than 1 or less less than 1? Explain.

2. Will $\frac{3}{4} + \frac{3}{4}$ be greater than 1 or less than 1? Explain.

3. Will $2\frac{7}{8} - 1\frac{5}{8}$ be closer to 1 or to 2? Explain.

4. Will $4\frac{7}{8} - 1\frac{5}{8}$ be closer to 3 or to 4? Explain.

Complete by writing the missing numerators.

5.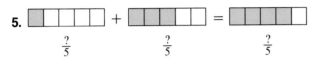

$\frac{?}{5}$ $\qquad$ $\frac{?}{5}$ $\qquad$ $\frac{?}{5}$

6.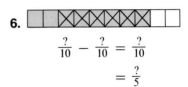

$\frac{?}{10} - \frac{?}{10} = \frac{?}{10}$

$= \frac{?}{5}$

Draw fraction models like those at the right for each of the following exercises. Shade the models to show each sum or difference.

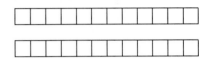

7. $\frac{1}{12} + \frac{4}{12}$

8. $2\frac{5}{12} + 1\frac{6}{12}$

9. $\frac{10}{12} - \frac{7}{12}$

10. $4\frac{11}{12} - 2\frac{5}{12}$

Add or subtract. Write each answer in lowest terms.

11. $\frac{1}{5} + \frac{1}{5}$

12. $\frac{3}{4} - \frac{2}{4}$

13. $\frac{4}{5} - \frac{3}{5}$

14. $\frac{7}{8} + \frac{3}{8}$

15. $\frac{9}{10} - \frac{8}{10}$

16. $\frac{3}{8} + \frac{3}{8}$

17. $\frac{3}{20} + \frac{7}{20}$

18. $\frac{9}{10} - \frac{4}{10}$

19. $3\frac{1}{5} + 2\frac{2}{5}$

20. $9\frac{2}{3} - 6\frac{1}{3}$

21. $1\frac{1}{4} + 5\frac{2}{4}$

22. $10\frac{7}{10} - 2\frac{2}{10}$

23. $4\frac{3}{6} - 1\frac{1}{6}$

24. $4\frac{5}{8} + 1\frac{4}{8}$

25. $9\frac{7}{12} - 5\frac{5}{12}$

26. $8\frac{1}{4} + 9\frac{1}{4}$

27. $\frac{11}{12} + \frac{9}{12}$

28. $7\frac{3}{4} - 4\frac{2}{4}$

29. $\frac{4}{6} - \frac{1}{6}$

30. $\frac{4}{5} + \frac{2}{5}$

31. $20\frac{2}{5} - 11\frac{1}{5}$

32. $8\frac{1}{6} + 7$

33. $13 + 5\frac{6}{7}$

34. $7\frac{2}{5} - 3\frac{1}{5}$

35. $\frac{4}{5} - \frac{2}{5}$

36. $\frac{5}{8} - \frac{3}{8}$

37. $12\frac{1}{3} + 5$

38. $\frac{1}{8} + \frac{2}{8}$

Complete.

39. $\frac{7}{12} + \frac{?}{\_} = \frac{11}{12}$

40. $\frac{?}{\_} + \frac{3}{8} = 1$

41. $\frac{?}{\_} - \frac{5}{6} = \frac{1}{6}$

42. $10 + \frac{?}{\_} = 11\frac{5}{6}$

43. $9\frac{9}{10} - \frac{?}{\_} = 4\frac{3}{10}$

44. $\frac{?}{\_} - 5\frac{1}{8} = 2\frac{1}{4}$

45. Find the missing numbers. (There is more than one answer.)

$\frac{?}{10} + \frac{?}{10} = \frac{8}{10}$

46. Find the missing numbers. (There is more than one answer.)

$\frac{?}{12} - \frac{?}{12} = \frac{5}{12}$

Mid-Chapter Review

Write the fraction that represents the point at X. (Pages 127–128)

1.

2.

3. Which of the fractions $\frac{1}{6}$, $\frac{4}{6}$, or $\frac{3}{6}$ is closest to 1? (Pages 127–128)

4. Which of the fractions $\frac{0}{3}$, $\frac{3}{4}$, $\frac{8}{9}$, $\frac{6}{6}$ is equal to 1? (Pages 127–128)

Write in lowest terms. (Pages 130–131)

5. $\frac{6}{24}$ 6. $\frac{4}{12}$ 7. $\frac{8}{36}$ 8. $\frac{14}{16}$ 9. $\frac{6}{4}$ 10. $\frac{24}{15}$

11. $\frac{36}{16}$ 12. $\frac{7}{42}$ 13. $\frac{40}{24}$ 14. $\frac{105}{70}$ 15. $\frac{52}{56}$ 16. $\frac{49}{63}$

Replace each ● *with* <, =, *or* >. (Pages 132–133)

17. $\frac{5}{6}$ ● 1   18. $\frac{10}{10}$ ● 1   19. $\frac{9}{8}$ ● 1   20. $\frac{7}{5}$ ● 1   21. $\frac{1}{2}$ ● 1

Write a mixed number for each fraction. (Pages 132–133)

22. $\frac{3}{2}$ 23. $\frac{7}{3}$ 24. $\frac{23}{4}$ 25. $\frac{29}{5}$ 26. $\frac{19}{10}$ 27. $\frac{13}{4}$

28. A meat loaf is cut into twelve slices. Five slices have been eaten. What fraction has not been eaten? (Pages 127–128)

29. Harvey completed 32 miles of a 50-mile bicycle ride. What fraction has he completed? Write the answer in lowest terms. (Pages 127–128)

MAINTENANCE

Multiply or divide as indicated. (Pages 41–42 and 68–69)

30. 4.7×0.2 31. 4.26×1.8

32. $0.6 \overline{)1.38}$ 33. $0.12 \overline{)10.08}$

34. At a certain store, a 20-ounce jar of jam costs $1.83, a 12-ounce jar of jam costs $1.09, and an 8-ounce jar of jam costs $0.82. Which is the best buy? (Pages 66–67)

35. Jeans cost $12.95 a pair and socks cost $3.50 a pair. What is the total cost of 3 pairs of jeans and 4 pairs of socks? (Pages 41–42)

For Exercises 36–37, refer to the bar graph. (Pages 84–85)

36. How much faster is a hyena than a cat?

37. How much faster than a rabbit is a lion?

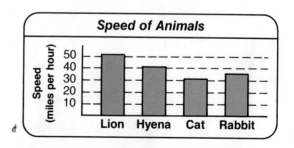

Estimation and Carpeting

Estimate the cost of tiling a floor that is $11\frac{3}{4}$ feet long and $13\frac{1}{4}$ feet wide. A package of tiles costs $11.90.

Room Length	Room Width									
	2	4	6	8	10	12	14	16	18	20
4	2	3	4	4	4	5	6	7	8	8
6	3	4	5	6	6	8	9	10	11	12
8	4	5	7	8	8	10	12	13	15	16
9	4	6	8	9	9	11	13	15	17	18
10	4	6	8	9	10	12	14	16	18	20
12	5	8	10	11	12	15	17	20	22	24

READ

What are the facts?

Length: $11\frac{3}{4}$ feet   **Width:** $13\frac{1}{4}$ feet   **Cost:** $11.90 per package

PLAN

Use the table to find the number of packages.
Multiply this number by the estimated cost.

SOLVE

1. Round the length and width to the <u>next largest</u> foot.

$11\frac{3}{4}$ feet ⟶ 12 feet $13\frac{1}{4}$ feet ⟶ 14 feet

Why do you think $13\frac{1}{4}$ is <u>not</u> rounded to 13?

2. Read the table. ⟶ 17 packages are needed.

3. Estimate the cost.
$11.90 is about $12. Therefore, $17 \times \$12 = \$204$. ◀ **Estimated cost**

CHECK

Compare the steps in the *Solve* with the facts.

EXERCISES   *Complete the table.*

	Floor		Number of Packages	Cost per Package of 10	Estimated Cost
	Length	Width			
1.	$11\frac{1}{2}$	$11\frac{1}{4}$	?	$10.80	?
2.	$9\frac{3}{4}$	$13\frac{1}{2}$	?	$11.75	?
3.	$11\frac{1}{4}$	$15\frac{1}{4}$	?	$12.50	?

In one state you pay 5¢ sales tax for each dollar you spend on certain items. Find the tax on:

8. Find the sales tax on $204.

9. Find the sales tax for the estimated costs in Exercises 1–3.

4. $2   5. $10   6. $100   7. $200

Least Common Denominator

Celia's dog, Domino, followed her $\frac{1}{4}$ of a mile to Janet's house, then $\frac{3}{8}$ of a mile to the park. You can use a number line as a model to determine how far they walked.

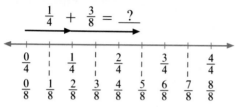

1. How many eighths equal $\frac{1}{4}$?

2. What is $\frac{2}{8} + \frac{3}{8}$?

3. *Complete:* To add $\frac{1}{4}$ and $\frac{3}{8}$, first write an __?__ fraction for $\frac{1}{4}$ having a denominator of __?__.

> Fractions such as $\frac{1}{4}$ and $\frac{3}{8}$ are **unlike fractions**.
> Their denominators are <u>not the same</u>.

Before adding or subtracting unlike fractions, you first find a common denominator. Then you write equivalent fractions having the <u>same</u> denominator.

> The **least common denominator** (LCD) of two or more fractions is the <u>smallest number</u> that is a multiple of the denominators of the fractions.

EXAMPLE 1

Find the LCD of $\frac{5}{6}$ and $\frac{7}{10}$.

1. Select the denominators. Choose 6.

2. Write multiples of 6 until you find a number that is <u>also</u> a multiple of 10.

Multiples of 6	Also a Multiple of 10?
$6 \times 1 = 6$	No
$6 \times 2 = 12$	No
$6 \times 3 = 18$	No
$6 \times 4 = 24$	No
$6 \times 5 = 30$	Yes. Stop!

The LCD is **30**, because 30 is the <u>smallest number</u> that is also a multiple of 10.

EXAMPLE 2 Write like fractions for $\frac{5}{6}$ and $\frac{7}{10}$.

1 Find the LCD for $\frac{5}{6}$ and $\frac{7}{10}$. ⟶ LCD: 30 ◀ **From Example 1**

2 Multiply the numerator and denominator of each fraction by a number that will make the denominator equal 30, the LCD.

For $\frac{5}{6}$: $\frac{5}{6} = \frac{5 \times ?}{6 \times ?} = \frac{?}{30}$ For $\frac{7}{10}$: $\frac{7}{10} = \frac{7 \times ?}{10 \times ?} = \frac{?}{30}$

$\frac{5}{6} = \frac{5 \times 5}{6 \times 5}$ ◀ $\frac{5}{5} = 1$ $\frac{7}{10} = \frac{7 \times 3}{10 \times 3}$ ◀ $\frac{3}{3} = 1$

$\frac{5}{6} = \frac{25}{30}$ $\frac{7}{10} = \frac{21}{30}$

↑———— **Like fractions** ————↑

EXERCISES

Write the first 5 multiples of each number.

1. 3   **2.** 6   **3.** 13   **4.** 11   **5.** 20   **6.** 15   **7.** 100

8. Which of the following is the LCD for $\frac{5}{8}$ and $\frac{1}{6}$?

8, 12, 18, 24, 36, 48

9. Which of the following is equivalent to $\frac{5}{8}$?

$\frac{5}{16}$ $\frac{1}{4}$ $\frac{15}{24}$ $\frac{5}{12}$ $\frac{3}{24}$

10. Which of the following is equivalent to $\frac{1}{6}$?

$\frac{5}{24}$ $\frac{5}{6}$ $\frac{4}{24}$ $\frac{4}{6}$ $\frac{1}{2}$

11. Which pair of like fractions shows the LCD for $\frac{5}{8}$ and $\frac{1}{6}$?

$\frac{30}{48}$ and $\frac{8}{48}$ $\frac{15}{24}$ and $\frac{4}{24}$

Find the LCD of each pair of fractions.

12. $\frac{1}{2}$ and $\frac{5}{6}$ **13.** $\frac{2}{3}$ and $\frac{1}{6}$ **14.** $\frac{3}{8}$ and $\frac{1}{2}$ **15.** $\frac{4}{5}$ and $\frac{3}{10}$ **16.** $\frac{1}{3}$ and $\frac{1}{12}$

17. $\frac{1}{2}$ and $\frac{1}{3}$ **18.** $\frac{1}{2}$ and $\frac{3}{5}$ **19.** $\frac{2}{3}$ and $\frac{1}{6}$ **20.** $\frac{1}{3}$ and $\frac{1}{4}$ **21.** $\frac{3}{4}$ and $\frac{2}{3}$

Write like fractions for each pair.

22. $\frac{1}{4}$ and $\frac{2}{3}$ **23.** $\frac{2}{5}$ and $\frac{1}{6}$ **24.** $\frac{1}{2}$ and $\frac{3}{5}$ **25.** $\frac{1}{2}$ and $\frac{1}{3}$ **26.** $\frac{3}{4}$ and $\frac{1}{5}$

27. $\frac{1}{6}$ and $\frac{3}{4}$ **28.** $\frac{5}{6}$ and $\frac{4}{5}$ **29.** $\frac{1}{3}$ and $\frac{4}{5}$ **30.** $\frac{1}{8}$ and $\frac{1}{4}$ **31.** $\frac{1}{8}$ and $\frac{1}{2}$

32. $\frac{3}{8}$ and $\frac{2}{3}$ **33.** $\frac{7}{8}$ and $\frac{1}{5}$ **34.** $\frac{3}{10}$ and $\frac{2}{5}$ **35.** $\frac{1}{10}$ and $\frac{3}{4}$ **36.** $\frac{7}{10}$ and $\frac{1}{6}$

37. The numerator of a fraction is 3. The denominator is a multiple of 3. Is the fraction in lowest terms? Explain.

38. The numerator of a fraction is a multiple of 2. The denominator is 10. Is the fraction in lowest terms? Explain.

Addition and Subtraction: UNLIKE FRACTIONS

Abigail plans the layout for the school newspaper. This week, she uses $\frac{1}{4}$ of the sports page for advertisements and $\frac{2}{3}$ of the page for sports articles.

1. Would Abigail add or subtract to find how much of the page is filled?

2. Will $\frac{1}{4} + \frac{2}{3}$ be closer to $\frac{1}{2}$ or to 1?

3. Would Abigail add or subtract to find out how much more of the page is used for sports articles than for ads?

4. Will $\frac{2}{3} - \frac{1}{4}$ be closer to $\frac{1}{2}$ or to 1?

EXAMPLE 1

a. $\frac{1}{4} + \frac{2}{3} = \underline{\ ?\ }$

   LCD: 12

$$\frac{1}{4} = \frac{3}{12}$$
$$+\frac{2}{3} = +\frac{8}{12}$$
$$\frac{11}{12}$$

◀ $\frac{11}{12}$ **of the page is filled.**

b. $\frac{2}{3} - \frac{1}{4} = \underline{\ ?\ }$

   LCD: 12

$$\frac{2}{3} = \frac{8}{12}$$
$$-\frac{1}{4} = -\frac{3}{12}$$
$$\frac{5}{12}$$

◀ $\frac{5}{12}$ **more is used for sports.**

Adding or subtracting with mixed numbers is similar to adding or subtracting with fractions.

EXAMPLE 2

a. $2\frac{2}{3} + 5\frac{3}{7} = \underline{\ ?\ }$

   LCD: 21

$$2\frac{2}{3} = 2\frac{14}{21}$$
$$+5\frac{3}{7} = +5\frac{9}{21}$$
$$7\frac{23}{21} = 7 + 1\frac{2}{21}$$

◀ $\frac{23}{21} = 1\frac{2}{21}$

$$= 8\frac{2}{21}$$

b. $5\frac{7}{8} - 4\frac{7}{12} = \underline{\ ?\ }$

   LCD: 24

$$5\frac{7}{8} = 5\frac{21}{24}$$
$$-4\frac{7}{12} = -4\frac{14}{24}$$
$$1\frac{7}{24}$$

EXERCISES

Complete. Choose your answers from the box at the right.

1. The LCD of $\frac{2}{5}$ and $\frac{1}{4}$ is ___?___ .

2. To find the sum or difference of $\frac{2}{5}$ and $\frac{1}{4}$, use the like fractions ___?___ and ___?___ .

3. The sum of $\frac{2}{5}$ and $\frac{1}{4}$ is ___?___ .

4. The difference between $\frac{2}{5}$ and $\frac{1}{4}$ is ___?___ .

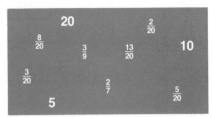

5. Will $\frac{2}{8} + \frac{5}{8}$ be less than 1 or greater than 1? Explain.

6. Will $\frac{5}{12} - \frac{1}{12}$ be greater than $\frac{1}{2}$ or less than $\frac{1}{2}$? Explain.

7. Will $\frac{4}{5} - \frac{6}{10}$ be closer to 0 or to $\frac{1}{2}$? Explain.

8. Will $\frac{9}{10} + \frac{3}{10}$ be closer to 1 or to 2? Explain.

Add or subtract. Write each answer in lowest terms.

9. $\frac{2}{3}$ $+\frac{5}{8}$

10. $\frac{2}{3}$ $-\frac{1}{6}$

11. $\frac{3}{4}$ $-\frac{1}{5}$

12. $\frac{5}{6}$ $+\frac{2}{5}$

13. $\frac{2}{3}$ $-\frac{1}{9}$

14. $\frac{3}{7}$ $+\frac{11}{14}$

15. $\frac{5}{8} + \frac{3}{16}$

16. $\frac{5}{12} - \frac{1}{4}$

17. $\frac{4}{5} - \frac{6}{10}$

18. $\frac{3}{11} + \frac{8}{22}$

19. $5\frac{1}{2}$ $+2\frac{1}{4}$

20. $5\frac{3}{4}$ $-2\frac{1}{2}$

21. $11\frac{1}{2}$ $-11\frac{6}{12}$

22. $8\frac{4}{5}$ $+5\frac{1}{3}$

23. $5\frac{11}{12}$ $-4\frac{3}{8}$

24. $9\frac{3}{8}$ $+8\frac{1}{6}$

25. $24\frac{3}{8} + 22\frac{1}{3}$

26. $28\frac{2}{3} - 26\frac{4}{9}$

27. $33\frac{1}{2} - 32\frac{1}{4}$

28. $12\frac{1}{12} + 13\frac{1}{4}$

29. $\frac{1}{5} + \frac{2}{3}$

30. $2\frac{5}{8} + 1\frac{1}{2}$

31. $\frac{6}{10} - \frac{3}{20}$

32. $5\frac{2}{3} + 5\frac{2}{3}$

33. $\frac{5}{6} - \frac{1}{2}$

34. $8\frac{7}{8} - 6\frac{3}{4}$

35. $\frac{1}{4} + \frac{5}{6}$

36. $46\frac{8}{15} - \frac{2}{5}$

Use the fractions in the box to complete Exercises 37–40.

$$\frac{1}{3}, \frac{1}{5}, \frac{1}{2}, \frac{3}{8}, \frac{1}{4}$$

37. $\frac{3}{10} + \frac{\square}{\square} = \frac{1}{2}$

38. $\frac{\square}{\square} - \frac{1}{6} = \frac{1}{3}$

39. $\frac{\square}{\square} + \frac{5}{12} = \frac{3}{4}$

40. $\frac{3}{4} - \frac{\square}{\square} = \frac{1}{2}$

Use the fractions in the box to complete Exercises 41–43.

$$\frac{1}{4}, 5\frac{3}{4}, \frac{1}{2}, 1\frac{1}{2}, 3\frac{3}{4}, 1\frac{1}{6}, 5\frac{1}{4}$$

41. $2\frac{3}{4} - \underline{\ ?\ } = 2\frac{1}{2}$

42. $\underline{\ ?\ } + 1\frac{1}{6} = 2\frac{2}{3}$

43. $\underline{\ ?\ } - 2\frac{1}{2} = 3\frac{1}{4}$

Use the numbers in parentheses to complete each problem so that the answer is the largest sum or difference.

44. $\frac{\square}{\square} + \frac{\square}{6} = \underline{\ ?\ }$ (3, 4, 2)

45. $\frac{12}{\square} - \frac{\square}{\square} = \underline{\ ?\ }$ (3, 4, 6)

Subtraction: MIXED NUMBERS

When subtracting with mixed numbers, you often have to rename a whole number.

EXAMPLE 1

Rename: $7\frac{3}{12} = 6\frac{\square}{12}$

1 Think: $7\frac{3}{12} = 6 + 1 + \frac{3}{12}$

2 $\qquad 7\frac{3}{12} = 6 + \frac{12}{12} + \frac{3}{12}$ ◀ $1 = \frac{12}{12}$

3 $\qquad 7\frac{3}{12} = 6 + \frac{15}{12}$, or $6\frac{15}{12}$

1. In step **2**, why is the number 1 renamed as $\frac{12}{12}$?

EXAMPLE 2

$7\frac{1}{4} - 4\frac{2}{3} = $ _____?_____

LCD: 12

$$7\frac{1}{4} = \quad 7\frac{3}{12}$$
$$-4\frac{2}{3} = -4\frac{8}{12}$$

◀ *Since $\frac{8}{12}$ is greater than $\frac{3}{12}$, rename $7\frac{3}{12}$ as $6\frac{15}{12}$.*

$$6\frac{15}{12}$$
$$-4\frac{8}{12}$$
$$\overline{\quad 2\frac{7}{12}}$$

◀ **See Example 1.**

2. How can you check the answer to Example 2?

3. To solve $4 - 2\frac{3}{4}$, how would you rename 4?

4. To solve $1 - \frac{5}{6}$, how would you rename 1?

EXAMPLE 3

a. $4 - 2\frac{3}{4} = $ _____?_____

$$4 \quad = \quad 3\frac{4}{4}$$
$$-2\frac{3}{4} = -2\frac{3}{4}$$
$$\overline{\quad\quad 1\frac{1}{4}}$$

b. $1 - \frac{5}{6} = $ _____?_____

$$1 \quad = \quad \frac{6}{6}$$
$$-\frac{5}{6} = -\frac{5}{6}$$
$$\overline{\quad\quad \frac{1}{6}}$$

EXERCISES

Complete.

1. $4\frac{2}{3} = 4 + \frac{\square}{\square}$

$= 3 + \frac{\square}{3} + \frac{2}{3}$

$= 3 + \frac{\square}{3}$

$= 3\frac{\square}{\square}$

2. $7\frac{3}{8} = 7 + \frac{\square}{\square}$

$= 6 + \frac{\square}{8} + \frac{3}{8}$

$= 6 + \frac{\square}{8}$

$= 6\frac{\square}{\square}$

3. $6\frac{1}{4} = 5\frac{\square}{4}$

4. $3\frac{5}{6} = 2\frac{\square}{6}$

5. $8\frac{3}{10} = 7\frac{\square}{10}$

6. $\begin{array}{r} 10\frac{1}{2} = \\ -\ 8\frac{5}{6} = \\ \hline \end{array} \quad \begin{array}{r} 10\frac{\square}{6} = \\ -\ 8\frac{5}{6} = \\ \hline \end{array} \quad \begin{array}{r} 9\frac{\square}{6} \\ -\ 8\frac{5}{6} \\ \hline ? \end{array}$

7. $\begin{array}{r} 5\frac{1}{3} = \\ -\ 2\frac{3}{4} = \\ \hline \end{array} \quad \begin{array}{r} 5\frac{\square}{12} = \\ -\ 2\frac{\square}{12} = \\ \hline \end{array} \quad \begin{array}{r} 4\frac{\square}{12} \\ -\ 2\frac{\square}{12} \\ \hline ? \end{array}$

8. Will $5 - \frac{1}{3}$ be closer to 4 or to 5? Explain.

9. Will $1 - \frac{3}{4}$ be closer to 0 or to 1? Explain.

10. Will $11 - \frac{5}{6}$ be closer to 10 or to 11? Explain.

11. Will $1\frac{5}{6} - \frac{1}{2}$ be greater than 1 or less than 1? Explain.

12. Will $8\frac{1}{4} - \frac{3}{4}$ be greater than 8 or less than 8?

13. Will $5\frac{9}{10} - \frac{1}{10}$ be closer to 6 or to 5? Explain.

Subtract. Write each answer in lowest terms.

14. $\begin{array}{r} 5\frac{1}{4} \\ -\ 1\frac{3}{4} \\ \hline \end{array}$

15. $\begin{array}{r} 3\frac{1}{8} \\ -\ 1\frac{3}{8} \\ \hline \end{array}$

16. $\begin{array}{r} 7\frac{1}{6} \\ -\ 4\frac{1}{2} \\ \hline \end{array}$

17. $\begin{array}{r} 10\frac{3}{10} \\ -\ 5\frac{3}{5} \\ \hline \end{array}$

18. $\begin{array}{r} 8\frac{7}{30} \\ -\ 2\frac{7}{10} \\ \hline \end{array}$

19. $\begin{array}{r} 9\frac{1}{12} \\ -\ 3\frac{2}{3} \\ \hline \end{array}$

20. $21\frac{3}{8} - 3\frac{11}{12}$

21. $6\frac{1}{2} - 4\frac{5}{7}$

22. $11\frac{1}{8} - 3\frac{2}{3}$

23. $1\frac{5}{6} - \frac{8}{9}$

24. $\begin{array}{r} 5 \\ -\ \frac{1}{3} \\ \hline \end{array}$

25. $\begin{array}{r} 1 \\ -\ \frac{1}{2} \\ \hline \end{array}$

26. $\begin{array}{r} 1 \\ -\ \frac{1}{3} \\ \hline \end{array}$

27. $\begin{array}{r} 11 \\ -\ 8\frac{1}{4} \\ \hline \end{array}$

28. $\begin{array}{r} 1 \\ -\ \frac{3}{8} \\ \hline \end{array}$

29. $\begin{array}{r} 8 \\ -\ 2\frac{9}{10} \\ \hline \end{array}$

30. $14 - 12\frac{1}{2}$

31. $1 - \frac{1}{5}$

32. $1 - \frac{3}{4}$

33. $6 - 5\frac{2}{3}$

34. $34 - 33\frac{1}{8}$

35. $17\frac{1}{2} - \frac{2}{3}$

36. $1 - \frac{4}{10}$

37. $11\frac{1}{5} - 6\frac{1}{3}$

38. $7\frac{1}{12} - 3\frac{2}{9}$

39. $8\frac{1}{8} - 2\frac{1}{4}$

40. $1 - \frac{3}{12}$

41. $11 - \frac{5}{6}$

Use the mixed numbers in the box to complete Exercises 42–44.

$$2\frac{2}{3},\ 2\frac{1}{2},\ 6\frac{1}{4},\ 3\frac{4}{5},\ 7\frac{1}{4}$$

42. $5\frac{1}{3} - \square = 2\frac{5}{6}$

43. $\square - 2\frac{3}{8} = 4\frac{7}{8}$

44. $8\frac{1}{3} - \square = 4\frac{8}{15}$

Strategy: MAKING A MODEL

There are 5 contestants in the 50-yard dash at the County Fair. Laura came in first. Tom finished last. Luke finished ahead of Janet. Sally finished after Janet. Who came in second?

READ   What are the facts?
Laura: Came in first.   Tom: Came in last.

Luke: Came in before Janet.   Janet: Came in before Sally.

PLAN   Make a line drawing to show the facts.

SOLVE

1. You know Laura came in first and Tom came in last. Show this on the line.

First ●————————————● Last
Laura   Tom

2. Luke came in before Janet. Show this on the line.

●——●——●————————●
Laura Luke Janet   Tom

3. Janet came in before Sally. Show this on the line.

●——●——●——●——●
Laura Luke Janet Sally Tom

Luke came in second.

CHECK   Does the drawing agree with the facts?

EXERCISES

Solve. The first two are started for you.

1. Adam's calf won second prize in the County Fair. Sue's entry came in next to last. Jane's calf placed just ahead of Sue's. Chen's entry came in right after Adam's. Flora's calf did <u>not</u> place sixth. Where did Nona's entry place?

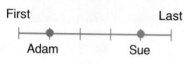

First ———————————— Last
     Adam   Sue

2. At the County Fair's baking contest, Entry A won fifth prize, Entry C came in ahead of Entry A and after Entry F. Entry F won third prize. D came in first and E placed seventh. Entry G came in after A. Where did B finish?

First ———————————— Last
D   E

3. The Top Fifty Records Shop at the Fair sold more Murky Mills records than Breezy Blues. They sold one fewer Thunder Tots record than Breezy Blues. More Cloud Trio records were sold than Murky Mills records. Which records had the greatest sales?

4. At the Fair's track meet, the javelin throw comes after the pole-climbing event. The 50-yard dash comes before pole climbing. The relay race comes between the 50-yard dash and the pole climbing. Which event is first?

5. At the Fair's softball games, the Reapers won more games than the Plows. The Harvesters won more games than the Balers and 2 fewer games than the Plows. Which two teams had the best records?

For Exercises 6–9, choose a strategy from the box at the right that you can use to solve each problem.

  a. *Name the strategy.*
  b. *Solve the problem.*

> **Guess and check**
> **Making a model**
> **More than one step**
> **Using patterns**

6. Alice won a prize at the fair by finding the missing numbers.

$\frac{1}{8}$ $\frac{1}{4}$ $\frac{3}{8}$ $\frac{1}{2}$ $\frac{5}{8}$ $\frac{3}{4}$ $\underline{\ ?\ }$ $\underline{\ ?\ }$

What are the numbers?

7. What does $\square \times \triangle = ?$
  Clue A: $\square + \triangle = 17$
  Clue B: $\square = \triangle + 1$

8. Alice bought two different souvenir shirts at the fair. The total cost for both was $18. The difference in the cost of the two shirts was $2. What did she pay for each shirt?

9. Keith has a part-time job at the fair. He uses $\frac{3}{10}$ of his pay for school expenses and $\frac{1}{5}$ of his pay for recreation. He saves the rest. What fractional part of his earnings does he save?

10. **Write your own problem** about this situation at the fair.

Frank, Meagan, Juan, and Wendy are in the archery competition. Here are their scores: Frank 96, Wendy 92, Juan 89, Meagan, 88. However, you cannot use these scores in your problem.

Math in Music

Annette wants to write a song for a victory celebration. She knows that measures in music are divided by **bars** and that notes of varying values are used in each measure.

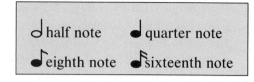

The **time signature** at the beginning of a piece of music tells how many beats per measure and the kind of note that receives one beat. For example, in a piece of music with a time signature of $\frac{3}{4}$, there are three beats in a measure and a quarter note receives one beat.

EXERCISES

1. For a time signature of $\frac{1}{2}$, what kind of note receives one beat?

2. For a time signature of $\frac{2}{4}$, how many sixteenth notes can there be in one measure?

3. For a time signature of $\frac{6}{8}$, how many beats will a quarter note receive?

4. For a time signature of $\frac{4}{4}$ with one half note in the measure, how many sixteenth notes can there be?

5. What note must be added to complete this measure? The time signature is $\frac{4}{4}$.

For each of these measures, a quarter note receives one beat. What is the time signature for each measure?

6.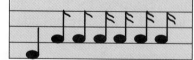

7.

PROJECT Write six measures having a time signature of $\frac{4}{4}$. Each measure should contain a different combination of notes. Practice clapping the rhythm of the measures you have written.

Chapter Summary

1. To write a fraction in lowest terms:

   [1] Divide the numerator <u>and</u> the denominator by a number that will divide evenly into both.

   [2] Repeat Step 1 until the numerator and denominator cannot be divided evenly by the same number except 1.

2. To determine whether a fraction is greater than 1, less than 1, or equal to 1, compare the numerator with the denominator.

3. To write a mixed number for a fraction greater than 1:

   [1] Divide the numerator by the denominator.

   [2] Write a fraction for the remainder.

4. To add or subtract like fractions:

   [1] Add or subtract the numerators.

   [2] Write the sum or difference over the common denominator.

   [3] Write the answer in lowest terms.

5. To add or subtract mixed numbers:

   [1] Add or subtract the fractional parts.

   [2] Add or subtract the whole numbers.

6. To find the LCD of two fractions:

   [1] Write the multiple of one of the denominators.

   [2] Stop when you find a common multiple of both denominators.

7. To write like fractions for unlike fractions:

   [1] Find the LCD.

   [2] Multiply <u>both</u> the numerator and denominator of <u>each</u> fraction by a number that will make the denominator equal to the LCD.

8. To add or subtract unlike fractions:

   [1] Find the LCD.

   [2] Use the LCD to write like fractions.

   [3] Add or subtract.

   [4] Write the answer in lowest terms.

9. To subtract with mixed numbers:

   [1] Find the LCD.

   [2] Use the LCD to write like fractions. Rename the whole number when necessary.

   [3] Subtract.

   [4] Write the answer in lowest terms.

---

1. $\dfrac{24}{30} = \dfrac{24 \div 2}{30 \div 2}$

   $= \dfrac{12}{15}$

   $= \dfrac{12 \div 3}{15 \div 3}$

   $= \dfrac{4}{5}$

2. Since $5 < 6$, $\dfrac{5}{6} < 1$.

   Since $4 = 4$, $\dfrac{4}{4} = 1$.

   Since $4 > 3$, $\dfrac{4}{3} > 1$.

3. $\dfrac{12}{5} \longrightarrow 5\overline{)12}\ ^{2\frac{2}{5}}$

4. $\dfrac{3}{10} + \dfrac{1}{10} = \dfrac{4}{10}$

   $= \dfrac{2}{5}$

5. $8\dfrac{5}{6} - 2\dfrac{1}{6} = 6\dfrac{4}{6}$

   $= 6\dfrac{2}{3}$

6. Given: $\dfrac{1}{2}$ and $\dfrac{1}{3}$

   Multiples of 2: 2, 4, ⑥

   LCD: **6**

7. Given: $\dfrac{1}{2}$ and $\dfrac{1}{3}$ LCD: 6

   $\dfrac{1}{2} = \dfrac{1 \times 3}{2 \times 3}$ $\dfrac{1}{3} = \dfrac{1 \times 2}{3 \times 2}$

   $\quad = \dfrac{3}{6}$ $\quad = \dfrac{2}{6}$

8. $\dfrac{1}{2} + \dfrac{1}{3} = \underline{\ ?\ }$ LCD: 6

   $\begin{aligned} \dfrac{1}{2} &= \dfrac{3}{6} \\ +\dfrac{1}{3} &= +\dfrac{2}{6} \\ \hline &\dfrac{5}{6} \end{aligned}$

9. $5\dfrac{1}{8} - 3\dfrac{3}{4} = \underline{\ ?\ }$ LCD: 8

   $\begin{aligned} 5\dfrac{1}{8} &= 5\dfrac{1}{8} = 4\dfrac{9}{8} \\ -3\dfrac{3}{4} &= -3\dfrac{6}{8} = -3\dfrac{6}{8} \\ \hline & \phantom{-3\dfrac{6}{8} = {}} 1\dfrac{3}{8} \end{aligned}$

Chapter Review

Part 1: VOCABULARY

For Exercises 1–6, choose from the box at the right the word(s) that complete(s) each statement.

1. The numerator and denominator are called the __?__ of a fraction.   (Page 127)

2. $\frac{1}{2}$ and $\frac{2}{4}$ are __?__ fractions.   (Page 129)

3. A fraction is greater than 1 when the numerator is __?__ than the denominator.   (Page 132)

4. A fraction greater than 1 can be written as a __?__ . (Page 132)

5. $\frac{2}{3}$ and $\frac{3}{4}$ are __?__ fractions.   (Page 138)

6. The abbreviation for least common denominator is __?__ . (Page 138)

> unlike
> terms
> like
> mixed number
> equivalent
> LCD
> greater
> whole number
> less

Part 2: SKILLS

Write each quotient as a fraction or as a whole number. (Pages 127–128)

7. $1 \div 5$ 8. $6 \div 6$ 9. $7 \div 8$ 10. $15 \div 3$ 11. $9 \div 5$

Write in lowest terms.   (Pages 130–131)

12. $\frac{12}{9}$ 13. $\frac{6}{9}$ 14. $\frac{14}{28}$ 15. $\frac{16}{14}$ 16. $\frac{24}{36}$ 17. $\frac{5}{25}$ 18. $\frac{18}{15}$

Replace each ● with $<$, $=$, or $>$.   (Pages 132–133)

19. $\frac{5}{8}$ ● 1   20. $\frac{3}{3}$ ● 1   21. $\frac{14}{11}$ ● 1   22. $\frac{9}{7}$ ● 1   23. $\frac{1}{6}$ ● 1

24. $\frac{4}{13}$ ● 1   25. $\frac{10}{4}$ ● 1   26. $\frac{3}{4}$ ● 1   27. $\frac{4}{4}$ ● 1   28. $\frac{9}{3}$ ● 1

Write a mixed number for each fraction.   (Pages 132–133)

29. $\frac{9}{4}$ 30. $\frac{20}{17}$ 31. $\frac{11}{2}$ 32. $\frac{8}{3}$ 33. $\frac{10}{7}$ 34. $\frac{24}{7}$ 35. $\frac{17}{4}$

Add or subtract. Write each answer in lowest terms.   (Pages 134–135)

36. $\begin{array}{r} \frac{3}{7} \\ +\frac{4}{7} \end{array}$ 37. $\begin{array}{r} \frac{5}{8} \\ +\frac{7}{8} \end{array}$ 38. $\begin{array}{r} \frac{10}{12} \\ -\frac{8}{12} \end{array}$ 39. $\begin{array}{r} \frac{12}{20} \\ -\frac{6}{20} \end{array}$ 40. $\begin{array}{r} 2\frac{3}{8} \\ +1\frac{3}{8} \end{array}$ 41. $\begin{array}{r} 6\frac{7}{10} \\ -4\frac{2}{10} \end{array}$

42. $5\frac{18}{24} - 1\frac{6}{24}$ 43. $7\frac{14}{18} - 3\frac{12}{18}$ 44. $7\frac{2}{3} + 1\frac{1}{3}$ 45. $9\frac{1}{2} + 3\frac{1}{2}$

46. $32\frac{10}{18} - 22\frac{7}{18}$ 47. $6\frac{5}{8} + 4\frac{3}{8}$ 48. $13\frac{3}{5} + 9\frac{1}{5}$ 49. $22\frac{2}{3} - 17\frac{1}{3}$

Find the LCD for each pair of fractions. (Pages 138–139)

50. $\frac{1}{3}$ and $\frac{3}{6}$ **51.** $\frac{2}{5}$ and $\frac{7}{10}$ **52.** $\frac{2}{3}$ and $\frac{3}{4}$ **53.** $\frac{1}{2}$ and $\frac{3}{5}$ **54.** $\frac{1}{7}$ and $\frac{3}{8}$

Write like fractions for each pair. (Pages 138–139)

55. $\frac{1}{4}$ and $\frac{1}{2}$ **56.** $\frac{3}{10}$ and $\frac{1}{5}$ **57.** $\frac{6}{8}$ and $\frac{4}{5}$ **58.** $\frac{2}{5}$ and $\frac{3}{4}$ **59.** $\frac{1}{3}$ and $\frac{3}{8}$

Add or subtract. Write each answer in lowest terms. (Pages 140–141)

60. $\frac{1}{2}$
$+\frac{1}{4}$

61. $\frac{2}{3}$
$+\frac{1}{6}$

62. $23\frac{3}{4}$
$-11\frac{1}{2} = \frac{2}{4}$

63. $4\frac{5}{8}$
$-2\frac{1}{4}$

64. $6\frac{2}{3} = \frac{16}{24}$
$+3\frac{1}{8} = \frac{3}{24}$

65. $17\frac{14}{15}$
$-4\frac{2}{3}$

66. $6\frac{2}{5}$
$+4\frac{1}{4}$

67. $2\frac{5}{8}$
$-1\frac{1}{6}$

68. $16\frac{4}{5} = \frac{28}{25}$
$-12\frac{2}{7} = \frac{10}{35}$

69. $21\frac{2}{5}$
$+5\frac{3}{4}$

70. $19\frac{1}{2} = \frac{9}{?}$
$-3\frac{1}{9} = \frac{2}{18}$

71. $17\frac{2}{7}$
$+6\frac{7}{8}$

72. $22\frac{3}{4} + 5\frac{2}{3}$ **73.** $16\frac{1}{8} + 15\frac{5}{6}$ **74.** $27\frac{3}{4} - 14\frac{1}{5} - \frac{4}{20}$ **75.** $15\frac{7}{8} - 12\frac{3}{5}$

Subtract. Write each answer in lowest terms. (Pages 142–143)

76. $14\frac{1}{3}$
$-2\frac{2}{3}$

77. $6\frac{1}{5}$
$-4\frac{3}{5}$

78. $7\frac{2}{3} = \frac{8}{6}$
$-4\frac{5}{6}$

79. $10\frac{1}{2}$
$-5\frac{3}{4}$

80. 6
$-3\frac{1}{8}$

81. 19
$-2\frac{7}{10}$

82. $16\frac{1}{4}$
$-3\frac{2}{5}$

83. $7\frac{3}{8}$
$-4\frac{4}{5}$

84. $3\frac{1}{2}$
$-\frac{7}{9}$

85. $24\frac{2}{3}$
$-\frac{7}{8}$

86. $4\frac{1}{10}$
$-3\frac{3}{4}$

87. $6\frac{1}{2}$
$-5\frac{5}{7}$

88. $5\frac{1}{3} - 4\frac{3}{8}$ **89.** $1\frac{5}{18} - \frac{3}{4}$ **90.** $3\frac{1}{6} - 1\frac{1}{5}$ **91.** $12\frac{3}{10} - 2\frac{1}{2}$

Part 3: PROBLEM SOLVING AND APPLICATIONS

92. There are 6 contestants in the carry-the-egg race at the County Fair. Corey finished second, and Luis finished next to last. Mai finished just before Luis, and Ben finished right behind Corey. Yoki did not finish last. In what place did Alix finish? (Pages 144–145)

93. a. Use the table on page 137 to find the number of packages of tiles needed to cover a floor that is $9\frac{3}{4}$ feet long and $13\frac{2}{3}$ feet wide.

b. One package of tiles costs $14.85. Estimate the cost of tiling the floor. (Page 137)

Chapter Test

Write in lowest terms. Then write a mixed number for any fraction greater than 1.

1. $\frac{2}{4}$ **2.** $\frac{6}{4}$ **3.** $\frac{3}{9}$ **4.** $\frac{14}{21}$ **5.** $\frac{16}{12}$ **6.** $\frac{18}{14}$ **7.** $\frac{15}{20}$

For Exercises 8–10, write the fractions in order from greatest to smallest.

8. $\frac{3}{4}, \frac{0}{6}, \frac{2}{2}, \frac{7}{6}$ **9.** $\frac{7}{9}, \frac{0}{8}, \frac{4}{2}, \frac{4}{4}$ **10.** $\frac{5}{5}, \frac{1}{4}, \frac{9}{3}, \frac{9}{5}$

Add or subtract. Write each answer in lowest terms.

11. $\begin{array}{r} \frac{5}{8} \\ +\frac{7}{8} \\ \hline \end{array}$ **12.** $\begin{array}{r} \frac{10}{12} \\ -\frac{7}{12} \\ \hline \end{array}$ **13.** $\begin{array}{r} 3\frac{3}{4} \\ +1\frac{2}{4} \\ \hline \end{array}$ **14.** $\begin{array}{r} 5\frac{7}{8} \\ +6\frac{4}{8} \\ \hline \end{array}$ **15.** $\begin{array}{r} 13\frac{12}{16} \\ -9\frac{4}{16} \\ \hline \end{array}$ **16.** $\begin{array}{r} 25\frac{16}{24} \\ -9\frac{7}{24} \\ \hline \end{array}$

Write like fractions for each pair.

17. $\frac{3}{4}$ and $\frac{1}{2}$ **18.** $\frac{2}{3}$ and $\frac{5}{6}$ **19.** $\frac{1}{6}$ and $\frac{3}{8}$ **20.** $\frac{3}{5}$ and $\frac{1}{4}$ **21.** $\frac{7}{8}$ and $\frac{1}{7}$

Add or subtract. Write each answer in lowest terms.

22. $\begin{array}{r} \frac{1}{2} \\ +\frac{3}{4} \\ \hline \end{array}$ **23.** $\begin{array}{r} 4\frac{1}{3} \\ +2\frac{4}{6} \\ \hline \end{array}$ **24.** $\begin{array}{r} 3\frac{5}{6} \\ -1\frac{1}{2} \\ \hline \end{array}$ **25.** $\begin{array}{r} \frac{7}{8} \\ -\frac{3}{4} \\ \hline \end{array}$ **26.** $\begin{array}{r} \frac{2}{3} \\ +\frac{3}{4} \\ \hline \end{array}$ **27.** $\begin{array}{r} 14\frac{2}{5} \\ +7\frac{1}{4} \\ \hline \end{array}$

28. $\begin{array}{r} 16\frac{2}{3} \\ -6\frac{1}{5} \\ \hline \end{array}$ **29.** $\begin{array}{r} \frac{4}{5} \\ -\frac{2}{7} \\ \hline \end{array}$ **30.** $\begin{array}{r} 14 \\ -1\frac{1}{3} \\ \hline \end{array}$ **31.** $\begin{array}{r} 23 \\ -12\frac{3}{8} \\ \hline \end{array}$ **32.** $\begin{array}{r} 14\frac{1}{2} \\ -12\frac{3}{4} \\ \hline \end{array}$ **33.** $\begin{array}{r} 7\frac{1}{3} \\ -4\frac{4}{5} \\ \hline \end{array}$

Solve.

34. Lou and Gretchen Reese jogged $2\frac{7}{10}$ miles on Monday and $3\frac{2}{5}$ miles on Wednesday. How many total miles did they jog on Monday and Wednesday?

35. Of the top five prize winners at the dog show, a beagle won third prize. A German shepherd came in one place ahead of a Labrador retriever. A cocker spaniel came in fifth place. Where did the bloodhound place?

Cumulative Maintenance Chapters 1–6

Choose the correct answer. Choose a, b, c, or d.

1. Add.

$$\begin{array}{r} 736 \\ 2952 \\ + 83 \\ \hline \end{array}$$

  a. 3761   **b.** 3671
  c. 3771   **d.** 3661

2. Divide:

$$14\,)\,\overline{3822}$$

  a. 273   **b.** 270   **c.** 274   **d.** 27

3. Subtract.

$$\begin{array}{r} 7003 \\ -876 \\ \hline \end{array}$$

  a. 7027   **b.** 6127
  c. 7127   **d.** 7023

4. This bar graph shows Jeff's earnings in tips for four weeks. During which week did he earn between $15 and $20?

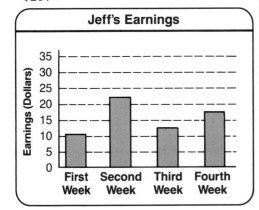

Jeff's Earnings

  a. First   **b.** Second
  c. Third   **d.** Fourth

5. Add.

$$4\tfrac{2}{15} + 3\tfrac{8}{15}$$

  a. $6\tfrac{9}{15}$ **b.** $7\tfrac{2}{3}$ **c.** $7\tfrac{6}{15}$ **d.** $8\tfrac{1}{15}$

6. Multiply.

$$\begin{array}{r} 309 \\ \times 76 \\ \hline \end{array}$$

  a. 23,474   **b.** 23,484
  c. 26,574   **d.** 24,384

7. The sum of the number of experiments performed by each of two students is 15. The difference in the number performed by each is 3. How many were performed by each student?

  a. 7, 8   **b.** 6, 9
  c. 5, 10   **d.** 4, 11

8. Which decimal has the smallest value?

  a. 0.306   **b.** 0.00036
  c. 3.6   **d.** 0.30006

9. Write a mixed number for $\tfrac{53}{9}$.

  a. $6\tfrac{1}{9}$ **b.** $5\tfrac{7}{9}$ **c.** $5\tfrac{1}{9}$ **d.** $5\tfrac{8}{9}$

10. Subtract.

$$\begin{array}{r} 6\tfrac{7}{8} \\ -2\tfrac{5}{8} \\ \hline \end{array}$$

  a. $4\tfrac{1}{4}$ **b.** $4\tfrac{3}{8}$ **c.** $3\tfrac{1}{4}$ **d.** $4\tfrac{1}{2}$

11. Multiply 0.56 by 0.07.

    a. 0.392   **b.** 0.0392
    c. 0.00392   **d.** 392

12. Soup sells at 4 cans for $1.59.
What is the cost of one can?

    a. 20¢   **b.** 45¢   **c.** 30¢   **d.** 40¢

13. Multiply: 8.73×1000

    a. 8730   **b.** 87.3
    c. 8.73   **d.** 87300

14. Subtract: $\frac{5}{8} - \frac{1}{2}$

    a. $\frac{3}{8}$ **b.** $\frac{1}{8}$ **c.** $\frac{4}{6}$ **d.** $\frac{5}{6}$

15. Add.

$$
\begin{array}{r}
6.21 \\
13.5 \\
3.963 \\
+\,354.7 \\
\hline
\end{array}
$$

    a. 366.743   **b.** 367.843
    c. 377.734   **d.** 378.373

16. Divide: $7\,\overline{)\,0.252}$

    a. 0.036 **b.** 3.6   **c.** 36   **d.** 0.36

17. Write the numeral for ten thousand
two.

    a. 10,000.2   **b.** 1002
    c. 10,200   **d.** 10,002

18. What is $\frac{63}{72}$ in lowest terms?

    a. $\frac{7}{8}$ **b.** $\frac{6}{7}$ **c.** $\frac{3}{4}$ **d.** $\frac{8}{9}$

19. Which of these represents thirteen
and seven hundredths?

    a. 13.07   **b.** 13.7
    c. 137   **d.** 13.007

20. Clara bought a pen for $4.68
and a writing tablet for $1.59. She
paid with a $10-bill. How much
change should she receive?

    a. $5.32   **b.** $6.91
    c. $3.73   **d.** $6.27

21. This graph shows the length of
the first day of every month for
a year. For which months do the
first days have the same length?

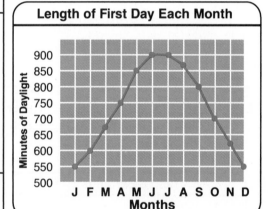

Length of First Day Each Month

    a. June and July
    b. March and October
    c. May and August
    d. January and November

22. In the last 5 years, the Blues
won 12, 8, 13, 10, and 17 games.
What is the median number of
games won?

    a. 12   **b.** 17   **c.** 13   **d.** 8

Fractions: Multiplication/Division

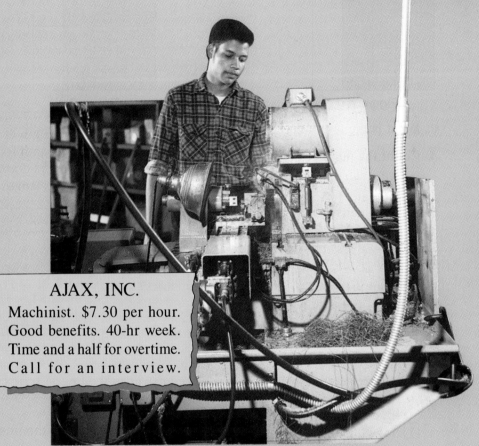

AJAX, INC.
Machinist. $7.30 per hour.
Good benefits. 40-hr week.
Time and a half for overtime.
Call for an interview.

- How much would a machinist earn for a 40-hour week at Ajax, Inc.?

- How much would a machinist earn for a 45-hour week at Ajax, Inc?

- Estimate how many 45-hour weeks it would take a machinist to earn $7,000 at Ajax, Inc.

- A machinist earned $109.50 in overtime pay in one week at Ajax, Inc. How many overtime hours did the employee work?

Meaning of Multiplication: FRACTIONS

You can fold a sheet of paper to solve this problem: $\frac{1}{3}$ of $\frac{3}{4}$ = _?_ .

ACTIVITY

1. Fold a sheet of paper into **fourths**. Shade 3 parts to show $\frac{3}{4}$.

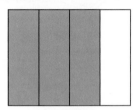

Figure A

2. Now fold the paper into **thirds** as shown in Figure B.

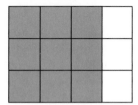

Figure B

3. What is the total number of parts in Figure B?

4. Count the number of parts in Figure B that shows $\frac{1}{3}$ of $\frac{3}{4}$.

   Complete: $\frac{1}{3}$ of $\frac{3}{4}$ = $\frac{\blacksquare}{12}$

5. Write the answer to $\frac{1}{3}$ of $\frac{3}{4}$ in lowest terms.

6. For the fractions $\frac{1}{3}$ and $\frac{3}{4}$:

   a. What is the product of the numerators?
   b. What is the product of the denominators?
   c. Write $\frac{3}{12}$ in lowest terms.

7. Compare the answer to Exercise **6c** with the answer to Exercise **5**.

   Complete: $\frac{1}{3}$ of $\frac{3}{4}$ means $\frac{1}{3}$ _?_ $\frac{3}{4}$.

8. Use Figure B to find $\frac{2}{3} \times \frac{3}{4}$. Write the answer in the lowest terms.

9. *Complete:* To multiply with fractions:

   ① Multiply the _?_ .   ② Multiply the _?_ .   ③ Write the answer in _?_ .

EXERCISES

Use the figures below to solve Exercises 1–5. Write answers in lowest terms.

   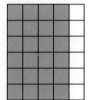

1. $\frac{2}{5} \times \frac{3}{4}$ **2.** $\frac{3}{5} \times \frac{1}{2}$ **3.** $\frac{1}{5} \times \frac{1}{2}$ **4.** $\frac{1}{6} \times \frac{4}{5}$ **5.** $\frac{5}{6} \times \frac{4}{5}$

Multiplication: FRACTIONS

Kevin Hall, a gourmet chef, prepares meals for customers in their own homes. The recipe for one of Kevin's special dishes requires $\frac{3}{4}$ of a cup of vegetable oil.

Kevin pours $\frac{1}{2}$ of $\frac{3}{4}$ cup into a second cup.

$\frac{3}{4}$ **of a cup** $\frac{1}{2}$ **of** $\frac{3}{4}$ **cup = ?**

1. *Complete:* Since $\frac{3}{4} = \frac{6}{8}$, $\frac{1}{2}$ of $\frac{3}{4} = \frac{1}{2}$ of $\frac{6}{8}$, or $\frac{\blacksquare}{8}$.

   Since $\frac{1}{2}$ of $\frac{3}{4}$ means $\frac{1}{2} \times \frac{3}{4}$, you can arrive at the same answer by multiplying.

2. *Complete:* $\frac{1}{2} \times \frac{3}{4} = \frac{1 \times \blacksquare}{2 \times \blacksquare} = \frac{\blacksquare}{8}$

   This example shows two methods for multiplying fractions.

EXAMPLE $\frac{4}{9} \times \frac{3}{8} = \underline{\quad ? \quad}$

Method 1: Multiply first. Then write in lowest terms.

$$\frac{4}{9} \times \frac{3}{8} = \frac{4 \times 3}{9 \times 8}$$

$$= \frac{12}{72} \quad \blacktriangleleft \text{ Not in lowest terms}$$

$$= \frac{6}{36} \quad \blacktriangleleft \text{ Not in lowest terms}$$

$$= \frac{1}{6} \quad \blacktriangleleft \text{ Lowest terms}$$

Method 2: Simplify by identifying common factors. Then multiply.

$$\frac{4}{9} \times \frac{3}{8} = \frac{\overset{1}{\cancel{4}}}{\underset{3}{\cancel{9}}} \times \frac{\overset{1}{\cancel{3}}}{\underset{2}{\cancel{8}}}$$

$$= \frac{1 \times 1}{3 \times 2}$$

$$= \frac{1}{6}$$

3. In Method 1, why does $\frac{12}{72} = \frac{6}{36}$?

4. In Method 1, why does $\frac{6}{36} = \frac{1}{6}$?

5. In Method 2, what is the common factor of 4 and 8? of 3 and 9?

EXERCISES

Complete. Choose your answers from the box at the right.

1. To multiply fractions, you multiply the __?__ and multiply the __?__ .

2. When you multiply fractions, you write the product in __?__ .

3. To compute $\frac{5}{6}$ of $\frac{8}{15}$, you __?__ $\frac{5}{6}$ by $\frac{8}{15}$.

4. To write $\frac{5}{8} \times \frac{4}{7}$ as $\frac{5}{2} \times \frac{1}{7}$, you __?__ 4 and 8 by a common factor, 4.

Multiply. Use the method you prefer. Write answers in lowest terms.

5. $\frac{3}{5} \times \frac{2}{7}$ 6. $\frac{7}{8} \times 3$ 7. $\frac{2}{9} \times \frac{4}{5}$ 8. $\frac{3}{11} \times \frac{5}{7}$ 9. $\frac{1}{2} \times \frac{1}{3}$ 10. $\frac{1}{2} \times \frac{3}{5}$

11. $\frac{1}{4} \times \frac{1}{2}$ 12. $\frac{1}{3} \times \frac{1}{5}$ 13. $\frac{3}{4} \times \frac{1}{4}$ 14. $\frac{3}{5} \times 9$ 15. $\frac{4}{5} \times \frac{1}{7}$ 16. $9 \times \frac{2}{3}$

17. $\frac{5}{9} \times \frac{3}{7}$ 18. $\frac{3}{4} \times \frac{2}{5}$ 19. $\frac{4}{5} \times 10$ 20. $\frac{2}{3} \times \frac{3}{8}$ 21. $\frac{2}{9} \times \frac{3}{4}$ 22. $\frac{9}{10} \times \frac{5}{6}$

23. $\frac{2}{3} \times \frac{9}{10}$ 24. $8 \times \frac{3}{4}$ 25. $\frac{3}{4} \times \frac{8}{15}$ 26. $\frac{7}{16} \times \frac{4}{21}$ 27. $\frac{4}{9} \times \frac{3}{8}$ 28. $\frac{12}{15} \times \frac{3}{4}$

29. $\frac{3}{5} \times \frac{1}{8}$ 30. $7 \times \frac{1}{5}$ 31. $\frac{1}{5} \times \frac{8}{9}$ 32. $\frac{3}{4} \times \frac{3}{5}$ 33. $6 \times \frac{5}{6}$ 34. $\frac{4}{5} \times \frac{2}{3}$

35. $\frac{5}{9} \times \frac{4}{7}$ 36. $\frac{1}{6} \times 3$ 37. $\frac{5}{8} \times \frac{1}{12}$ 38. $8 \times \frac{2}{3}$ 39. $\frac{2}{5} \times \frac{1}{6}$ 40. $\frac{2}{7} \times \frac{1}{6}$

For Exercises 41–46, find the missing factors.

41. $\frac{1}{8} \times \frac{\blacksquare}{\blacksquare} = \frac{3}{40}$ 42. $\frac{\blacksquare}{\blacksquare} \times \frac{2}{3} = \frac{8}{15}$ 43. $\frac{2}{5} \times \frac{\blacksquare}{\blacksquare} = \frac{4}{15}$

44. $\frac{9}{10} \times \frac{\blacksquare}{\blacksquare} = \frac{1}{2}$ 45. $\frac{1}{3} \times \frac{\blacksquare}{\blacksquare} = \frac{2}{9}$ 46. $\frac{\blacksquare}{\blacksquare} \times \frac{3}{8} = \frac{5}{24}$

47. Kevin's recipe for orange custard serves 12 people. The recipe calls for $\frac{2}{3}$ of a cup of orange juice. Kevin wants to make $\frac{1}{4}$ of the regular recipe. How much orange juice will he need?

48. Kevin allows $\frac{3}{4}$ of an hour to prepare a special sauce. One-third of that time is spent in cooking the sauce. What fraction of an hour is spent in cooking?

Math in Health

Dan's health class is studying about foods and their effect on good health. He learned that potatoes are starchy foods and that, for many years, people believed that potatoes were a high-calorie food. Recent studies, however, show that a potato of medium size contains only about 150 calories. Dan found this recipe for preparing potatoes using low-calorie and low-fat ingredients.

Potato Casserole	
4 large potatoes, peeled and cubed	$\frac{1}{2}$ teaspoon parsley
$\frac{3}{4}$ cup low-fat cottage cheese	$\frac{3}{4}$ cup low-fat yogurt
$\frac{2}{3}$ cup of onions, chopped	2 cloves of garlic, minced

Mix all ingredients together. Put in a casserole. Bake at 350° for 45 minutes. (Serves 4)

EXERCISES

Refer to the recipe for Exercises 1–5.

1. How much cottage cheese should Dan use if he prepares $\frac{1}{2}$ of the recipe?

2. How much parsley should Dan use if he prepares $\frac{1}{2}$ of the recipe?

3. How much yogurt should Dan use if he plans to prepare a recipe that serves 8 people?

4. Dan has $\frac{2}{3}$ of a cup of yogurt. How much more does he need to prepare a recipe that serves 8 people?

5. How many cups of onions will Dan need to prepare a recipe that serves 12 people?

6. Dan dropped the cup containing $\frac{3}{4}$ of a cup of cottage cheese and spilled $\frac{1}{3}$ of it. How much is left in the cup?

PROJECT Find a recipe for preparing a starchy vegetable. Rewrite the recipe to show the amount of each ingredient you would use to prepare $\frac{1}{2}$ of the recipe.

Multiplication: MIXED NUMBERS

Rounding mixed numbers will help you to estimate products.

Rules for Rounding Mixed Numbers

To round a mixed number, look at the fractional part.

a. If the fraction is <u>less than</u> $\frac{1}{2}$, round <u>down</u> to the nearest whole number.

b. If the fraction is <u>greater than or equal to</u> $\frac{1}{2}$, round <u>up</u> to the next whole number.

---

1. Round to the nearest whole number.

    a. $5\frac{3}{4}$ **b.** $2\frac{2}{7}$ **c.** $6\frac{1}{4}$ **d.** $\frac{7}{8}$

2. Estimate these products.

    a. $5\frac{3}{4} \times 2\frac{2}{7}$ **b.** $6\frac{1}{4} \times \frac{7}{8}$

To find the exact product, you first write a fraction for each mixed number. The table shows you how to do this.

Mixed Number	**1** Multiply the denominator and the whole number.	**2** Add this product to the numerator.	**3** Write this sum over the denominator.
$5\frac{3}{4}$	$4 \times 5 = 20$	$20 + 3 = 23$	$\frac{23}{4}$ $5\frac{3}{4} = \frac{23}{4}$

3. Write fractions for $2\frac{2}{7}$ and $6\frac{1}{4}$.

EXAMPLE 1

a. $5\frac{3}{4} \times 2\frac{2}{7} = \underline{\quad ? \quad}$

$$5\frac{3}{4} \times 2\frac{2}{7} = \frac{23}{\overset{1}{\cancel{4}}} \times \frac{\overset{4}{\cancel{16}}}{7}$$

$$= \frac{92}{7} \quad \frac{23 \times 4}{1 \times 7}$$

$$= 13\frac{1}{7}$$

b. $6\frac{1}{4} \times \frac{7}{8} = \underline{\quad ? \quad}$

$$6\frac{1}{4} \times \frac{7}{8} = \frac{25}{4} \times \frac{7}{8}$$

$$= \frac{175}{32} \quad \frac{25 \times 7}{4 \times 8}$$

$$= 5\frac{15}{32}$$

4. Compare the answers in the Example with your estimates in Exercise 2. Are the answers reasonably close to the estimates?

EXERCISES

Round to the nearest whole number.

1. $1\frac{1}{16}$ 2. $4\frac{5}{7}$ 3. $8\frac{9}{10}$ 4. $6\frac{1}{4}$ 5. $\frac{5}{6}$ 6. $15\frac{1}{3}$ 7. $10\frac{5}{8}$

8. $6\frac{1}{25}$ 9. $7\frac{11}{12}$ 10. $21\frac{7}{9}$ 11. $11\frac{2}{5}$ 12. $\frac{7}{10}$ 13. $5\frac{1}{6}$ 14. $44\frac{3}{4}$

Estimate each product.

15. $3\frac{1}{2} \times 2$ 16. $5\frac{1}{6} \times 4$ 17. $4 \times 1\frac{7}{8}$ 18. $\frac{9}{10} \times 3\frac{1}{10}$ 19. $8\frac{8}{9} \times \frac{11}{12}$

20. $3\frac{1}{2} \times 6\frac{1}{12}$ 21. $8\frac{3}{4} \times 3\frac{2}{3}$ 22. $7\frac{1}{4} \times 6\frac{1}{5}$ 23. $9\frac{1}{8} \times 4\frac{5}{6}$ 24. $10\frac{2}{5} \times 8\frac{5}{8}$

Write a fraction for each mixed number.

25. $2\frac{1}{2}$ 26. $4\frac{1}{3}$ 27. $6\frac{1}{5}$ 28. $5\frac{3}{4}$ 29. $3\frac{3}{4}$ 30. $9\frac{3}{8}$ 31. $11\frac{5}{8}$

32. $5\frac{1}{6}$ 33. $2\frac{2}{3}$ 34. $3\frac{4}{7}$ 35. $7\frac{1}{8}$ 36. $8\frac{7}{9}$ 37. $16\frac{1}{12}$ 38. $12\frac{11}{10}$

For Exercises 39–68:

a. *Estimate the answer.*

b. *Find the exact answer.*

c. *Compare the exact answer with the estimate to determine whether your answer is reasonable.*

39. $\frac{5}{6} \times 3\frac{3}{10}$ 40. $\frac{3}{4} \times 2\frac{2}{5}$ 41. $\frac{9}{10} \times 4$ 42. $2\frac{1}{6} \times 3$ 43. $5\frac{2}{3} \times 2\frac{1}{5}$

44. $5\frac{3}{4} \times 5\frac{1}{3}$ 45. $4\frac{2}{3} \times 5\frac{1}{6}$ 46. $8\frac{2}{3} \times 4\frac{3}{8}$ 47. $7\frac{1}{3} \times 8\frac{1}{4}$ 48. $1\frac{1}{6} \times 1\frac{4}{5}$

49. $2\frac{4}{5} \times \frac{6}{7}$ 50. $1\frac{7}{8} \times 6\frac{2}{5}$ 51. $1\frac{7}{8} \times 3\frac{5}{6}$ 52. $3\frac{1}{3} \times 1\frac{1}{8}$ 53. $1\frac{1}{4} \times 1\frac{1}{5}$

54. $8 \times 1\frac{8}{9}$ 55. $4 \times 2\frac{1}{3}$ 56. $3\frac{1}{8} \times 3\frac{1}{10}$ 57. $\frac{7}{8} \times 6\frac{1}{6}$ 58. $3\frac{3}{4} \times 8\frac{5}{6}$

59. $9\frac{1}{8} \times 4\frac{1}{5}$ 60. $3\frac{4}{5} \times \frac{9}{10}$ 61. $6\frac{5}{6} \times 3$ 62. $1\frac{1}{12} \times 1\frac{1}{3}$ 63. $5\frac{1}{4} \times 3\frac{5}{6}$

64. $7\frac{3}{10} \times 5$ 65. $4\frac{1}{8} \times 4$ 66. $2\frac{1}{10} \times 9\frac{7}{12}$ 67. $6\frac{2}{5} \times 4$ 68. $3\frac{15}{16} \times 3\frac{11}{12}$

69. In Exercises 49, 51, 58, 60, and 68, you rounded both numbers up to the next whole number. Was each exact answer greater or less than each estimate? Explain.

70. In Exercises 47, 52, 53, 56, 59, and 62, you rounded both numbers down to the nearest whole number. Was each exact answer greater or less than each estimate? Explain.

Division: FRACTIONS

Anna has a cloth remnant $\frac{3}{4}$ of a yard long. She wants to cut it into pieces $\frac{1}{8}$ of a yard long.

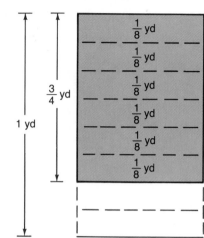

1. How many $\frac{1}{8}$'s are there in $\frac{3}{4}$? (See the figure at the right.)

   The figure shows that there are six $\frac{1}{8}$'s in $\frac{3}{4}$, or that
   $$\frac{3}{4} \div \frac{1}{8} = 6.$$

2. What is $\frac{3}{4} \times 8$?

3. Does $\frac{3}{4} \div \frac{1}{8} = \frac{3}{4} \times 8$?

4. *Complete:* $42 \div 7 = \underline{\quad?\quad}$ and $42 \times \frac{1}{7} = \underline{\quad?\quad}$

> Numbers such as 6 and $\frac{1}{6}$ are **reciprocals** because their product is 1. The number 0 has no reciprocal because you cannot divide by 0.

5. *Complete:* In $\frac{3}{4} \times \frac{4}{3} = 1$, $\frac{3}{4}$ and $\frac{4}{3}$ are $\underline{\quad?\quad}$ of each other.

6. *Complete:* In $8 \times \frac{1}{8} = 1$, 8 and $\frac{1}{8}$ are $\underline{\quad?\quad}$ of each other.

   Look at your answers to Exercises 2, 3, and 4 above.

7. *Complete:* Dividing by a fraction is the same as multiplying by its $\underline{\quad?\quad}$.

EXAMPLE

a. $\frac{3}{8} \div \frac{3}{4}$

$$\frac{3}{8} \div \frac{3}{4} = \frac{3}{8} \times \frac{4}{3}$$
$$= \frac{\overset{1}{\cancel{3}}}{\underset{2}{\cancel{8}}} \times \frac{\overset{1}{\cancel{4}}}{\underset{1}{\cancel{3}}}$$
$$= \frac{1 \times 1}{2 \times 1}, \text{ or } \frac{1}{2}$$

Multiply by the reciprocal

b. $\frac{2}{3} \div 9$

$$\frac{2}{3} \div 9 = \frac{2}{3} \times \frac{1}{9}$$
$$= \frac{2 \times 1}{3 \times 9}$$
$$= \frac{2}{27}$$

8. Will $\frac{3}{8} \div 3$ be greater or less than $\frac{3}{8}$? Explain.

9. Will $\frac{3}{8} \div \frac{1}{2}$ be greater or less than $\frac{3}{8}$? Explain.

10. Will $\frac{3}{8} \div \frac{5}{4}$ be greater or less than $\frac{3}{8}$? Explain.

EXERCISES

Write the reciprocal of each number.

1. $\frac{4}{7}$ **2.** 8   **3.** 4   **4.** $\frac{1}{4}$ **5.** $\frac{1}{3}$ **6.** 1   **7.** $\frac{21}{4}$ **8.** $\frac{11}{5}$

Complete.

9. $4 \div \frac{1}{2} = 4 \times \underline{\ ?\ }$ **10.** $\frac{1}{5} \div 2 = \frac{1}{5} \times \underline{\ ?\ }$ **11.** $8 \div \frac{2}{3} = 8 \times \underline{\ ?\ }$

12. $\frac{3}{5} \div \frac{2}{5} = \frac{3}{5} \times \underline{\ ?\ }$ **13.** $\frac{1}{6} \div \frac{3}{4} = \frac{1}{6} \times \underline{\ ?\ }$ **14.** $16 \div \frac{1}{4} = 16 \times \underline{\ ?\ }$

Divide. Write each answer in lowest terms.

15. $\frac{2}{3} \div \frac{4}{5}$ **16.** $\frac{7}{8} \div \frac{3}{4}$ **17.** $\frac{4}{7} \div \frac{7}{12}$ **18.** $\frac{1}{3} \div \frac{5}{9}$ **19.** $\frac{1}{8} \div \frac{1}{4}$

20. $\frac{5}{8} \div \frac{5}{6}$ **21.** $\frac{1}{9} \div \frac{8}{21}$ **22.** $\frac{3}{5} \div \frac{5}{12}$ **23.** $\frac{6}{7} \div \frac{9}{14}$ **24.** $\frac{3}{8} \div \frac{1}{6}$

25. $\frac{7}{9} \div \frac{14}{15}$ **26.** $\frac{5}{9} \div \frac{11}{12}$ **27.** $\frac{3}{7} \div \frac{3}{7}$ **28.** $\frac{1}{2} \div 6$ **29.** $\frac{1}{2} \div 2$

30. $\frac{1}{5} \div 2$ **31.** $\frac{1}{5} \div 6$ **32.** $\frac{4}{5} \div \frac{4}{5}$ **33.** $7 \div \frac{1}{3}$ **34.** $9 \div \frac{1}{3}$

35. $3 \div \frac{1}{4}$ **36.** $8 \div \frac{1}{4}$ **37.** $4 \div \frac{1}{7}$ **38.** $\frac{1}{6} \div 5$ **39.** $\frac{2}{3} \div 9$

40. $4 \div \frac{2}{3}$ **41.** $\frac{5}{6} \div \frac{2}{3}$ **42.** $1 \div \frac{4}{5}$ **43.** $1 \div \frac{2}{3}$ **44.** $\frac{1}{8} \div \frac{3}{4}$

45. $\frac{5}{6} \div 10$ **46.** $\frac{8}{27} \div 16$ **47.** $\frac{3}{5} \div \frac{5}{6}$ **48.** $\frac{3}{8} \div \frac{2}{3}$ **49.** $\frac{5}{6} \div \frac{3}{5}$

50. $5 \div \frac{2}{3}$ **51.** $\frac{5}{9} \div \frac{5}{12}$ **52.** $\frac{3}{8} \div \frac{3}{10}$ **53.** $1 \div \frac{3}{8}$ **54.** $2 \div \frac{3}{8}$

Garret A. Morgan invented the first traffic light in 1914. There were two lights, red and green, and they had to be changed by hand.

55. A traffic light at an intersection changes every $\frac{5}{6}$ of a minute.
  a. How many times will it change in 10 minutes?
  b. How many times will it change in 10 hours?

56. A traffic light changes every $\frac{3}{4}$ of a minute from 6 A.M. to 6 P.M. and every $\frac{2}{3}$ minute from 6 P.M. to 6 A.M. How many times does it change in 24 hours?

Division: MIXED NUMBERS

You have a six-mile paper route. It takes you one hour to cover $1\frac{1}{2}$ miles. You can use a drawing as a **model** to find how long it will take you to cover the 6 miles.

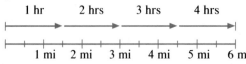

The drawing shows that there are **four** $1\frac{1}{2}$'s in 6.

1. To compute how many $1\frac{1}{2}$'s there are in 6, would you multiply 6 by $1\frac{1}{2}$, or divide 6 by $1\frac{1}{2}$? Why?

EXAMPLE 1

$$6 \div 1\frac{1}{2} = \underline{\ ?\ }$$

$$6 \div 1\frac{1}{2} = \frac{6}{1} \div \frac{3}{2} \quad \longleftarrow \text{ \textbf{Write as fractions.}}$$

$$= \frac{6}{1} \times \frac{2}{3} \quad \longleftarrow \text{ \textbf{The reciprocal of } } \frac{3}{2} \text{ \textbf{ is } } \frac{2}{3}.$$

$$= \frac{\overset{2}{6}}{1} \times \frac{2}{\underset{1}{3}}$$

$$= \frac{4}{1}, \text{ or \textbf{4}}$$

You can use rounding to help you estimate quotients.

$$1\frac{1}{2} \div 1\frac{1}{4} \approx 2 \div 1, \text{ or } 2$$

$$10\frac{1}{4} \div 1\frac{7}{8} \approx 10 \div 2, \text{ or } 5$$

$$8\frac{5}{6} \div 2\frac{1}{12} \approx 9 \div 2, \text{ or } 4\frac{1}{2}$$

> $\approx$ **means "is approximately equal to."**

$$\frac{1}{8} \div 1\frac{1}{12} \approx 0 \div 1, \text{ or } 0$$

EXAMPLE 2

$$5\frac{1}{6} \div 2\frac{1}{4} = \underline{\ ?\ } \qquad \textbf{Estimate: } 5 \div 2 = 2\frac{1}{2}$$

$$5\frac{1}{6} \div 2\frac{1}{4} = \frac{31}{6} \div \frac{9}{4}$$

$$= \frac{31}{\underset{3}{6}} \times \frac{\overset{2}{4}}{9}$$

$$= \frac{31 \times 2}{3 \times 9} = \frac{62}{27}, \text{ or } 2\frac{8}{27}$$

> **Compare with the estimate.**

EXERCISES

For Exercises 1–8, estimate each quotient. Choose your answer from the list below.

$\frac{1}{2}$ 1   2   3   4   5   6   7   8

1. $3\frac{1}{8} \div 2\frac{5}{8}$ **2.** $6\frac{1}{12} \div 2\frac{1}{4}$ **3.** $11\frac{7}{8} \div 6$ **4.** $10\frac{1}{4} \div 1\frac{7}{8}$

5. $6\frac{1}{12} \div 2\frac{1}{4}$ **6.** $8\frac{1}{6} \div \frac{7}{8}$ **7.** $1\frac{1}{6} \div 2$ **8.** $5\frac{1}{4} \div 9\frac{5}{6}$

9. How many $1\frac{1}{2}$'s are there in 12?   **10.** How many $1\frac{1}{4}$'s are there in 60?

11. How many $1\frac{1}{8}$'s are there in 72?   **12.** How many $1\frac{1}{5}$'s are there in 120?

Divide. Write each answer in lowest terms.

13. $\frac{1}{3} \div 3\frac{3}{5}$ **14.** $\frac{2}{5} \div 7\frac{1}{8}$ **15.** $7\frac{1}{8} \div \frac{2}{5}$ **16.** $4\frac{1}{3} \div 4$ **17.** $7\frac{1}{2} \div 1\frac{1}{4}$

18. $3\frac{1}{4} \div 6\frac{1}{2}$ **19.** $24 \div 5\frac{1}{3}$ **20.** $8 \div 4\frac{1}{3}$ **21.** $5\frac{7}{8} \div 6\frac{2}{5}$ **22.** $9\frac{3}{4} \div 3\frac{1}{4}$

23. $6\frac{1}{3} \div 1\frac{1}{3}$ **24.** $8\frac{1}{5} \div 2\frac{1}{2}$ **25.** $3\frac{1}{3} \div 1\frac{1}{5}$ **26.** $2\frac{1}{4} \div 3\frac{3}{8}$ **27.** $2 \div 1\frac{1}{4}$

28. $6 \div 4\frac{1}{2}$ **29.** $5\frac{2}{3} \div 2$ **30.** $3\frac{1}{2} \div 4$ **31.** $6\frac{1}{4} \div 25$ **32.** $3\frac{7}{8} \div 4$

33. $8\frac{1}{3} \div \frac{5}{12}$ **34.** $1\frac{1}{2} \div 1\frac{1}{4}$ **35.** $1\frac{1}{3} \div 2\frac{2}{3}$ **36.** $1\frac{7}{8} \div 1\frac{2}{3}$ **37.** $1\frac{7}{8} \div 1\frac{7}{8}$

For Exercises 38–47:

a. *Estimate the quotient.*

b. *Find the exact answer.*

c. *Compare the estimate with the quotient to determine whether your answer is reasonable.*

38. $1\frac{11}{12} \div 2$ **39.** $1\frac{1}{8} \div 4$ **40.** $2 \div 4\frac{1}{8}$ **41.** $10 \div 4\frac{7}{8}$ **42.** $2\frac{1}{8} \div 1\frac{1}{12}$

43. $5\frac{5}{6} \div 2\frac{1}{3}$ **44.** $3\frac{1}{3} \div 6$ **45.** $4\frac{4}{5} \div 10$ **46.** $24 \div 5\frac{2}{3}$ **47.** $8 \div 4\frac{1}{6}$

48. How many slices, each $1\frac{1}{4}$ inches thick, can be cut from a loaf of bread 15 inches long?

49. A model train circled the track $7\frac{1}{2}$ times in 10 minutes. How long will it take the train to go around the track once?

50. Costumes for a school play each require $3\frac{1}{3}$ yards of cloth. How many costumes can be made from a bolt of material that contains 40 yards?

Mid-Chapter Review

Multiply. Write each answer in lowest terms. (Pages 155–156)

1. $\frac{5}{8} \times \frac{11}{25}$ 2. $\frac{5}{6} \times \frac{4}{5}$ 3. $\frac{10}{21} \times \frac{14}{15}$ 4. $\frac{9}{10} \times \frac{8}{15}$ 5. $\frac{4}{3} \times \frac{3}{4}$

6. $9 \times \frac{3}{5}$ 7. $\frac{1}{9} \times \frac{3}{8}$ 8. $\frac{1}{9} \times \frac{6}{7}$ 9. $\frac{3}{8} \times \frac{6}{7}$ 10. $\frac{8}{9} \times \frac{9}{8}$

For Exercises 11–25:

a. *Estimate the answer.*

b. *Find the exact answer.*

c. *Compare the exact answer with the estimate to determine whether your answer is reasonable.*
(Pages 158–159 and 162–163)

11. $1\frac{9}{10} \times 2\frac{3}{4}$ 12. $1\frac{1}{4} \times 1\frac{1}{4}$ 13. $2\frac{1}{5} \times \frac{4}{5}$ 14. $\frac{7}{8} \times 3\frac{1}{5}$ 15. $3\frac{1}{3} \times 1\frac{1}{10}$

16. $3\frac{4}{5} \times 1\frac{7}{8}$ 17. $6\frac{1}{4} \times 2$ 18. $8\frac{3}{4} \times 2\frac{1}{5}$ 19. $5\frac{7}{8} \times 2\frac{1}{4}$ 20. $5\frac{1}{3} \times \frac{5}{6}$

21. $8\frac{2}{3} \div 6$ 22. $6\frac{2}{3} \div 1\frac{3}{4}$ 23. $4\frac{4}{5} \div 6$ 24. $9\frac{4}{5} \div 2\frac{1}{6}$ 25. $36 \div 4\frac{1}{10}$

Divide. Write each answer in lowest terms. (Pages 160–161)

26. $\frac{1}{3} \div \frac{3}{8}$ 27. $\frac{3}{10} \div \frac{3}{8}$ 28. $\frac{3}{8} \div 64$ 29. $6 \div \frac{3}{8}$ 30. $\frac{8}{9} \div \frac{3}{4}$

31. $\frac{7}{8} \div \frac{1}{6}$ 32. $\frac{3}{8} \div \frac{3}{4}$ 33. $\frac{24}{36} \div 12$ 34. $10 \div \frac{1}{2}$ 35. $\frac{3}{7} \div \frac{3}{7}$

36. It takes Angela $2\frac{1}{2}$ hours to cook dinner in a conventional oven. A microwave oven takes $\frac{1}{4}$ of that time. How much time does it take in microwave oven? (Pages 158–159)

37. A passenger jet circled an airport $5\frac{1}{2}$ times in 20 minutes. How long did it take the jet to circle the airport once? (Pages 162–163)

MAINTENANCE

Add or subtract. Write each answer in lowest terms.
(Pages 134–135 and 140–141)

38. $\frac{2}{3} - \frac{1}{3}$ 39. $\frac{1}{6} + \frac{3}{6}$ 40. $\frac{3}{5} + \frac{1}{3}$ 41. $\frac{7}{8} - \frac{2}{3}$ 42. $\frac{1}{4} + \frac{1}{2}$

43. $1\frac{2}{5} + 3\frac{1}{5}$ 44. $4\frac{9}{10} - 3\frac{7}{10}$ 45. $6\frac{3}{5} - 2\frac{1}{2}$ 46. $2\frac{1}{6} + 4\frac{3}{4}$ 47. $4\frac{3}{4} - 1\frac{1}{5}$

48. Choose the set of numbers at the right that are correctly ordered from least to greatest. Choose a, b, c, or d.

a. $\frac{3}{4}, \frac{7}{8}, \frac{2}{3}, \frac{4}{5}$ b. $\frac{7}{8}, \frac{3}{4}, \frac{4}{5}, \frac{2}{3}$

c. $\frac{2}{3}, \frac{3}{4}, \frac{4}{5}, \frac{7}{8}$ d. $\frac{2}{3}, \frac{4}{5}, \frac{3}{4}, \frac{7}{8}$

Math in Selling DISCOUNT

Sales clerks often answer customers' questions about items on sale. Such items are sold at a **discount** from the regular price. Sometimes the rate of discount is given as a fraction such as "$\frac{1}{3}$ off." The amount the customer actually pays is the **net price.**

Sale! ⅓ OFF
Baseball Gloves
REGULARLY $24.90

EXAMPLE What is the net price of the baseball gloves in the advertisement?

READ What are the facts?   Regular price: $24.90   Discount rate: $\frac{1}{3}$ off

PLAN First answer the "hidden question":

What is the amount of discount?

SOLVE 1 Amount of discount:

$\frac{1}{\overset{1}{\cancel{3}}} \times \$24.\overset{8.30}{\cancel{90}} = \$8.30$ ◀ **Rate of Discount × Regular Price = Discount**

2 Net price: $24.90 − $8.30 = **$16.60** ◀ **Regular Price − Discount = Net Price**

CHECK Did you use all the facts correctly in the solution?

EXERCISES

1. A camera is on sale for $\frac{1}{5}$ off the regular price of $165.00. What is the discount?

2. The regular price of a cassette is $7.20. It is on sale for $\frac{1}{3}$ off the regular price. What is the amount of discount?

3. A clock radio is on sale for $\frac{1}{4}$ off the regular price of $73.96. What will the customer pay?

4. A piano with a regular price of $1200 is sold for $800. Was the discount rate "$\frac{1}{2}$ off" or "$\frac{1}{3}$ off"? Explain.

5. Jane paid 68¢ for a piece of ribbon regularly sold for 85¢. Was the discount "$\frac{1}{5}$ off" or "$\frac{1}{4}$ off"? Explain.

Comparing Fractions and Decimals

During track practice, Vernon ran $\frac{5}{8}$ of a mile while Ed ran $\frac{3}{8}$ of a mile. You can use a number line to compare distances expressed as fractions.

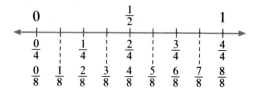

a. $\frac{5}{8} > \frac{3}{8}$ **b.** $\frac{1}{4} < \frac{3}{4}$ ◀ **The tip of > or < points to the smaller number.**

1. Are the fractions in **a** and **b** like or unlike fractions? Explain.

2. Use > or < to complete each of the following:

   a. $5 \underline{\ ?\ } 3$ and $\frac{5}{8} \underline{\ ?\ } \frac{3}{8}$ **b.** $1 \underline{\ ?\ } 3$ and $\frac{1}{4} \underline{\ ?\ } \frac{3}{4}$

3. State a rule for comparing like fractions.

4. *Complete:* To add or subtract unlike fractions, you first find the $\underline{\ ?\ }$ of the fractions. Then you write $\underline{\ ?\ }$ fractions.

5. *Complete:* To compare unlike fractions, you first find the $\underline{\ ?\ }$ of the fractions. Then you write $\underline{\ ?\ }$ fractions.

EXAMPLE 1 Compare the fractions. Write <, =, or >: $\frac{5}{6} \bullet \frac{7}{8}$

1 Find the LCD.   LCD: 24

2 Write equivalent fractions. $\frac{5}{6} = \frac{20}{24}$ $\frac{7}{8} = \frac{21}{24}$

3 Compare.   Since $20 < 21$, $\frac{20}{24} < \frac{21}{24}$. Thus, $\frac{5}{6} < \frac{7}{8}$.

You can also use decimal equivalents to compare fractions.

EXAMPLE 2 Compare the fractions. Write <, =, or >: $\frac{3}{4} \bullet \frac{7}{8}$

Plan: Write a decimal for each fraction.

$$\frac{3}{4} \longrightarrow 4\overline{)3.00} \quad \begin{array}{r} 0.75 \\ \underline{2\,8} \\ 20 \\ \underline{20} \\ 0 \end{array}$$

$$\frac{7}{8} \longrightarrow 8\overline{)7.000} \quad \begin{array}{r} 0.875 \\ \underline{6\,4} \\ 60 \\ \underline{56} \\ 40 \\ \underline{40} \\ 0 \end{array}$$

Since $0.75 < 0.875$,

$\frac{3}{4} < \frac{7}{8}$.

EXERCISES

Use the number line at the right to compare each pair of fractions. Write <, =, or >.

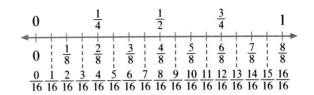

1. $\frac{1}{8}$ ● $\frac{3}{8}$ **2.** $\frac{7}{8}$ ● $\frac{3}{8}$ **3.** $\frac{2}{4}$ ● $\frac{1}{4}$ **4.** $\frac{3}{16}$ ● $\frac{5}{16}$ **5.** $\frac{15}{16}$ ● $\frac{11}{16}$

For Exercises 6–15, write like fractions to compare. Write <, =, or >.

6. $\frac{5}{6}$ ● $\frac{3}{8}$ **7.** $\frac{1}{2}$ ● $\frac{2}{3}$ **8.** $\frac{1}{3}$ ● $\frac{3}{4}$ **9.** $\frac{5}{6}$ ● $\frac{7}{9}$ **10.** $\frac{4}{5}$ ● $\frac{14}{15}$

11. $\frac{3}{5}$ ● $\frac{2}{3}$ **12.** $\frac{7}{10}$ ● $\frac{17}{20}$ **13.** $\frac{5}{12}$ ● $\frac{1}{3}$ **14.** $\frac{4}{5}$ ● $\frac{12}{15}$ **15.** $\frac{4}{9}$ ● $\frac{1}{2}$

For Exercises 16–25, write a decimal for each fraction. Then compare. Write <, =, or >.

16. $\frac{3}{4}$ ● $\frac{7}{8}$ **17.** $\frac{2}{5}$ ● $\frac{1}{2}$ **18.** $\frac{5}{8}$ ● $\frac{2}{5}$ **19.** $\frac{7}{20}$ ● $\frac{3}{10}$ **20.** $\frac{1}{2}$ ● $\frac{5}{10}$

21. $\frac{3}{10}$ ● $\frac{2}{5}$ **22.** $\frac{1}{4}$ ● $\frac{3}{10}$ **23.** $\frac{3}{5}$ ● $\frac{7}{10}$ **24.** $\frac{3}{5}$ ● $\frac{3}{8}$ **25.** $\frac{3}{4}$ ● $\frac{7}{8}$

Write the fractions in order from least to greatest.

26. $\frac{2}{3}, \frac{5}{6}, \frac{5}{7}, \frac{7}{8}$ **27.** $\frac{7}{12}, \frac{3}{4}, \frac{8}{9}, \frac{5}{6}$ **28.** $\frac{11}{12}, \frac{9}{10}, \frac{7}{8}, \frac{5}{6}$

29. $\frac{3}{4}, \frac{4}{5}, \frac{1}{2}, \frac{3}{10}, \frac{7}{8}$ **30.** $\frac{2}{3}, \frac{5}{6}, \frac{4}{5}, \frac{1}{2}, \frac{3}{4}$ **31.** $\frac{2}{5}, \frac{1}{8}, \frac{3}{7}, \frac{3}{10}, \frac{1}{3}$

Compare. Write <, =, or >.

32. $\frac{1}{5}$ ● $\frac{1}{4}$ **33.** $\frac{1}{8}$ ● $\frac{1}{9}$ **34.** $\frac{1}{5}$ ● $\frac{1}{6}$ **35.** $\frac{1}{3}$ ● $\frac{1}{2}$ **36.** $\frac{1}{7}$ ● $\frac{1}{10}$

For Exercises 37–39, replace each box with 1, 3, or 9 to make each comparison correct. You can use each number only once for each fraction. Write as many answers as possible.

37. $\frac{\blacksquare}{\blacksquare} < \frac{3}{8}$ **38.** $\frac{\blacksquare}{\blacksquare} > \frac{3}{7}$ **39.** $\frac{\blacksquare}{\blacksquare} < \frac{2}{5}$

40. Mary has two recipes for macaroni and cheese. The first calls for $\frac{5}{8}$ of a cup of cheese and the second calls for $\frac{5}{6}$ of a cup. Which amount is greater?

41. Pierre and Ellen are reading the same book. Ellen has finished $\frac{7}{8}$ of the book and Pierre has finished $\frac{11}{12}$ of the book. Who has more pages left to read?

42. Sam and Walter start a one-kilometer race together. At a particular moment, Sam has run $\frac{15}{16}$ of the distance. Walter has run $\frac{9}{10}$. Who is ahead?

43. A swimmer completed the 200-meter freestyle event in 2.15 minutes and the 200-meter backstroke in $2\frac{2}{5}$ minutes. Which time was faster?

FRACTIONS: MULTIPLICATION/DIVISION   **167**

Income and Wages

The Seaside Resort pays Linda $5 an hour for a 40-hour week.

She receives **overtime pay** when she works more than a 40-hour week. She receives $1\frac{1}{2}$, or 1.5, times the hourly pay for each hour of overtime. This is called **time and a half.**

1. How would you find her overtime pay per hour?

2. How would you find her overtime pay for the week?

EXAMPLE

Linda worked 42.5 hours last week. Find her total pay.

READ

What are the facts?

She earns $5 an hour for a 40-hour week. She worked 42.5 hours.

PLAN

To find her total income, first answer these "hidden questions."

What is her regular pay for a 40-hour week?

What is her overtime pay per hour?

What is her overtime pay?

SOLVE

1. Find the pay for 40 hours.

   $5 × 40 = __?__

   $\begin{array}{c} \text{Rate Per} \\ \text{Hour} \end{array} × \begin{array}{c} \text{Hours} \\ \text{Worked} \end{array} = \begin{array}{c} \text{Regular} \\ \text{Pay} \end{array}$

2. Find the overtime pay per hour.

   $5 × 1.5 = __?__

   $\begin{array}{c} \text{Regular} \\ \text{Pay Per} \\ \text{Hour} \end{array} × 1.5 = \begin{array}{c} \text{Overtime} \\ \text{Pay Per} \\ \text{Hour} \end{array}$

3. Find the overtime pay for $2\frac{1}{2}$ hours.

   $7.50 × 2.5 = __?__

   $42\frac{1}{2} - 40 = 2\frac{1}{2}$, or 2.5

4. Now you can find the total income.

   $200.00 + $18.75 = __?__

   $\begin{array}{c} \text{Regular} \\ \text{Pay} \end{array} + \begin{array}{c} \text{Overtime} \\ \text{Pay} \end{array} = \begin{array}{c} \text{Total} \\ \text{Pay} \end{array}$

CHECK

Did you use all the facts correctly in the formulas?

Linda's total pay for last week was **$218.75.**

EXERCISES   *For Exercises 1–6, find the weekly pay.*

1. 40 hours at $4.00 per hour

2. 40 hours at $5.35 per hour

3. 38 hours at $5.38 per hour

4. 25 hours at $3.85 per hour

5. An auto mechanic earns $8.60 per hour and works 35 hours per week. Find the weekly pay.

6. A computer repair technician earns $11 per hour and works 40 hours per week. Find the weekly pay.

For Exercises 7–10, find the number of hours of overtime work. Overtime is paid for all hours worked over 37$\frac{1}{2}$ hours.

7. 39

8. 41$\frac{1}{2}$

9. 43$\frac{1}{4}$

10. 45$\frac{3}{4}$

Complete this table. The regular work week is 38 hours.

	Hourly Wage	Hours Worked	Regular Pay	Overtime Pay per Hour	Overtime Pay	Total Pay
11.	$6.00	42$\frac{1}{2}$	?	?	?	?
12.	$7.50	44	?	?	?	?

For each of Exercises 13–14, use the ad to find each of the following.

a. *Find the regular pay.*
b. *Find the overtime pay per hour.*
c. *Find the overtime pay.*
d. *Find the total pay.*

DRIVER

with own late model 4-door car or station wagon. $8 per hour. 35 hr per wk. Time and a half for overtime. Will train. 622-0500

13. Jane McGuire works as a driver. One week she worked 40 hours.

14. Bill Mann also works as a driver. One week he worked 41$\frac{3}{4}$ hours.

15. On a certain day, Rosita worked 7$\frac{1}{2}$ hours. She began work that day at 8:15 A.M. and took an hour off for lunch. What time did she leave work?

16. Thomas works from 8:30 A.M. to 5:00 P.M. with 1 hour off for lunch. One day he leaves work at 4:15 P.M. planning to make up the time lost on the next day. Describe 3 ways in which he can do this.

17. These two want ads are for the same type of company. Which company would you work for? Why?

AJAX, INC.

Machinist. $7.25 per hour. Good benefits. 40-hr week. Time and a half for overtime. Call for an interview.

ACME, INC.

Machinist. $7.00 per hour. Good benefits. 35-hr week. Time and a half for overtime. Guaranteed 5 hours overtime per week.

Payroll Deductions

Linda receives a **statement of earnings** with her paycheck.

HOURS	HOURLY RATE	REGULAR PAY	OVERTIME PAY	GROSS PAY	NET PAY
42.5	5.00	200.00	18.75	218.75	?

TAX DEDUCTIONS				PERSONAL DEDUCTIONS		
FEDERAL	FICA	STATE	LOCAL	INSURANCE	RETIREMENT	OTHER
32.70	16.43	7.25	- - - -	5.85	- - - - - -	6.00

Gross pay is the amount of money earned for the pay period.

1. How do you find the gross pay?

Federal income tax, social security tax (FICA), state, and local taxes are **deducted** (subtracted) from the gross pay. **Personal deductions** are also deducted from gross pay.

2. What amounts do you add to find the following?

   a. Total taxes   **b.** Total personal deductions   **c.** Total deductions

After the total deductions are deducted from the gross pay, what remains is the **net pay**, or **take-home pay**.

EXAMPLE   Find Linda's net pay.

READ   What are the facts?

Taxes paid: $32.70, $16.43, $7.25

Personal deductions: $5.85, $6.00

Gross pay: $218.75

PLAN   To find Linda's net pay, first answer this "hidden question."
What are the total deductions?

SOLVE   ☐1 Find the total deductions.

Total taxes: $32.70 + $16.43 + $7.25 = $56.38
Total personal deductions: $5.85 + $6.00 $11.85
Total deductions: **$68.23**

☐2 Find the net pay.
$218.75 - $68.23 = **$150.52**   ◀ $\dfrac{Gross}{Pay} - \dfrac{Total}{Deductions} = \dfrac{Net}{Pay}$

CHECK   Did you use all the facts correctly in the solution?

EXERCISES

For Exercises 1–6, find the total deductions and the net pay.

	Gross Pay	Federal Tax	FICA Tax	State Tax	Local Tax	Insurance	Retirement	Total Deductions	Net Pay
1.	$250	$ 45.05	$18.78	$ 9.98	—	—	$12.50	?	?
2.	$300	$ 59.21	$22.53	$12.05	$4.12	$1.50	—	?	?
3.	$180	$ 19.10	$13.52	$ 7.20	$3.10	$0.75	—	?	?
4.	$520	$137.02	$39.05	$20.85	—	—	$35.00	?	?
5.	$600	$168.93	$45.06	$24.10	—	$3.00	$45.00	?	?
6.	$430	$102.27	$32.29	$17.15	$7.27	$4.00	$25.00	?	?

7. Joseph Ortega earned $580 last week. The deductions were: federal tax – $73.80; state tax – $14.65; FICA tax – $43.52; life insurance – $2.50. Find the net pay.

8. Marie Kelly earned $440 last week. The deductions were: federal tax – $33.00; state tax – $12.90; FICA tax – $33.04; pension plan – $20.00. Find the net pay.

Use the statement of earnings below to find the following.

9. Gross pay

10. Total taxes

11. Total personal deductions

12. Net pay

HOURS	HOURLY RATE	REGULAR PAY	OVERTIME PAY	GROSS PAY	NET PAY
40.0	5.50	?	- - - - - -	?	?

TAX DEDUCTIONS				PERSONAL DEDUCTIONS		
FEDERAL	FICA	STATE	LOCAL	INSURANCE	RETIREMENT	OTHER
23.80	16.52	2.64	- - - -	1.16	- - - - - -	25.30

13. Clyde's paycheck for last week was $480.25. His gross pay was $657.82. His personal deductions were $17.37. How much were the total taxes?

14. Sue's net pay last week was $364.72. Her total taxes were $175.20. Her personal deductions were $25.50. How much was her gross pay?

15. Marie's paycheck for last week was $282.50. Her gross pay was $362.50. Her total taxes were $62.00. How much were the personal deductions?

16. Harry's gross pay last week was $270.25 and his net pay was $210.25. The total personal deductions were $15.25. How much was deducted for taxes?

FRACTIONS: MULTIPLICATION/DIVISION **171**

Rational Numbers

A **rational number** is any number that can be written as a fraction in the form $\frac{a}{b}$. Remember that the denominator, b, cannot equal zero. Why?

Complete the following.

1. Since 152 can be written as $\frac{152}{\blacksquare}$, 152 is a rational number.

2. Since $1\frac{7}{8}$ can be written as $\frac{\blacksquare}{8}$, $1\frac{7}{8}$ is a rational number.

3. Since 3.5 can be written as $\frac{\blacksquare}{2}$, 3.5 is a rational number.

Between any two rational numbers there is always another rational number.

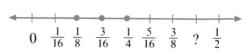

4. What rational number is halfway between $\frac{1}{8}$ and $\frac{1}{4}$?

5. What rational number is halfway between $\frac{3}{8}$ and $\frac{1}{2}$?

Another method of finding a rational number halfway between $\frac{3}{8}$ and $\frac{1}{2}$ is to add the two numbers. Then divide the sum by 2.

EXAMPLE Find the rational number halfway between $\frac{3}{8}$ and $\frac{1}{2}$.

$(\frac{3}{8} + \frac{1}{2}) \div 2 = (\frac{3}{8} + \frac{4}{8}) \div 2$ ◀ **Do the work inside parentheses first.**

$= \frac{7}{8} \div 2$

$= \frac{7}{8} \times \frac{1}{2}$

$= \frac{7}{16}$ So $\frac{7}{16}$ is halfway between $\frac{3}{8}$ and $\frac{1}{2}$.

6. How would you find the rational number halfway between $\frac{3}{8}$ and $\frac{7}{16}$?

EXERCISES

Write each rational number in the form $\frac{a}{b}$.

1. 14   2. $2\frac{1}{8}$ 3. 4.5   4. 0.9   5. $8\frac{1}{3}$ 6. 0

Find the rational number halfway between each pair of rational numbers.

7. $\frac{1}{2}$ and $\frac{5}{12}$ 8. $\frac{2}{3}$ and $\frac{3}{4}$ 9. $\frac{1}{2}$ and $\frac{1}{3}$ 10. $\frac{1}{8}$ and $\frac{5}{32}$

11. 0.4 and 0.5   12. $\frac{9}{10}$ and 1   13. $\frac{9}{10}$ and 2   14. $1\frac{1}{5}$ and $1\frac{2}{5}$

Chapter Summary

1. To multiply with fractions:
   - [1] Multiply the numerators and multiply the denominators.
   - [2] Write the answer in lowest terms.

2. To multiply with mixed numbers:
   - [1] Write a fraction for each mixed number.
   - [2] Multiply the fractions.
   - [3] Write the answer in lowest terms.

3. To divide with fractions:
   - [1] Use the reciprocal of the divisor to write the corresponding multiplication problem.
   - [2] Multiply.
   - [3] Write the answer in lowest terms.

4. To divide with mixed numbers:
   - [1] Write a fraction for each mixed number.
   - [2] Divide the fractions.
   - [3] Write the answer in lowest terms.

5. To compare fractions:
   - [1] Find the LCD.
   - [2] Write equivalent fractions.
   - [3] Compare.

6. To compare fractions:
   - [1] Write a decimal for each fraction.
   - [2] Compare.

---

1. $\dfrac{5}{6} \times \dfrac{7}{8} = \dfrac{5 \times 7}{6 \times 8}$

   $= \dfrac{35}{48}$ ◀ **Lowest terms**

2. $4\dfrac{2}{3} \times 3\dfrac{1}{7} = \dfrac{14}{3} \times \dfrac{22}{7}$

   $= \dfrac{\overset{2}{\cancel{14}}}{3} \times \dfrac{22}{\underset{1}{\cancel{7}}}$

   $= \dfrac{44}{3} = 14\dfrac{2}{3}$

3. $\dfrac{7}{8} \div \dfrac{1}{6} = \dfrac{7}{8} \times \dfrac{6}{1}$

   $= \dfrac{7}{\underset{4}{\cancel{8}}} \times \dfrac{\overset{3}{\cancel{6}}}{1}$

   $= \dfrac{21}{4} = 5\dfrac{1}{4}$

4. $6\dfrac{1}{4} \div 1\dfrac{1}{2} = \dfrac{25}{4} \div \dfrac{3}{2}$

   $= \dfrac{25}{\underset{2}{\cancel{4}}} \times \dfrac{\overset{1}{\cancel{2}}}{3}$

   $= \dfrac{50}{12} = 4\dfrac{1}{6}$

5. Compare $\dfrac{2}{3}$ and $\dfrac{3}{4}$.

   LCD: 12

   $\dfrac{2}{3} = \dfrac{8}{12}$ $\dfrac{3}{4} = \dfrac{9}{12}$

   Since $8 < 9$, $\dfrac{8}{12} < \dfrac{9}{12}$.

   Thus, $\dfrac{2}{3} < \dfrac{3}{4}$.

6. Compare $\dfrac{1}{4}$ and $\dfrac{2}{5}$.

   $\dfrac{1}{4} \rightarrow 4\overline{)1.00}\;^{0.25}$ $\dfrac{2}{5} \rightarrow 5\overline{)2.0}\;^{0.4}$

   Since $0.25 < 0.4$, $\dfrac{1}{4} < \dfrac{2}{5}$.

Chapter Review

Part 1: VOCABULARY

For Exercises 1–6, choose from the box at the right the word(s) that complete(s) each statement.

1. To multiply fractions, you multiply the __?__ and multiply the __?__ .   (Page 155)

2. Two fractions whose product is one are __?__ of each other.   (Page 160)

3. Dividing by a fraction is the same as __?__ by its reciprocal.   (Page 160)

4. You can write __?__ fractions to compare two unlike fractions.   (Page 166)

5. When you work longer than the regular work week, you may receive __?__ .   (Page 168)

6. After the total deductions are deducted from the gross pay, what remains is the __?__ .   (Page 170)

> multiplying
> net pay
> equivalent
> numerators
> overtime pay
> denominators
> reciprocals

Part 2: SKILLS

Multiply. Write each answer in lowest terms.   (Pages 155–156)

7. $\frac{2}{3} \times \frac{4}{9}$ 8. $\frac{3}{4} \times \frac{5}{7}$ 9. $4 \times \frac{1}{3}$ 10. $\frac{2}{7} \times 8$ 11. $\frac{4}{7} \times \frac{5}{12}$ 12. $\frac{5}{8} \times \frac{8}{9}$

13. $\frac{1}{9} \times 12$ 14. $15 \times \frac{2}{3}$ 15. $\frac{8}{15} \times \frac{3}{4}$ 16. $\frac{2}{3} \times \frac{6}{8}$ 17. $\frac{21}{30} \times \frac{5}{7}$ 18. $\frac{2}{15} \times \frac{3}{4}$

Write a fraction for each mixed number.   (Pages 158–159)

19. $6\frac{2}{3}$ 20. $5\frac{4}{7}$ 21. $3\frac{4}{5}$ 22. $7\frac{9}{10}$ 23. $4\frac{1}{15}$ 24. $2\frac{7}{15}$ 25. $7\frac{1}{6}$

Round to the nearest whole number.   (Pages 158–159)

26. $1\frac{1}{4}$ 27. $3\frac{3}{4}$ 28. $5\frac{7}{8}$ 29. $6\frac{1}{2}$ 30. $4\frac{3}{7}$ 31. $13\frac{1}{3}$ 32. $20\frac{5}{9}$

33. $1\frac{7}{9}$ 34. $1\frac{1}{5}$ 35. $5\frac{1}{4}$ 36. $2\frac{5}{6}$ 37. $4\frac{4}{5}$ 38. $3\frac{3}{10}$ 39. $8\frac{2}{9}$

For Exercises 40–47:   **a.** *Estimate the answer.*   **b.** *Find the exact answer.*   **c.** *Compare the exact answer with the estimate to determine whether your answer is reasonable.*   (Pages 158–159)

40. $5\frac{3}{4} \times 2$ 41. $3\frac{3}{5} \times 1\frac{1}{10}$ 42. $2\frac{3}{5} \times 1\frac{1}{6}$ 43. $1\frac{1}{5} \times 4\frac{3}{8}$

44. $4\frac{1}{4} \times 2\frac{2}{3}$ 45. $2\frac{1}{10} \times 3\frac{3}{4}$ 46. $8\frac{1}{8} \times 2\frac{1}{5}$ 47. $5\frac{2}{3} \times 2\frac{1}{4}$

Write the reciprocal of each number.   (Pages 160–161)

48. $\frac{2}{3}$ **49.** $\frac{1}{6}$ **50.** 7   **51.** 5   **52.** $\frac{21}{4}$ **53.** $\frac{5}{8}$ **54.** $\frac{16}{15}$ **55.** 3

Divide. Write each answer in lowest terms.   (Pages 160–161)

56. $\frac{3}{4} \div \frac{3}{5}$ **57.** $\frac{2}{3} \div \frac{2}{5}$ **58.** $\frac{2}{5} \div 4$ **59.** $\frac{9}{10} \div 3$ **60.** $3 \div \frac{6}{10}$ **61.** $8 \div \frac{4}{5}$

62. $\frac{5}{12} \div \frac{1}{3}$ **63.** $\frac{3}{16} \div \frac{3}{4}$ **64.** $\frac{3}{5} \div \frac{9}{10}$ **65.** $\frac{3}{7} \div \frac{9}{14}$ **66.** $\frac{4}{5} \div \frac{8}{15}$ **67.** $\frac{3}{4} \div \frac{1}{8}$

Divide. Write each answer in lowest terms.
(Pages 162–163)

68. $\frac{1}{3} \div 4\frac{2}{3}$ **69.** $\frac{3}{7} \div 6\frac{3}{7}$ **70.** $2\frac{2}{3} \div \frac{10}{21}$ **71.** $3\frac{3}{4} \div \frac{5}{8}$ **72.** $6\frac{1}{8} \div 7$

73. $5\frac{5}{6} \div 14$ **74.** $3\frac{1}{2} \div 1\frac{3}{4}$ **75.** $9\frac{3}{7} \div 5\frac{1}{2}$ **76.** $4\frac{2}{3} \div 1\frac{3}{4}$ **77.** $2\frac{1}{2} \div 3\frac{1}{8}$

For Exercises 78–82, write like fractions to compare.
Write <, =, or >.   (Pages 166–167)

78. $\frac{1}{2}$ ● $\frac{3}{8}$ **79.** $\frac{5}{8}$ ● $\frac{3}{4}$ **80.** $\frac{2}{3}$ ● $\frac{2}{5}$ **81.** $\frac{2}{5}$ ● $\frac{1}{2}$ **82.** $\frac{5}{6}$ ● $\frac{7}{9}$

For Exercises 83–87, write a decimal for each fraction. Then compare.
Write <, =, or >.   (Pages 166–167)

83. $\frac{4}{5}$ ● $\frac{3}{4}$ **84.** $\frac{1}{4}$ ● $\frac{1}{10}$ **85.** $\frac{7}{10}$ ● $\frac{17}{20}$ **86.** $\frac{9}{20}$ ● $\frac{1}{2}$ **87.** $\frac{3}{5}$ ● $\frac{3}{8}$

Part 3: APPLICATIONS

88. Each costume for a play requires $2\frac{1}{2}$ yards of material. How many costumes can be made from 30 yards of material?   (Pages 162–163)

89. Miguel has paddled his canoe $\frac{2}{3}$ of the length of the river. Rosa has paddled $\frac{7}{12}$ of the river's length. Who has the farthest left to paddle? (Pages 166–167)

90. Heather is paid $4.50 per hour for a 36-hour workweek. One week she worked $40\frac{1}{2}$ hours. Find her total pay for the week.   (Pages 168–169)

91. Heathcliff earned $390 last week. The deductions were: federal tax: $32.00; state tax: $9.60; FICA tax: $29.29; insurance: $8.72. Find the net pay.   (Pages 170–171)

Chapter Test

Multiply. Write each answer in lowest terms.

1. $\frac{1}{2} \times \frac{3}{5}$ **2.** $\frac{2}{3} \times 4$ **3.** $5 \times \frac{3}{4}$ **4.** $\frac{2}{3} \times \frac{9}{10}$ **5.** $\frac{5}{7} \times \frac{14}{15}$

6. $\frac{1}{5} \times \frac{3}{8}$ **7.** $\frac{2}{5} \times \frac{5}{8}$ **8.** $6 \times \frac{3}{5}$ **9.** $\frac{4}{5} \times \frac{1}{2}$ **10.** $\frac{4}{21} \times \frac{7}{8}$

Round to the nearest whole number.

11. $7\frac{2}{3}$ **12.** $9\frac{1}{2}$ **13.** $4\frac{1}{16}$ **14.** $10\frac{3}{4}$ **15.** $10\frac{3}{14}$ **16.** $17\frac{6}{11}$ **17.** $24\frac{8}{9}$

For Exercises 18–22:

a. *Estimate the answer.*

b. *Find the exact answer.*

c. *Compare the exact answer with the estimate to determine whether your answer is reasonable.*

18. $1\frac{1}{12} \times 2\frac{4}{5}$ **19.** $4\frac{2}{3} \times 6\frac{3}{7}$ **20.** $1\frac{2}{3} \times 3\frac{1}{5}$ **21.** $1\frac{1}{4} \times 3\frac{1}{2}$ **22.** $4 \times 3\frac{7}{8}$

Divide. Write each answer in lowest terms.

23. $\frac{2}{3} \div \frac{4}{9}$ **24.** $3 \div \frac{3}{8}$ **25.** $\frac{9}{10} \div \frac{3}{5}$ **26.** $\frac{6}{7} \div 12$ **27.** $1\frac{1}{2} \div 2\frac{1}{4}$

28. $6\frac{1}{8} \div 3\frac{1}{16}$ **29.** $2\frac{2}{3} \div 4\frac{1}{8}$ **30.** $5\frac{3}{5} \div 2\frac{1}{3}$ **31.** $12\frac{1}{2} \div 2\frac{1}{2}$ **32.** $4\frac{5}{7} \div 5\frac{1}{2}$

Write the fractions in order from least to greatest.

33. $\frac{1}{3}, \frac{1}{2}, \frac{1}{4}, \frac{1}{6}$ **34.** $\frac{2}{3}, \frac{7}{8}, \frac{4}{5}, \frac{3}{4}$ **35.** $\frac{9}{10}, \frac{5}{6}, \frac{1}{2}, \frac{7}{12}$

Solve each problem.

36. A carpenter wishes to cut 5 strips, each $3\frac{1}{8}$ inches wide, from a plank. The plank is $16\frac{3}{4}$ inches wide. Can the carpenter cut the 5 strips?

37. Henry's net pay last week was $428.55. His total taxes were $115.42. His personal deductions were $11.63. Find Henry's gross pay.

Cumulative Maintenance Chapters 1–7

Choose the correct answer. Choose a, b, c, or d.

1. Add.

$$2.26 + 52.8 + 0.007$$

a. 55.076   b. 55.067
c. 55.167   d. 55.087

2. Subtract.

$$56.007 - 3.87$$

a. 52.137   b. 52.037
c. 50.137   d. 52.130

3. Which number represents thirty and forty-two thousandths?

a. 3042   b. 30.42
c. 30.042   d. 30.0042

4. Multiply.

$$\begin{array}{r} 0.836 \\ \times\ 0.53 \\ \hline \end{array}$$

a. 44.038   b. 44.208
c. 43.208   d. 0.44308

5. Add.

$$\tfrac{1}{3} + \tfrac{2}{5}$$

a. $\frac{3}{8}$ b. $\frac{14}{15}$ c. $\frac{2}{8}$ d. $\frac{11}{15}$

6. For which group of numbers is the mean closest to most of the numbers?

a. 15, 11, 14, 42, 10, 12
b. 10, 12, 9, 11, 1, 10
c. 10, 8, 12, 10, 9, 11
d. 20, 18, 19, 5, 2, 4

7. Divide.

$$38\,\overline{)\,239.4}$$

a. 6.5   b. 63
c. 630   d. 6.3

8. Subtract.

$$\begin{array}{r} 7\frac{5}{11} \\ -\,3\frac{3}{11} \\ \hline \end{array}$$

a. $3\frac{3}{11}$ b. $4\frac{2}{11}$

c. $4\frac{3}{11}$ d. $3\frac{2}{11}$

9. Subtract.

$$\tfrac{7}{15} - \tfrac{1}{3}$$

a. $\frac{4}{15}$ b. $\frac{1}{2}$

c. $\frac{2}{15}$ d. $\frac{6}{15}$

10. Choose the best estimate.

$$10 \div \tfrac{11}{12}$$

a. 8   b. 10
c. 13   d. 15

11. Corn is selling at 6 ears for 96¢. What does one ear cost?

    a. 20¢

    b. 18¢

    c. 16¢

    d. 14¢

12. Multiply.
$$\frac{2}{5} \times \frac{3}{7}$$

    a. $\frac{5}{35}$ **b.** $\frac{6}{35}$

    c. $\frac{5}{12}$ **d.** $\frac{6}{12}$

13. Divide.
$$42 \div 100$$

    a. 0.42   **b.** 4200

    c. 4.2   **d.** 0.042

14. Each apple below represents 25,000 apples. What is the total number of apples represented?

    a. 125,000   **b.** 125,000,000

    c. 12,500   **d.** 1,250,000

15. Subtract.
$$7\tfrac{2}{3}$$
$$-4\tfrac{1}{6}$$

    a. $3\tfrac{1}{3}$ **b.** $3\tfrac{1}{6}$

    c. $3\tfrac{1}{2}$ **d.** $2\tfrac{1}{6}$

16. Which shows the numbers 0.7, 0.007, 7.7, and 0.07 listed in order from least to greatest?

    a. 0.007,   0.7,   7.7,   0.07

    b. 7.7,   0.07,   0.7,   0.007

    c. 0.007,   0.07,   0.7,   7.7

    d. 0.07,   0.7,   0.007,   7.7

17. Choose the best estimate.
$$8.3 \times 11.7$$

    a. 96   **b.** 90

    c. 88   **d.** 99

18. In the last 4 years, the Jets won 12, 13, 10, and 17 games. What is the mean number of games won?

    a. 12   **b.** 13

    c. 17   **d.** 8

19. Stan is paid $7.50 per hour for a 40-hour week. One week he worked 46 hours. What was his total pay for the week?

    a. $345.00   **b.** $300.00

    c. $367.50   **d.** $517.50

Measurement: Perimeter/Area

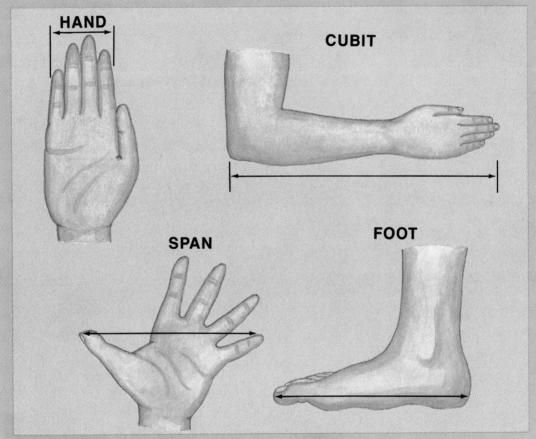

CHAPTER 8

HAND

CUBIT

SPAN

FOOT

Compare each measurement with the results obtained by your classmates. Are the measurements the same?

- Measure the length of your desk in **hands.**

- Measure the height of your desk from the floor in **spans.**

- Use your **feet** to measure the length of your classroom.

- A span, a foot, a hand, and a cubit are called "nonstandard" units of measure. Why are they called "nonstandard"?

Standard Units

In early times, people used parts of the body to measure lengths.

1. Is the board 4 feet long or 6 feet long? Explain.

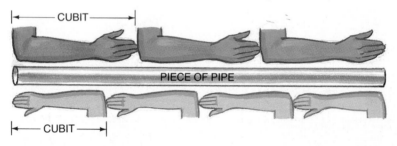

2. Is the pipe 3 cubits long or 4 cubits long? Explain.

ACTIVITY *Use these three different units to measure the length of your desk.*

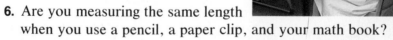

3. Use a pencil.
   Length of desk: __?__ pencils

4. Use a paper clip.
   Length of desk: __?__ paper clips

5. Use your math book.
   Length of desk: __?__ math books

6. Are you measuring the same length when you use a pencil, a paper clip, and your math book?

7. Which unit, the pencil, the paper clip, or your math book, gives the fewest number of units for the length of your desk?

8. Which unit, the pencil, the paper clip, or your math book, gives the greatest number of units for the length of your desk?

9. Why is it important to use **standard units,** such as the inch or the centimeter, to measure length?

10. *Complete:* To find the __?__ of an object, compare its __?__ to the length of a standard unit such as the inch or the centimeter.

EXERCISES

The smallest violin that can be played was made by Morris Samskin of Brooklyn, New York. Find each indicated length for this violin.

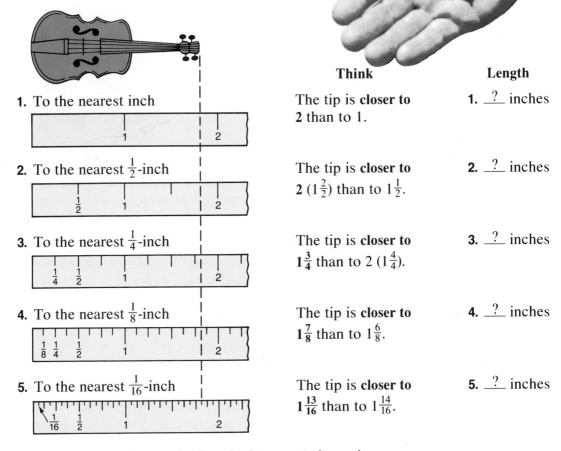

		Think	Length
1. To the nearest inch		The tip is **closer to** 2 than to 1.	**1.** _?_ inches
2. To the nearest $\frac{1}{2}$-inch		The tip is **closer to** 2 ($1\frac{2}{2}$) than to $1\frac{1}{2}$.	**2.** _?_ inches
3. To the nearest $\frac{1}{4}$-inch		The tip is **closer to** $1\frac{3}{4}$ than to 2 ($1\frac{4}{4}$).	**3.** _?_ inches
4. To the nearest $\frac{1}{8}$-inch		The tip is **closer to** $1\frac{7}{8}$ than to $1\frac{6}{8}$.	**4.** _?_ inches
5. To the nearest $\frac{1}{16}$-inch		The tip is **closer to** $1\frac{13}{16}$ than to $1\frac{14}{16}$.	**5.** _?_ inches

Give the length of each object as indicated.

6. To the nearest $\frac{1}{4}$-inch

7. To the nearest $\frac{1}{8}$-inch

Measure the length of each object to the nearest inch, to the nearest $\frac{1}{2}$-inch, to the nearest $\frac{1}{8}$-inch, and to the nearest $\frac{1}{16}$-inch.

8. Nail

9. Bolt

10.

11. Tack

Customary Measures of Length

Some standard **customary units** for measuring length are the inch (in), foot (ft), yard (yd), and mile (mi).

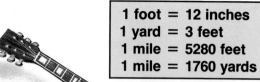

1 foot = 12 inches	
1 yard = 3 feet	
1 mile = 5280 feet	
1 mile = 1760 yards	

1. How many inches are there in one yard?

2. Which customary unit would you use to measure these?

   a. Distance to the sun

   b. Your waist

The world's largest guitar is 8 ft 10 in long.

EXAMPLE 1   How many inches long is this?

Think: Since 1 foot = 12 inches, 8 ft = 8 × 12, or 96 in.

So 8 ft 10 in = 8 ft + 10 in = (96 + 10) in, or 106 in.

The world's tallest guitar is **106 inches** tall.

The world's largest harp is 13 feet 4 inches tall.

3. Is this taller or shorter than the world's tallest guitar?

4. What operation would you use to find how much taller the harp is?

EXAMPLE 2   How much taller is the harp?

$$\begin{array}{cc} 13 \text{ ft} \quad 4 \text{ in} \\ - \ 8 \text{ ft } 10 \text{ in} \end{array} \longrightarrow \begin{array}{c} 12 \text{ ft } 16 \text{ in} \\ - \ 8 \text{ ft } 10 \text{ in} \\ \hline 4 \text{ ft} \quad 6 \text{ in} \end{array}$$

13 ft = 12 ft 12 in
12 in + 4 in = 16 in

$4 \text{ ft} \quad 6 \text{ in} = 4 \text{ ft} + \frac{6}{12} \text{ ft} = 4\frac{1}{2} \text{ ft}$

The harp is 4 ft 6 in, or **$4\frac{1}{2}$ feet** taller than the guitar.

EXERCISES

Which standard customary unit would you use to measure these?

1. Width of a room   **2.** Length of a football field   **3.** Distance to Mars

Which is greater?

4. 15 in or 1 ft

5. 6 ft or 80 in

6. 10 ft or 3 yd

7. 2 mi or 10,000 ft

8. 6000 yd or 5 mi

9. 7 yd or 400 in

Copy and complete.

10. 2 ft 9 in = <u> ? </u> ft

11. 96 in = <u> ? </u> ft

12. 73 ft = <u> ? </u> yd <u> ? </u> ft

13. 9 yd 2 ft = <u> ? </u> ft

14. 3 mi = <u> ? </u> yd

15. 7000 ft = <u> ? </u> mi <u> ? </u> ft

16. 59 in = <u> ? </u> yd <u> ? </u> in

17. 2 yd 5 in = <u> ? </u> in

18. 10 mi = <u> ? </u> ft

Perform the indicated operations.

19.   1 ft 10 in
    $+$ 3 ft 2 in

20.   5 yd 2 ft
    $+$ 1 yd 1 ft

21.   4 yd 1 ft
    $-$ 1 yd 2 ft

22.   2 yd
    $-$ 1 yd 2 ft

23.   2 ft 8 in
    $\times$ 4

24.   46 mi 680 ft
    $\times$ 8

25.   3 yd 1 ft 4 in
    $-$ 1 yd 1 ft 8 in

26.   4 yd 2 ft 9 in
    $+$ 2 yd 2 ft 11 in

27.   1 ft ? in
    3) 4 ft 3 in
         3 ft
         1 ft 3 in $\longrightarrow$ 15 in

28. 3) 29 ft 9 in

29. 5) 13 yd 1 ft

30. 6) 7 mi 4 yd

31. Mauna Loa, a volcano in Hawaii, is 2 miles 3117 feet above sea level. Mount St. Helens, a volcano in Washington, is 1 mile 3084 feet above sea level. How much higher than Mount St. Helens is Mauna Loa?

32. The largest recorded brass instrument is a tuba $7\frac{1}{2}$ feet tall. How much taller is the world's largest harp, which is 13 ft 4 in tall?

33. The traditional **Roman mile** was 1000 double paces of 5 feet each. How many feet were there in 3 "Roman miles"?

|$\leftarrow$————Single Pace————$\rightarrow$|

34. Gray whales generally grow no longer than 16 yards 2 feet. Blue whales can grow up to twice that long. To what length can Blue whales grow?

35. A cheetah can run 121 yards 1 foot in 4 seconds. At this rate, how far can the cheetah run in 1 second?

Metric Units of Length

The **meter** (m) is the base unit in the metric system. One meter is about the width of a door. The height of a kitchen counter is about one meter. This table shows how other metric units of length are related to the meter.

1 kilometer (km) = 1000 meters (m)	1 m = 0.001 km
1 meter = 100 centimeters (cm)	1 cm = 0.01 m
1 centimeter = 10 millimeters (mm)	1 mm = 0.01 cm
1 meter = 1000 millimeters	1 mm = 0.001 m

1. How many meters are there in 1 kilometer? in 2 kilometers?

2. How many millimeters are there in 1 centimeter? in 2 centimeters?

3. Which metric unit would you use to measure these?
   a. Distance to the North Pole
   b. Width of a dime

4. Find the length in metric units of the smallest violin that can be played.

   a. Since the tip of the violin is closer to 5 cm than to 4 cm, its length is _?_ cm (nearest centimeter).

   b. Since the tip of the violin is closer to 47 mm than to 48 mm, its length is _?_ mm (nearest millimeter).

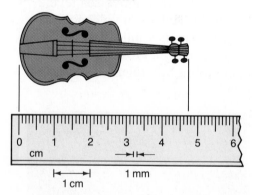

EXERCISES

Which metric unit would you use to measure these? Use km, m, cm, or mm.

1. Length of a dollar bill

2. Height of a building

3. Width of a quarter

4. Length of a toothbrush

5. Your height

6. Distance from Dallas to Chicago

Give the length of each object as indicated.

7. Nearest centimeter

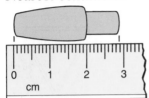

8. Nearest millimeter

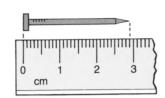

9. Nearest millimeter

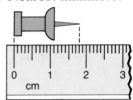

Measure the length of each object to the nearest centimeter and to the nearest millimeter.

10. Chain

11. Spike

12. Bolt

*Choose the most suitable measure. Choose **a, b,** or **c.***

13. The length of a car	**a.** 3 cm	**b.** 3 m	**c.** 3 km
14. The width of a button	**a.** 20 mm	**b.** 20 cm	**c.** 0.5 m
15. The width of a dollar bill	**a.** 6.6 mm	**b.** 6.6 cm	**c.** 0.75 m
16. The distance from Chicago to Los Angeles	**a.** 3400 cm	**b.** 3400 m	**c.** 3400 km

17. A penny is 1.8 centimeters wide. Can you place 14 pennies touching each other in a row so that the total length is 27 centimeters? Explain.

18. Joanne has 300 centimeters of ribbon to decorate 16 flower pots. Each pot will require a ribbon 18.6 centimeters long. Does she have enough ribbon?

Explain these statements in your own words. Give examples to illustrate your explanation.

19. No matter how carefully you measure something, the measurement is never <u>exact</u>.

20. When you give a measurement to the nearest unit, you are actually <u>rounding</u> the measurement.

Perimeter

The **perimeter** of a figure is the distance around it.

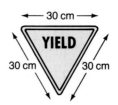

1. What is the perimeter, P, of this traffic sign?
   Complete: $P = 30 + 30 + \underline{\ ?\ }$
   $ = \underline{\ ?\ }$ cm

2. *Complete:* To find the perimeter of a figure, add the lengths of the __?__.

The figure at the right is a **rectangle.**

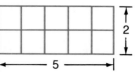

3. *Complete:* The opposite sides of a rectangle have equal __?__.

4. Which of these can you use to find the perimeter of the rectangle?
   - **a.** $P = 5 + 2 + 5 + 2$
   - **b.** $P = 5 + 5 + 2 + 2$
   - **c.** $P = (2 \times 5) + (2 \times 2)$
   - **d.** $P = 2 \times (5 + 2)$

The length of the rectangle at the right is l and the width is w.

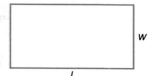

5. Which of these can you use to find the perimeter?
   - **a.** $P = l + w + l + w$
   - **b.** $P = l + l + w + w$
   - **c.** $P = (2 \times l) + (2 \times w)$
   - **d.** $P = 2 \times (l + w)$

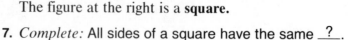

Expressions such as $P = 2 \times (l + w)$ are called **formulas.**

6. Which formula in Exercise 5 would be easiest for you to use? Why?

The figure at the right is a **square.**

7. *Complete:* All sides of a square have the same __?__.

8. Which of these can you use to find the perimeter of the square?
   - **a.** $P = 3 + 3 + 3 + 3$
   - **b.** $P = 4 \times 3$
   - **c.** $P = 2 \times (3 + 3)$
   - **d.** $P = 3 \times 3$

The length of each side of this square is s units.

9. *Complete:* $P = s + s + s + \underline{\ ?\ }$
   $ P = 4 \times \underline{\ ?\ }$

EXERCISES

Find the perimeter of each figure.

1. Postage Stamp

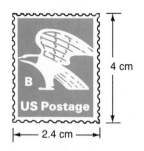

4 cm

2.4 cm

2. Pendant

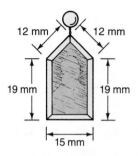

12 mm   12 mm

19 mm   19 mm

15 mm

3. Swimming Pool

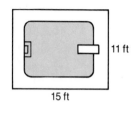

11 ft

15 ft

4. Baseball Diamond

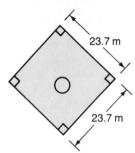

23.7 m

23.7 m

5. City Square

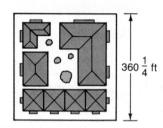

$360\frac{1}{4}$ ft

6. Garden

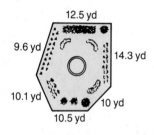

12.5 yd

9.6 yd   14.3 yd

10.1 yd   10 yd

10.5 yd

Complete.

	Length	Width	Perimeter
7.	$6\frac{1}{2}$ in	3 in	?
8.	8.2 cm	5.1 cm	?
9.	9 yd	?	24 yd

	Length	Width	Perimeter
10.	?	75 m	500 m
11.	12 ft	?	35 ft
12.	?	2.8 m	13.6 m

13. Find the perimeter of the roof shown below.

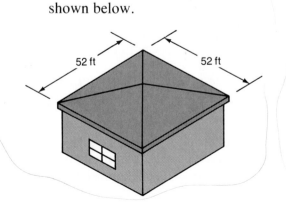

52 ft   52 ft

14. Two tennis courts are laid out side-by-side as shown below. The perimeter is 116.6 meters. Find the width.

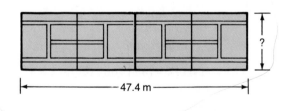

?

47.4 m

The plan below shows the dimensions of some of the rooms of the first floor of a house. For example, the dimensions 4.2 m × 4 m for the dining room mean that it is 4.2 meters long and 4 meters wide. Use this plan for Exercises 15–18.

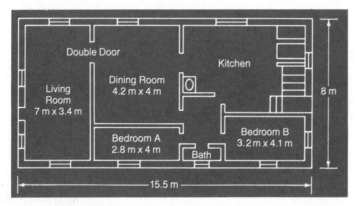

15. What is the perimeter of the living room?

16. What is the perimeter of the outside walls?

17. How much greater is the perimeter of the living room than the perimeter of bedroom B?

18. How much greater is the perimeter of the dining room than the perimeter of bedroom A?

For Exercises 19–22, first draw a rectangle or a square. Use the information in the problem to mark lengths of the sides. Then solve the problem.

19. The rectangular building in which spacecraft are assembled at Cape Canaveral in Florida is 199.33 meters long and 147.47 meters wide. Find the perimeter.

20. The base of the Great Pyramid of Egypt is a square. The distance around the base is 945.6 meters. How long is each side?

21. A square rug has a perimeter of 30 yards. How much will it cost to repair the fringe on one side of the rug at a cost of $15.70 per yard?

22. A school's rectangular swimming pool is 540 feet around and 90 feet wide. How many feet would be traveled by a swimmer who swims 5 laps (1 lap is 2 lengths of the pool)?

For Exercises 23–24, use graph paper to make drawings that illustrate each answer. Use whole numbers only for the lengths of the sides of each figure.

23. Draw as many different rectangles as you can that have a perimeter of 24 units.

24. Draw as many figures as you can that have a perimeter of 60 units. In each figure, the sides must be equal in length.

Mid-Chapter Review

Perform the indicated operations.   (Pages 182–183)

1.   7 ft 9 in
     + 3 ft 4 in

2.   6 yd 1 ft
     − 2 yd 2 ft

3.   1 ft 7 in
     ×   3

Measure the length of each object to the nearest centimeter and to the nearest millimeter.   (Pages 184–185)

4. Screwdriver

5. Straw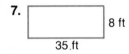

Find the perimeter of each figure.   (Pages 186–188)

6. 15.2 cm
     15.2 cm

7. 8 ft
     35 ft

8.
4.0 m
8.4 m
3.9 m
3.6 m
5.3 m

For Exercises 9–10, first draw a rectangle or square. Use the information in the problem to mark lengths of the sides. Then solve the problem.   (Pages 186–188)

9. A rectangular parking garage is 102.4 meters long and 48.6 meters wide. Find the perimeter.

10. The base of a monument is a square. The distance around the base is 23.2 meters. How long is each side?

MAINTENANCE

11. Add. (Pages 28–29)
43.6 + 13.5 + 11.02

12. Multiply. (Pages 41–42)
6.21 × 13.4

13. Divide. (Pages 68–69)
$0.06 \overline{)\, 2.4}$

Multiply or divide.   (Pages 155–156 and 160–163)

14. $\frac{1}{4} \times \frac{4}{10} = \underline{\ ?\ }$

15. $\frac{3}{4} \div \frac{1}{8} = \underline{\ ?\ }$

16. $4\frac{1}{2} \div 3 = \underline{\ ?\ }$

17. Two posters of football teams cost $6.75. One poster costs $0.75 more than the other. How much does each poster cost?
(Pages 28–29)

18. The distance from Ron's office to the home of a client is 142 miles. He makes a round trip to visit the client once each month. How many miles does he travel in one year going to and from his client's home?
(Pages 32–33)

Math and Formulas

Top-Drive rentals uses this formula to compute the cost, C, of renting a car.

$$C = 0.22 \times n + 52.50$$

The formula shows the sum of two costs.

a. A **base fee** ($52.50) that varies with the size of the car you decide to rent.

b. A **mileage fee** that varies with the number of miles you drive ($0.22 \times n$).

EXAMPLE Dan rented a car from Top-Drive Rentals to travel to a tennis tournament. The base fee was $52.50, and Dan drove 512 miles. How much did he owe?

Think: First write the formula.

$$C = 0.22 \times n + 52.50$$

Then replace n with 512.

$$C = 0.22 \times 512 + 52.50$$

 165.14

Dan paid **$165.14** to rent the car.

EXERCISES

Use the formula $C = 0.22 \times n + 52.50$ to find the cost of renting a car to travel each number of miles.

1. 400

2. 155

3. 1000

4. 362

5. Gregg rented a car from Top-Drive Rentals and drove it 715 miles. He paid $50 in advance. How much did he owe when he returned the car?

6. Pam drove the car she rented from Top-Drive Rentals a distance of 530 miles. Because she rented a large car, she was charged a base fee of $60. What formula should she use to compute the total charges?

Area: RECTANGLES AND SQUARES

The world's smallest republic is the country of Nauru (NAH roo) in the South Pacific Ocean. Nauru has an area of about 8 square miles (mi²).

Area is the number of square units needed to cover a surface. For example, each side of this square is 1 unit long. Its area is 1 **square unit**.

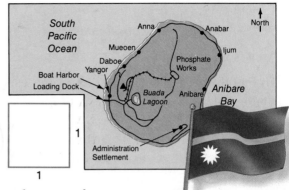

Activity
Copy the table. Draw each rectangle or square on graph paper. Then complete the table.

Figure	Length	Width	Count the number of square units.	Multiply the length and width.
Rectangle	8	2	_?_ square units	_?_ square units
Rectangle	3	?	15 square units	_?_ square units
Square	4	?	_?_ square units	_?_ square units
Square	?	6	_?_ square units	_?_ square units

1. Complete this formula for the area of a rectangle.

   $A = \underline{\ ?\ } \times \underline{\ ?\ }$

   A = area
   l = length
   w = width

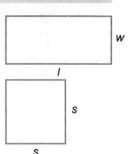

2. Complete this formula for the area of a square.

   $A = \underline{\ ?\ } \times \underline{\ ?\ }$

   A = area
   s = side of the square

EXAMPLE | Find the area of each figure.

a. 20 yd, 119 yd

b. 1.2 m, 1.2 m

$A = l \times w$ ⟵ Formulas ⟶ $A = s \times s$ $s \times s = s^2$

$A = 119 \times 20$ $A = 1.2 \times 1.2$

$A = 2380 \text{ yd}^2$ $A = 1.44 \text{ m}^2$

NOTE: Square yards is written as yd^2, square meters is written as m^2, square inches is written as in^2, and so on.

EXERCISES

Complete. Choose your answers from the box at the right.

1. Area is measured in __?__.

2. To find the area of rectangles and squares, you __?__ the length and the width.

3. The formula for the area of a square with sides of length, s, is __?__.

4. The formula for the area of a rectangle is __?__.

5. The perimeter, P, of a square with each side of length, s, is __?__ units.

6. The perimeter, P, of a rectangle with length, l, and width, w, is __?__.

> add
> $4 \times s$
> $2 \times (l + w)$
> multiply
> square units
> $A = l \times w$
> $A = s \times s$

Find the area of each figure.

7. Gym Floor

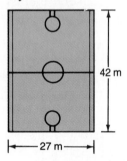

42 m

27 m

8. Solar Panel

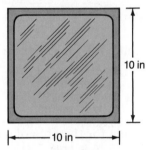

10 in

10 in

9. Rug

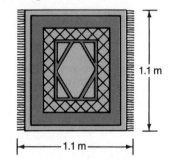

1.1 m

1.1 m

10. Dollar Bill

$2\frac{1}{2}$ in

6 in

11. Business Card

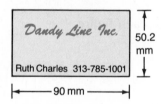

Dandy Line Inc.

Ruth Charles 313-785-1001

50.2 mm

90 mm

12. Stamp

USA   32¢

$\frac{7}{8}$ in

$\frac{7}{8}$ in

Complete.

	Length	Width	Area
13.	4 cm	3 cm	_?_ cm²
14.	5 m	4.6 m	_?_ m²
15.	8 mi	_?_ mi	64 mi²
16.	3 yd	_?_ yd	7 yd²
17.	_?_ km	1.5 km	6 km²

	Length	Width	Area
18.	2 yd	2 yd	_?_ yd²
19.	13 mm	2.8 mm	_?_ mm²
20.	_?_ in	2.8 in	21 in²
21.	5 mm	_?_ mm	7.5 mm²
22.	$2\frac{1}{2}$ ft	$\frac{4}{5}$ ft	_?_ ft²

For Exercises 23–24, draw a rectangle to fit each description.
Use graph paper if you wish.

23. Perimeter: 16 units
Area: 15 square units

24. Perimeter: 12 units
Area: 8 square units

Complete. In these exercises, s and w represent whole numbers.

25.

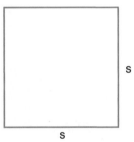

s

s

Area: 16 m² $s =$ _?_ m

26.

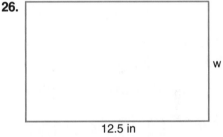

w

12.5 in

Area: 100 in² $w =$ _?_ in

27. The flag that hangs from the George Washington Bridge on holidays is 18 meters wide and 27 meters long. Find its area.

28. The top of a rectangular table is 1 meter long and has an area of 0.8 square meters. What is the width?

29. Floor tiles are sold in boxes containing 10 tiles. Each tile is a square with sides that are 12 inches long. Find the number of square inches that can be covered by one box of floor tiles.

Strategy: SOLVING A SIMPLER PROBLEM

The students at Alamo High School are holding a raffle to raise funds for their school library. The four winners will share $3000 according to these rules.

Second Prize: Three–fifths of the first prize
Third Prize: One–half of the second prize
Fourth Prize: One–third of the third prize

EXAMPLE How much will each winner receive?

READ What are the facts?

Total for prizes: $3000
The total will be shared according to the rules.

PLAN Solve a simpler problem; that is, use smaller numbers.

Then use the results to solve the original problem.

SOLVE 1 Assume the first prize is $100.

First prize:	$100	
Second prize:	60	← $\frac{3}{5} \times \$100$
Third prize:	30	← $\frac{1}{2} \times \$60$
Fourth prize:	10	← $\frac{1}{3} \times \$30$
Total:	**$200**	

2 Compare the total with $3000.

Think: Since $3000 \div \$200 = 15$, multiply each prize in Step 1 by 15.

First prize:	$1500	← $15 \times \$100$
Second prize:	900	← $15 \times \$60$
Third prize:	450	← $15 \times \$30$
Fourth prize:	150	← $15 \times \$10$
Total:	**$3000**	

CHECK Did you use the facts correctly in the problem?

EXERCISES

Last year, Alamo High School also held a raffle for the library fund. The four winners shared $4500 according to the same rules as this year. How much was each prize?

1. First prize   **2.** Second prize   **3.** Third prize   **4.** Fourth Prize

Use the rules on page 194 to determine each prize if the four winners share $2500.

5. First prize   **6.** Second prize   **7.** Third prize   **8.** Fourth Prize

These rules were used by Alamo's town library to raise funds for the science-fiction section.

> **Second Prize:** Two–thirds of the first prize
> **Third Prize:** One–half of the second prize
> **Fourth Prize:** One–fourth of the third prize

9. If the first prize is $600, what is the total prize money?

10. If the fourth prize is $500, what is the first prize?

11. If the third prize is $1200, what is the second prize?

12. If the second prize is $8,000, what is the total prize money?

13. The four winners will share $10,000. How much will each receive?

14. The four winners will share $15,000. How much will each receive?

There are 10,000 different names in a drawing. Yours is one of them. The chances of your winning a prize when the first name is picked are 1 out of 10,000, or $\frac{1}{10,000}$.

15. Your name is not picked on the first drawing. What are your chances of winning a prize when the second name is picked?

PROJECT

a. Work in small groups or as your teacher directs to write rules for a drawing having five prizes.

b. Use the rules to write three problems for a drawing for prizes.

Area: PARALLELOGRAMS/TRIANGLES

A **parallelogram** can be changed into a rectangle of equal area.

Parallelogram

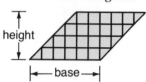

Cut off this ▶ part

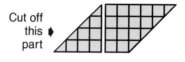

Rectangle

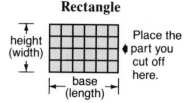

Place the ◀ part you cut off here.

ACTIVITY Copy the table below.

  a. Use graph paper to draw each parallelogram.

  b. Count the squares to estimate the area of each parallelogram. Round the area to the nearest whole number when necessary.

  c. Change each parallelogram into a rectangle of equal area.

  d. Complete the table.

Parallelogram				Rectangle			
	Base	Height	Estimated Area (square units)		Base	Height	Area (square units)
I.	7	5	?	I.	?	?	?
II.	4	9	?	II.	?	?	?
III.	10	8	?	III.	?	?	?

1. Complete this formula for the area of a parallelogram.

$$A = \underline{\ ?\ } \times \underline{\ ?\ }$$

A = area
b = base
h = height

EXAMPLE 1 | Find the area of this parallelogram.

$A = b \times h$ {
b = 5 in
h = 9 in
A = ?

$A = 5 \times 9$
$A = 45$ The area is **45 in²**.

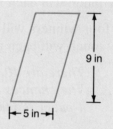

9 in

5 in

A parallelogram can be changed into two triangles of equal area.

2. Estimate the area of the parallelogram.

3. Estimate the area of Triangle I. Count the units of area.

4. Estimate the area of Triangle II. Count the units of area.

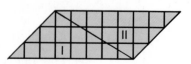

5. Complete the formula for the area of a triangle.

$$A = \frac{1}{2} \times \underline{\ ?\ } \times \underline{\ ?\ }$$

A = **area**
b = **base**
h = **height**

EXAMPLE 2 | Find the area of this sail.

$$A = \frac{1}{2} \times b \times h$$

b = **6 m**
h = **8.6 m**

$$A = \frac{1}{2} \times 6 \times 8.6$$

$$A = 25.8 \qquad \text{The area is } \textbf{25.8 m}^2.$$

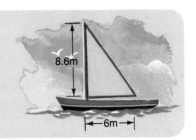

EXERCISES

Complete. Choose your answers from the box at the right.

1. Rectangle

$P = \underline{\ ?\ } ; A = \underline{\ ?\ }$

2. Square

$P = \underline{\ ?\ } ; A = \underline{\ ?\ }$

$P = 4 \times a$
$P = 2 \times (a + b)$
$P = a + b + c$
$P = a + b$
$A = a \times a$
$A = \frac{1}{2} \times a \times h$
$A = a \times b$
$A = a \times h$

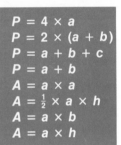

3. Parallelogram

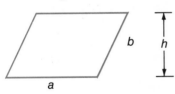

$P = \underline{\ ?\ } ; A = \underline{\ ?\ }$

4. Triangle

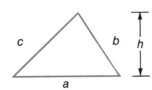

$P = \underline{\ ?\ } ; A = \underline{\ ?\ }$

Find the area of each parallelogram or triangle.

5. Flag

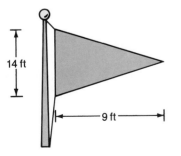

6. Parking Lot

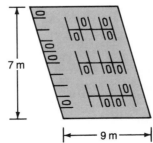

7. Gable of a Roof

MEASUREMENT: PERIMETER/AREA   **197**

8. Tie Back

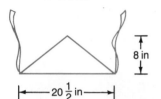

8 in

20 ½ in

9.

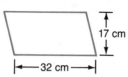

17 cm

32 cm

10.

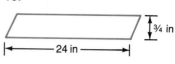

¾ in

24 in

11. Fish Net

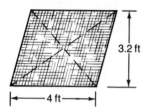

3.2 ft

4 ft

12.

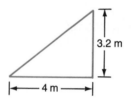

3.2 m

4 m

13.

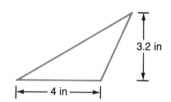

3.2 in

4 in

Figure	Base	Height	Area
14. Parallelogram	4 cm	6 cm	?
15. Triangle	3 in	4 in	?
16. Triangle	6 m	4.2 m	?

Figure	Base	Height	Area
17. Parallelogram	5 ft	?	30 ft²
18. Triangle	$3\frac{1}{2}$ yd	2 yd	?
19. Parallelogram	7.9 m	4.6 m	?

20. The sides of the Transamerica Building in San Francisco are triangular in shape. The base of each triangle is 34.7 meters and the height is 257 meters. What is the area of one side of the building?

21. Each side of a steeple has the shape of a triangle. The base of each triangle is 3.6 meters long and the height is 55 meters. Find the total area of the four sides of the steeple.

22. Mr. and Mrs. Sands installed out-door carpet on their boat dock as shown at the right. The carpet costs $9 per square foot. What is the total cost?

23. Samantha paid $344 to have tile installed on the floor of her patio. Each tile was a square, 12 inches on a side. How much did she pay for each tile?

Transamerica Building
San Francisco

3 ft

6 ft

12 ft

20 ft

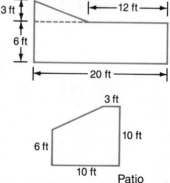

3 ft

10 ft

6 ft

10 ft

Patio

Math in Language Arts

Joyce spent her summer vacation with a team of archeologists in Illinois. This is an entry from her diary.

Week of July 24

Our team's digging site is rectangular in shape. It's dimensions are 2.2 meters by 1.4 meters. A metal detector located artifacts at a depth of 1.7 meters. After finding nothing for 3 days, I uncovered an arrowhead that is almost a perfect triangle at a depth of 2.1 meters. The base of the arrowhead is about 6 centimeters long. The height of the arrowhead is about 3.2 centimeters.

EXERCISES

1. What was the perimeter of the digging site?

2. What was the area of the digging site?

3. How much deeper was the arrowhead than the artifacts located by the detector?

4. What was the area of the arrowhead?

5. The arrowhead Joyce found was made of stone. Joyce knew that native Americans made almost no stone arrowheads before 500 A.D. What is the earliest century in which the maker of the arrowhead could have lived?

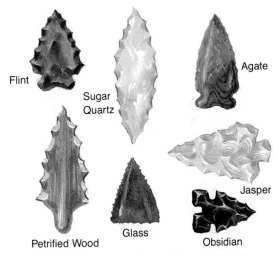

Flint

Sugar Quartz

Agate

Petrified Wood

Glass

Jasper

Obsidian

Native Americans found countless materials for their arrowheads. They used such hard materials as slate, basalt, quartz, glass, agate, and copper.

PROJECT Do research on one of these archeological sites or on a site of your own choice.

Valley of the Kings   Mesopotamia   Mayan Indians

Write a report on some of the artifacts found. Make drawings or a poster to illustrate your report.

Area: TRAPEZOIDS

You can use the formula for the area of a triangle to find the area of this **trapezoid.**

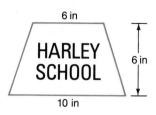

1. Draw one line to separate the trapezoid into 2 triangles.

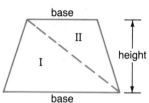

2. What is the area of Triangle I?

$A = \frac{1}{2} \times b \times h$ ◀ $\begin{array}{l} b = 10 \\ h = 6 \\ A = ? \end{array}$

$A = \frac{1}{2} \times 10 \times 6$

$A = \underline{\ ?\ } \ \text{in}^2$

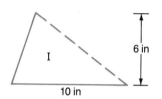

3. What is the area of Triangle II?

$A = \frac{1}{2} \times b \times h$ ◀ $\begin{array}{l} b = 6 \\ h = 6 \\ A = ? \end{array}$

$A = \frac{1}{2} \times \underline{\ ?\ } \times 6$

$A = \underline{\ ?\ } \ \text{in}^2$

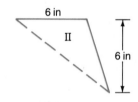

4. What is the total area?

$30 \ \text{in}^2 + 18 \ \text{in}^2 = \underline{\ ?\ } \ \text{in}^2$ ◀ **Area of the trapezoid**

EXERCISES

Complete. Choose your answers from the box at the right.

1. A trapezoid has __?__ bases.

2. The height of a trapezoid is the distance between the __?__ .

3. You can draw a line that separates a trapezoid into __?__ triangles.

4. Each triangle into which the trapezoid is divided has the same __?__ .

5. Each triangle into which the trapezoid is divided will <u>not</u> have the __?__ area.

6. The bases of a trapezoid have different __?__ .

7. Area of a Trapezoid = Area of __?__ + Area of __?__ .

bases
height
lengths
Triangle I
Triangle II
two
same

Find the area of each trapezoid.

8. Side of a Wheelbarrow

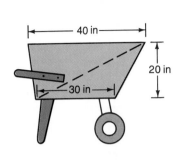

9. Window Box

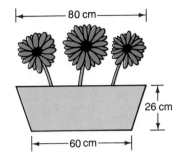

10. Side of a Tool Shed

	Upper Base	Lower Base	Height	Area
11.	4 in	6 in	4 in	?
12.	2 cm	4 cm	5 cm	?
13.	9 yd	6 yd	2 yd	?

	Upper Base	Lower Base	Height	Area
14.	9.4 m	10.3 m	9.2 m	?
15.	$16\frac{1}{2}$ ft	$21\frac{1}{2}$ ft	10 ft	?
16.	6.3 cm	8.5 cm	4 cm	?

17. Use graph paper to draw a trapezoid and a triangle each having an area of 60 square units.

18. Use graph paper to draw a rectangle and a trapezoid each having an area of 80 square units.

19. Find the amount of waste in the piece of metal shown below. Round your answer to the nearest whole number.

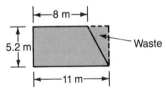

20. The Pooles are building a swimming pool. The shaded area will be a deck. Find the area of the deck.

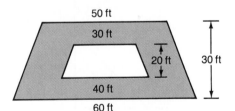

21. A landscape architect made this design for a garden. Find the total area of the garden.

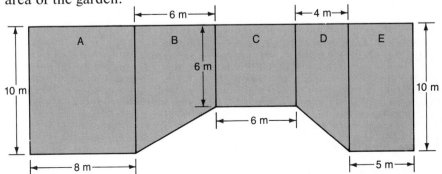

Squares in Design

Geometric figures such as the square often appear in designs of floor tiles, bridges, windows, cloth patterns, wallpaper patterns, and so on.

How would you find the area of the shaded part of this square?

1. *Complete:*

$$\begin{array}{ccc} \text{Area of Shaded} \\ \text{Part} \end{array} = \begin{array}{c} \text{Area of Larger} \\ \text{Square} \end{array} - \begin{array}{c} \text{Area of} \\ \underline{\quad?\quad} \end{array}$$

2. Area of larger square: $4 \times 4 = \underline{\ ?\ }$ cm^2

3. Area of smaller square: $2 \times 2 = \underline{\ ?\ }$ cm^2

4. Area of shaded part: 16 cm$^2 - 4$ cm$^2 = \underline{\ ?\ }$

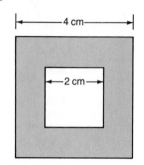

EXERCISES

Find the area of the shaded part of each figure. Round answers to the nearest square unit.

1.

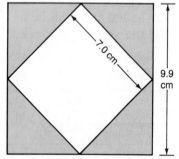

2.

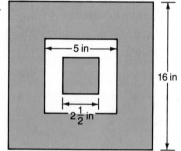

3.

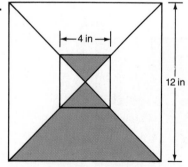

4.

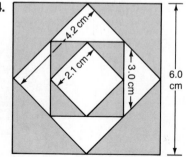

Chapter Summary

1. To find the perimeter, P, of a rectangle:
   1. Write the formula $P = 2 \times (l + w)$.
   2. Add the length, l, and width, w. Then multiply the sum by 2.

2. To find the perimeter, P, of a square:
   1. Write the formula $P = 4 \times s$.
   2. Multiply the length of a side, s, by 4.

3. To find the area, A, of a rectangle:
   1. Write the formula $A = l \times w$.
   2. Multiply the length, l, and the width, w.

4. To find the area, A, of a square:
   1. Write the formula $A = s \times s$ or $A = s^2$.
   2. Multiply the length of any two sides, s.

5. To find the area, A, of a parallelogram:
   1. Write the formula $A = b \times h$.
   2. Multiply the base, b, and the height, h.

6. To find the area, A, of a triangle:
   1. Write the formula $A = \frac{1}{2} \times b \times h$.
   2. Multiply the base, b, and the height, h. Then multiply by $\frac{1}{2}$.

7. To find the area of a trapezoid:
   1. Separate the trapezoid into two triangles. Find the area of each triangle.
   2. Find the sum of the areas of the triangles.

1.
6 m
20 m
$P = 2 \times (l + w)$
$P = 2 \times (20 + 6) =$ **52 m**

2.
$P = 4 \times s$
$P = 4 \times 3 =$ **12 cm**
3 cm

3.
5 m
8 m
$A = l \times w$
$A = 8 \times 5$
$A =$ **40 m²**

4.
4 mm
$A = s \times s$
$A = 4 \times 4$
$A =$ **16 mm²**

5.
5 m
6 m
$A = b \times h$
$A = 6 \times 5$
$A =$ **30 m²**

6.
4 cm
3 cm
$A = \frac{1}{2} \times 3 \times 4$
$A = \frac{1}{2} \times 12$
$A =$ **6 cm²**

7.
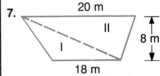
20 m
II
8 m
I
18 m
Triangle I: $A =$ **72 m²**
Triangle II: $A =$ **80 m²**
Trapezoid: $A = 72 + 80$
$A =$ **152 m²**

Chapter Review

Part 1: VOCABULARY

For Exercises 1–10, choose from the box at the right the word(s) that complete(s) each statement.

1. A centimeter is a unit of __?__ .   (Page 184)

2. The __?__ of a figure is the distance around it.   (Page 186)

3. The opposite sides of a __?__ have equal lengths. (Page 186)

4. All sides of a square have the __?__ length.   (Page 186)

5. A formula for the perimeter of a rectangle is __?__ . (Page 186)

6. The formula for the perimeter of a square is __?__ . (Page 186)

7. Area is measured in __?__ .   (Page 191)

8. The formula for the area of rectangle is __?__ .   (Page 191)

9. The formula for the area of a triangle is __?__ .   (Page 197)

10. To find the area of a trapezoid, separate it into __?__ triangles.   (Page 200)

> two
> perimeter
> length
> $P = 4 \times s$
> $A = \frac{1}{2} \times b \times h$
> same
> $A = l \times w$
> rectangle
> square units
> $P = 2 \times (l + w)$

Part 2: SKILLS

Perform the indicated operations.   (Pages 182–183)

11.
$$\begin{array}{r} 1 \text{ ft } 9 \text{ in} \\ + 5 \text{ ft } 3 \text{ in} \\ \hline \end{array}$$

12.
$$\begin{array}{r} 6 \text{ yd } 1 \text{ ft} \\ - 3 \text{ yd } 2 \text{ ft} \\ \hline \end{array}$$

13.
$$\begin{array}{r} 2 \text{ ft } 5 \text{ in} \\ \times \quad 4 \\ \hline \end{array}$$

14. $5 \overline{) 12 \text{ mi } 420 \text{ ft}}$

15.
$$\begin{array}{r} 5 \text{ yd } 1 \text{ ft } 6 \text{ in} \\ - 1 \text{ yd } 2 \text{ ft } 9 \text{ in} \\ \hline \end{array}$$

16.
$$\begin{array}{r} 8 \text{ yd } 2 \text{ ft } 10 \text{ in} \\ + 5 \text{ yd } 2 \text{ ft } 11 \text{ in} \\ \hline \end{array}$$

Measure the length of each object to the nearest centimeter and to the nearest millimeter.   (Pages 184–185)

17. Match

18. Key

Choose the most suitable measure. Choose a, b, or c.   (Pages 184–185)

19. The diameter of a baseball   **a.** 8 mm   **b.** 8 cm   **c.** 8 m

20. The height of a tall building   **a.** 381 mm   **b.** 381 cm   **c.** 381 m

Find the perimeter.   (Pages 186–188)

21. Book Cover

Health
and
Fitness

11 in

6 in

22. Table Top

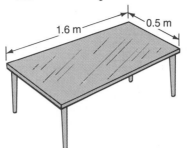

1.6 m   0.5 m

23. Checker Board

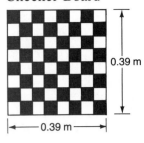

0.39 m

0.39 m

Find the area of each figure.   (Pages 191–193 and 196–198)

	Figure	Length	Width
24.	Rectangle	9 cm	5 cm
25.	Rectangle	26 ft	14 ft
26.	Square	23 in	23 in
27.	Square	5.7 cm	5.7 cm

	Figure	Base	Height
28.	Parallelogram	9 ft	3 ft
29.	Parallelogram	9.3 cm	12.5 cm
30.	Triangle	3.6 m	4.2 m
31.	Triangle	22 in	11 in

Find the area of each trapezoid.   (Pages 200–201)

32. Piece of Slate

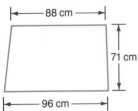

88 cm

71 cm

96 cm

33. Trough

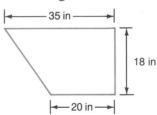

35 in

18 in

20 in

34. Side of a Stairway

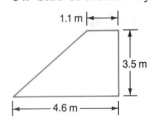

1.1 m

3.5 m

4.6 m

Part 3: APPLICATIONS

35. A carpet is 180 centimeters long and 135 centimeters wide. Find the perimeter of the carpet. (Pages 186–188)

36. One side of a square pool is 5.4 meters long. Find the area of the pool.   (Pages 191–193)

37. Four winners in a contest share $15,000. The second prize is three-fifths of the first prize. The third prize is one-half of the second prize. The fourth prize is one-third of the third prize. How much will each winner receive? (Pages 194–195)

38. Each side of a steeple has the shape of a triangle. The base of the triangle is 4.8 meters long and the height is 45 meters. Find the area of one side of the steeple.   (Pages 196–198)

Chapter Test

Perform the indicated operations.

1. 8 yd 4 ft
      + 2 yd 9 ft

2. 3 ft 6 in
      ×   3

3. 6 yd 1 ft 5 in
      − 2 yd 2 ft 8 in

Measure the length of each object to the nearest centimeter and to the nearest millimeter.

4. Thermometer

5. Eyedropper

Find the perimeter of each rectangle or square.

Figure	Length	Width
6. Rectangle	1.4 m	0.8 m
7. Square	30 in	30 in

Figure	Length	Width
8. Square	24 yd	24 yd
9. Rectangle	15 cm	2 cm

Find the area.

Figure	Length	Width
10. Rectangle	2.3 mm	1.9 mm
11. Square	7 ft	7 ft

Figure	Length	Width
12. Square	3.1 m	3.1 m
13. Rectangle	10 mm	6.2 mm

Find each area.

Figure	Base	Height
14. Triangle	6 yd	5 yd
15. Parallelogram	4.1 cm	5.6 cm

Figure	Base	Height
16. Parallelogram	1.9 m	2 m
17. Triangle	4.2 m	2.9 m

18. Use graph paper to draw a rectangle having a perimeter of 26 units and an area of 40 square units.

19. Three winners share $3,500. Each receives one-half the amount of the next higher prize. What is the third prize?

20. A patio having the shape of a trapezoid has a height of 2 meters. The bases of the trapezoid are 3 meters and 2.5 meters. Find the area of the patio.

Cumulative Maintenance   Chapters 1–8

Choose the correct answer. Choose a, b, c, or d.

1. Add.

$$83 + 296 + 2352$$

a. 2831   b. 2741
c. 2731   d. 2841

2. Subtract.

$$56.007 - 3.87$$

a. 52.137   b. 52.037
c. 50.137   d. 52.130

3. Multiply 0.56 by 0.07.

a. 0.392   b. 0.0392
c. 0.00392   d. 392

4. Divide.

$$1.8\overline{)5.508}$$

a. 30.6   b. 3.006
c. 306   d. 3.06

5. Add.

$$\frac{1}{3} + \frac{2}{5}$$

a. $\frac{3}{8}$ b. $\frac{3}{15}$ c. $\frac{2}{8}$ d. $\frac{11}{15}$

6. Sam gathered 12 buckets of scallops, Kirstin gathered 9 buckets of scallops, and José gathered 6 buckets. They sold the scallops for $4 a bucket and split the money evenly. How much did each person earn?

a. $27   b. $36
c. $108   d. $30

7. At which price will one pear cost the least?

a. 3 for 35¢   b. 4 for 50¢
c. 5 for 65¢   d. 6 for 75¢

8. Which shows the fractions listed in order from least to greatest?

a. $\frac{2}{3}, \frac{3}{8}, \frac{4}{5}, \frac{5}{6}$

b. $\frac{2}{3}, \frac{4}{5}, \frac{5}{6}, \frac{3}{8}$

c. $\frac{3}{8}, \frac{4}{5}, \frac{5}{6}, \frac{2}{3}$

d. $\frac{3}{8}, \frac{2}{3}, \frac{4}{5}, \frac{5}{6}$

9. Robin earns $6 per hour. How much does she earn by working $7\frac{1}{2}$ hours?

a. $45   b. $40
c. $42   d. $48

10. Multiply: $\frac{3}{4} \times \frac{8}{9}$

a. $\frac{1}{3}$ b. $\frac{3}{4}$ c. $\frac{2}{3}$ d. $\frac{1}{2}$

11. How many millimeters long is this tie bar (nearest millimeter)?

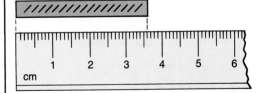

a. 40 mm   b. 36 mm
c. 3.6 mm   d. 35 mm

12. Write a mixed number for $\frac{57}{12}$.

a. $4\frac{3}{4}$ **b.** $4\frac{7}{12}$

c. $5\frac{3}{12}$ **d.** $5\frac{1}{12}$

---

13. What is the area in square meters of this rectangular pool?

a. 40
b. 45
c. 14
d. 28

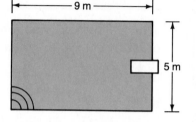

←——— 9 m ———→

5 m

---

14. Subtract.
$$7\frac{2}{3} - 2\frac{1}{6}$$

a. $5\frac{1}{2}$ **b.** $5\frac{1}{3}$

c. $5\frac{1}{6}$ **d.** $4\frac{1}{6}$

---

15. Divide.
$$16 \div \frac{1}{4}$$

a. 24   **b.** 4
c. 64   **d.** 12

---

16. What is the perimeter of a triangle whose sides are 8, 6, and 11?

a. 25   **b.** 24
c. 60   **d.** 42

---

17. Written in words, 7.08 is

a. Seven and eight tenths
b. Seven and eight hundredths
c. Seven and eight thousandths
d. Seven hundred eight

---

18. Multiply.
$$3\frac{1}{2} \times 3\frac{1}{7}$$

a. $5\frac{1}{2}$ **b.** 10

c. 11   **c.** $9\frac{1}{14}$

---

19. In the last 5 years, the Reds won 9, 6, 14, 12, and 19 games. What is the mean number of games won?

a. 10   **b.** 15
c. 12   **d.** 8

---

20. One Thursday, Bob worked from 7:30 A.M. until 6:00 P.M. How many hours did he work?

a. $13\frac{1}{2}$ **b.** $12\frac{1}{2}$

c. $10\frac{1}{2}$ **d.** $11\frac{1}{2}$

---

21. This bar graph shows the number of tickets sold for a school dance. Who sold more than 15 tickets?

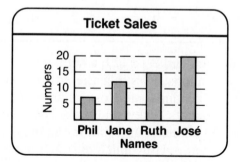

Ticket Sales

Numbers

20
15
10
5

Phil Jane Ruth José
Names

a. José   **b.** Jane
c. Phil   **d.** Ruth

Equations/Ratio and Proportion

A survey showed that 3 out of every 5 teen-age TV viewers watch the Burt Silver Comedy Show on Thursday nights.

- Does this mean that only 5 teen-age TV viewers were surveyed?

- Does this mean that only 3 teen-age TV viewers watch the Burt Silver Comedy Show on Thursday nights?

- Suppose that 500 teen-agers were actually surveyed. How many watched the Burt Silver Show?

- There are 30 students in your class who watch TV on Thursday nights. About how many would you expect to watch the Burt Silver Comedy Show?

Solving Equations: ADDITION

Jeff had several baseball cards. After getting two more from a friend he had a total of 7. How many did he have originally?

You can represent the problem in two ways.

	Number Jeff had	+	Two more	=	Seven in all
Model	■	+	△△	=	△△△△ △△△
Equation	n	+	2	=	7

In $n + 2 = 7$, the letter n is a variable. A **variable** is a letter used to represent a number.

EXAMPLE

Find how many cards Jeff had originally.

Method 1: *Draw a model.*

Let ■ represent how many cards Jeff had and let △ represent one card.

$$■ + △△ = △△△△ △△△$$

Method 2: *Use an equation.*

An **equation** is a mathematical sentence that uses " = ." Let n represent how many cards Jeff had originally.

$$n + 2 = 7$$

In order to get ■ or n alone, you remove two △'s, or 2, from the left side of the model or of the equation. To keep both sides equal, you also remove two △'s, or 2, from the right side of the model or of the equation.

$$■ + △\!\!\!/△\!\!\!/ = \frac{△△△△}{△△\!\!\!/ △\!\!\!/}$$

$$■ = △△△△△$$

$$n + 2 - 2 = 7 - 2$$

$$n = 5$$

Check: $n + 2 = 7$ ⟵ **Replace n with 5.**

$5 + 2 \overset{?}{=} 7$

$7 \overset{?}{=} 7$ Yes ✓   It checks!

EXERCISES

Write the number you would subtract from each side of the equation in order to solve for n.

1. $n + 2 = 15$ **2.** $n + 17 = 20$ **3.** $n + 8 = 18$ **4.** $n + 15 = 26$

Solve for n by drawing a model. Let ■ *= n and* △ *= 1. Check each answer.*

5. $n + 3 = 6$ **6.** $n + 6 = 8$ **7.** $n + 8 = 9$ **8.** $n + 2 = 9$

9. $n + 5 = 10$ **10.** $n + 4 = 7$ **11.** $n + 7 = 11$ **12.** $n + 3 = 8$

13. $n + 4 = 12$ **14.** $n + 9 = 15$ **15.** $n + 5 = 14$ **16.** $n + 9 = 16$

Solve and check each equation.

17. $n + 5 = 7$ **18.** $n + 4 = 10$ **19.** $n + 16 = 31$ **20.** $n + 11 = 22$

21. $n + 13 = 72$ **22.** $n + 25 = 51$ **23.** $n + 20 = 93$ **24.** $n + 32 = 45$

25. $n + 26 = 50$ **26.** $n + 17 = 28$ **27.** $n + 45 = 90$ **28.** $n + 30 = 86$

29. $n + 1.6 = 9.6$ **30.** $n + 3.2 = 8.2$ **31.** $n + \frac{2}{9} = \frac{8}{9}$ **32.** $n + \frac{3}{8} = \frac{7}{8}$

33. $n + 1.8 = 2.5$ **34.** $n + 1.2 = 7.5$ **35.** $n + \frac{1}{4} = \frac{1}{2}$ **36.** $n + \frac{1}{3} = \frac{5}{6}$

For Exercises 37–40:
a. *Write an equation to represent each problem. Tell what n represents.*
b. *Solve the equation.*

37. Denise had several shells in her collection. After finding six more shells at the beach, she had a total of 18. How many shells did she have originally?

38. Pancho drove a certain number of miles in the morning. He drove another 85 miles in the afternoon for a total of 142 miles. How many miles did Pancho drive in the morning?

39. If Robert saves 37 more dollars, he will have $100. How much does Robert have now?

40. Louise wants to increase the number of fish in her aquarium. If Louise adds 8 new fish to her aquarium, she will have a total of 22. How many fish does she have now?

Solving Equations: SUBTRACTION

Ruth went shopping and spent $7. She has $1 left. How much did she have before going shopping?

1. When shopping, do you add or subtract the amount spent from the amount you had?

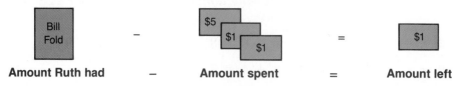

| Amount Ruth had | – | Amount spent | = | Amount left |

2. In order to <u>return</u> to the amount Ruth had <u>before</u> shopping, would you add $7 to, or subtract $7 from, the amount left? Why?

But, to keep the model balanced, you would also add $7 to the right side.

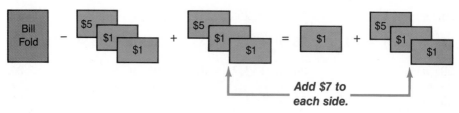

Add $7 to each side.

3. *Complete:* $7 − $7 = $ _?_

4. *Complete:* Before going shopping, Ruth had $ _?_ .

You can also use an equation to solve the problem.

5. What will the variable n represent?

EXAMPLE

Solve: $n − 7 = 1$

Think: To get n alone, add 7 to each side of the equation.

$$n − 7 + 7 = 1 + 7$$
$$n = 8$$

Check: $n − 7 = 1$ ⟵ **Replace n with 8.**

$$8 − 7 \overset{?}{=} 1$$
$$1 \overset{?}{=} 1 \quad \text{Yes} \checkmark \quad \text{It checks!}$$

6. What number would you add to each side of $n − 0.5 = 2.5$ to get n alone?

7. What number would you add to each side of $n - \frac{1}{4} = \frac{3}{4}$ to get n alone?

EXERCISES

For Exercises 1–12, choose from the box below the value of n that makes each equation true.

| 1 | 2 | 3 | 5 | 6 | 7 | 8 | 9 | 12 | 14 |

1. $n - 2 = 3$ **2.** $n + 2 = 3$ **3.** $n - 5 = 2$ **4.** $n + 1 = 2$

5. $n - 7 = 2$ **6.** $n - 8 = 6$ **7.** $n + 2 = 7$ **8.** $n + 6 = 8$

9. $n - 1 = 4$ **10.** $n + 1 = 4$ **11.** $n + 3 = 6$ **12.** $n - 6 = 3$

For Exercises 13–20, write the number you would add to each side of the equation in order to solve for n.

13. $n - 2 = 15$ **14.** $n - 17 = 20$ **15.** $n - 16 = 9$ **16.** $n - 17 = 50$

17. $n - 8 = 10$ **18.** $n - 12 = 35$ **19.** $n - 2.1 = 4$ **20.** $n - \frac{1}{2} = \frac{3}{4}$

Solve each equation for n. Check each answer.

21. $n - 9 = 10$ **22.** $n - 14 = 25$ **23.** $n - 6 = 65$ **24.** $n - 13 = 7$

25. $n - 11 = 46$ **26.** $n - 7 = 18$ **27.** $n - 9 = 36$ **28.** $n - 50 = 40$

29. $n - 10 = 34$ **30.** $n - 15 = 63$ **31.** $n - 42 = 21$ **32.** $n - 35 = 15$

33. $n - 12 = 8$ **34.** $n - 20 = 29$ **35.** $n - 7 = 38$ **36.** $n - 17 = 82$

37. $n - 4.9 = 6.3$ **38.** $n - 7.5 = 7.5$ **39.** $n - 0.5 = 1$ **40.** $n - 3.6 = 8.5$

41. $n - \frac{1}{2} = \frac{1}{2}$ **42.** $n - \frac{1}{3} = \frac{2}{3}$ **43.** $n - \frac{3}{4} = \frac{1}{2}$ **44.** $n - \frac{1}{6} = \frac{5}{6}$

For Exercises 45–49, match each problem with the correct equation. Then solve the equation.

45. If George had $7 more, he could buy a shirt that costs $19.

46. After Maria spent $19, she had $7 left.

47. Willie loaned $7 to his friend. Now he has $19 left.

48. If you subtract 19 from Sue's present age, the answer will be 7.

49. Larry jogged 7 miles more than Sam. Larry jogged a total of 19 miles.

a. $n - 7 = 19$

b. $n + 7 = 19$

c. $n - 19 = 7$

d. $n + 19 = 7$

Using Equations: ADDITION/SUBTRACTION

To use equations to solve problems, it is important to be able to identify the equation that represents the information given in the problem.

EXAMPLE

The smallest airplane ever flown was called Sky Baby. Its wingspan increased by 28 feet equals 35 feet. What is the wingspan of Sky Baby?

READ   What are the facts?

Sky Baby's wingspan increased by 28 feet equals 35 feet.

PLAN   Write an equation for the problem.

SOLVE   1 Let n = Sky Baby's wingspan.

2 Write an equation.

Think: Sky Baby's wingspan increased by 28 ft is 35 feet.

Translate: n $+$ 28 $= 35$

3 Solve the equation. $n + 28 - 28 = 35 - 28$

$n = 7$

CHECK   Does $7 + 28 = 35$?   Yes ✓

Sky Baby has a wingspan of **7 feet.**

EXERCISES

Choose the equation you would use to solve the problem.
Then solve and check.

1. The weight of a textbook plus 4.5 pounds is 7 pounds, the weight of a goliath frog. What is the weight of the textbook?

   a. $n + 4.5 = 7$

   b. $4.5 + 7 = n$

   c. $n - 4.5 = 7$

2. Ninety feet less than the length of a ribbon worm is 90 feet, the length of a blue whale. What is the length of a ribbon worm?

   a. $n + 90 = 90$

   b. $n - 90 = 90$

   c. $90 - 90 = n$

3. "The American Dream," the world's longest car, contains a swimming pool and a helicopter landing pad. The length of "The American Dream" minus 46 feet is 14 feet, the length of an ordinary car. What is the length of "The American Dream"?

a. $n + 46 = 14$ **b.** $46 - 14 = n$ **c.** $n - 46 = 14$

4. The number of golf balls balanced one on top of the other by Lang Martin increased by 8 equals 15. How many golf balls did Lang balance?

a. $n - 8 = 15$

b. $n + 8 = 15$

c. $8 + 15 = n$

5. At Mt. Washington, New Hampshire, the highest windspeed decreased by 157 is 74 miles per hour. What was the highest windspeed at Mt. Washington?

a. $n - 157 = 74$

b. $n + 74 = 157$

c. $157 - 74 = n$

6. The sum of the width of the smallest book ever published and 194 millimeters is 195 millimeters, the width of this textbook. What is the width of the smallest book?

a. $n - 194 = 195$

b. $194 + 195 = n$

c. $n + 194 = 195$

7. The boiling point of water at sea level minus 52°F is 160°F, the boiling point of water atop Mt. Everest. What is the boiling point of water at sea level?

a. $160 - 52 = n$

b. $n + 52 = 160$

c. $n - 52 = 160$

8. The blood temperature of a golden hamster during hibernation minus 6°F is 32°F, the freezing point of water. What is the golden hamster's blood temperature during hibernation?

a. $n - 6 = 32$

b. $n + 6 = 32$

c. $32 - 6 = n$

9. John Howard's speed record on a bicycle plus 48 miles per hour is 200 miles per hour, the speed of a Formula 1 race car. What is John's speed record on a bicycle?

a. $n - 48 = 200$

b. $200 + 48 = n$

c. $n + 48 = 200$

Solving Equations: MULTIPLICATION/DIVISION

Yoshi has $12. If this amount will pay for four movie tickets, how much does each cost?

Model

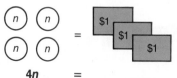

Equation   4n   = $12 ◄——— *4n means 4 × n.*

Since there are four tickets, divide each side into four equal groups.

1. How many *n*'s will there be in each group?

2. How many one-dollar bills will there be in each group?

3. *Complete:* With each *n*, you can match $_?_. So *n* = $_?_.

You can also use an equation to solve the problem.

EXAMPLE 1

Solve and check: $4n = 12$

Think: To get *n* alone, divide each side of the equation by 4.

$$4n = 12$$
$$\frac{4n}{4} = \frac{12}{4} \quad \blacktriangleleft \quad \frac{4n}{4} = \frac{\overset{1}{4} \times n}{\underset{1}{4}} = n$$
$$n = 3$$

Check: $4n = 12$ ◄ **Replace *n* with 3.**
$$4 \times 3 \overset{?}{=} 12$$
$$12 \overset{?}{=} 12 \quad \text{Yes} \checkmark$$

In Example 2, recall that $\frac{n}{2}$ means $n \div 7$.

EXAMPLE 2

Solve and check: $\frac{n}{7} = 18$

Think: To get *n* alone, multiply each side by 7.

$$\frac{n}{7} = 18$$
$$7 \times \frac{n}{7} = 7 \times 18 \quad \blacktriangleleft \quad \overset{1}{7} \times \frac{n}{\underset{1}{7}} = n$$
$$n = 126$$

Check: $\frac{n}{7} = 18$ ◄ **Replace *n* with 126.**
$$\frac{126}{7} \overset{?}{=} 18$$
$$18 \overset{?}{=} 18 \quad \text{Yes} \checkmark$$

EXERCISES

For Exercises 1–5, write the number by which you would divide each side of the equation in order to solve for n.

1. $6n = 30$ **2.** $9n = 45$ **3.** $7n = 49$ **4.** $1.2n = 72$ **5.** $0.3n = 2.7$

Solve and check.

6. $5n = 60$ **7.** $9n = 45$ **8.** $4n = 32$ **9.** $20n = 180$ **10.** $7n = 140$

11. $5n = 45$ **12.** $4n = 16$ **13.** $7n = 105$ **14.** $20n = 1800$ **15.** $3n = 48$

16. $4n = 116$ **17.** $7n = 91$ **18.** $9n = 729$ **19.** $6n = 132$ **20.** $8n = 96$

21. $3.2n = 16$ **22.** $1.8n = 9$ **23.** $2.4n = 36$ **24.** $1.5n = 24$ **25.** $4.5n = 27$

In Exercises 26–30, write the number by which you would multiply each side of the equation in order to get n alone.

26. $\frac{n}{6} = 14$ **27.** $\frac{n}{9} = 13$ **28.** $\frac{n}{15} = 18$ **29.** $\frac{n}{13} = 4.1$ **30.** $\frac{n}{3} = 9.1$

Solve and check.

31. $\frac{n}{5} = 6$ **32.** $\frac{n}{8} = 2$ **33.** $\frac{n}{20} = 9$ **34.** $\frac{n}{4} = 36$ **35.** $\frac{n}{10} = 3$

36. $\frac{n}{9} = 8$ **37.** $\frac{n}{6} = 12$ **38.** $\frac{n}{5} = 11$ **39.** $\frac{n}{7} = 5$ **40.** $\frac{n}{15} = 4$

41. $\frac{n}{3} = 7$ **42.** $\frac{n}{9} = 10$ **43.** $\frac{n}{7} = 8$ **44.** $\frac{n}{2} = 3$ **45.** $\frac{n}{11} = 6$

46. $\frac{n}{5} = 1.1$ **47.** $\frac{n}{7} = 4.1$ **48.** $\frac{n}{6} = 7.5$ **49.** $\frac{n}{4} = 0.4$ **50.** $\frac{n}{10} = 10.9$

51. $2.6n = 13$ **52.** $\frac{n}{16} = 4$ **53.** $3.6n = 5.4$ **54.** $3.8n = 26.6$ **55.** $\frac{n}{12} = 0.6$

56. $\frac{n}{15} = 7.5$ **57.** $7.2n = 36$ **58.** $\frac{n}{12} = 24$ **59.** $\frac{n}{18} = 7$ **60.** $7.2n = 27$

For Exercises 61–64:
a. *Choose the equation below that represents each problem.*
b. *Solve the equation.*

$$n + 4 = 12 \qquad n - 4 = 12 \qquad 4n = 12 \qquad \frac{n}{4} = 12$$

61. Six pizzas were divided equally among 12 people. Each person received 4 slices. How many slices of pizza were there?

62. After spending $4, Joe had $12 left. How much money did he originally have?

63. Lunch costs $12. Each of 4 friends pays an equal amount. How much does each pay?

64. A concert ticket costs $12. Rosa needs $4 more to pay for the ticket. How much money does Rosa actually have?

Using the *d* = *rt* Formula

This formula shows how distance, rate, and time are related.

> **distance = rate × time, or d = rt** ◀ *rt means r × t.*

The table below shows some units that are used to measure distance, rate, and time.

Distance	Time	Rate
miles	hours	miles per hour
feet	seconds	feet per second
kilometers	hours	kilometers per hour
meters	seconds	meters per second

1. How many miles will you travel in 8 hours at an average rate of 50 miles per hour?

EXAMPLE 1

In 1932, Amelia Earhart Putnam was the first woman to fly an airplane across the Atlantic Ocean alone. She crossed the ocean in 14.9 hours at an average speed of 135.9 miles per hour. How many miles was this?

READ What are the facts?

Time (t): 14.9 hours
Rate (r): 135.9 miles per hour
Distance (d): __?__

PLAN Use the formula $d = rt$ to find d.

SOLVE

$d = rt$ ◀——— **Replace r with 135.9 and t with 14.9.**
$d = 135.9 \times 14.9$

[1] [3] [5] [·] [9] [×] [1] [4] [·] [9] [=]   ⬛ 2024.91 ⬛

She traveled **2024.91 miles.**

CHECK Did you use the facts correctly in the formula?

You can also use the distance formula to find the time or the rate.

| **EXAMPLE 2** | In 1978, three Americans were the first to cross the Atlantic Ocean in a balloon. They traveled about 5202 kilometers at an average rate of 38.6 kilometers per hour. To the nearest hour, how long did the crossing take? |

READ   What are the facts?

Distance (*d*): 5202 km
Rate (*r*): 38.6 kilometers per hour
Time (*t*): __?__

PLAN   Use the formula $d = rt$ to find *t*.

SOLVE

$$d = rt$$

$$5202 = 38.6t \longleftarrow \textbf{Divide each side by 38.6.}$$

$$\frac{5202}{38.6} = \frac{38.6t}{38.6} \longleftarrow \textbf{Use a calculator to find } \textbf{5202} \div \textbf{38.6.}$$

 $\langle$ *t* ≈ **135**

The crossing took **135 hours** (nearest hour).

CHECK   Did you use the facts correctly in the formula?

EXERCISES

Use the distance formula to complete each of the following.

1. *d* = 1080 miles, *t* = 18 hours, *r* = __?__ miles per hour

2. *t* = 3 hours, *r* = 96 kilometers per hour, *d* = __?__ kilometers

3. *d* = 305 meters, *t* = 10 seconds, *r* = __?__ meters per second

4. *d* = 1260 yards, *r* = 28 yards per minute, *t* = __?__ minutes

5. In the first nonstop balloon crossing of North America, Maxie Anderson and his son traveled about 4500 kilometers in 100 hours. What was the average speed?

6. In 1903, Wilbur Wright flew the Kitty Hawk for 59 seconds at an average speed of 14 feet per second. How far did he travel?

7. On December 3, 1980, Janice Brown made the first solar-powered flight in the Solar Challenger. The flight lasted 20 minutes and covered 6 miles. How many miles per minute did the plane travel?

On the same day as his brother, Orville Wright flew the Kitty Hawk 120 feet at an average rate of 10 feet per second.

8. Whose flight lasted longer, Orville's or Wilbur's?

9. How much longer did it last?

10. In December, 1986, Dick Rutan and Jeana Yeager flew the Voyager 26,000 miles around the world without landing or refueling. The flight lasted 216 hours. Find the speed (nearest whole number).

11. The Voyager has a top speed of about 120 miles per hour. At this rate, how many hours (nearest whole number) would it take to fly 2000 miles?

12. In a recent "speed test," a giant tortoise traveled 5 yards in 43.5 seconds. What was its rate in yards per second? Round your answer to the nearest tenth.

13. The average snail travels at a **snail's pace** of 0.5 meters per hour. At this rate, how many hours would it take a snail to travel 1.6 kilometers (about 1 mile)?

The graph at the right shows the distance covered by a car traveling at an average rate.

14. What information is given on the horizontal scale?

15. What information is given on the vertical scale?

16. How far does the car travel in $\frac{1}{2}$ hour?

17. How far does the car travel in $4\frac{1}{2}$ hours?

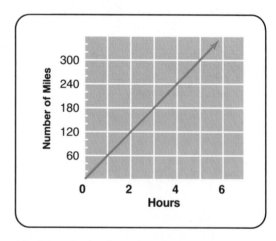

18. How long does it take the car to travel 300 miles?

19. How long does it take the car to travel 200 miles?

Use the graph to complete this table.

20.

Hours	0	1	?	3	4	5	?
Distance	?	?	120	?	?	?	330

21. Use the table in Exercise 20 to find the speed of the car in miles per hour.

22. Predict how far the car will travel in 9 hours.

More on Solving Equations

Jessica said that four times her age increased by 15 would equal 75. How old is Jessica?

1. What variable can you use to represent Jessica's age?

2. If n represents Jessica's age, how can you represent 4 times her age?

3. How would you represent "increased by 15"?

Four times Jessica's age **increased by** **15** **equals** **75.**

$$4n \qquad + \qquad 15 \quad = \quad 75$$

EXAMPLE

Solve and check: $4n + 15 = 75$

Think: To get $4n$ alone, subtract 15 from each side.

$$4n + 15 = 75$$
$$4n + 15 - \mathbf{15} = 75 - \mathbf{15}$$
$$4n = 60 \qquad \blacktriangleleft \begin{array}{l} \textbf{Divide each} \\ \textbf{side by 4.} \end{array}$$
$$\frac{4n}{4} = \frac{60}{4}$$
$$n = \mathbf{15}$$

Check: Does 4 times Jessica's age increased by 15 equal 75?

$$\text{Does} \qquad 4 \quad \times \quad 15 \qquad + \qquad 15 = 75?$$

Does $60 + 15 = 75$?

Does $75 = 75$?   Yes ✓

Jessica is **15 years old.**

When equations involve more than one operation, you usually add or subtract first. Then you multiply or divide.

EXERCISES

For Exercises 1–9, write the first step you would use to solve the equation. Write "Subtract 5 from each side," "Add 3 to each side," and so on.

1. $4n + 7 = 15$

2. $3n + 18 = 21$

3. $5n - 2 = 38$

4. $31n - 29 = 23$

5. $12n - 0.4 = 2$

6. $27n - 1.8 = 0.36$

7. $\frac{n}{6} + \frac{1}{3} = \frac{5}{6}$

8. $\frac{n}{4} - \frac{1}{2} = \frac{1}{4}$

9. $31n + 2.1 = 23.8$

Solve and check.

10. $21n + 6 = 48$

11. $8n + 25 = 57$

12. $6n - 3 = 15$

13. $8n - 7 = 41$

14. $12n - 6 = 30$

15. $4n - 6 = 34$

16. $3n + 1 = 19$

17. $7n + 2 = 51$

18. $4n + 4 = 40$

19. $\frac{n}{6} - 19 = 3$

20. $\frac{n}{5} - 8 = 42$

21. $\frac{n}{3} + 6 = 9$

22. $\frac{n}{8} + 1 = 15$

23. $\frac{n}{6} + 1 = 17$

24. $\frac{n}{5} - 4 = 4$

25. $\frac{n}{30} + 62 = 65$

26. $\frac{n}{28} + 16 = 25$

27. $\frac{n}{7} - 6 = 21$

28. $2n - 27 = 9$

29. $\frac{n}{5} + 1 = 3$

30. $4n + 13 = 77$

31. $\frac{n}{4} - 6 = 1$

32. $\frac{n}{6} + 8 = 23$

33. $9n + 16 = 81.7$

Choose the equation you would use to solve the problem. Then solve the problem.

34. Twice the cost of a bicycle increased by $50 is $164. Find the cost of the bicycle.
    a. $2n + 50 = 164$
    b. $2n - 164 = 50$
    c. $2n - 50 = 164$

35. Five times Debbie's age decreased by 9 is 41. Find Debbie's age.
    a. $9 - 5n = 41$
    b. $5n - 9 = 41$
    c. $5n + 41 = 9$

36. Three times the number of cars in a parking lot decreased by 15 is 27. Find the number of cars in the parking lot.
    a. $27 - 3n = 15$
    b. $3n + 27 = 15$
    c. $3n - 15 = 27$

37. Four times the number of students at Lincoln High School increased by 8 is 1200. How many students are at Lincoln High School?
    a. $4n - 8 = 1200$
    b. $4n + 8 = 1200$
    c. $1200 - 4n = 8$

Mid-Chapter Review

Solve and check.   (Pages 210–211)

1. $n + 4 = 9$ **2.** $n + 10 = 19$ **3.** $n + 23 = 38$ **4.** $n + 16 = 25$

5. $n + 27 = 45$ **6.** $n + 35 = 60$ **7.** $n + 2.5 = 8.7$ **8.** $n + \frac{1}{4} = \frac{2}{3}$

(Pages 212–213)

9. $n - 1 = 7$ **10.** $n - 8 = 12$ **11.** $n - 11 = 21$ **12.** $n - 15 = 30$

13. $n - 34 = 63$ **14.** $n - 46 = 51$ **15.** $n - 0.7 = 3$ **16.** $n - \frac{1}{8} = \frac{1}{4}$

(Pages 216–217)

17. $2n = 12$ **18.** $4n = 20$ **19.** $5n = 35$ **20.** $7n = 63$

21. $\frac{n}{2} = 2$ **22.** $\frac{n}{6} = 1$ **23.** $\frac{n}{4} = 18$ **24.** $\frac{n}{9} = 21$

Solve and check.   (Pages 221–222)

25. $3n + 4 = 16$ **26.** $6n - 1 = 17$ **27.** $15n + 3 = 93$

28. $\frac{n}{4} + 2 = 4$ **29.** $\frac{n}{3} - 2 = 3$ **30.** $\frac{n}{7} + 10 = 16$

31. The length of a flight for a paper airplane over level ground plus 5 seconds is 20 seconds.

    a. Which equation could you use to find the length of the flight, $n + 5 = 20$ or $n - 5 = 20$?

    b. Solve the equation.   (Pages 214–215)

32. The Gossamer Albatross, a human powered airplane, flew 22 miles in 169 minutes. What was the rate of the Albatross in miles per minute? Round your answer to the nearest hundredth.   (Pages 218–220)

MAINTENANCE   *Add or subtract.*   (Pages 140–141)

33. $\frac{1}{2} + \frac{1}{3}$ **34.** $\frac{5}{8} - \frac{1}{4}$ **35.** $\frac{5}{6} - \frac{2}{3}$ **36.** $\frac{2}{5} + \frac{1}{2}$

37. $1\frac{3}{4} + 1\frac{1}{10}$ **38.** $3\frac{1}{4} + 2\frac{1}{3}$ **39.** $6\frac{1}{6} - 2\frac{1}{8}$ **40.** $5\frac{4}{5} - 1\frac{1}{3}$

41. Find the area of this rectangle. (Pages 191–193)

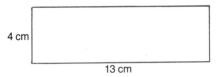

4 cm

13 cm

42. Find the area of this triangle. (Pages 196–198)

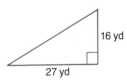

16 yd

27 yd

EQUATIONS/RATIO AND PROPORTION   **223**

Ratio

The results of a survey showed that, on Thursday nights, 3 out of every 5 TV viewers watch the Burt Silver Comedy Show.

1. Does this mean that only 5 TV viewers were surveyed?

2. Does this mean that only 3 viewers watch the Comedy Show?

A **ratio** is a way to compare numbers. You can write the ratio "3 out of 5" in several ways.

$$3 \text{ to } 5 \qquad 3:5 \qquad \frac{3}{5}$$

Ratios such as $\frac{3}{5}$ and $\frac{6}{10}$ are *equivalent ratios*. **Equivalent ratios** are equal.

$$\frac{3}{5} = \frac{6}{10}$$

3. In the survey above, could there have been 5000 TV viewers who took part in the survey and 3000 viewers who said that they watch the Burt Silvers Show? Explain your answer.

You can use **cross-products** to determine whether two ratios are equivalent.

EXAMPLE

Determine whether the ratios are equivalent.

a. $\frac{3}{5}$ and $\frac{6}{10}$

$$\frac{3}{5} \stackrel{?}{=} \frac{6}{10}$$ *The arrows show the cross-products*

$$3 \times 10 \stackrel{?}{=} 5 \times 6$$

$$30 \stackrel{?}{=} 30 \quad \text{Yes} \checkmark$$

Since the cross-products are equal, the ratios are **equivalent.**

b. $\frac{7}{3}$ and $\frac{28}{12}$

$$\frac{7}{3} \stackrel{?}{=} \frac{28}{12}$$

$$7 \times 12 \stackrel{?}{=} 3 \times 28$$

$$84 \stackrel{?}{=} 84 \quad \text{Yes} \checkmark$$

Since the cross-products are equal, the ratios are **equivalent.**

---

5. *Complete:* When cross-products are __?__, the ratios are equivalent.

6. *Complete:* When cross-products are <u>not</u> equal, the ratios are __?__ equivalent.

EXERCISES

Write each ratio in three different ways.

1. Circles to triangles

2. Triangles to circles

3. Circles to all the figures

4. Triangles to all the figures

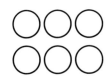

Write each ratio in the form $\frac{a}{b}$.

5. 3 to 7	**6.** 4 to 20	**7.** 14 to 42	**8.** 16 to 25
9. 7 to 21	**10.** 30 to 80	**11.** 5 to 25	**12.** 15 to 60
13. 16 to 64	**14.** 5 to 9	**15.** 10 to 100	**16.** 4 to 11

For Exercises 17–44, determine whether the ratios are equivalent. Answer <u>Yes</u> or <u>No</u>.

17. $\frac{3}{4}$ and $\frac{15}{22}$ **18.** $\frac{1}{2}$ and $\frac{6}{12}$ **19.** $\frac{4}{5}$ and $\frac{18}{25}$ **20.** $\frac{2}{3}$ and $\frac{14}{21}$

21. $\frac{2}{7}$ and $\frac{21}{63}$ **22.** $\frac{5}{8}$ and $\frac{30}{32}$ **23.** $\frac{1}{6}$ and $\frac{10}{72}$ **24.** $\frac{1}{4}$ and $\frac{25}{100}$

25. $\frac{2}{5}$ and $\frac{20}{45}$ **26.** $\frac{3}{10}$ and $\frac{22}{80}$ **27.** $\frac{5}{7}$ and $\frac{25}{35}$ **28.** $\frac{2}{15}$ and $\frac{10}{75}$

29. $\frac{11}{6}$ and $\frac{66}{36}$ **30.** $\frac{12}{5}$ and $\frac{60}{25}$ **31.** $\frac{3}{7}$ and $\frac{20}{49}$ **32.** $\frac{7}{20}$ and $\frac{40}{120}$

33. $\frac{9}{7}$ and $\frac{36}{28}$ **34.** $\frac{8}{3}$ and $\frac{18}{48}$ **35.** $\frac{1}{5}$ and $\frac{4}{20}$ **36.** $\frac{3}{5}$ and $\frac{21}{35}$

37. $\frac{4}{1}$ and $\frac{28}{7}$ **38.** $\frac{12}{1}$ and $\frac{132}{13}$ **39.** $\frac{3}{16}$ and $\frac{9}{48}$ **40.** $\frac{3}{2}$ and $\frac{27}{18}$

41. $\frac{7}{10}$ and $\frac{54}{80}$ **42.** $\frac{7}{12}$ and $\frac{36}{60}$ **43.** $\frac{6}{5}$ and $\frac{18}{12}$ **44.** $\frac{1}{3}$ and $\frac{16}{51}$

45. For every 10 columns of news stories in a newspaper, 4 columns are advertisements. What is the ratio of advertisements to news stories?

46. To make orange juice, Laura mixed 3 cans of water with 1 can of orange juice concentrate. What is the ratio of water to concentrate?

47. *Complete:* When writing a ratio in the form $\frac{a}{b}$ for 2:7, *a* equals __?__ and *b* equals __?__ .

48. Write the ratio of the number of dimes in a dollar to the number of nickels in a quarter.

Math in Photography: RATIOS

Photographers can use *"f-stops"* on a camera to control the overall sharpness of photographs.

Light enters a camera through a small hole called an **aperture.** The size of the aperture is related to numbers on a ring that can be turned to change the size. The numbers are called **"f-stops."** The largest f-stop has the number 2. Each stop after f 2 allows $\frac{1}{2}$ as much light into the camera as the previous f-stop.

EXAMPLE

Write the ratio of f 8 to f 2.8.

1 Count the number of f-stops from f 8 to f 2.8.

f 8 → f 5.6 → f 4 → f 2.8

      1   2   3 ⟶ **3 stops**

2 Since there are 3 stops, multiply $\frac{1}{2}$ by $\frac{1}{2}$ by $\frac{1}{2}$.

$$\frac{1}{2} \times \frac{1}{2} \times \frac{1}{2} = \frac{1}{8}$$ Thus, the ratio of f 8 to f 2.8 is $\frac{1}{8}$ or **1 to 8.**

---

1. *Complete:* An f-stop of 2.8 allows __?__ times as much light to enter a camera than an f-stop of 8.

EXERCISES

Write the ratio of each pair of apertures.

1. f 11 to f 8
2. f 16 to f 8
3. f 5.6 to f 2

4. f 8 to f 2
5. f 4 to f 2
6. f 8 to f 16

7. *Complete:* An f-stop of 2.8 allows __?__ times as much light to enter a camera as an f-stop of 5.6.

8. *Complete:* An f-stop of 2 allows __?__ times as much light to enter a camera as an f-stop of 16.

9. *Complete:* An f-stop of 16 allows __?__ as much light to enter a camera as an f-stop of f 8.

Math in Music: RATIOS

Mel interviewed the conductor of a local symphony orchestra.
He made a table to show the number of instruments in four sections
of the orchestra.

Dale City Symphony Orchestra

I. String Section
- **A.** 20 violins
  - **a.** 12 first violins
  - **b.** 8 second violins
- **B.** 8 violas
- **C.** 10 cellos
- **D.** 6 string basses
- **E.** 1 harp

II. Woodwind Section
- **A.** 3 flutes   **C.** 2 clarinets
- **B.** 2 oboes   **D.** 2 bassoons

III. Brass Section
- **A.** 3 trumpets
- **B.** 2 French horns
- **C.** 2 trombones
- **D.** 1 tuba

IV. Percussion Section
- **A.** 2 kettle drums
- **B.** 1 bass drum
- **C.** 1 set of bells
- **D.** 1 set of cymbals

EXERCISES

1. What is the ratio of first violins to second violins?

2. What is the ratio of cellos to the total number of instruments in the string section?

3. What is the ratio of trumpets to the total number of french horns and trombones?

4. What is the ratio of percussion instruments to the total number of instruments in the orchestra?

5. If the number of first violins is increased to 18, how many second violins must be added to maintain the present ratio of first to second violins?

6. If the number of cellos is increased to 12, how must the number of violins be changed for the ratio of cellos to violins to be 2 to 1.

PROJECT   Interview a member of your local high school or community orchestra. Make a table to show the number of instruments in each section of the orchestra. Use the information to write four problems involving ratios.

Proportion

Sandra's art teacher told her that she could obtain the shade of orange paint she wanted by mixing yellow paint with red paint in the ratio of 4:3.

How much yellow paint should Sandra buy to mix with 12 quarts of red paint?

1. Will Sandra use more yellow paint or more red paint?

2. Complete this ratio: $\frac{n}{\blacksquare}$ ◄——— **Number of quarts of yellow paint**
   ◄——— **Number of quarts of red paint**

A **proportion** is an equation that shows equivalent ratios.

Yellow paint ——► $\frac{4}{3} = \frac{n}{12}$ ◄——— **Number of quarts of yellow paint**
Red paint ——► ◄——— **Number of quarts of red paint**

EXAMPLE Solve $\frac{4}{3} = \frac{n}{12}$

Think: Equivalent ratios have equal cross-products.

$$\frac{4}{3} \times \frac{n}{12}$$

$$3n = 48$$

$$\frac{3n}{3} = \frac{48}{3}$$

$$n = 16$$

Check: $\frac{4}{3} = \frac{n}{12}$ ◄ **Replace n with 16.**

$$\frac{4}{3} \overset{?}{\times} \frac{16}{12}$$

$$3 \times 16 \overset{?}{=} 4 \times 12$$

$$48 \overset{?}{=} 48 \quad \text{Yes } ✓$$

3. How many quarts of yellow paint should Sandra buy?

4. Determine whether you can solve the problem by using each of these proportions. Give a reason for each answer.

   a. $\frac{12}{n} = \frac{3}{4}$ **b.** $\frac{4}{3} = \frac{n}{12}$ **c.** $\frac{3}{4} = \frac{12}{n}$ **d.** $\frac{4}{3} = \frac{n}{12}$

EXERCISES

Determine whether each pair of ratios is a proportion.
Write = or ≠ (is not equal to).

1. $\frac{4}{7} \bullet \frac{12}{21}$ **2.** $\frac{2}{6} \bullet \frac{5}{15}$ **3.** $\frac{3}{5} \bullet \frac{6}{8}$ **4.** $\frac{6}{30} \bullet \frac{8}{40}$ **5.** $\frac{6}{8} \bullet \frac{5}{9}$

6. $\frac{4}{10}$ ● $\frac{8}{20}$ 7. $\frac{10}{12}$ ● $\frac{25}{30}$ 8. $\frac{9}{12}$ ● $\frac{10}{11}$ 9. $\frac{3}{15}$ ● $\frac{7}{35}$ 10. $\frac{15}{12}$ ● $\frac{50}{45}$

Solve each proportion for n.

11. $\frac{n}{36} = \frac{1}{4}$ 12. $\frac{6}{27} = \frac{n}{9}$ 13. $\frac{65}{100} = \frac{13}{n}$ 14. $\frac{20}{n} = \frac{4}{7}$ 15. $\frac{n}{98} = \frac{13}{14}$

16. $\frac{12}{18} = \frac{n}{42}$ 17. $\frac{25}{30} = \frac{5}{n}$ 18. $\frac{9}{n} = \frac{1}{3}$ 19. $\frac{n}{72} = \frac{1}{12}$ 20. $\frac{33}{36} = \frac{n}{12}$

21. $\frac{20}{70} = \frac{6}{n}$ 22. $\frac{72}{n} = \frac{9}{8}$ 23. $\frac{n}{60} = \frac{3}{20}$ 24. $\frac{72}{80} = \frac{n}{10}$ 25. $\frac{48}{56} = \frac{6}{n}$

26. $\frac{14}{n} = \frac{10}{15}$ 27. $\frac{n}{24} = \frac{5}{4}$ 28. $\frac{18}{84} = \frac{n}{14}$ 29. $\frac{28}{64} = \frac{7}{n}$ 30. $\frac{5}{n} = \frac{3}{15}$

31. $\frac{n}{8} = \frac{15}{24}$ 32. $\frac{9}{21} = \frac{n}{7}$ 33. $\frac{42}{48} = \frac{7}{n}$ 34. $\frac{16}{n} = \frac{2}{9}$ 35. $\frac{n}{12} = \frac{21}{36}$

36. $\frac{5}{7} = \frac{n}{14}$ 37. $\frac{3}{4} = \frac{9}{n}$ 38. $\frac{4}{n} = \frac{16}{60}$ 39. $\frac{n}{3} = \frac{20}{60}$ 40. $\frac{16}{100} = \frac{n}{25}$

41. $\frac{9}{57} = \frac{3}{n}$ 42. $\frac{12}{n} = \frac{3}{5}$ 43. $\frac{n}{20} = \frac{9}{60}$ 44. $\frac{5}{6} = \frac{n}{30}$ 45. $\frac{9}{42} = \frac{3}{n}$

46. $\frac{13}{n} = \frac{5}{10}$ 47. $\frac{n}{32} = \frac{3}{8}$ 48. $\frac{3}{10} = \frac{n}{20}$ 49. $\frac{7}{10} = \frac{35}{n}$ 50. $\frac{6}{n} = \frac{18}{33}$

For Exercises 51–55:
a. *Write a proportion for the problem.*
b. *Solve the problem.*

51. Three out of every five freshmen voted for Gregg for class president. There are 500 freshmen. How many voted for Gregg? (HINT: $\frac{3}{5} = \frac{n}{\blacksquare}$)

52. In a store, the ratio of hamsters to parakeets is 4 to 3. If there are 28 hamsters, how many parakeets are there?

GRANDMA MOSES (1860–1961) MIXING COLORS FOR A PAINTING.

53. Louis mixed blue paint with red in the ratio 5:2 to obtain purple paint. How much red paint should he buy to mix with 10 pints of blue paint?

54. A painter mixed yellow paint with blue in the ratio of 2:3 to obtain a shade of green. How many gallons of blue paint are needed to mix with 12 gallons of yellow paint?

55. A certain shade of paint can be obtained by mixing yellow with red in the ratio of 1:2. How many gallons of red paint are needed to mix with 5 gallons of yellow?

Strategy: USING A DRAWING

The drawing below shows the plan for a house. The **scale** of the drawing is given at the left below.

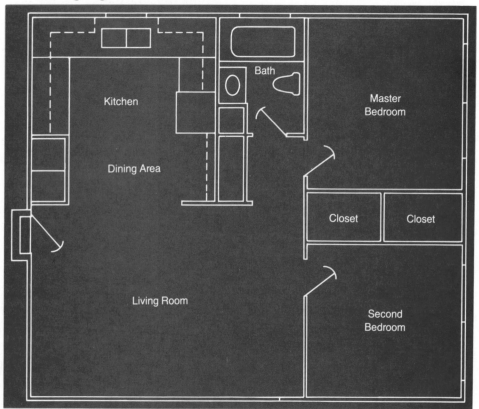

Kitchen

Bath

Master Bedroom

Dining Area

Closet

Closet

Living Room

Second Bedroom

One inch represents 8 feet.

1. What ratio represents $\dfrac{\text{Length on the Drawing}}{\text{Actual Length}}$?

EXAMPLE On the drawing, the width of the house is 4.75 inches. What is the actual width?

READ What are the facts?

Width of house on drawing: 4.75   Scale: 1 to 8, or 1:8

PLAN Write a proportion. Let *n* represent the actual width.

SOLVE

Width on drawing ⟶ $\dfrac{1}{8}$ = $\dfrac{4.75}{n}$ ⟵ Width on drawing
Actual width ⟶   ⟵ Actual width

$$1 \times n = 8 \times 4.75$$
$$n = 38 \quad \text{The actual width is } \textbf{38 feet.}$$

CHECK Did you use the facts correctly in the problem? Is your answer reasonable?

EXERCISES

1. On the drawing, the length of the living room is 3 inches. Find the actual length.

2. The length of the two closets on the drawing is 1.75 inches. Find the actual length.

3. On the drawing, the kitchen has a width of 2 inches. What is the actual width?

4. The length of the stone step on the drawing is about 0.56 inch. What is actual length?

For Exercises 5–8, choose a strategy from the box at the right that you can use to solve each problem.

a. Name the strategy.

b. Solve the problem.

> **Choosing a computation method**
> **More than one step**
> **Guess and check**
> **Making a model**

5. One third of the floor space of an apartment is for the living room. One twelfth of the floor space is for the kitchen. What fractional part of the apartment remains?

6. Five architects competed in an office building design contest. Waldo came in first. Ralph finished last. Erma finished ahead of Mary. Victor finished after Mary. Who came in second?

7. A plumber and a carpenter each compute their earnings for a job. Use the table at the right to answer each question.

Worker	Hours	Rate Per Hour
Plumber	24.2	$12.75
Carpenter	18.8	$10.50

   a. Should each worker estimate personal earnings or is an exact answer needed? Why?

   b. What method or methods of computation would you use to compute the earnings? Give reasons for your answer.

   c. Find the earnings for each worker.

8. The Carltons ordered two optional features with their new home. The total cost for both features is $9000. The difference in their cost is $5000. How much will the Carltons pay for each optional feature?

9. **Write your own problem** about this situation.

   The drawing at the right shows a plan for a patio deck. The scale on the drawing is: One-half inch represents 8 feet.

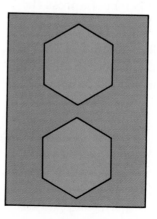

Golden Ratio

For thousands of years, architects and artists have believed that the most pleasing shape of a rectangle is one where the ratio of the width to the length is about $\frac{618}{1000}$, or 0.618. This is known as the **Golden Ratio.**

EXERCISES

The Parthenon, one of the finest buildings of ancient Greece, has a length of 101 feet and a height of 65 feet.

1. Find the ratio of the height to the length. (HINT: Use a calculator.) Round to the nearest ten-thousandth.

2. How close is the ratio you found in Exercise 1 to the Golden Ratio?

3. Write a ratio that compares the height of your waist from the ground to your total height. Express the ratio as a decimal.

4. How close is the ratio you found in Exercise 3 to the Golden Ratio?

5. Draw a Golden Rectangle whose width is 6 millimeters. Round the length to the nearest millimeter.

6. Draw a Golden Rectangle whose length is 9 inches. Round the width to the nearest inch.

PROJECT Measure art and photographs in magazines and newspapers. Use a calculator to compute the ratio of width to length. Display on the class bulletin board any examples with ratios close to the Golden Ratio.

Chapter Summary

1. To solve an addition equation:

   1 Subtract the number being added from each side of the equation.

   2 Write the solution.

   3 Check by replacing the variable with the solution. Does a correct statement result?

2. To solve a subtraction equation:

   1 Add the number being subtracted to each side of the equation.

   2 Write the solution.

   3 Check by replacing the variable with the solution. Does a correct statement result?

3. To solve a multiplication equation:

   1 Divide each side of the equation by the multiplier of the variable.

   2 Write the solution.

   3 Check by replacing the variable with the solution. Does a correct statement result?

4. To solve a division equation:

   1 Multiply each side of the equation by the divisor of the variable.

   2 Write the solution.

   3 Check by replacing the variable with the solution. Does a correct statement result?

5. To solve an equation with more than one operation:

   1 Add or subtract the same number from each side of the equation.

   2 Multiply or divide each side of the equation by the same number (except zero).

   3 Check.

6. To determine whether two ratios are equivalent:

   1 Find the cross-products.

   2 Compare the cross-products. Equivalent ratios have equal cross-products.

7. To solve a proportion for n:

   1 Write the cross-products.

   2 Solve the equation for n.

---

1. Solve: $n + 5 = 19$

$$n + 5 = 19$$
$$n + 5 - \mathbf{5} = 19 - \mathbf{5}$$
$$n = \mathbf{14}$$

2. Solve: $n - 4 = 12$

$$n - 4 = 12$$
$$n - 4 + \mathbf{4} = 12 + \mathbf{4}$$
$$n = \mathbf{16}$$

3. Solve: $6n = 42$

$$6n = 42$$
$$\frac{\mathbf{6}n}{\mathbf{6}} = \frac{\mathbf{42}}{\mathbf{6}}$$
$$n = \mathbf{7}$$

4. Solve: $\frac{n}{3} = 5$

$$\frac{n}{3} = 5$$
$$\mathbf{3} \times \frac{n}{3} = \mathbf{3} \times 5$$
$$n = \mathbf{15}$$

5. Solve: $2n + 1 = 7$

$$2n + 1 = 7$$
$$2n + 1 - \mathbf{1} = 7 - \mathbf{1}$$
$$2n = 6$$
$$\frac{2n}{2} = \frac{6}{2}$$
$$n = \mathbf{3}$$

6.

$$\frac{8}{12} \overset{?}{=} \frac{10}{15}$$
$$8 \times 15 \overset{?}{=} 12 \times 10$$
$$120 \overset{?}{=} 120 \quad \textbf{Yes} \checkmark$$

7. Solve: $\frac{2}{3} = \frac{n}{18}$

$$3n = 36$$
$$\frac{3n}{3} = \frac{36}{3}$$
$$n = \mathbf{12}$$

Chapter Review

Part 1: VOCABULARY

For Exercises 1–7, choose from the box at the right the word(s) that complete(s) each statement.

1. To solve an addition equation, you __?__ the same number from each side of the equation. (Page 210)

2. To solve a subtraction equation, you __?__ the same number to each side of the equation. (Page 212)

3. To solve a multiplication equation, you __?__ each side of the equation by the same number. (Page 216)

4. To solve a division equation, you __?__ each side of the equation by the same number. (Page 216)

5. Distance equals rate times __?__. (Page 218)

6. Equivalent ratios are __?__. (Page 224)

7. An equation that shows equivalent ratios is called a __?__. (Page 228)

> divide
> equal
> add
> time
> proportion
> subtract
> multiply

Part 2: SKILLS

Solve for n. Check each answer. (Pages 210–211)

8. $n + 5 = 21$ 9. $n + 6 = 19$ 10. $n + 7 = 20$ 11. $n + 3 = 12$

12. $n + 4 = 15$ 13. $n + 12 = 24$ 14. $n + 20 = 87$ 15. $n + 32 = 79$

16. $n + 6.2 = 9.8$ 17. $n + 7.1 = 8.5$ 18. $n + 2.9 = 12.3$ 19. $n + 1.8 = 7.2$

(Pages 212–213)

20. $n - 8 = 20$ 21. $n - 5 = 16$ 22. $n - 4 = 13$ 23. $n - 7 = 21$

24. $n - 1 = 1$ 25. $n - 16 = 35$ 26. $n - 17 = 22$ 27. $n - 20 = 40$

28. $n - 1.1 = 6.5$ 29. $n - 3.2 = 4.7$ 30. $n - 3.4 = 5.6$ 31. $n - 4.9 = 7.3$

(Pages 216–217)

32. $3n = 21$ 33. $5n = 40$ 34. $10n = 90$ 35. $20n = 160$

36. $12n = 108$ 37. $11n = 132$ 38. $4n = 48$ 39. $25n = 350$

40. $\frac{n}{6} = 18$ 41. $\frac{n}{4} = 13$ 42. $\frac{n}{12} = 10$ 43. $\frac{n}{13} = 14$

44. $\frac{n}{6} = 21$ 45. $\frac{n}{9} = 17$ 46. $\frac{n}{11} = 64$ 47. $\frac{n}{12} = 39$

(Pages 221–222)

48. $14n + 9 = 51$ **49.** $12n + 8 = 32$ **50.** $5n - 7 = 28$

51. $6n - 4 = 20$ **52.** $2n + 12 = 48$ **53.** $4n - 3 = 21$

54. $\frac{n}{6} - 7 = 10$ **55.** $\frac{n}{7} - 9 = 18$ **56.** $\frac{n}{5} + 8 = 14$

Write each ratio in the form $\frac{a}{b}$. (Pages 224–225)

57. 2 to 4   **58.** 7 to 14   **59.** 36 to 9   **60.** 84 to 36

61. 18 to 45   **62.** 15 to 45   **63.** 60 to 48   **64.** 9 to 6

Determine whether the ratios are equivalent. Answer <u>Yes</u> or <u>No</u>.
(Pages 224–225)

65. $\frac{1}{2}$ and $\frac{4}{6}$ **66.** $\frac{2}{3}$ and $\frac{16}{24}$ **67.** $\frac{4}{5}$ and $\frac{8}{20}$ **68.** $\frac{2}{7}$ and $\frac{6}{14}$

69. $\frac{5}{7}$ and $\frac{35}{49}$ **70.** $\frac{2}{15}$ and $\frac{12}{60}$ **71.** $\frac{8}{3}$ and $\frac{24}{9}$ **72.** $\frac{9}{5}$ and $\frac{18}{12}$

Solve each proportion for n. (Pages 228–229)

73. $\frac{n}{14} = \frac{2}{7}$ **74.** $\frac{n}{12} = \frac{5}{6}$ **75.** $\frac{n}{8} = \frac{3}{4}$ **76.** $\frac{n}{21} = \frac{5}{7}$ **77.** $\frac{n}{16} = \frac{3}{8}$

78. $\frac{n}{20} = \frac{3}{5}$ **79.** $\frac{n}{13} = \frac{8}{52}$ **80.** $\frac{4}{9} = \frac{20}{n}$ **81.** $\frac{6}{5} = \frac{24}{n}$ **82.** $\frac{9}{7} = \frac{81}{n}$

83. $\frac{7}{12} = \frac{14}{n}$ **84.** $\frac{1}{3} = \frac{12}{n}$ **85.** $\frac{n}{16} = \frac{21}{28}$ **86.** $\frac{n}{12} = \frac{10}{15}$ **87.** $\frac{n}{8} = \frac{5}{10}$

Part 3: APPLICATIONS

88. The top speed of a lion decreased by 23 miles per hour is 27 miles per hour, the top speed of a human. Which equation would you use to find a lion's top speed? Solve the equation.   (Pages 214–215)

    a. $n + 23 = 27$ **b.** $n - 23 = 27$

89. A passenger jet flew at a speed of 475 miles per hour for 3 hours. How far did it travel?   (Pages 218–220)

90. To make grapefruit juice, Charles mixed 4 cans of water with 1 can of concentrate. What is the ratio of concentrate to water? (Pages 224–225)

91. A painter mixed blue and red paint in the ratio of 3:4. How many gallons of red paint will be needed to mix with 9 gallons of blue paint? (Pages 228–229)

92. On a drawing, one inch represents 5 feet. The width of a patio is 2.75 inches. What is the actual width? (Pages 230–231)

Chapter Test

Solve and check.

1. $n + 7 = 23$

2. $n - 8 = 16$

3. $n + 21 = 35$

4. $n - 12 = 27$

5. $2n = 24$

6. $13n = 26$

7. $\frac{n}{5} = 15$

8. $\frac{n}{6} = 12$

9. $2n + 8 = 14$

10. $3n - 9 = 21$

11. $\frac{n}{4} - 5 = 2$

12. $\frac{n}{20} + 3 = 10$

Write each ratio in the form $\frac{a}{b}$.

13. 5 to 16

14. 27 to 3

15. 14 to 49

16. 60 to 20

Determine whether the ratios are equivalent. Answer <u>Yes</u> or <u>No</u>.

17. $\frac{2}{3}$ and $\frac{9}{12}$

18. $\frac{9}{20}$ and $\frac{44}{100}$

19. $\frac{3}{10}$ and $\frac{27}{90}$

20. $\frac{5}{4}$ and $\frac{25}{20}$

Solve each proportion.

21. $\frac{2}{3} = \frac{n}{9}$

22. $\frac{3}{5} = \frac{9}{n}$

23. $\frac{2}{n} = \frac{6}{21}$

24. $\frac{3}{8} = \frac{24}{n}$

25. $\frac{3}{10} = \frac{21}{n}$

26. $\frac{24}{27} = \frac{n}{9}$

27. $\frac{24}{56} = \frac{3}{n}$

28. $\frac{9}{20} = \frac{n}{100}$

29. $\frac{n}{5} = \frac{18}{30}$

30. $\frac{15}{1} = \frac{n}{2}$

31. George Blanda's record number of attempted passes in a football game minus 31 is 37, his number of completed passes in that game. Which equation would you use to find the number of passes attempted? Solve the equation.

    a. $37 - 31 = n$

    b. $n + 31 = 37$

    c. $n - 31 = 37$

32. Susan drove 450 miles at an average speed of 50 miles per hour. How many hours did she drive?

33. On a drawing, one inch represents 6 feet. The length of a hallway is 2.25 inches. What is the actual length?

Cumulative Maintenance Chapters 1–9

*Choose the correct answer. Choose **a, b, c, or d**.*

1. Add.

$$24 + 0.86 + 5.308$$

a. 29.168   **b.** 29.394
c. 30.168   **d.** 31.394

2. Choose the best estimate.

$$39.7 \div 7.9$$

a. 3   **b.** 5
c. 4   **d.** 6

3. Multiply.

$$\begin{array}{r} 0.718 \\ \times\ 0.64 \\ \hline \end{array}$$

a. 0.45952   **b.** 45.952
c. 4.5952   **d.** 459.52

4. Which unit would you use to measure a door frame?

a. millimeters

b. kilometers

c. kilograms

d. meters

5. A rectangular garden is 12.3 meters long and 6.8 meters wide. Find the perimeter in meters.

a. 72.24   **b.** 72
c. 19.1   **d.** 38.2

6. Twelve pounds of ground meat costs $24.30. Estimate the cost of one pound.

a. $3   **b.** $4
c. $2   **d.** $1.75

7. Divide. Round the quotient to the nearest tenth.

$$4.3\overline{)7.31}$$

a. 1.4   **b.** 1.6
c. 1.7   **d.** 1.5

8. Subtract.

$$\begin{array}{r} 7 \\ -\ 2.4 \\ \hline \end{array}$$

a. 4.6   **b.** 9.4
c. 5.4   **d.** 4.4

9. Give the length of the eraser to the nearest millimeter.

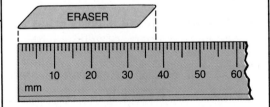

a. 3.8   **b.** 43
c. 0.38   **d.** 38

10. In five games, a football team scored 21, 10, 24, and 3 points. How many points in the fifth game would the team have had to score to have a mean of 14 points scored?

a. 10   **b.** 9
c. 12   **d.** 7

11. Choose the best estimate.

$$8.3 \times 11.7$$

a. 96   **b.** 90
c. 88   **d.** 99

12. Subtract.

$$\frac{6}{7}$$
$$-\frac{2}{3}$$

a. $\frac{7}{21}$ **b.** $\frac{4}{21}$

c. $\frac{5}{21}$ **d.** $\frac{4}{7}$

13. Multiply.

$$2\frac{2}{3} \times 1\frac{3}{8}$$

a. $2\frac{1}{4}$ **b.** $3\frac{1}{4}$

c. $3\frac{2}{3}$ **d.** $2\frac{2}{3}$

14. A certain color of paint is made by mixing 5 parts of yellow with 3 parts of blue. How much blue paint should be mixed with 15 parts of yellow paint?

a. 12   **b.** 25
c. 9   **d.** 15

15. Laurie drove to the lake at an average rate of 46 miles per hour. The trip took $3\frac{1}{2}$ hours. How many miles did she travel?

a. 128   **b.** 147.2
c. 167   **d.** 161

16. What is the area of this figure in square centimeters?

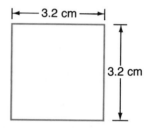

a. 10.24   **b.** 9.4
c. 1.024   **d.** 0.94

17. Add.

$$4\frac{1}{3}$$
$$+2\frac{2}{5}$$

a. $6\frac{3}{8}$ **b.** $6\frac{2}{15}$

c. $6\frac{11}{15}$ **d.** $7\frac{1}{3}$

18. On their vacation, the Martin family traveled 314.6 kilometers in 6.5 hours. Find the rate in kilometers per hour.

a. 484   **b.** 4840
c. 48.4   **d.** 4.84

SALE!
3 DAYS ONLY!

40% off!
Reg. $32.95
Trail Pack

25% off!
Reg. $60
Binoculars

33% off!
Reg. $80
Boots

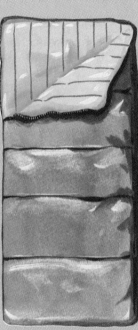

15% off!
Reg. $140
Sleeping Bag

- How much will you save on the binoculars?

- How much will you pay for the trail pack?

- Which amount is greater, the discount on the sleeping bag or the discount on the trail pack?

- Which has the lower sale price, the binoculars or the boots?

Meaning of Percent

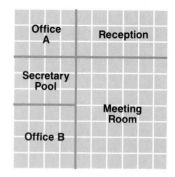

On the plans for a new office suite, an architect divided the total area into 100 equal square units.

1. How many square units will Office A occupy?

2. *Complete:* $\dfrac{\text{Area of Office A}}{\text{Total Area}} = \dfrac{\square}{\square}$

You can write a *percent* for a ratio (or fraction) with a denominator of 100. **Percent** means per hundred or hundredths. Thus, Office A occupies $\frac{12}{100}$, or 12% of the total area.

3. What ratio compares the area of Office B to the entire area?

4. What percent of the total area is occupied by Office B?

5. What is the ratio of the areas of all the rooms in the office suite to the total area?

6. Write a percent for the ratio in Exercise 5.

> $\frac{100}{100}$ means 100 parts out of 100 parts or 100%.
> 100% represents the total area, or the whole.

Since percent means per hundred or hundredths, you can write both a fraction and a decimal for a percent.

$$12\% = \frac{12}{100} = 0.12 \qquad\qquad 20\% = \frac{20}{100} = 0.20, \text{ or } 0.2$$

$$1\% = \frac{1}{100} = 0.01 \qquad\qquad 97\% = \frac{97}{100} = 0.97$$

$$5\% = \frac{5}{100} = 0.05 \qquad\qquad 100\% = \frac{100}{100} = 1.00, \text{ or } 1$$

EXERCISES

Use the house design at the right for Exercises 1–7.

1. How many square units are there in all?

2. How many square units does the living room occupy?

3. Write a ratio to compare the number of square units in the living room to the total number of square units.

4. What percent of the total area does the living room occupy?

5. What percent of the total area is each closet?

6. Which rooms occupy 16% of the total area?

7. Which two rooms occupy 52% of the total area?

For Exercises 8–11:
  a. *Determine what percent of each figure is shaded.*
  b. *Determine what percent of each figure is not shaded.*

You may have to give approximate answers for Exercises 10–11.

8.

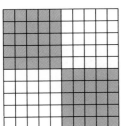

9.

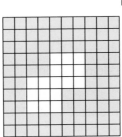

10.

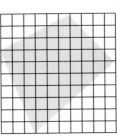

11.

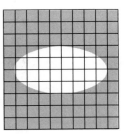

Write a percent for each fraction.

12. $\frac{30}{100}$ **13.** $\frac{40}{100}$ **14.** $\frac{16}{100}$ **15.** $\frac{21}{100}$ **16.** $\frac{19}{100}$ **17.** $\frac{11}{100}$

18. $\frac{9}{100}$ **19.** $\frac{7}{100}$ **20.** $\frac{5}{100}$ **21.** $\frac{3}{100}$ **22.** $\frac{8}{100}$ **23.** $\frac{1}{100}$

24. $\frac{51}{100}$ **25.** $\frac{83}{100}$ **26.** $\frac{90}{100}$ **27.** $\frac{77}{100}$ **28.** $\frac{45}{100}$ **29.** $\frac{99}{100}$

Write a percent for each decimal.

30. 0.15   **31.** 0.35   **32.** 0.71   **33.** 0.83   **34.** 0.09   **35.** 0.06

36. 0.02   **37.** 0.07   **38.** 0.70   **39.** 0.30   **40.** 0.03   **41.** 0.50

Complete. Write a two-place decimal and a percent for each fraction.

		Two-Place Decimal	Percent			Two-Place Decimal	Percent
42.	$\frac{19}{100}$	?	?	**48.**	$\frac{25}{100}$	?	?
43.	$\frac{11}{100}$	?	?	**49.**	$\frac{75}{100}$	?	?
44.	$\frac{6}{100}$	?	?	**50.**	$\frac{42}{100}$	?	?
45.	$\frac{9}{100}$	?	?	**51.**	$\frac{83}{100}$	?	?
46.	$\frac{20}{100}$	?	?	**52.**	$\frac{29}{100}$	?	?
47.	$\frac{80}{100}$	?	?	**53.**	$\frac{13}{100}$	?	?

Percents and Decimals

Since percent means hundredths, you can write a decimal for a percent.

1. *Complete:* 50% means __?__ parts out of 100 parts.

2. *Complete:* 50 parts out of 100 parts equals $\frac{\square}{\square}$.

3. *Complete:* Written as a decimal, $\frac{50}{100}$ = __?__

   Thus, $50\% = \frac{50}{100} = 0.5$

   You can use a shortcut method to write a decimal for a percent.

4. *Complete:*
   To divide by 100, move the decimal point __?__ places to the left.

EXAMPLE 1 Write a decimal for each percent.

a.	b.	c.
13% = 13%	147% = 147%	18.5% = 18.5%
= 0.13	= 1.47	= 0.185

Sometimes you have to insert zeros.

EXAMPLE 2 Write a decimal for each percent.

a.
3% = 03% ◀ **Insert one zero.**
  = 0.03

b.
8.5% = 08.5% ◀ **Insert one zero.**
  = 0.085

5. *Complete:* To write a decimal for a percent, move the decimal point two places to the __?__.

   You can also use a shortcut to write a percent for a decimal.

EXAMPLE 3 Write a percent for each decimal.

a.	b.	c.
0.35 = 0.35	0.08 = 0.08	0.875 = 0.875
= 35%	= 8%	= 87.5%

6. *Complete:* To write a percent for a decimal, move the decimal point two places to the __?__.

EXERCISES

For Exercises 1–6, choose from the box at the right a fraction and a decimal equal to each percent.

$\frac{1.7}{100}$	$\frac{170}{100}$	$\frac{7.5}{100}$	$\frac{700}{100}$
$\frac{7}{1}$	$\frac{70}{100}$	$\frac{7}{100}$	$\frac{17}{100}$
0.75	7	0.7	0.17
0.075	0.07	1.7	700

1. 7%   **2.** 17%   **3.** 7.5%

4. 170%   **5.** 700%   **6.** 70%

For Exercises 7–24, write a fraction and a decimal for each percent.

7. 13%   **8.** 29%   **9.** 3%   **10.** 6%   **11.** 60%   **12.** 10%

13. 112%   **14.** 130%   **15.** 370%   **16.** 106%   **17.** 208%   **18.** 37.5%

19. 87.5%   **20.** 16.8%   **21.** 52.9%   **22.** 5%   **23.** 9%   **24.** 110%

For Exercises 25–42, write a percent for each decimal.

25. 0.21   **26.** 0.93   **27.** 0.04   **28.** 0.09   **29.** 0.805   **30.** 0.609

31. 0.575   **32.** 0.865   **33.** 1.75   **34.** 3.6   **35.** 0.3   **36.** 0.1

37. 0.90   **38.** 1.82   **39.** 0.008   **40.** 0.005   **41.** 3.9   **42.** 1.7

For Exercises 43–48, write a decimal for each percent.

43. Paul Rios bought a suit at a 30% off sale.

44. In a class of 210 students, 65% voted for Teresa for class president.

45. Okra National Bank pays $5\frac{1}{2}\%$ interest on savings accounts.

46. On Tuesday, 5% of the students were absent.

47. Katie has saved 60% of the amount needed to buy a camera.

48. About $87\frac{1}{2}\%$ of a melon is water.

For Exercises 49–52, use the survey at the right below. The survey shows the percent of people (written as a decimal) who chose each sport as their favorite.

49. What percent of the people surveyed preferred football?

50. What percent preferred swimming?

51. What percent preferred sports other than golf?

52. How much greater was the percent who preferred baseball than the percent who preferred tennis?

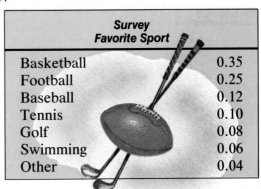

**Survey
Favorite Sport**

Basketball	0.35
Football	0.25
Baseball	0.12
Tennis	0.10
Golf	0.08
Swimming	0.06
Other	0.04

Writing Fractions for Percents

This bar graph shows which kinds of television programs are preferred by top American executives.

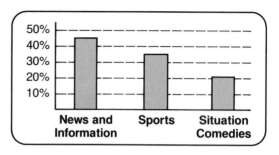

1. What percent of the executives prefer sports?

EXAMPLE 1

a. What fraction of the executives prefer news programs?

$$45\% = \frac{45}{100}$$

$$= \frac{9}{20} \quad \blacktriangleleft \textbf{ Lowest terms}$$

$\frac{9}{20}$ of the executives prefer news programs.

b. What fraction represents how many more executives prefer news and information than sports?

$$45\% - 35\% = 10\% \quad \blacktriangleleft \begin{array}{l}\textbf{\textit{Percent who}} \\ \textbf{\textit{prefer news}}\end{array} - \begin{array}{l}\textbf{\textit{Percent who}} \\ \textbf{\textit{prefer sports}}\end{array}$$

$$10\% = \frac{10}{100}$$

$$10\% = \frac{1}{10}$$

$\frac{1}{10}$ of the executives prefer news and information over sports.

---

2. *Complete:* Dividing by 100 is the same as multiplying by $\frac{1}{?}$.

3. *Complete:* Since $\frac{3}{4}\%$ is less than 1% and $1\% = \frac{1}{?}$, $\frac{3}{4}\%$ will be less than $\frac{1}{?}$.

EXAMPLE 2

Write a fraction for each percent.

a. $\frac{3}{4}\%$ b. $33\frac{1}{3}\%$

a. $\frac{3}{4}\% = \frac{\frac{3}{4}}{100}$ b. $33\frac{1}{3}\% = \frac{33\frac{1}{3}}{100}$

$\frac{3}{4}\% = \frac{3}{4} \div 100$ $33\frac{1}{3}\% = 33\frac{1}{3} \div 100$

$\frac{3}{4}\% = \frac{3}{4} \times \frac{1}{100}$ $33\frac{1}{3}\% = \frac{\overset{1}{\cancel{100}}}{3} \times \frac{1}{\underset{1}{\cancel{100}}}$

$\frac{3}{4}\% = \frac{3}{400}$ $33\frac{1}{3}\% = \frac{1}{3}$

EXERCISES

For Exercises 1–6, copy and complete.

1. $35\% = \frac{\square}{100} = \frac{\square}{\square}$

2. $90\% = \frac{\square}{100} = \frac{\square}{\square}$

3. $4\% = \frac{\square}{100} = \frac{\square}{\square}$

4. $8\% = \frac{\square}{100} = \frac{\square}{\square}$

5. $4\frac{1}{2}\% = \frac{\square}{100} = \frac{\square}{200}$

6. $9\frac{1}{4}\% = \frac{\square}{100} = \frac{\square}{400}$

Write each percent as a fraction in lowest terms.

7. 40% **8.** 90% **9.** 50% **10.** 20% **11.** 75% **12.** 25%

13. 35% **14.** 15% **15.** 44% **16.** 66% **17.** 98% **18.** 87%

19. 43% **20.** 21% **21.** 72% **22.** 62% **23.** 88% **24.** 67%

25. $1\frac{1}{2}\%$ **26.** $1\frac{1}{5}\%$ **27.** $\frac{3}{4}\%$ **28.** $\frac{3}{8}\%$ **29.** $\frac{5}{8}\%$ **30.** $66\frac{2}{3}\%$

31. $62\frac{1}{2}\%$ **32.** $5\frac{1}{4}\%$ **33.** $87\frac{1}{2}\%$ **34.** $16\frac{2}{3}\%$ **35.** $83\frac{1}{3}\%$ **36.** $12\frac{1}{2}\%$

37. 5% **38.** $37\frac{1}{2}\%$ **39.** $2\frac{1}{2}\%$ **40.** $\frac{1}{2}\%$ **41.** $3\frac{1}{3}\%$ **42.** 6%

For Exercises 43–50, replace each ● with <, =, or >.

43. 30% ● $\frac{7}{10}$ **44.** 85% ● $\frac{13}{20}$ **45.** 10% ● $\frac{1}{12}$ **46.** 80% ● $\frac{4}{5}$

47. 50% ● $\frac{1}{3}$ **48.** 15% ● $\frac{1}{10}$ **49.** $66\frac{2}{3}\%$ ● $\frac{3}{4}$ **50.** $\frac{1}{2}\%$ ● $\frac{1}{2}$

This bar graph shows which kinds of radio programs are preferred by top American executives. Refer to this graph for Exercises 51–54.

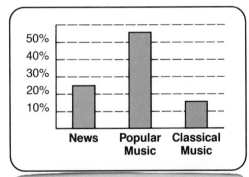

51. What fraction of the executives prefer classical music?

52. Write a fraction to represent how many more executives prefer popular music than classical music?

·53. Write a fraction to represent how many executives prefer popular and classical music.

54. Write a fraction to represent how many more executives prefer news programs over classical music.

Percents/Decimals/Fractions

About $\frac{4}{5}$ of all American households own a car. What percent is this?

1. *Complete:*

$$\frac{4}{5} = \frac{\square}{10} = \frac{\square}{20} = \frac{\square}{40} = \frac{\square}{50} = \frac{\square}{100}$$

EXAMPLE 1

Write a percent for $\frac{4}{5}$.

Method 1

Think: $\frac{4}{5} = \frac{\square}{100}$ ◀ **Let n = □**

$$\frac{4}{5} = \frac{n}{100}$$

$$5 \times n = 4 \times 100$$

$$5n = 400$$

$$\frac{5n}{5} = \frac{400}{5}$$

$$n = 80$$

So $\frac{4}{5} = \frac{80}{100} = \mathbf{80\%}.$

Method 2

Think: $\frac{4}{5} \longrightarrow 5\overline{)4}$

$$\begin{array}{r} 0.8 \\ 5\overline{)4.0} \\ \underline{4\,0} \\ 0 \end{array}$$ ◀ **0.8 = 0.80**

$$0.80 = \frac{80}{100}$$

$$= 80\%$$

So $\frac{4}{5} = 0.80 = \mathbf{80\%}.$

Sometimes you may have to divide beyond the hundredths place.

2. Write a percent for 0.625.

3. Write a percent for $0.66\frac{2}{3}$.

EXAMPLE 2

Write a percent for each fraction.

a. $\frac{5}{8}$

b. $\frac{2}{3}$

a. $\frac{5}{8} \longrightarrow \begin{array}{r} 0.625 \\ 8\overline{)5.000} \\ \underline{4\,8} \\ 20 \\ \underline{16} \\ 40 \\ \underline{40} \\ 0 \end{array}$

b. $\frac{2}{3} \longrightarrow \begin{array}{r} 0.66\frac{2}{3} \\ 3\overline{)2.00} \\ \underline{1\,8} \\ 20 \\ \underline{18} \\ 2 \end{array}$ ◀ **The remainder keeps repeating.**

So $\frac{5}{8} = 0.625 = \mathbf{62.5\%}.$

So $\frac{2}{3} = 0.66\frac{2}{3} = \mathbf{66\frac{2}{3}\%}.$

EXERCISES

Complete.

1. $\frac{1}{2} = \frac{\square}{100} = \underline{?}\%$ **2.** $\frac{1}{4} = \frac{\square}{100} = \underline{?}\%$ **3.** $\frac{2}{5} = \frac{\square}{100} = \underline{?}\%$

4. $\frac{3}{10} = \frac{\square}{100} = \underline{?}\%$ **5.** $\frac{7}{20} = \frac{\square}{100} = \underline{?}\%$ **6.** $\frac{3}{25} = \frac{\square}{100} = \underline{?}\%$

Use a proportion to write a percent for each fraction.

7. $\frac{3}{4}$ **8.** $\frac{1}{10}$ **9.** $\frac{1}{5}$ **10.** $\frac{7}{10}$ **11.** $\frac{1}{20}$ **12.** $\frac{1}{25}$

13. $\frac{1}{50}$ **14.** $\frac{8}{25}$ **15.** $\frac{3}{20}$ **16.** $\frac{29}{50}$ **17.** $\frac{4}{25}$ **18.** $\frac{11}{20}$

Write a decimal for each fraction. Then write a percent.

19. $\frac{1}{3}$ **20.** $\frac{3}{8}$ **21.** $\frac{5}{8}$ **22.** $\frac{1}{12}$ **23.** $\frac{5}{12}$ **24.** $\frac{1}{40}$

25. $\frac{9}{40}$ **26.** $\frac{1}{6}$ **27.** $\frac{5}{6}$ **28.** $\frac{1}{16}$ **29.** $\frac{2}{9}$ **30.** $\frac{4}{7}$

Complete.

	Fraction	Per 100	Decimal	Percent
31.	$\frac{17}{20}$	?	?	?
32.	?	$\frac{15}{100}$	?	?
33.	?	?	0.65	?
34.	?	?	?	30%
35.	$\frac{3}{5}$	?	?	?
36.	?	$\frac{40}{100}$	?	?
37.	?	?	0.875	?
38.	$\frac{1}{3}$	?	?	$33\frac{1}{3}\%$
39.	$\frac{8}{25}$	?	?	?

40. Optional features made up $\frac{11}{50}$ of the price of a new car. What percent is this?

41. Loretta's car uses $\frac{5}{6}$ of a tank of gasoline each week. What percent is this?

Write a fraction that is approximately equal to each percent. For example, $24\frac{1}{2}\% \approx 25\% = \frac{1}{4}$.

42. 9.7%   **43.** $19\frac{1}{2}\%$ **44.** 33.1%   **45.** 79.8%   **46.** 51%   **47.** 68%

48. 10.3%   **49.** 29%   **50.** $74\frac{1}{2}\%$ **51.** 81%   **52.** 26%   **53.** 34%

Math in Social Studies

Vanessa's social studies class is studying about the government of the United States. She made this table to show the political parties to which the presidents belonged.

Political Party	Number of Presidents
Federalist	2
Democrat/Republican	4
Whig	4
Union	1
Democrat	13
Republican	16

EXERCISES

1. What is the total number of presidents?

2. What fraction of the presidents were Republicans?

3. Use your answer to Exercise 2 to write what percent of the presidents were Republicans.

4. What fraction of the presidents belonged to the Whig Party?

5. What percent of the presidents belonged to the Whig party?

6. What percent of the presidents belonged to the Federalist party?

7. What percent of the presidents belonged to the Union party?

8. Write a percent for the ratio of Democrat/Republican presidents to Whig presidents.

9. Write a percent for the ratio of Whig presidents to Republican presidents.

10. Write a percent for the ratio of Republican presidents to Federalist presidents.

11. Construct a bar graph to show the percent of the number of presidents that belonged to each political party.

PROJECT Do research to determine the political parties of the last ten governors of your state. Construct a bar graph to show the results of your research.

Mid-Chapter Review

Write a percent for each fraction. (Pages 240–241)

1. $\frac{5}{100}$　　**2.** $\frac{12}{100}$　　**3.** $\frac{40}{100}$　　**4.** $\frac{17}{100}$　　**5.** $\frac{25}{100}$　　**6.** $\frac{78}{100}$

Write a decimal for each percent. (Pages 242–243)

7. 18%　　**8.** 20%　　**9.** 104%　　**10.** 7%　　**11.** 9%　　**12.** 12.8%

Write a percent for each decimal. (Pages 242–243)

13. 0.24　　**14.** 0.46　　**15.** 0.05　　**16.** 0.01　　**17.** 0.721　　**18.** 4.2

Write each percent as a fraction in lowest terms. (Pages 244–245)

19. 50%　　**20.** 75%　　**21.** 57%　　**22.** $\frac{1}{4}$%　　**23.** $\frac{7}{8}$%　　**24.** $33\frac{1}{3}$%

Use a proportion to write a percent for each fraction. (Pages 246–247)

25. $\frac{1}{2}$　　**26.** $\frac{7}{10}$　　**27.** $\frac{3}{20}$　　**28.** $\frac{6}{25}$　　**29.** $\frac{13}{50}$　　**30.** $\frac{49}{50}$

Write a decimal for each fraction. Then write a percent. (Pages 246–247)

31. $\frac{7}{8}$　　**32.** $\frac{1}{3}$　　**33.** $\frac{4}{9}$　　**34.** $\frac{5}{6}$　　**35.** $\frac{11}{12}$　　**36.** $\frac{3}{40}$

37. About 36% of the students surveyed prefer country music. What fraction of the students did not prefer country music? (Pages 244–245)

38. Mr. Steer has used $\frac{3}{4}$ of his vacation days. What percent of his vacation days has he used? (Pages 246–247)

MAINTENANCE

Multiply. Write the answer in lowest terms. (Pages 155–156 and 158–159)

39. $\frac{1}{2} \times \frac{1}{4}$　　**40.** $\frac{2}{3} \times \frac{9}{20}$　　**41.** $1\frac{1}{4} \times 2\frac{2}{5}$　　**42.** $2\frac{1}{8} \times 3\frac{1}{6}$　　**43.** $1\frac{3}{4} \times 4\frac{1}{7}$

The graph at the right shows the results of a student survey. (Pages 84–85)

44. How many students responded to the survey?

45. Which subject was preferred by 30 of the 200 students?

46. Which subject was preferred by 13 out of every 40 students?

47. Which subject was preferred by 30 out of every 100 students?

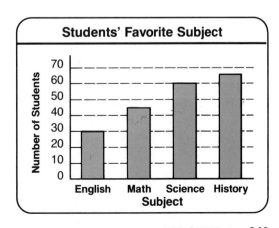

Students' Favorite Subject

Number of Students (vertical axis: 0, 10, 20, 30, 40, 50, 60, 70)

Subject (horizontal axis: English, Math, Science, History)

PERCENT　　**249**

Finding a Percent of a Number

There are eight different types of blood. This table shows what percent of the American population has each type.

Blood Type	Percent of Population (approximate)
O Positive	39
O Negative	7
A Positive	30
A Negative	6
B Positive	11
B Negative	2
AB Positive	4
AB Negative	1

1. What is the most common type?
   O
2. What is the rarest type?
3. *Complete:* For every 100 people, __?__ can be expected to have AB positive blood.

EXAMPLE 1
There are 850 ninth graders at Emerson High. How many can be expected to have AB positive blood?

$\boxed{1}$ **Think:** 4% of 850 can be expected to have AB positive blood. 4% of 850 means 4% × 850.

$\boxed{2}$ 4% × 850 = 0.04 × 850
$\qquad$ = **34**

So **34 students** can be expected to have AB positive blood.

Sometimes you can use a fraction for a percent.

4. *Complete:* $10\% = \frac{\square}{100} = \frac{\square}{10}$

5. *Complete:* $30\% = \frac{\square}{100} = \frac{\square}{\square}$

EXAMPLE 2
How many ninth graders at Emerson High can be expected to have A positive blood?

Think: 30% of 850 = __?__

$\boxed{1}$ Write a fraction for 30%. $\longrightarrow$ $30\% = \frac{30}{100} = \frac{3}{10}$

$\boxed{2}$ $\frac{3}{10}$ of 850 = $\frac{3}{10} \times \overset{85}{850}$
$\qquad \qquad \underset{1}{}$
$\qquad$ = **255**

About **255 ninth graders** can be expected to have A positive blood.

The table at the top of the next page can help you in writing equivalent fractions for some percents. It may be helpful if you memorize these equivalent fractions and percents.

Equivalent Fractions and Percents			
$\frac{1}{4} = 25\%$	$\frac{1}{2} = 50\%$	$\frac{3}{4} = 75\%$	
$\frac{1}{5} = 20\%$	$\frac{2}{5} = 40\%$	$\frac{3}{5} = 60\%$	$\frac{4}{5} = 80\%$
$\frac{1}{6} = 16\frac{2}{3}\%$	$\frac{1}{3} = 33\frac{1}{3}\%$	$\frac{2}{3} = 66\frac{2}{3}\%$	$\frac{5}{6} = 83\frac{1}{3}\%$
$\frac{1}{8} = 12\frac{1}{2}\%$	$\frac{3}{8} = 37\frac{1}{2}\%$	$\frac{5}{8} = 62\frac{1}{2}\%$	$\frac{7}{8} = 87\frac{1}{2}\%$

EXERCISES

For Exercises 1–4, refer to the table on page 250.

1. For every 100 people, __?__ can be expected to have A positive blood.

2. For every 200 people, __?__ can be expected to have A positive blood.

3. For every 300 people, __?__ can be expected to have A positive blood.

4. For every 650 people, __?__ can be expected to have A positive blood.

Find each answer. Write a decimal for the percent.

5. 12% of 900

6. 65% of 500

7. 8% of 80

8. 7% of 160

9. 28% of 75

10. 36% of 85

11. 3% of 27

12. 1% of 69

13. 95% of 188

14. 36% of 25

15. 18% of 215

16. 81% of 52

17. 16% of 200

18. 35% of 246

19. 87.5% of 912

20. 62.5% of 480

Find each answer. Write a fraction for the percent.

21. 25% of 16

22. 75% of 20

23. 80% of 60

24. 40% of 35

25. 10% of 516

26. 30% of 505

27. 60% of 325

28. 50% of 812

29. 40% of 324

30. 70% of 910

31. $37\frac{1}{2}\%$ of 56

32. $12\frac{1}{2}\%$ of 24

Find each answer. Use the method you prefer.

33. 25% of 160

34. 75% of 200

35. 1% of 27

36. 8% of 55

37. $37\frac{1}{2}\%$ of 64

38. $62\frac{1}{2}\%$ of 568

39. 30% of 60

40. 40% of 18

41. $16\frac{2}{3}\%$ of 420

42. $33\frac{1}{3}\%$ of 522

43. 2% of 90

44. 7% of 20

45. 50% of 456

46. 80% of 900

47. 35% of 200

48. 60% of 92

49. Which is greater, 30% of 40 or 40% of 32?

50. Which is greater, 20% of 50 or 25% of 45?

51. Which is greater, 50% of 600 or 40% of 550?

52. Which is greater, 90% of 36 or 80% of 45?

Discount

SALE!
3 DAYS ONLY!

40% off!
Reg. $32.95
Trail Pack

25% off!
Reg. $60
Binoculars

33% off!
Reg. $80
Boots

15% off!
Reg. $140
Sleeping Bag

The **discount** is the amount you save when you buy an item on sale. The **sale price** is the amount you actually pay. To find the sale price, you subtract the amount of discount from the regular price.

1. *Complete:* Sale price = Regular price − <u>?</u>

EXAMPLE	How much will you pay for a sleeping bag advertised on sale at 15% off?
READ	What are the facts? Regular price: $140 Rate of discount: 15%
PLAN	First answer the "hidden question": **What is the amount of discount?**
SOLVE	**1** Find the amount of discount. 15% of 140 = 0.15 × 140 ⟵ $\frac{\textbf{Rate of}}{\textbf{Discount}} \times \frac{\textbf{Regular}}{\textbf{Price}} = \textbf{Discount}$ = **21** The discount is **$21.** **2** Find the sale price. 140 − 21 = **119** ⟵ $\frac{\textbf{Regular}}{\textbf{Price}} - \textbf{Discount} = \frac{\textbf{Sale}}{\textbf{Price}}$ The sale price is **$119.**
CHECK	Did you use the facts correctly in the problem?

EXERCISES

Complete. Choose your answers from the box at the right.

1. The price of an item before it is marked down is called the __?__ .

2. The amount a customer saves by buying items on sale is called the __?__ .

3. To find the sale price of an item on sale, you first find the __?__ .

4. Amount of discount = Rate of discount × __?__ .

5. Sale price = Regular price − __?__ .

> discount
> rate of discount
> regular price
> sale price

For Exercises 6–8, find the discount on each item.

6.

CEILING FANS
52″

50% OFF
Regularly $79

7.

EVENING BAGS
25% OFF
Regularly $36

8.

CLOSE-OUT
Warm-up Suits

Originally $56.95
Now
40% OFF

Find the selling price of each item. Use the advertisements for Exercises 6–8.

9. Ceiling fans

10. Evening bags

11. Warm-up suits

For Exercises 12–15, use the ads on page 252.

12. How much will you pay for a trail pack advertised on sale at 40% off?

13. How much will you pay for binoculars on sale at 25% off?

14. How much will you pay for boots on sale at 33% off?

15. Mr. Hand bought the binoculars on sale. He also gets a 10% discount off the sale price. How much did he pay?

16. These ads show the same pair of shoes on sale at two different stores. At which store can you buy the shoes for less? Explain.

Sale
$6.00 Off
regular price $28.95
CROWNOVER'S SHOE STORE

Sale
15% Off
regular price $28.95
SOFT SOLE SHOE STORE

Sales Tax

Many cities and states have a **sales tax** on certain items or services. The sales tax rate is usually expressed as a percent.

1. What is the sales tax rate in your state?

2. Do all states have the same sales tax rate?

EXAMPLE

Carla Silverheels bought a pair of shoes for $42.50 and a pair of boots for $68. The sales tax rate is $4\frac{1}{2}\%$. How much will she pay in all?

READ

What are the facts?

Purchases: $42.50 and $68   Sales tax rate: 4.5%

PLAN

To find the total cost, first answer the "hidden questions."

What is the total purchase price?

What is the amount of tax?

SOLVE

☐1 Find the total purchase price.
$42.50 + $68.00 = **$110.50**

☐2 Find the amount of tax. ⟶ 4.5% of $110.50 = __?__

| 1 | 1 | 0 | . | 5 | 0 | × | 4 | . | 5 | % | | 4.9725 |

OR

| 1 | 1 | 0 | . | 5 | 0 | × | . | 0 | 4 | 5 | = | 4.9725 |

Amount of tax: **$4.97** ◀── *Nearest cent*

☐3 Find the total cost.
$110.50 + $4.97 = **$115.47**   The total cost is **$115.47.**

CHECK

Did you use the facts correctly in the problem?
Is the answer reasonable?

---

3. What are some advantages of a sales tax?

4. What are some disadvantages of a sales tax?

> **Excise taxes** are collected by the federal government on such items as gasoline and airplane tickets. Taxes such as these are sometimes called **"hidden taxes"** because they are "hidden" in what the customer pays for the item.

EXERCISES

Find the sales tax on each purchase. The sales tax rate is 5%.

1. $19.15   **2.** $6.85   **3.** $14.50   **4.** $21.00   **5.** $12.95

Find the sales tax on each purchase. The sales tax rate is 4%.

6. $9.00   **7.** $17.50   **8.** $12.42   **9.** $53.41   **10.** $35.99

Find the sales tax and the total cost for each purchase. The sales tax rate is $4\frac{1}{2}\%$.

11. $7.15   **12.** $10.58   **13.** $15.00   **14.** $18.85   **15.** $21.25

Find the total cost for each purchase. The sales tax rate is 8%.

16. $10.50   **17.** $15.75   **18.** $20.11   **19.** $23.84   **20.** $32.05

21. At Hightop Pharmacy, Mrs. Carr paid $2.25 for a toothbrush, $1.50 for a comb, and $3.89 for shampoo. The sales tax rate is 3%. Find the sales tax and the total cost.

22. Louis bought a shirt for $9.95 and a pair of blue jeans for $15.50. The sales tax rate is 4%. Find the sales tax and the total cost.

23. Deanna bought a bicycle for $159.95, an air pump for $17.65, and a rack for $10.99. The sales tax is $4\frac{1}{2}\%$. Find the sales tax and the total cost.

24. At the Record Ranch, Juan bought 2 albums for $6.99 each, and a cassette tape for $5.59. The sales tax rate is 5%. Find the total cost.

25. A round-trip airline ticket for a flight from Orlando, Florida to Portland, Oregon costs $452 without taxes. The federal excise tax is 8% of the cost. How much is the excise tax on this ticket?

26. Mr. and Mrs. Van filled the fuel tank of their recreational vehicle with 52 gallons of unleaded gasoline. The excise tax is 9.1¢ per gallon. How much did the Vans pay in excise tax?

Savings Accounts/Interest

Money deposited in a savings account earns *interest*. **Interest** is money earned on the funds or *principal* in the account.

> Interest = Principal × Rate × Time (in years)
> i = p × r × t, or $i = prt$

EXAMPLE 1

Jill Kiley deposited $800 in a savings account. The account pays a yearly interest rate of $5\frac{1}{2}\%$. How much interest did the $800 earn at the end of 9 months?

$$i = p \times r \times t \begin{cases} p = 800 \\ r = 5\frac{1}{2}\%, \text{ or } 0.055 \\ t = 9 \text{ months} = \frac{9}{12}, \text{ or } \frac{3}{4} \text{ years} \end{cases}$$

$$i = \overset{200}{\cancel{800}} \times 0.055 \times \underset{1}{\frac{3}{4}}$$

$$i = 33 \qquad \text{The interest is } \mathbf{\$33.}$$

When you have a savings account, your bank may mail you a monthly or **quarterly** (every 3 months) account statement. The statement shows all deposits, withdrawals, and interest credited to your account. The statement also shows the **new balance,** or adjusted amount, in the account.

> **New Balance** = **Previous Balance** + Interest + Deposit − Withdrawals

EXAMPLE 2

Find the new balance in Stephen Salk's account.

Depositor: Stephen Salk			Acc. No. 10-376-54	
Date	Withdrawal	Deposit	Interest	Balance
1/12		50.00		834.15
1/29		24.50		858.65
2/8	20.00			838.65
3/20		162.25		1000.90
4/01			13.93	?

Previous Statement		This Statement	
Date	Balance	Date	Balance
01-01	$784.15	04-01	?

$$\begin{aligned}\frac{\text{New}}{\text{Balance}} &= \frac{\text{Previous}}{\text{Balance}} + \text{Interest} + \text{Deposits} - \text{Withdrawals} \\ &= \$784.15 + \$13.93 + \$236.75 - \$20.00 \\ &= \mathbf{\$1014.83}\end{aligned}$$

EXERCISES

Complete. Choose your answers from the box at the right.

1. A sum of money that you take out of your savings account is called a __?__ .

2. Money deposited in a savings account earns __?__ .

3. To find the new balance on a bank statement, the deposits and interest are __?__ the previous balance.

deposit
interest
withdrawal
added to
subtracted from
principal

For Exercises 4–9, find the interest.

	Principal	Rate	Time
4.	$800	5%	4 months
5.	$1500	6%	9 months
6.	$2500	5%	6 months

	Principal	Rate	Time
7.	$7200	5%	10 months
8.	$1200	8%	3 months
9.	$10,000	8%	7 months

10. Mike has $600 in an account that pays a yearly interest rate of 6%. How much interest will the account earn in 6 months?

11. Wilma's investment account pays a yearly interest rate of 15%. She has $840 in the account. Find the interest after 4 months.

12. Find the new balance in Selena's account.

13. Find the new balance in Thomas' account.

Depositor: Selena Mink Acc. No. 10-451-66				
Date	Withdrawal	Deposit	Interest	Balance
5/02		150.00		565.07
5/28		62.40		627.47
6/5	30.00			597.47
6/29	25.00			572.47
7/01			9.83	?

Previous Statement		This Statement	
Date	Balance	Date	Balance
05-01	$415.07	07-01	?

Depositor: Thomas Kinder Acc. No. 10-355-41				
Date	Withdrawal	Deposit	Interest	Balance
1/10		100.00		788.72
1/27	50.00			738.72
2/5	25.50			713.22
2/22		195.25		908.47
3/01			11.87	?

Previous Statement		This Statement	
Date	Balance	Date	Balance
01-01	$688.72	03-01	?

Strategy: USING MENTAL COMPUTATION

Sometimes the percent of a number can be computed mentally. It is both useful and convenient to be able to determine the amount for a tip or to check that the amount of sales tax is correct without using paper and pencil.

Study the Examples.

EXAMPLES

a. 10% of $85 = __?__

Think: Move the decimal point one place to the left.
10% of $85 = **$8.50**

b. 20% of $98 = __?__

Think: Since 20% = 2 × 10%, find 10% of $98 first.
Then multiply by 2.
10% of $98 = $9.80
20% of $98 = 2 × $9.80, or **$19.60**

c. 5% of $36.20 = __?__

Think: Since 5% = $\frac{1}{2}$ of 10%, find 10% of $36.20 first.
Then multiply by $\frac{1}{2}$, or divide by 2.
10% of $36.20 = $3.62
5% of $36.20 = $3.62 ÷ 2, or **$1.81**

d. 15% of $154 = __?__

Think: Since 15% = 10% + 5%, add 10% of $154 and 5% of $154.
10% of $154 = $15.40
5% of $154 = $7.70
15% of $154 = $15.40 + $7.70, or **$23.10**

EXERCISES

Do each computation mentally. Write only the answers.

1. 10% of $105

2. 10% of $89

3. 10% of $125.40

4. 20% of $68

5. 20% of $126

6. 20% of $540

7. 5% of $65

8. 5% of $24

9. 5% of $92

10. 15% of $80

11. 15% of $120

12. 15% of $16

For the remaining exercises, use mental computation and shortcuts when you can. For Exercises 13–16, find the amount of discount.

13. Jeans marked at a 20% discount from $24

14. A radio regularly priced at $55 at a 15%-off sale

15. A television regularly priced at $220 at a 15%-off sale

16. A chair marked at a 5% discount from $45

Find the amount of sales tax if the sales tax rate is 5%.

17. Golf clubs: $280

18. Tape deck: $58

19. Running shoes: $42

The usual tip for restaurant service is 15% of the total cost of the food._ For Exercises 20–22, find the amount of the tip.

20. Total cost: $21

21. Total cost: $75

22. Total cost: $128

23. What is the regular price of the jogging suit?

24. What is the sale price of the suit?

25. If the sales tax is 5%, what will be the total cost of the suit, including sales tax?

> **Jogging Suit**
> Regularly $50
> **20% OFF**

26. What is the regular price of the grill?

27. What is the sale price of the grill?

28. If the sales tax is 6%, what will be the total cost?

> **Outdoor Grill**
> Regularly $200
> **10% OFF**

29. What is the original price of the racquet?

30. What is the sale price of the racquet?

31. The sales tax is 5%. What will be the total cost?

> **TENNIS RACQUET**
> Regularly $60
> **SALE 5% OFF**

For each of Exercises 32–35, determine the better buy.

32. A $20 lawn chair at 10% off or at $10 off

33. A $90 coat at 10% off or at $10 off

34. Hiking boots regularly priced at $75 at $10 off or at 10% off

35. A $45 tape deck at $10 off or at 10% off

36. For items with a regular price less than $100, which is the better buy, 10% off or $10 off? Explain.

Compound Interest

Compound interest is interest earned not only on the original principal but also on the interest earned during previous interest periods. You can use a table to compute compound interest.

Compound Interest for $1.00			
Quarters	1%	1.25%	1.5%
1	1.0100	1.0125	1.0150
2	1.0201	1.0252	1.0302
3	1.0303	1.0380	1.0457
4	1.0406	1.0509	1.0614
5	1.0510	1.0641	1.0773
6	1.0615	1.0774	1.0934
7	1.0721	1.0909	1.1098

EXAMPLE A bank offers savings accounts at 5% interest compounded quarterly (every 3 months). Find how much $10,000 will amount to (new balance) after 6 months.

1. Find the quarterly rate.
5% ÷ 4 = **1.25%**

2. Find the number of quarters.
6 mo = $\frac{1}{2}$ yr = **2 quarters**

3. Use the table to find how much $1.00 will amount to at a rate of 1.25% for 2 quarters. ⟶ **1.0252**

4. Multiply to find how much $10,000 will amount to at 1.25% for 2 quarters.
1.0252 × 10,000 = **$10,252** ⟵ *New balance*

EXERCISES

Find the new balance for interest compounded quarterly.

	Present Balance	Yearly Rate	Quarterly Rate	Time	Number of Quarters	New Balance
1.	$750	6%	?	6 mo	?	?
2.	$2000	5%	?	1 yr	?	?

3. Suppose you deposited $5000 at 6% interest compounded quarterly and your friend deposited $5000 at 5% interest compounded quarterly. How much more money would you have than your friend at the end of 18 months?

Chapter Summary

1. To write a decimal for a percent:
   1. Move the decimal point two places to the left.
   2. Drop the percent symbol.

2. To write a percent for a decimal:
   1. Move the decimal point two places to the right.
   2. Write the percent symbol.

3. To write a fraction for a percent:
   1. Write a fraction with a denominator of 100 for the percent.
   2. Write the fraction in lowest terms.

4. To write a percent for a fraction:
   1. Divide the numerator of the fraction by the denominator. Carry the division to two decimal places.
   2. Write a percent for the decimal.

5. To find a percent of a number:
   1. Write a decimal or a fraction for the percent.
   2. Multiply.

6. To find the sale price of an item:
   1. Multiply the rate of discount and the regular price to find the discount.
   2. Subtract the discount from the regular price.

7. To find the total cost of an item:
   1. Find the total purchase price.
   2. Multiply the sales tax rate and the total purchase price to find the amount of tax.
   3. Add the total purchase price and the amount of tax.

8. To find the amount of interest, multiply the principal, the rate of interest, and the time in years.

   $$i = p \times r \times t$$ *i = interest; p = principal*
   r = rate; t = time in years

---

1. $52\% = 52\%$
   $= \mathbf{0.52}$

2. $0.25 = 0.25$
   $= \mathbf{25\%}$

3. $60\% = \frac{60}{100}$
   $= \frac{3}{5}$

4. $\frac{3}{4} \longrightarrow$ $\begin{array}{r} 0.75 = \mathbf{75\%} \\ 4\overline{)3.00} \end{array}$

5. $5\% \times 40 = 0.05 \times 40$
   $= \mathbf{2}$

6. 20% of $\$120 = 0.20 \times 120$
   $= \$24$ *Dis-count*
   $\$120 - \$24 = \mathbf{\$96}$

7. $\$15 + \$45 = \$60$
   $5\% \times \$60 = 0.05 \times 60$
   $= \$3$
   $\$60 + \$3 = \mathbf{\$63}$

8. $p = \$600; r = 6\%;$
   $t = 9$ months
   $i = p \times r \times t$
   $i = \$600 \times 0.06 \times \frac{3}{4}$
   $i = \mathbf{\$27}$

Chapter Review

Part 1: VOCABULARY

For Exercises 1–9, choose from the box at the right the word(s) that complete(s) each statement.

1. Percent means per hundred or ___?___ .   (Page 240)

2. To write a decimal for a percent, move the decimal point ___?___ places to the left.   (Page 242)

3. To write a percent for a decimal, move the decimal point two places to the ___?___ .   (Page 242)

4. To write a percent for a fraction, first ___?___ the numerator of the fraction by the denominator.   (Page 246)

5. The amount you save when you buy an item on sale is the ___?___ .   (Page 252)

6. To find the sale price of an item, ___?___ the amount of discount from the regular price.   (Page 252)

7. In the formula $i = p \times r \times t$, time is expressed in ___?___ .
(Page 256)

> right
> subtract
> years
> hundredths
> divide
> two
> discount
> denominator
> excise

Part 2: SKILLS

Write a percent for each fraction.   (Pages 240–241)

8. $\frac{2}{100}$ 9. $\frac{4}{100}$ 10. $\frac{17}{100}$ 11. $\frac{35}{100}$ 12. $\frac{60}{100}$ 13. $\frac{18}{100}$

Write a decimal for each percent.   (Pages 242–243)

14. 43%   15. 54%   16. 321%   17. 280%   18. 61.2%   19. 17.9%

20. 5%   21. 3%   22. 2.1%   23. 7.9%   24. 0.8%   25. 0.5%

Write a percent for each deimal.   (Pages 242–243)

26. 0.26   27. 0.31   28. 0.73   29. 0.46   30. 0.09   31. 0.06

32. 0.431   33. 0.792   34. 0.068   35. 0.071   36. 0.007   37. 0.003

Write a fraction in lowest terms for each percent.   (Pages 244–245)

38. 16%   39. 24%   40. 5%   41. 4%   42. $\frac{7}{8}$%   43. $\frac{5}{7}$%

44. $\frac{7}{9}$%   45. $\frac{5}{9}$%   46. $6\frac{1}{2}$%   47. $5\frac{3}{4}$%   48. $12\frac{2}{3}$%   49. $15\frac{1}{2}$%

Write a percent for each fraction. (Pages 246–247)

50. $\frac{3}{5}$ **51.** $\frac{7}{10}$ **52.** $\frac{9}{20}$ **53.** $\frac{19}{20}$ **54.** $\frac{11}{25}$ **55.** $\frac{39}{50}$

56. $\frac{6}{7}$ **57.** $\frac{7}{9}$ **58.** $\frac{2}{3}$ **59.** $\frac{7}{8}$ **60.** $\frac{5}{6}$ **61.** $\frac{7}{16}$

Find each answer. (Pages 250–251)

62. 15% of 300 **63.** 23% of 200 **64.** 9% of 700 **65.** 7% of 210

66. 3% of 65 **67.** 1% of 82 **68.** 43% of 221 **69.** 51% of 145

70. 25% of 20 **71.** 50% of 14 **72.** $37\frac{1}{2}$% of 64 **73.** $66\frac{2}{3}$% of 39

74. 40% of 55 **75.** 75% of 28 **76.** $33\frac{1}{3}$% of 120 **77.** $12\frac{1}{2}$% of 72

Find the sales tax and total cost for each purchase. The sales tax rate is 7%. (Pages 254–255)

78. $7.95 **79.** $18.25 **80.** $32.00 **81.** $55.05 **82.** $137.21

For Exercises 83–86, find the interest. (Pages 256–257)

	Principal	Rate	Time			Principal	Rate	Time
83.	$500	5%	4 months		**85.**	$1800	6%	10 months
84.	$900	6%	9 months		**86.**	$3200	8%	3 months

Do each computation mentally. Write only the answer. (Pages 258–259)

87. 10% of $80 **88.** 10% of $75 **89.** 10% of $130.45

90. 20% of $34 **91.** 5% of $66 **92.** 15% of $420.00

Part 3: APPLICATIONS

93. About 2% of the population has B negative blood. Of 500 eleventh graders at Emerson High, how many can be expected to have B negative blood? (Pages 250–251)

95. The regular price of a hardwood table at Mason's Furniture is $300. Display models are sold at 40% off. Find the sale price. (Pages 252–253)

96. Raymond bought a puzzle for $5.67 and a magazine for $2.50. The sales tax rate is 6%. Find the total cost. (Pages 254–255)

94. Find the new balance in Ramona's account. (Pages 256–257)

Depositor: *Ramona Razon* Acc. No. 10 - 411 - 37				
Date	Withdrawal	Deposit	Interest	Balance
1/10		400.00		698.03
1/27		78.50		776.53
2/5	48.00			728.53
2/22	55.20			673.33
3/01			8.31	?

Previous Statement		This Statement	
Date	Balance	Date	Balance
01-01	$298.03	04-01	?

Chapter Test

Write a decimal for each percent.

1. 25%　　**2.** 63%　　**3.** 181%　　**4.** 4%　　**5.** 3%　　**6.** 6.2%

Write a percent for each decimal.

7. 0.32　　**8.** 0.41　　**9.** 0.58　　**10.** 0.046　　**11.** 0.073　　**12.** 0.007

Write a fraction in lowest terms for each percent.

13. 23%　　**14.** 44%　　**15.** 5%　　**16.** 6%　　**17.** $\frac{5}{6}$%　　**18.** $3\frac{1}{2}$%

Write a percent for each fraction.

19. $\frac{4}{5}$　　**20.** $\frac{3}{4}$　　**21.** $\frac{7}{20}$　　**22.** $\frac{5}{7}$　　**23.** $\frac{8}{9}$　　**24.** $\frac{3}{8}$

Find each answer.

25. 10% of 40　　**26.** $12\frac{1}{2}$% of 24　　**27.** 23% of 60　　**28.** 1% of 810

For Exercises 29–30, find the interest.

29. Principal: $600; Rate: 6%;
Time: 3 months

30. Principal: $10,000; Rate: 8%;
Time: 9 months

31. What is the sale price of the
calculator? Use mental computation.
Write only the answer.

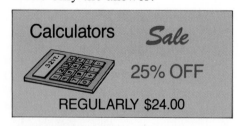

Calculators **Sale**

25% OFF

REGULARLY $24.00

32. Find the new balance in Paula's
account.

Depositor: *Paula Print* Acc. No. *10-001-02*				
Date	Withdrawal	Deposit	Interest	Balance
5/02		231.00		881.00
5/28	78.25			802.75
6/5	100.10			702.65
6/29		80.50		783.15
7/01			12.51	?

Previous Statement		This Statement	
Date	Balance	Date	Balance
05-01	*$650.00*	*07-01*	?

33. Cid bought a BMX bicycle for
$249.99 and a helmet for $32.00.
The sales tax rate is 5%. Find the
total cost.

Cumulative Maintenance Chapters 1–10

Choose the correct answer. Choose a, b, c, or d.

1. Add.

4.32 + 13.7 + 6.845 + 327.6

  a. 352.465   **b.** 351.365
  c. 352.365   **d.** 351.465

2. Which is equal to 70%?

  a. $\frac{7}{100}$ **b.** $\frac{3}{4}$

  c. $\frac{7}{10}$ **d.** $\frac{3}{5}$

3. Which shows the numbers listed in order from least to greatest?

  a. 0.32, 0.65, $\frac{9}{40}$, $\frac{17}{50}$

  b. $\frac{9}{40}$, 0.32, $\frac{17}{50}$, 0.65

  c. $\frac{17}{50}$, $\frac{9}{40}$, 0.65, 0.32

  d. 0.65, $\frac{17}{50}$, $\frac{9}{40}$, 0.32

4. Write a decimal for 8.3%.

  a. 8.3   **b.** 0.83
  c. 0.083   **d.** $0.8\frac{1}{3}$

5. Find the missing terms.

2, 7, __?__ , __?__ , 22, 27, 32

  a. 5, 5   **b.** 12, 17
  c. 13, 18   **d.** 8, 9

6. Solve the proportion for n.

$$\frac{13}{15} = \frac{n}{105}$$

  a. 5.4   **b.** 91
  c. 54   **d.** 9.1

7. Solve for n.

$$n + 8 = 32$$

  a. 30   **b.** 40
  c. 4   **d.** 24

8. Larry earns $7 per hour and works 38 hours per week. Deductions amount to $58. Find the net pay.

  a. $208   **b.** $205
  c. $266   **d.** $218

9. Use the graph below to find how many houses were sold in 1987.

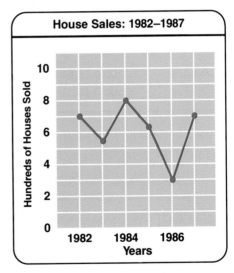

House Sales: 1982–1987

  a. 800   **b.** 700
  c. 900   **d.** 500

10. Which ratios are equivalent?

a. $\frac{3}{5}$ and $\frac{9}{25}$ b. $\frac{8}{15}$ and $\frac{4}{5}$

c. $\frac{30}{50}$ and $\frac{9}{15}$ d. $\frac{5}{8}$ and $\frac{50}{85}$

11. Multiply.
$$\frac{3}{4} \times \frac{8}{9}$$

a. $\frac{1}{3}$ b. $\frac{3}{4}$

c. $\frac{2}{3}$ d. $\frac{1}{2}$

12. Solve the proportion for n.
$$\frac{8}{9} = \frac{40}{n}$$

a. 5   b. 45

c. $3\frac{1}{2}$ d. 54

13. Rob bought a pair of boots that were on sale for $\frac{1}{4}$ off the regular price of $87.52. What was the amount of discount?

a. $32.50   b. $21.88
c. $20   d. $25

14. Give the length of this pencil to the nearest $\frac{1}{4}$–inch.

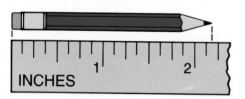

INCHES

a. $2\frac{2}{4}$ b. $2\frac{1}{4}$

c. $2\frac{3}{4}$ d. 2

15. Sam bought two shirts for $11.00 each. He bought four pairs of socks for $2.00 each. How much did he pay in all?

a. $9   b. $13
c. $24   d. $30

16. Find 30% of 90.

a. 270   b. 300
c. 30   d. 27

17. Solve for n.
$$\frac{n}{1.7} = 3.2$$

a. 7.12   b. 5.44
c. 71.2   d. 54.4

18. The scale on a map is:

1 cm represents 200 km

The distance between two cities is $7\frac{1}{2}$ centimeters. Find the actual distance in kilometers.

a. 1500   b. 1450
c. 7500   d. 750

19. John spent $36.50 for bowling shoes. He paid 4% sales tax. What is the sales tax?

a. $1.46   b. $1.58
c. $5.16   d. $1.64

20. Solve for n.
$$3n + 1.2 = 7.5$$

a. 2.1   b. 3.23
c. 18.9   d. 3.6

More on Percent

CHAPTER 11

Name _Tom Russell_   Number correct __36__
                       Grade __90%__

Match the traffic warning sign with its corresponding description.

A.

B.

C.

D.

E.

1. You are approaching a point where the traffic lanes come together.

2. You are near a school. Slow down.

3. A sharp turn to the right. Slow down.

4. A road crosses the main highway ahead. Look right and left for other traffic.

5. You are approaching a railroad crossing. Slow down, look, and listen.

6. A gradual curve to the left. Slow down and keep well to the right.

7. The highway on which you are traveling is divided ahead. You will be entering a one-way roadway.

1. __B__

2. __a__

3. __E__

4. __D̸__

5. __H__

6. __K__

7. __R__

- What is 90% of the total number of questions?

- What will 10% of the total number of questions equal?

- What will 100% of the total number of questions equal?

- How many questions were there on the test?

Finding The Percent

Apple growers in the United States produce about 2 million bushels of apples every year. Most of the apples grown commercially and in home gardens are of three varieties, Delicious, Golden Delicious, and McIntosh.

EXAMPLE 1

An apple weighing 160 grams contains 136 grams of water. What percent of the apple is water?

1 Write an equation.

What percent of 160 is 136?

$n \times 160 = 136$, or

$160n = 136$

2 Solve the equation.

$160n = 136$

$\dfrac{160n}{160} = \dfrac{136}{160}$

$n = \dfrac{136}{160}$

3 Write a percent for $\dfrac{136}{160}$.

$$\dfrac{136}{160} \longrightarrow 160 \overline{)\, 136.00} \quad \begin{array}{c} 0.85 = 85\% \end{array}$$

So **85%** of the apple is water.

NOTE: In Step 3 of Example 1, you could write $\dfrac{136}{160}$ in lowest terms.

$$\dfrac{\overset{17}{\cancel{136}}}{\underset{20}{\cancel{160}}} = \dfrac{17}{20} \longrightarrow 20 \overline{)\, 17.00} \quad \overset{0.85}{} \longrightarrow 85\%$$

Sometimes the percent may be greater than 100%.

1. When the numerator of a fraction is greater than the denominator, will the fraction be greater than 1 or less than 1?

2. When you write a percent for a fraction whose numerator is greater than the denominator, will it be greater or less than 100%? Explain.

EXAMPLE 2

Nine is what percent of 6?

1 $9 = n \times 6$, or

2 $6n = 9$ ← Same as $9 = 6n$.

$\dfrac{6n}{6} = \dfrac{9}{6}$

$n = \dfrac{9}{6}$, or $\dfrac{3}{2}$

3 $\dfrac{3}{2} \longrightarrow 2 \overline{)\, 3.0} \quad \overset{1.5}{}$

$1.50 = 150\%$

So **150%** of 6 is 9.

EXERCISES

Write "greater than 100%" or "less than 100%" for each fraction.

1. $\frac{3}{4}$ 2. $\frac{4}{3}$ 3. $\frac{1}{2}$ 4. $\frac{2}{1}$ 5. $\frac{196}{75}$ 6. $\frac{96}{95}$

Write an equation for each exercise. Do NOT solve.

7. What percent of 20 is 15?

8. 15 is what percent of 20?

9. What percent of 56 is 60?

10. What percent of 60 is 56?

11. Thirty is what percent of 66?

12. Fifty is what percent of 2?

Solve.

13. What percent of 80 is 20?

14. What percent of 25 is 8?

15. What percent of 15 is 5?

16. What percent of 4 is 12?

17. What percent of 16 is 48?

18. What percent of 72 is 36?

19. Three is what percent of 4?

20. Two is what percent of 5?

21. Seven is what percent of 8?

22. One is what percent of 4?

23. Twenty is what percent of 16?

24. Ninety is what percent of 18?

25. What percent of 60 is 50?

26. Sixty-three is what percent of 70?

27. What percent of 180 is 30?

28. Twenty-five is what percent of 10?

29. Seventy-five is what percent of 200?

30. What percent of 8 is 12?

31. There are 30 varieties of wild apples. Seven of these varieties are grown in the United States. What percent of the varieties can be found in the United States?

32. Of 120 apple trees, 72 produce Delicious, Golden Delicious, or McIntosh apples. What percent of the total number of trees is this?

33. Of 280 apple trees, 56 older trees were replaced. What percent were not replaced?

34. John Chapman (Johnny Appleseed) planted apple orchards along the western frontier from 1797 to 1845. He spent 25 of these years in Ohio. What percent of the planting years did he spend in Ohio (nearest 10%)?

35. What percent of the area of the large rectangle is the area of the small rectangle?

9 in
4 in
2 in
3 in

Percent of Increase or Decrease

The cost of apples in a supermarket rose from 45¢ a pound to 60¢ a pound.

1. By how much per pound did the cost of the apples increase?

2. Write a fraction for this ratio:
$$\frac{\text{Amount of Increase}}{\text{Original Amount}}$$

EXAMPLE 1 Find the percent of increase in the cost of the apples.

① Amount of increase: 60¢ − 45¢ = **15¢**

② Percent of increase = $\dfrac{\text{Amount of Increase}}{\text{Original Amount}}$

$$= \frac{15}{45} \longrightarrow 45 \overline{)\begin{array}{l} .33\frac{1}{3} = 33\frac{1}{3}\% \\ 15.00 \end{array}}$$

$$\begin{array}{r} 13\ 5 \\ \hline 1\ 50 \\ 1\ 35 \\ \hline 15 \end{array} \longleftarrow \frac{15}{45} = \frac{1}{3}$$

The cost of apples increased **33⅓%**.

Finding the percent of decrease is similar to finding the percent of increase.

3. *Complete:* Percent of decrease = $\dfrac{?}{\text{Original Amount}}$

EXAMPLE 2 In a year in which a disease called "apple scab" attacked the trees, the United States production of apples fell from 190 million bushels to 114 million bushels. What was the percent of decrease?

① Amount of decrease: 190 − 114 = **76** ◀ **In millions of bushels**

② Percent of decrease = $\dfrac{\text{Amount of Decrease}}{\text{Original Amount}}$

$$= \frac{76}{190} \longrightarrow 190 \overline{)\begin{array}{l} 0.4 \\ 76.0 \end{array}}$$

$$= 0.4, \text{ or } \mathbf{40\%} \quad ◀ \text{ \textbf{Percent of decrease}}$$

4. Why can you use 190 and 114 in place of 190,000,000 and 114,000,000 in step ☐1 of Example 2?

EXERCISES

For Exercises 1–6, write the fraction you could use to find the percent of increase or decrease.

Item	Original Cost	Present Cost
1. Rent	$320/mo	$368/mo
2. Electric Bill	$60/mo	$55/mo
3. Car	$4650	$3720

Item	Original Cost	Present Cost
4. Horse	$88,000	$110,000
5. Groceries	$24.20	$25.41
6. Motorcycle	$2400	$1500

Find the percent of increase or decrease.

7. From 8 to 14   **8.** From 15 to 6   **9.** From 16 to 8   **10.** From 100 to 5

11. From 500 to 1500   **12.** From 100 to 90   **13.** From 100 to 20   **14.** From 90 to 100

15. A bakery's sales of apple pies rose from $2\frac{1}{2}$ dozen per day to 4 dozen per day. What was the percent of increase?

16. The price of a home video rental was $3 in April. In May the price was $2. What was the percent of decrease?

17. On weekdays during the day, teenagers watch about 3 hours of TV. During the evening, they watch about 4 hours of TV. What is the percent of increase?

18. During the football strike in 1987, attendance at some stadiums fell from about 60 thousand to about 8 thousand. What was the percent of decrease?

The broken line graph at the right shows the approximate U.S. population growth from 1870 through 1990 (estimated).

19. What was the percent of increase in population from 1870 to 1890?

20. How much greater is the percent of increase in population from 1950 to 1970 than the estimated percent of increase from 1970 to 1990?

21. During which 40–year period was the <u>increase</u> the greatest?

22. During which 40–year period was the <u>percent of increase</u> in population the greatest? Why is this not the same as the answer to Exercise 21?

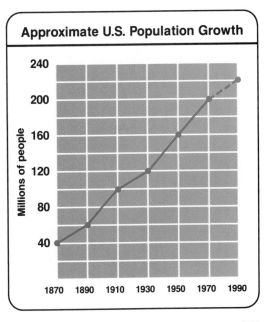

Approximate U.S. Population Growth

Finding a Number Given the Percent

The United States produces about 60% of the world's grapefruit. Grapefruit trees were first planted in Florida around 1820. By the year 1900, Florida was shipping grapefruit throughout the United States.

EXAMPLE 1 | Florida produces about 75% of the grapefruit grown in the United States. In a year in which 60 million boxes of grapefruit were produced in Florida, how many boxes were produced in the United States?

Think: 60 million is 75% of how many million?

1. Write an equation.

$$60 = 75\% \times n, \quad \text{or}$$

$$0.75n = 60 \quad \begin{array}{l} \textbf{0.75 n = 60 is the} \\ \textbf{same as 60 = 0.75n.} \end{array}$$

2. Solve.

$$\frac{0.75n}{0.75} = \frac{60}{0.75}$$

$$n = 80 \quad \begin{array}{l} \textbf{60 million is 75\%} \\ \textbf{of 80 million.} \end{array}$$

About **80 million boxes of grapefruit** were produced.

1. If 27 is 30% of some number n, will n be greater or less than n?

EXAMPLE 2 | 19 is 20% of what number?

$$19 = 20\% \times n, \quad \text{or} \quad \begin{array}{l} \textit{Write an} \\ \textit{equation.} \end{array}$$

$$0.20n = 19$$

$$\frac{0.20n}{0.20} = \frac{19}{0.20}$$

$$n = 95 \qquad \text{So 19 is 20\% of } \textbf{95.}$$

EXERCISES

Write an equation for each exercise. Do NOT solve.

1. 18 is 10% of what number?

2. 15 is 75% of what number?

3. 6 is 15% of what number?

4. 12 is 16% of what number?

5. Don spent $2, which is 25% of his money. How much does he have?

6. A baseball team lost 8 games. This is 40% of the games played. How many games were played?

Find the number.

7. 9 is 2% of what number?

8. 18 is 60% of what number?

9. 5 is 25% of what number?

10. 52 is 80% of what number?

11. 7 is 20% of what number?

12. 81 is 27% of what number?

13. 16 is 80% of what number?

14. 18 is 10% of what number?

15. 23 is 50% of what number?

16. 42 is 25% of what number?

17. 14 is 70% of what number?

18. 12 is 30% of what number?

19. 15 is 1% of what number?

20. 4 is 2% of what number?

Solve.

21. The Emerson High baseball team lost 4 games this year. This is 16% of all the games they played. How many games did they play?

22. Hondo has saved $100 for a new bicycle. This is 40% of the price. What is the price of the bicycle?

23. Juan made 15 baskets in a basketball game. This was 75% of his free throws. How many free throws did Juan make?

24. The sale price of a television set is $120. The sales clerk says this is 80% of the regular price. What is the regular price?

25. Nora spent four vacation days at the beach. This was 20% of her total number of vacation days. How long was her vacation?

26. Tanya paid $13.50 for a one-year subscription to New Teen magazine. This price was 60% of the newsstand price. What would be the yearly cost of buying the magazine at the newsstand?

Mid-Chapter Review

Write an equation for each exercise. Then solve.   (Pages 268–269)

1. What percent of 70 is 14?

2. What percent of 50 is 5?

3. 3 is what percent of 15?

4. 54 is what percent of 27?

5. What percent of 30 is 25?

6. 9 is what percent of 12?

Find the percent of increase or decrease.   (Pages 270–271)

7. From 9 to 15   **8.** From 20 to 30   **9.** From 18 to 12   **10.** From 25 to 15

11. From 50 to 10   **12.** From 75 to 50   **13.** From 100 to 120   **14.** From 150 to 165

Write an equation for each exercise. Then solve.   (Pages 272–273)

15. 50 is 20% of what number?

16. 9 is 30% of what number?

17. 18 is 75% of what number?

18. 4 is 20% of what number?

19. 25 is 50% of what number?

20. 21 is 7% of what number?

Solve.

21. An orchard contains 120 apple trees. Ninety of the trees produce Golden Delicious apples. What percent of the total number of trees in the orchard is this?   (Pages 268–269)

22. In 1980, the value of an early comic book was $4000. In 1987, the value was $5200. What was the percent of increase?   (Pages 270-271)

MAINTENANCE

For Exercises 23–24, write a ratio in lowest terms.   (Pages 224–225)

23. Kiyo won 24 out of 30 games that he pitched. What is the ratio of games won to games played?

24. For every 10 columns of news stories in a newspaper, four columns are advertisements. What is the ratio of advertisements to news stories?

25. Bob Lightfoot deposits into his bank account checks of $25, $100, and $470. He receives $75 in cash. How much is his total deposit?
(Pages 36–37)

26. Janet has $80 for buying clothes. She wants to buy 2 pairs of jeans for $15 each and 3 blouses for $18 each. How much more money does she need to buy all five items?
(Pages 8–9)

Math in Business: COMMISSION

Sometimes a salesperson's earnings are based on a fixed percent of total sales. This percent is called the **rate of commission.**

EXAMPLE

Emily Coalter accepted the job advertised at the right.

In March, her sales total was $14,480. Find her total earnings for March.

> **SALESPERSON**
> Good opportunity to earn high pay. $400 per mo. + a 4% commission on sales. Car necessary. Will train. Good benefits. Equal Oppty Employer

[1] Find the amount of commission.

Total Sales × **Commission Rate** = **Amount of Commission**

$\boxed{1}\boxed{4}\boxed{4}\boxed{8}\boxed{0}$ $\boxed{\times}$ $\boxed{4}\boxed{\%}$ $\boxed{579.2}$

OR

$\boxed{1}\boxed{4}\boxed{4}\boxed{8}\boxed{0}$ $\boxed{\times}$ $\boxed{.}\boxed{0}\boxed{4}$ $\boxed{=}$ 579.2 = $579.20

[2] Find the total earnings.

$400 + 579.20 = **$979.20** ◀ Salary + Commission = **Total Earnings**

Emily's total earnings for March were **$979.20.**

EXERCISES

For Exercises 1–3, find the commission.

1. Sales: $25,000
   Rate of Commission: 5%

2. Sales: $6500
   Rate of Commission: 3%

3. Sales: $9000
   Rate of Commission: $7\frac{1}{2}\%$

4. A lawyer earned a 15% commission for collecting a debt of $20,000 for a client. What was the amount of commission?

5. A real estate agent received a 7% commission for selling a house for $138,000. How much did the agent receive?

6. An auto mechanic receives a weekly salary of $480 plus a commission of $33\frac{1}{3}\%$ on the service income brought in each week. What is the weekly income of a mechanic who brings in $1860 in service income one week?

7. Elena Carson receives a salary of $800 per month plus commission. The rate of commission is 5% of the first $2000 of sales, and 6% on sales over $2000. Find her total income on monthly sales of $4380.

Using Proportions to Solve Percent Problems

You can also use proportions to solve percent problems.

1. Write a ratio for 75%.

EXAMPLE 1

Eric answered 75% of the 40 questions on a test correctly. How many items did he solve correctly?

1 **Think:** 75% of 40 is what number?

          Rate   **Whole**   **Part**

2 Write a proportion.

Rate ⟶ $\frac{75}{100} = \frac{n}{40}$ ⟵ **Part**
                ⟵ **Whole**

3 Solve the proportion.

$$\frac{75}{100} \times \frac{n}{40}$$

$$75 \times 40 = 100 \times n, \text{ or}$$
$$100n = 75 \times 40$$
$$\frac{100n}{100} = \frac{3000}{100}$$
$$n = 30$$

Eric solved **30 items** correctly.

In Example 2, you use a proportion to find the rate.

EXAMPLE 2

In a science class, two out of every 5 students are boys. What percent are boys?

1 **Think:** Two is what percent of 5?

       Part   **Rate**   **Whole**

2 Proportion: **Rate** ⟶ $\frac{n}{100} = \frac{2}{5}$ ⟵ **Part**
                           ⟵ **Whole**

3 Solve: $\frac{n}{100} \times \frac{2}{5}$

$$n \times 5 = 100 \times 2, \text{ or}$$
$$5n = 200$$
$$\frac{5n}{5} = \frac{200}{5}$$
$$n = 40 \quad \frac{40}{100} = 0.40 = 40\%$$

So **40%** of the class are boys.

Example 3 shows how to find the whole, given the part and the rate.

EXAMPLE 3

Ellis left a tip of $2.40 at a restaurant. This was 15% of the bill. How much was the bill?

1. **Think:** $2.40 is 15% of what number?

Part	Rate	Whole

2. Rate → $\dfrac{15}{100} = \dfrac{2.40}{n}$ ← Part ← Whole

3. $15 \times n = 100 \times 2.40$

   $$\frac{15n}{15} = \frac{240}{15}$$

   $n = 16$ The bill was **$16.**

EXERCISES

For each of Exercises 1–8:

a. *Identify the rate, the whole, and the part.*

b. *Write the proportion you could use to solve the problem. Do NOT solve the proportion.*

1. What number is 30% of 60?

2. What percent of 100 is 20?

3. 12 is what percent of 75?

4. 15 is 20% of what number?

5. What percent of 32 is 8?

6. 36% of 68 is what number?

7. 63% of what number is 63?

8. What percent of 54 is 18?

For Exercises 9–30, use a proportion to solve.

9. 5% of 60 is what number?

10. 10% of 40 is what number?

11. 20% of 45 is what number?

12. 50% of 84 is what number?

13. 6 is what percent of 12?

14. 3 is what percent of 4?

15. 6 is what percent of 10?

16. 24 is what percent of 12?

17. 24 is 12% of what number?

18. 18 is 36% of what number?

19. 20 is 20% of what number?

20. 44 is 55% of what number?

21. 25% of 48 is what number?

22. 17 is 85% of what number?

23. 15 is what percent of 75?

24. 16% of 125 is what number?

Use a proportion to solve each problem.

25. Of 150 students in a high school band, 20% play brass instruments. How many students play brass instruments?

26. An auctioneer sold goods amounting to $5000. As payment, he received 6% of the total sales. How much did he receive?

27. Of 24 shots attempted by Sandra in a basketball game, 18 were baskets. What percent of her shots were baskets?

28. Bobbie Jo worked 16 math problems. She solved 2 problems incorrectly. What percent of the problems did she solve incorrectly?

29. At Thoreau High, 90 students participate in school clubs. This is 20% of the number of students attending the school. How many students attend Thoreau High?

30. In three hours, 75 passenger jets landed at an airport. Twelve percent of the jets belonged to Skyway Airlines. How many jets belonged to Skyway Airlines?

31. Mr. Kline's class collected 45 cans of food for a charity drive. This was 15% of the total number of cans collected at the school. How many cans of food were collected in all?

32. During the first 40 school days, Emilio was absent 2 days. What percent of the first 40 school days was he present?

33. Kay Wu spent 16 days on vacation. She spent 25% of her vacation at Padre Island and 50% of her vacation in New Orleans. The remaining days were spent traveling. How many days did she spend traveling?

Review of Percent

Each of the three types of percent problems involves three numbers. Two of these numbers are known. The third number is the one you have to find.

Type 1	Type 2	Type 3
What number is 60% of 98?	What percent of 40 is 18?	14 is 5% of what number?
Unknown Known Known	Unknown Known Known	Known Known Unknowm

When you write an equation (or a proportion) for a percent problem, let n represent the unknown.

EXAMPLE 1

What number is 60% of 98?

$$n = 60\% \times 98 \quad \longleftarrow \textbf{Write an equation.}$$

$$n = 0.60 \times 98$$

$$n = \mathbf{58.8}$$

So 60% of 98 is **58.8**.

1. How could you check your answer to Example 1?

EXAMPLE 2

What percent of 40 is 18?

$$n \times 40 = 18, \text{ or}$$

$$40n = 18$$

$$\frac{40n}{40} = \frac{18}{40}$$

$$n = \frac{9}{20} \longrightarrow 20 \overline{)9.00} \longrightarrow \mathbf{45\%}$$

$$\begin{array}{r} 0.45 \\ 20 \overline{)9.00} \\ \underline{8\ 0} \\ 1\ 00 \\ \underline{1\ 00} \end{array}$$

So **45%** of 40 is 18.

2. How could you check your answer to Example 2?

EXAMPLE 3

14   is   5% of <u>what number?</u>

$14 = 5\% \times n$

$14 = 0.05n,$ or $0.05n = 14$

$$\frac{0.05n}{0.05} = \frac{14}{0.05}$$

$$n = \mathbf{280}$$

So 14 is 5% of **280**.

3. How could you check your answer to Example 3?

EXERCISES

For Exercises 1–4, choose the correct equation.

1. What percent of 10 is 3?
   a. $n \times 3 = 10$
   b. $n \times 10 = 3$
   c. $n = 10 \times 3$

2. What is 3% of 24?
   a. $n \times 0.03 = 24$
   b. $n \times 24 = 0.03$
   c. $n = 0.03 \times 24$

3. 14 is 5% of what number?
   a. $14 \times n = 0.05$
   b. $14 = 0.05 \times n$
   c. $14 \times 0.5 = n$

4. 16 is 15% of what number?
   a. $16 = 0.15 \times n$
   b. $16 \times n = 0.15$
   c. $16 \times 0.15 = n$

Match each question with the correct equation.

	Problem	Equations
5.	What is 30% of 50?	**a.** $n \times 50 = 30$
6.	50 is 30% of what number?	**b.** $50 = 0.30 \times n$
7.	What percent of 50 is 30?	**c.** $n = 0.30 \times 50$
8.	What percent of 72 is 18?	**d.** $18 = 0.72 \times n$
9.	What is 72% of 18?	**e.** $n \times 72 = 18$
10.	18 is 72% of what number?	**f.** $n = 0.72 \times 18$

Solve.

11. What percent of 20 is 12?

12. What is 45% of 120?

13. 18 is 90% of what number?

14. What is 18% of 150?

15. 50 is 50% of what number?

16. What percent of 84 is 63?

17. What percent of 72 is 60?

18. 72 is 9% of what number?

Review: PROBLEMS INVOLVING PERCENT

You have studied various types of problems involving percent.
Before writing an equation (or a proportion) for a percent problem,
you can make the problem simpler to solve by restating the
question.

EXAMPLE 1

About 70% of a person's body weight is
water. For a person who weighs 125
pounds, how much of the weight is water?

1	Restate the question.	What is 70% of 125?
2	Write an equation.	$n = 70\% \times 125$
3	Solve.	$n = 0.70 \times 125$
		$n = \mathbf{87.5}$

About **87.5 pounds** are water.

1. In Example 1, how much of the person's weight is <u>not</u> water?

EXAMPLE 2

There are 5600 registered voters in
Lonetree County. Of them, 2100 voted in
the last election. What percent of registered
voters voted?

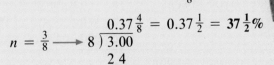

1	What percent of 5600 is 2100?

$$\begin{array}{c} 2 \\ 3 \end{array} \quad \begin{aligned} n \times 5600 &= 2100 \\ \frac{5600n}{5600} &= \frac{2100}{5600} \\ &= \frac{3}{8} \end{aligned}$$

$$n = \frac{3}{8} \longrightarrow \quad \begin{array}{r} 0.37\frac{4}{8} = 0.37\frac{1}{2} = 37\frac{1}{2}\% \\ 8\overline{\smash{)}3.00} \\ \underline{2\ 4} \\ 60 \\ \underline{56} \\ 4 \end{array}$$

Of 5600 registered voters,
$37\frac{1}{2}\%$ voted.

2. In Example 2, what percent of the registered voters did <u>not</u> vote in
the election?

EXAMPLE 3

Shirts'n Sweaters advertised that you could save $8 if you bought a sweater at a 20%-off sale. What was the regular price of the shirt?

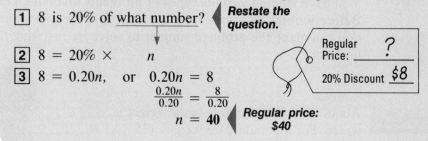

1 8 is 20% of _what number?_ ◀ *Restate the question.*

2 $8 = 20\% \times \quad n$

3 $8 = 0.20n, \quad$ or $\quad 0.20n = 8$

$$\frac{0.20n}{0.20} = \frac{8}{0.20}$$

$$n = 40$$ ◀ *Regular price: $40*

Regular Price: **?**

20% Discount **$8**

EXERCISES

For Exercises 1–7, choose the correct question for each problem. Choose **a, b,** *or* **c.**

1. Three hundred students participated. This is 80% of the students. How many students are there?
  a. What percent of 300 is 80?
  b. What is 80% of 300?
  c. 300 is 80% of what number?

2. One hundred fifty students took the test. This is 50% of the students. How many students are there?
  a. 150 is 50% of what number?
  b. What is 50% of 150?
  c. What percent of 150 is 50?

3. Five students live on South Street. There are 400 students in all. What percent live on South Street?
  a. What is 5% of 400?
  b. What percent of 400 is 5?
  c. 400 is 5% of what number?

4. One hundred eighty students voted. There are 200 students. What percent voted?
  a. What percent of 200 is 180?
  b. What is 180% of 200?
  c. 200 is 180% of what number?

5. Four percent of the students are absent. There are 900 students. How many are absent?
  a. What percent of 900 is 4?
  b. What is 4% of 900?
  c. 900 is 4% of what number?

6. Two hundred eighty-five students graduated. This is 95% of the seniors. How many seniors were there?
  a. 95% of 285 is what number?
  b. 285 is 95% of what number?
  c. What percent of 285 is 95?

7. Eighty students ride bicycles to school. There are 200 students. What percent ride bicycles to school?
  a. What percent of 80 is 200?
  b. 80% of 200 is what number?
  c. What percent of 200 is 80?

Solve each problem.

8. A real estate agent earns 4% on sales. How much does the agent earn on a house that sells for $80,000?

9. A family budgets 28% of take-home pay for food. The family's monthly take-home pay is $1600. How much is budgeted for food each month?

10. Jill has read 50 pages of a book that has 150 pages. What percent of the book has she read?

11. A waiter in a restaurant received a tip of $2.25 from a customer. The customer's bill for the meal was $15.00. What percent of the bill was the tip?

12. One season, a kicker scored a field goal on 80% of his attempts. He scored 20 field goals. How many attempts did he have?

13. Juanita paid a sales tax of $0.12 on a pair of socks. The rate of sales tax in her state is 5%. How much did she pay for the socks?

14. A certain automobile lost 14% of its original value in a year. When new, the car cost $8200. How much did the value of the car decrease in one year?

15. Ted, who earns $5.40 an hour, receives a raise of $0.75 an hour. What is the percent of increase? Round your answer to the nearest whole percent.

16. Lisa paid $67 on a bedroom set. This was 10% of the purchase price. How much does she have left to pay?

17. Bob has 25 dimes and 30 nickels. What percent of the total amount of money is in dimes?

18. A 9 foot by 12 foot rectangular rug is placed in a room 12 feet wide and 15 feet long. What percent of the floor is covered by the rug?

19. Carole bought one record at the regular price. She bought a second record at 50% off. She saved $3.01 on the second record. What was the cost for both records?

Strategy: USING ESTIMATION WITH PERCENTS

In everyday life, consumers make quick estimates to determine the discount or sale price, to compare prices, to compute the amount of tip for a restaurant meal, and so on. This often involves percents.

EXAMPLE 1

Estimate the sale price of the basketball.

Think: When you save 20%, you pay 80%.

$80\% = \frac{4}{5}$ $19.95 is about $20.

$\frac{4}{5} \times \$20 = \$16$

The sale price is about **$16.**

Basketball

REGULARLY $19.95
20% OFF

1. What percent do you pay if the discount is 45%?

2. What percent do you pay if the discount is $33\frac{1}{3}\%$?

3. What other method could you use to estimate the sale price in Example 1?

In Example 2, you use mental computation to estimate the amount of a tip and to estimate the total cost.

EXAMPLE 2

A restaurant bill totals $29.60 without tax. You want to leave a tip of 15% of the cost of the meal.

a. Estimate the tip.

Think: $29.60 is about $30   and   15% = 10% + 5%.

You know that 10% of $30 is **$3.**
Also, 5% is $\frac{1}{2}$ of 10%. So 5% of $30 = $\frac{1}{2} \times$ $3, or **$1.50.**
Then 15% of $30 is about $3 + $1.50, or **$4.50.**

b. Estimate the total cost of the meal if the sales tax is 4%.

Think: $29.60 is about $30.

You know that 1% of $30 is $0.30.
So 4% of $30 is 4 × $0.30, or **$1.20.** ◀ *About $1*

Estimated total cost:
$30 + $4.50 + $1 = **$35.50** ◀ *Cost of Meal* + *Tip* + *Sales Tax*

4. Why is it useful to estimate the total cost of a restaurant bill?

EXERCISES

Estimate the sale price.

1. $89.98 60% off!

2. $79.99 25% off!

3. $56.70 10% off!

4. $602.80 19% off!

5. Mr. Paul's Cookware
Regularly $89.99
30% OFF!

6. Sweaters
Regularly $37.89
60% OFF!

7. 14K Gold Necklace
Regularly $59.98
33⅓% OFF!

Estimate the tip of 15% for each restaurant meal.

8. $25.80 **9.** $9.05 **10.** $39.60 **11.** $9.75 **12.** $19.20

13. Dolores ate lunch at a sidewalk cafe. The cost of the meal amounted to $6.90. The rate of tax was 5%. Estimate the total cost, including a 15% tip.

14. The bill for Nick and Karen's dinner at the Seaside Inn Restaurant cost $19.67. They left a tip of $5. Is the tip more than 15% of the bill?

For Exercises 15–18, choose a strategy from the box at the right that you can use to solve each problem.

a. *Name the strategy.*

b. *Solve the problem.*

> Guess and check
> Making a model
> More than one step
> Using a drawing

15. Bill, Cathy, Jon, and Alix each left a tip of 20% for their meal. Cathy left $2 more than Alix. Alix's meal cost more than Jon's and less than Bill's. Whose tip was the smallest?

16. An architect's drawing uses this scale:

$$3 \text{ cm} = 5 \text{ m}$$

If the actual length of a dining room is 12 meters, what is the length on the drawing?

17. Which is more, a discount of $33\frac{1}{3}\%$ on a regular price of $69, or a discount of 30% on a regular price of $90?

18. Marlene earned $249 last week. She worked 37 hours at the regular rate per hour and 3 hours at time and a half. What is her regular hourly rate?

Averages and Percents

You have learned that *a percent is a ratio*. Because of this, computing averages involving percents is done in a special way.

EXAMPLE

In one basketball game, Adam took 15 shots and made 80% of them. In the next game, Adam took 35 shots and made 60% of them. What was his shooting average for the two games?

1 Total shots taken: $15 + 35 = $ **50**

2 Find the total shots made.

$$80\% \text{ of } 15 = \tfrac{4}{5} \times 15 = 12$$

$$60\% \text{ of } 35 = \tfrac{3}{5} \times 35 = \underline{21}$$

Total shots made: **33**

3 Write the ratio of total shots made to total shots taken.

$\dfrac{33}{50}$ ◄—— **Shots made**
◄—— **Shots taken**

4 Write a percent for the ratio.

$\dfrac{33}{50} = \dfrac{66}{100} = $ **66%**

Adam's shooting average was **66%**.

1. Why can't you find Adam's shooting average by finding the mean of 80% and 60%?

EXERCISES

1. In one week, a quarterback completed 70% of his 30 attempted passes. During the next week, he completed 40% of his 20 attempts. What was his completion average for both weeks?

2. Ada's average on 5 tests was 60%. On the next 10 tests, her average was 90%. What was her average for the 15 tests?

3. Marilyn averaged 91% on 10 tests. The final examination is counted as equal to five tests. On the final examination, Marilyn's grade is 64%. What is her average?

4. Suppose Marilyn had received a grade of 85% on the final examination. What would her average be?

Chapter Summary

1. To find what percent a number is of another:
   1. Write an equation.
   2. Solve the equation for n.
   3. Write a percent for n.

---

1. What percent of 85 is 17?

$$85n = 17$$
$$\frac{85n}{85} = \frac{17}{85}$$

$$\frac{17}{85} \longrightarrow 85 \overline{)\ 17.00}^{\ 0.20}$$

$0.20 = 20\%$
So **20%** of 85 is 17.

---

2. To find the percent of increase:
   1. Find the amount of increase.
   2. Write the ratio: $\dfrac{\text{Amount of Increase}}{\text{Original Amount}}$
   3. Write a percent for the ratio.

---

2. Find the percent of increase from 6 to 9.

$9 - 6 = 3$

Percent of increase $= \dfrac{3}{6}$
$= \textbf{50\%}$

---

3. To find the percent of decrease:
   1. Find the amount of decrease.
   2. Write the ratio: $\dfrac{\text{Amount of Decrease}}{\text{Original Amount}}$
   3. Write a percent for the ratio.

---

3. Find the percent of decrease from 20 to 15.

$20 - 15 = 5$

Percent of decrease $= \dfrac{5}{20}$
$= \textbf{25\%}$

---

4. To find a number when a percent of it is known:
   1. Write an equation.
   2. Solve the equation.

---

4. 20 is 80% of what number?

$20 = 80\% \times n$

$0.80n = 20$

$\dfrac{0.80n}{0.80} = \dfrac{20}{0.80}$

$n = 25$
So 20 is 80% of **25.**

---

5. To use a proportion to solve percent problems:
   1. Identify the rate, the whole, and the part.
   2. Write a proportion.
   3. Solve the proportion.

---

5. What number is 60% of 80?

Part   Rate Whole

Rate $\rightarrow \dfrac{60}{100} = \dfrac{n}{80}$ $\leftarrow$ Part $\leftarrow$ Whole

$100n = 4800$
$\dfrac{100n}{100} = \dfrac{4800}{100}$

$n = 48$
So **48** is 60% of 80.

Chapter Review

Part 1: VOCABULARY

For Exercises 1–4, choose from the box at the right the word(s) that complete(s) each statement.

1. The percent of increase is the ratio of the amount of __?__ to the original amount.   (Page 270)

2. The ratio of the amount of decrease to the original amount is the percent of __?__ .   (Page 270)

3. A fixed percent of total sales earned by a salesperson is called __?__ .   (Page 275)

4. The amount of money a diner leaves for the waiter is a __?__ . (Page 284)

> commission
> increase
> tip
> decrease
> mark-up

Part 2: SKILLS

First write an equation. Then solve the equation.   (Pages 268–269)

5. What percent of 8 is 2?

6. 9 is what percent of 81?

7. What percent of 25 is 5?

8. 20 is what percent of 70?

9. 9 is what percent of 15?

10. 81 is what percent of 90?

11. 12 is what percent of 30?

12. What percent of 30 is 27?

For Exercises 13–18, find the percent of increase or decrease. Identify each answer as a percent of increase or decrease. (Pages 270–271)

	Item	Original Value	Present Value		Item	Original Value	Present Value
13.	Computer	$4000	$5000	16.	Painting	$100	$120
14.	House	$70,000	$80,000	17.	Blouse	$ 30	$ 20
15.	Car	$12,000	$10,000	18.	Tape Player	$ 60	$ 50

For Exercises 19–26, first write an equation. Then solve the equation. (Pages 272–273)

19. 19 is 50% of what number?

20. 75 is 75% of what number?

21. 8 is 10% of what number?

22. 8 is 25% of what number?

23. 25 is 20% of what number?

24. 50 is 1% of what number?

25. 72 is 30% of what number?

26. 65 is 13% of what number?

For Exercises 27–32, use a proportion to solve. (Pages 276–278)

27. 10% of 50 is what number?

28. 8 is what percent of 20?

29. 12 is 15% of what number?

30. 20% of 150 is what number?

31. 312 is 40% of what number?

32. What percent of 56 is 14?

For Exercises 33–36, choose the correct equation. (Pages 279–280)

33. What is 16% of 18?
  a. $0.16 \times n = 18$
  b. $n = 0.16 \times 18$

34. 21 is 30% of what number?
  a. $21 = 0.30 \times n$
  b. $21 \times 0.30 = n$

35. 20 is what percent of 500?
  a. $n = 20 \times 500$
  b. $20 = n \times 500$

36. What is 40% of 72?
  a. $0.40 \times n = 72$
  b. $n = 0.40 \times 72$

First write an equation. Then solve the equation. (Pages 279–280)

37. What is 5% of 60?

38. 14 is what percent of 200?

39. 25 is $33\frac{1}{3}$% of what number?

40. 23% of what number is 10.35?

41. What percent of 200 is 180?

42. What is $66\frac{2}{3}$% of 60?

Estimate the sale price. (Pages 284–285)

43. 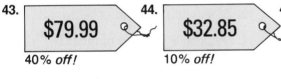 **$79.99** 40% off!

44. **$32.85** 10% off!

45. 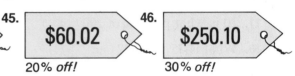 **$60.02** 20% off!

46. **$250.10** 30% off!

Part 3: APPLICATIONS

47. In 1987, a Firefly Turbo sports car cost $16,000. In 1988, the cost increased to $16,800. Find the percent of increase. (Pages 270–271)

48. A family budgets $160 for food. This is 8% of monthly take-home pay. How much is the monthly take-home pay? (Pages 272–273)

49. Cindy paid a sales tax of $0.96 on the purchase of several books. The rate of sales tax in her state is 4%. How much did she pay for the books? (Pages 272–273)

50. Caesar and Cleo's dinner check amounted to $20.18. They left a tip of 15% of the bill. Estimate the total cost, including the tip. (Pages 284–285)

Chapter Test

For Exercises 1–4, choose the correct equation.

1. 16 is what percent of 75?
    a. $16 = n \times 75$
    b. $n = 16 \times 75$
    c. $75 = 16 \times n$

2. 15 is 75% of what number?
    a. $n = 0.75 \times 15$
    b. $0.75 = 15 \times n$
    c. $15 = 0.75 \times n$

3. What is 45% of 600?
    a. $n = 0.45 \times 600$
    b. $0.45 = n \times 600$
    c. $600 = 0.45 \times n$

4. What percent of 200 is 100?
    a. $n \times 100 = 200$
    b. $n \times 200 = 100$
    c. $100 \times 200 = n$

For Exercises 5–10, first write an equation. Then solve the equation.

5. What is 60% of 400?

6. 300 is what percent of 500?

7. 16 is what percent of 64?

8. 64 is 25% of what number?

9. 75 is 30% of what number?

10. What is 16% of 80?

For Exercises 11–16, solve by the method you prefer.

11. 20 is what percent of 200?

12. 35 is what percent of 50?

13. What is 25% of 40?

14. 70 is 50% of what number?

15. 81 is 90% of what number?

16. What is 40% of 600?

Solve each problem.

17. Skates and Things advertised that you could save $6 if you bought a skateboard at a 15%-off sale. What was the regular price of the skateboard?

18. A personal computer that sold for $800 last year sells for $775 this year. Find the percent of decrease in the price of the computer. Round the percent to the nearest tenth.

19. A bicycle is on sale at 20% off the regular price. The regular price of the bicycle is $98.73. Estimate the sale price of the bicycle.

20. Denise's dinner check amounted to $9.85. She left a 15% tip. Estimate the total cost of the meal.

Cumulative Maintenance Chapters 1–11

Choose the correct answer. Choose **a**, **b**, **c**, or **d**.

1. Divide.

$$32 \overline{)\, 201.6}$$

  a. 73   **b.** 7.3
  c. 63   **d.** 6.3

2. Janine bought shoes at 40% off the regular price of $35. How much did she pay for them?

  a. $25   **b.** $30
  c. $14   **d.** $21

3. Richard paid $5.80 for lunch. A 15% tip is closest to:

  a. 90¢   **b.** 80¢
  c. $1.00   **d.** $1.20

4. A recipe for 4 quarts of lemonade uses 3 cups of sugar. How much sugar is needed for 12 quarts of lemonade?

  a. 9   **b.** 12
  c. 15   **d.** 8

5. What percent of 60 is 27?

  a. 30%   **b.** 45%
  c. 35%   **d.** 42%

6. Solve for n.

$$\frac{n}{4} + 3 = 15$$

  a. $42\frac{3}{4}$ **b.** $30\frac{1}{2}$
  c. 48   **d.** 50

7. Write a percent for 0.96.

  a. 9.6%   **b.** 0.0096%
  c. 96%   **d.** 0.96%

8. A radio is selling for 40% off its regular price of $15.00. Find the new selling price.

  a. $6   **b.** $9
  c. $7.50   **d.** $10.50

9. The number of home runs hit by six people are given below. Find the median.

    10   12   11   14   20   15

  a. 14   **b.** 13
  c. 12   **d.** 15

10. The table below shows the results of a survey of favorite books.

Type of Book	Count
Mystery	43
Science Fiction	25
Drama	10
Comedy	25
Adventure	40
Novel	56

Find the mode.

  a. Adventure   **b.** Novel
  c. Comedy   **d.** Drama

11. Write a percent for $\frac{21}{100}$.

a. 210% b. 21%

c. 2.1% d. 0.21%

---

12. Solve the proportion for n.

$$\frac{5}{6} = \frac{15}{n}$$

a. 5 b. $12\frac{1}{2}$

c. 2 d. 18

---

13. The distance from Sally's house to Scuba Cove is 80 miles. Sally makes a round trip once each month. How many miles does she travel in 18 months going to and from Scuba Cove?

a. 960 b. 2880

c. 1440 d. 1920

---

14. 72 is 12% of what number?

a. 6 b. 6000

c. 60 d. 600

---

15. Find the interest on $400 at 14% for 6 months. Use the formula:

$$i = p \times r \times t \ (t \text{ is in years})$$

a. $45 b. $45

c. $28 d. $36

---

16. LeRoy recorded the wind speeds below at 6:00 A.M. on seven consecutive days. Find the mean wind speed.

15 mph 12 mph 13 mph 20 mph
35 mph 17 mph 7 mph

a. 16 mph b. 17 mph

c. 13 mph d. 19 mph

---

17. At a softball tournament, the Aces won more games than the Blue Jays. The Reds won more games than the Smashers and 3 fewer games than the Blue Jays. Which two teams had the best records?

a. Aces, Blue Jays

b. Reds, Smashers

c. Blue Jays, Reds

d. Aces, Reds

---

18. Which group of fractions is arranged in order from least to greatest.

a. $\frac{1}{9}, \frac{1}{11}, \frac{1}{12}, \frac{1}{10}$

b. $\frac{1}{10}, \frac{1}{12}, \frac{1}{9}, \frac{1}{11}$

c. $\frac{1}{9}, \frac{1}{10}, \frac{1}{11}, \frac{1}{12}$

d. $\frac{1}{12}, \frac{1}{11}, \frac{1}{10}, \frac{1}{9}$

---

19. Multiply.

$$2\frac{2}{3} \times 1\frac{1}{8}$$

a. 4 b. $3\frac{1}{4}$ c. 3 d. $2\frac{2}{3}$

---

20. A refrigerator regularly sells for $696.50. It is now selling at a 25% discount. Choose the best estimate for the discount.

a. $175 b. $180

c. $150 d. $200

---

21. A rectangular garden is 10.7 meters long and 5.2 meters wide. Find the perimeter in meters.

a. 15.9 b. 31.8

c. 55.64 d. 111.28

Measurement: Surface Area/Volume

CHAPTER 12

Gregg's Photo Box

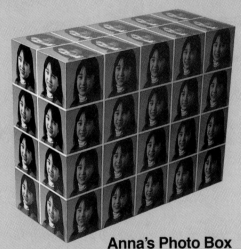

Anna's Photo Box

Sara's Photo Box

Each photo box has photos as shown on all six sides, or **faces.**

- Which photo box has the same number of photos on all faces?

- How many photos are there are Gregg's photo box?

- Which photo box has the most photos on it?

- How many more photos are there on Anna's box than on Sara's?

- Which photo box has the greatest number of photos on one side?

Metric System: CAPACITY/MASS

The **liter** (L) and the **milliliter** (mL) are commonly used metric units of capacity.

The **milligram** (mg), the **gram** (g), and the **kilogram** (kg) are commonly used metric units of mass.

This jar contains about 1 liter of orange juice.

An ant has a mass of about 200 milligrams.

This paper clip has a mass of about 1 gram.

An adult's pair of shoes have a mass of about 1 kilogram.

The glass contains about 250 milliliters of liquid.

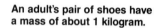

The eyedropper contains about 1 milliliter of liquid.

1. Which is more sensible for the mass of a shoelace, 1 gram or 1 kilogram?

2. Which is more sensible for the amount of soup in a cup, 150 liters or 150 milliliters?

EXERCISES

Copy the figure at the right below. Fill in the squares by choosing the most likely measure.

Across

1. Capacity of a perfume bottle

2. Mass of a truck

Down

1. Mass of a drop of water

3. Capacity of a gas tank (car)

4. Mass of a quarter

*Choose the best estimate. Choose **a** or **b.***

5.

a. 30 mL   **b.** 30 L

6.

a. 100 g   **b.** 100 kg

7.

a. 40 g   **b.** 40 kg

8.

a. 10 mL   **b.** 1 L

9.

a. 5 mL   **b.** 5 L

10.

a. 2 kg   **b.** 2 g

11. Jeff estimates that the mass of the family car is four times this.

   a. 545 g   **b.** 545 kg

12. Karen estimates that half of this amount of water will fill a teakettle.

   a. 4 L   **b.** 0.1 L

13. Pharmacists use a minim glass to measure liquids. A **minim** is about one drop. Estimate the number of minims in a milliliter.

   a. 16   **b.** 160

14. The Harveys' swimming pool is half full. Which is the better estimate for the amount of water it contains?

   a. 100,000 mL   **b.** 37,500 L

A Minim Glass

15. Claude charged these amounts of gasoline on a credit card:

   38.1 L   26.8 L   52.5 L
   44.9 L   32 L   41.7 L

   At a cost of 23 cents per liter, what was the cost of the gasoline?

16. Claude also has to pay a tax of 5% on the gasoline. How much did he pay for the gasoline plus tax?

Metric System: CHANGING UNITS

These tables show how the most commonly used metric units are related.

1000 Kilo-

Length
1 km = 1000 m
1 m = 100 cm
1 cm = 10 mm

Mass
1 kg = 1000 g
1 g = 1000 mg

Capacity
1 L = 1000 mL

Meter Gram liter

1. *Complete:* Since 1 meter = 100 centimeters,
250 meters = 250 × __?__ , or 25,000 centimeters.

2. *Complete:* Since 1 liter = 1000 milliliters,
3.8 liters = 3.8 × __?__ , or 3800 milliliters.

$\frac{1}{100}$ Centi-

> To change from larger to smaller metric units, you **multiply** by a number such as 10, 100, or 1000.

$\frac{1}{1000}$ Milli-

3. *Complete:* Since 1000 grams = 1 kilogram,
5000 grams = 5000 ÷ __?__ , or 5 kilograms.

4. *Complete:* Since 10 millimeters = 1 centimeter,
8 millimeters = 8 ÷ __?__ , or 0.8 centimeter.

> To change from smaller to larger metric units, you **divide** by a number such as 10, 100, or 1000.

EXAMPLE

In one day, Americans make 100,000 speeches. If the speakers were all waiting their turn at a soap box, the line would be about 45,000 meters long.

a. How many kilometers is this?
Think: Smaller to larger: Divide.
Since 1000 m = 1 km, 45,000 m = (45,000 ÷ 1000) km
= **45 km**

b. How many centimeters is this?
Think: Larger to smaller: Multiply.
Since 1 m = 100 cm, 45,000 m = (45,000 × 100) cm
= **4,500,000 cm**

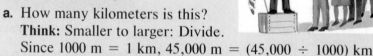

EXERCISES

For each change in metric units, write whether you would multiply or divide.

1. centimeters to kilometers

2. liters to milliliters

3. kilometers to meters

4. grams to kilograms

5. kilograms to grams

6. millimeters to centimeters

7. milliliters to liters

8. meters to millimeters

Complete.

9. 4 L = _?_ mL

10. 50,000 g = _?_ kg

11. 200 cm = _?_ m

12. 17 km = _?_ m

13. 0.5 L = _?_ mL

14. 4286 mg = _?_ g

15. 54 mm = _?_ cm

16. 680 mL = _?_ L

17. 5760 m = _?_ km

18. 36 g = _?_ kg

19. 3000 mL = _?_ L

20. 64 mg = _?_ g

21. 5.3 kg = _?_ g

22. 72 km = _?_ m

23. 5.8 cm = _?_ mm

24. 0.5 m = _?_ cm

25. 10 mg = _?_ g

26. 180 g = _?_ kg

27. A grocery shopper bought the following items.

tomatoes	754 g
soup	772 g
sugar	4.62 kg
potatoes	3.45 kg
raisins	425 g

How many kilograms was this in all?

28. Allen made this fruit punch recipe for a party of 32 people.

   a. How many liters of punch does the recipe make?

   b. How many milliliters of punch per person were there?

Fruit Punch

pineapple juice	3.5 L
orange juice	300 mL
ginger ale	4 L
soda water	2.5 L
strawberries (mashed)	500 mL
sugar syrup	800 mL

29. Clare made the same fruit punch for a fair. The punch was sold for 50¢ per 80-milliliter cup. What was the profit if the ingredients for the punch cost $36?

30. The **odometer** (mileage meter) on a car read 15468.7 kilometers before a trip and 15751.6 kilometers after a trip. If the car used 34.5 liters of gas for the trip, how many kilometers per liter was this?

MEASUREMENT: SURFACE AREA/VOLUME   **297**

Customary Measures: CHANGING UNITS

The longest loaf of bread ever baked was 15,075 inches long.

1. How many inches are there in one foot?

2. To change 15,075 inches to feet, would you multiply or divide 15,075 by 12? Why?

EXAMPLE

Complete: 15,075 in = __?__ ft

Think: Smaller to larger: Divide.

Since 12 in = 1 ft, 15,075 in = $\dfrac{15,075}{12}$ ft ⟶ $12\overline{)15075}$ $\begin{array}{c} 1256 \ \ \text{r } 3 \end{array}$

= 1256 ft 3 in

Thus, the loaf of bread was **1256 feet 3 inches** long.

3. What is 1256 ft 3 in expressed as a mixed number?

EXERCISES

The tables at the right show commonly used customary measures. Refer to these tables when necessary.

Complete.

1. 18 qt = __?__ gal __?__ qt

2. 17 ft = __?__ yd __?__ ft

3. 58 oz = __?__ lb __?__ oz

4. 11 c = __?__ pt __?__ c

5. 45 in = __?__ ft __?__ in

6. 3600 lb = __?__ T __?__ lb

7. 15,000 ft = __?__ mi __?__ ft

8. 54 oz = __?__ lb __?__ oz

Length
12 inches (in) = 1 foot (ft)
3 feet = 1 yard (yd)
36 inches = 1 yard
5280 feet = 1 mile (mi)

Capacity
2 cups (c) = 1 pint (pt)
2 pints = 1 quart (qt)
4 quarts = 1 gallon (gal)

Weight
16 ounces (oz) = 1 pound (lb)
2000 pounds = 1 ton (T)

Complete. Write each answer as a mixed number.

9. 3 lb 10 oz = _?_ lb

10. 1 yd 4 ft = _?_ yd

11. 3 qt 1 pt = _?_ qt

12. 8 yd 27 in = _?_ yd

13. 1 T 500 lb = _?_ T

14. 1 mi 528 ft = _?_ mi

Add or subtract as indicated. Some exercises are done for you.

15.
$$\begin{array}{r} 3 \text{ ft } \ 4 \text{ in} \\ + 1 \text{ ft } 11 \text{ in} \\ \hline 4 \text{ ft } \underline{15 \text{ in}} = 4 \text{ ft } + 1 \text{ ft } 3 \text{ in} \end{array}$$
$$= 5 \text{ ft } 3 \text{ in, or } 5\tfrac{1}{4} \text{ ft}$$

16.
$$\begin{array}{r} 2 \text{ ft } 7 \text{ in} \\ + 3 \text{ ft } 6 \text{ in} \\ \hline \end{array}$$

17.
$$\begin{array}{r} 12 \text{ ft } 6 \text{ in} \\ + \ 6 \text{ ft } 8 \text{ in} \\ \hline \end{array}$$

18.
$$\begin{array}{r} 1 \text{ yd } 2 \text{ ft} \\ + 8 \text{ yd } 1 \text{ ft} \\ \hline \end{array}$$

19.
$$\begin{array}{r} 9 \text{ yd } 2 \text{ ft} \\ + \qquad 3 \text{ ft} \\ \hline \end{array}$$

20.
$$\begin{array}{r} 4 \text{ gal } 2 \text{ qt} \\ - 1 \text{ gal } 3 \text{ qt} \\ \hline \end{array} \longrightarrow \begin{array}{r} 3 \text{ gal } 6 \text{ qt} \\ - 1 \text{ gal } 3 \text{ qt} \\ \hline 2 \text{ gal } 3 \text{ qt, or } 2\tfrac{3}{4} \text{ gal} \end{array}$$

4 gal 2 qt =
3 gal + 1 gal 2 qt

21.
$$\begin{array}{r} 7 \text{ gal } 2 \text{ qt} \\ - 1 \text{ gal } 3 \text{ qt} \\ \hline \end{array}$$

22.
$$\begin{array}{r} 9 \text{ gal } 1 \text{ qt} \\ - 6 \text{ gal } 2 \text{ qt} \\ \hline \end{array}$$

23.
$$\begin{array}{r} 7 \text{ lb } 5 \text{ oz} \\ - 3 \text{ lb } 7 \text{ oz} \\ \hline \end{array}$$

24.
$$\begin{array}{r} 5 \text{ ft } 3 \text{ in} \\ - 2 \text{ ft } 7 \text{ in} \\ \hline \end{array}$$

25.
$$\begin{array}{r} 7 \text{ lb } 15 \text{ oz} \\ + 3 \text{ lb } \ 5 \text{ oz} \\ \hline \end{array}$$

26.
$$\begin{array}{r} 98 \text{ lb } 10 \text{ oz} \\ - 17 \text{ lb } 11 \text{ oz} \\ \hline \end{array}$$

27.
$$\begin{array}{r} 3 \text{ pt } 1 \text{ c} \\ - 1 \text{ pt } 3 \text{ c} \\ \hline \end{array}$$

28.
$$\begin{array}{r} 1 \text{ qt } 1 \text{ pt} \\ + 2 \text{ qt } 1 \text{ pt} \\ \hline \end{array}$$

29.
$$\begin{array}{r} 1 \text{ T } 200 \text{ lb} \\ - \qquad 400 \text{ lb} \\ \hline \end{array}$$

30.
$$\begin{array}{r} 6 \text{ yd } 5 \text{ in} \\ - 2 \text{ yd } 8 \text{ in} \\ \hline \end{array}$$

31.
$$\begin{array}{r} 3 \text{ qt } 2 \text{ pt} \\ + 5 \text{ qt } 2 \text{ pt} \\ \hline \end{array}$$

32.
$$\begin{array}{r} 3 \text{ mi } 600 \text{ ft} \\ - \qquad 900 \text{ ft} \\ \hline \end{array}$$

33.
$$\begin{array}{r} 6 \text{ pt } 2 \text{ c} \\ + 2 \text{ pt } 2 \text{ c} \\ \hline \end{array}$$

The largest hamburger on record weighed 3,591 pounds. It was 16 feet in diameter and was $2\tfrac{1}{2}$ inches thick.

Complete each of Exercises 34–35.

34. 3591 lb = _?_ T _?_ lb

35. 16 ft = _?_ yd _?_ ft

36. If the hamburger was divided into 9-ounce portions, how many people could receive a portion?

37. If the hamburger had to be divided into 9000 portions, how large would each portion be (to the nearest ounce)?

Surface Area

The Twin Towers of the World Trade Center in New York City are the world's largest office buildings. The base of each tower is a square with sides 207 feet long. The height of one tower is 1350 feet.

1. How many sides (or **faces**) does one tower have? Include the bottom and top faces.

2. What is the shape of the bottom and top faces of each tower?

3. What is the shape of the side faces (front, back, left side, right side) of each tower?

Each of the Twin Towers is a **rectangular prism.** The six faces of a rectangular prism are squares or rectangles.

EXAMPLE What is the surface area of the Twin Tower that has a height of 1350 feet?

Think: The tower is like a box. Draw the box unfolded.

Top (207 ft × 207 ft)

| Left Side (207 ft × 1350 ft) | Front (207 ft × 1350 ft) | Right Side (207 ft × 1350 ft) | Back (207 ft × 1350 ft) |

| Bottom (207 ft × 207 ft) |

Use a calculator.

Area of front:	$207 \times 1350 =$	$279{,}450 \text{ ft}^2$
Area of back:	$207 \times 1350 =$	$279{,}450 \text{ ft}^2$
Area of top:	$207 \times 207 =$	$42{,}849 \text{ ft}^2$
Area of bottom:	$207 \times 207 =$	$42{,}849 \text{ ft}^2$
Area of left side:	$207 \times 1350 =$	$279{,}450 \text{ ft}^2$
Area of right side:	$207 \times 1350 =$	$279{,}450 \text{ ft}^2$
	Surface area:	$1{,}203{,}498 \text{ ft}^2$

The surface area of the Twin Tower is **1,203,498 ft².**

EXERCISES

Complete the table to find the surface area.

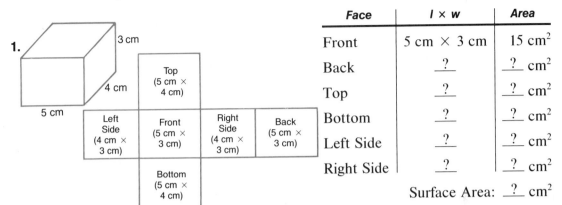

1.

Face	l × w	Area
Front	5 cm × 3 cm	15 cm²
Back	?	? cm²
Top	?	? cm²
Bottom	?	? cm²
Left Side	?	? cm²
Right Side	?	? cm²

Surface Area: ? cm²

Complete. Refer to your answers to Exercise 1.

2. The front of a rectangular prism has the same area as the _?_ .

3. The top of a rectangular prism has the same area as the _?_ .

4. The left side of a rectangular prism has the same area as the _?_ side.

5. Surface area = 2 × area of top + _?_ × area of front + _?_ × area of right side.

Find the surface area of each rectangular prism.

6. Book

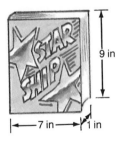

9 in
7 in
1 in

7. Stereo Speaker

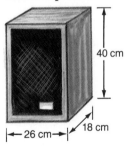

40 cm
26 cm
18 cm

8. Cereal Box

10 in
7½ in
2 in

9. Trunk

65 cm
80 cm
92 cm

10. Cabinet

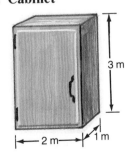

3 m
2 m
1 m

11. Packing Crate

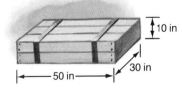

10 in
50 in
30 in

12.

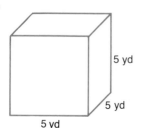

5 yd
5 yd
5 yd

13.

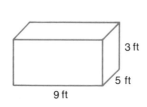

3 ft
5 ft
9 ft

14.

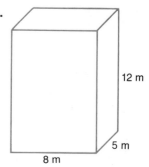

12 m
5 m
8 m

	Length	Width	Height
15.	15 in	5 in	3 in
16.	40 cm	29 cm	10.2 cm
17.	$11\frac{1}{2}$ ft	15 ft	3 ft

	Length	Width	Height
18.	3 ft	3 ft	3 ft
19.	1.1 m	2.4 m	8 m
20.	9 cm	9 cm	0.5 cm

A cube is a rectangular prism with six square faces.

21. What is the area of the front face of the cube at the right?

22. Does each face of the cube have the same area?

23. *Complete:* For a cube, surface area = ___?___ × area of one face.

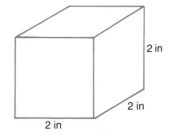

2 in
2 in
2 in

24. What is the surface area of the cube?

25. A rectangular pool is 8 meters long, 4.5 meters wide, and 2 meters deep. What is the surface area?

26. Which prism below has the smaller surface area?

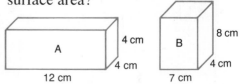

A
4 cm
4 cm
12 cm

B
8 cm
4 cm
7 cm

27. The surface area of a rectangular prism is 184 square yards. The length is 10 yards and the width is two yards. What is the height?

28. Before going camping, Susan plans to waterproof her new tent. She will not waterproof the base. How much surface area will be treated?

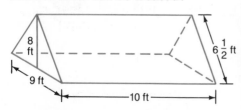

8 ft
$6\frac{1}{2}$ ft
9 ft
10 ft

Mid-Chapter Review

Choose the best estimate. Choose a or b.   (Pages 294–295)

1.   **a.** 3 g
   b. 3 kg

2.    **a.** 20 mL
   b. 10 L

3.    **a.** 1 L
   b. 10 mL

Complete.   (Pages 296–297)

4. 3 L = _?_ mL

5. 2.4 kg = _?_ g

6. 100 mg = _?_ g

7. 500 cm = _?_ m

8. 0.4 L = _?_ mL

9. 6250 m = _?_ km

Add or subtract as indicated.   (Pages 298–299)

10.   1 ft 7 in
      + 2 ft 7 in

11.   9 lb 4 oz
      − 4 lb 6 oz

12.   1 T 500 lb
      −   800 lb

13.   2 qt 1 pt
      + 3 qt 2 pt

Find the surface area of each rectangular prism.   (Pages 300–302)

	Length	Width	Height
14.	10 in	8 in	4 in
15.	25 cm	15 cm	9 cm

	Length	Width	Height
16.	7 m	7 m	2.4 m
17.	$4\frac{1}{2}$ ft	3 ft	2 ft

18. Which is the better estimate for the amount of water a bathtub will hold, 150 L or 150 mL?   (Pages 294–295)

19. A packing crate is 2 meters long, 1 meter wide, and 2 meters high. What is the surface area? (Pages 300–302)

MAINTENANCE

20. Mike is planning a banquet. He estimates that 2 pounds of cole slaw will feed 8 guests. How many pounds of cole slaw will he need to feed 60 guests?   (Pages 228–229)

21. In a student council election, Rosita received 60% of the votes cast. A total of 500 students voted. How many students voted for Rosita? (Pages 250–251)

22. Sheila ordered clothing by mail. She ordered 2 pairs of slacks for $15 each, 2 shirts for $10 each, and 3 pairs of socks for $2 each. She paid $3 in tax and $2 in postage. Find the total cost.   (Pages 8–9)

23. What is the area of the triangle below?   (Pages 196–197)

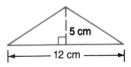

5 cm

12 cm

Volume: RECTANGULAR PRISMS

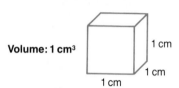

Americans buy about 428 bushels of paper clips per day. A **bushel** equals $1\frac{1}{4}$ cubic feet. Cubic feet (ft³) is a measure of *volume*.

1. How many cubic feet of paper clips do Americans buy per day?

Volume is the number of cubic units a solid contains. A cube with a 1-centimeter edge has a volume of 1 cubic centimeter (cm³); a cube with a 1-inch edge has a volume of 1 cubic inch (in³), and so on.

Volume: 1 cm³

2. Complete the table for rectangular prisms A, B, and C.

Prism	Number of cubes	l	w	h	$l \times w \times h$
A	_?_ cubic units	4	_?_	_?_	40 cubic units
B	_?_ cubic units	_?_	_?_	_?_	_?_ cubic units
C	_?_ cubic units	_?_	_?_	_?_	_?_ cubic units

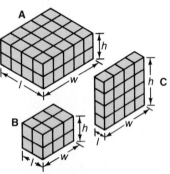
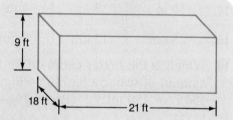

3. Volume of a rectangular prism = _?_ × _?_ × _?_

EXAMPLE

How many cubic feet of air are there in a room 21 feet long, 18 feet wide, and 9 feet high?

$V = l \times w \times h$ ◀ $l = ?$
$w = ?$
$h = ?$

$V = 21 \times 18 \times 9$

$V = 3402$ The room contains **3402 ft³** of air.

EXERCISES

1. *Complete:* Volume is measured in _?_ units.

2. Which formula can be used to find the volume of a rectangular prism?
   a. $V = l + w + h$ **b.** $V = l \times w \times h$ **c.** $V = l \times w \div h$

3. Each side of a cube has length s. Which formula can be used to find the volume?
   a. $V = s \times s \times s$ **b.** $V = 3 \times s$ **c.** $V = s + s + s$

Determine the number of cubic units in each figure.

4.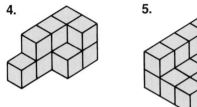

5.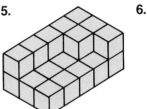

6.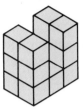

7.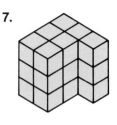

Find the volume.

8. Cereal Box

25 cm
20 cm
8 cm

9. Ice Cube Tray

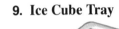

2 in
12 in
5 in

10. Suitcase

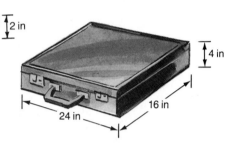

4 in
16 in
24 in

11.

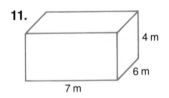

4 m
6 m
7 m

12.

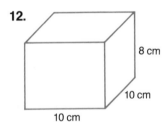

8 cm
10 cm
10 cm

13.

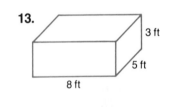

3 ft
5 ft
8 ft

	Length	Width	Height
14.	3 cm	4 cm	8 cm
15.	10 in	10 in	10 in
16.	4.5 m	6.8 m	3 m

	Length	Width	Height
17.	0.3 cm	0.3 cm	0.1 cm
18.	6 yd	$3\frac{1}{2}$ yd	$5\frac{1}{4}$ yd
19.	10 ft	8 ft	$3\frac{1}{2}$ ft

20. A desk drawer is 12 inches long, 18 inches wide, and 6 inches deep. Find its volume.

22. Joel wants to buy goldfish for his aquarium. A goldfish 5 centimeters long requires 3000 cm^3 of water to survive. How many goldfish should Joel buy?

21. The volume of a cube is 27 cubic meters. Find the height of the cube.

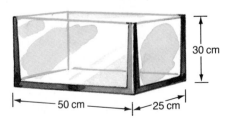

30 cm
50 cm
25 cm

Strategy: USING FORMULAS

This thermometer shows the two scales used to measure temperature, the **Celsius** scale and the **Fahrenheit** scale.

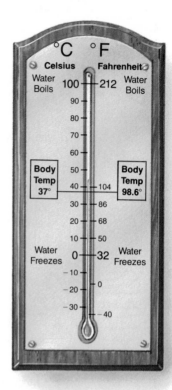

1. What is normal body temperature on the Fahrenheit scale? on the Celsius scale?

2. What is the boiling point of water on the Celsius scale? on the Fahrenheit scale?

3. *Complete:* $0°C = \underline{\ ?\ }°F$

   You can use this formula to change Celsius degrees to Fahrenheit degrees. In the formula, 1.8 C means $1.8 \times C$.

   $$F = 1.8C + 32 \quad \begin{cases} C = \text{degrees Celsius} \\ F = \text{degrees Fahrenheit} \end{cases}$$

EXAMPLE 1

At about 6000 feet above sea level, water will boil at 94°C. How many degrees Fahrenheit (to the nearest whole degree) is this?

Think: Use the formula to find F. Let C = 94.

   $F = 1.8C + 32 \longleftarrow$ **Replace C with 94.**
   $F = 1.8 \times 94 + 32$

   $\boxed{1}\ \boxed{\cdot}\ \boxed{8}\ \boxed{\times}\ \boxed{9}\ \boxed{4}\ \boxed{+}\ \boxed{3}\ \boxed{2}\ \boxed{=}$ $\boxed{\quad 201.2 \quad}$

So 94°C is about **201°F** (nearest whole degree).

You can use this formula to change Fahrenheit degrees to Celsius degrees.

$$C = \tfrac{5}{9}(F - 32) \quad \begin{cases} C = \text{degrees Celsius} \\ F = \text{degrees Fahrenheit} \end{cases}$$

4. To compute $C = \tfrac{5}{9}(79 - 32)$, do you subtract first or multiply first?

5. In the formula $C = \tfrac{5}{9}(F - 32)$, suppose that F = 32. What is the value of $\tfrac{5}{9}(F - 32)$?

EXAMPLE 2 The temperature on a winter day is 36°F. How many degrees Celsius (to the nearest whole degree) is this?

Think: Use the formula to solve for C. Let F = 36.

$$C = \frac{5}{9}(F - 32) \longleftarrow \textbf{Replace F with 36.}$$

$$C = \frac{5}{9}(36 - 32) \longleftarrow \textbf{Do the work in parentheses first.}$$

$$\boxed{3}\boxed{6}\ \boxed{-}\ \boxed{3}\boxed{2}\ \boxed{=}\ \boxed{\times}\ \boxed{5}\ \boxed{\div}\ \boxed{9}\ \boxed{=}\qquad \boxed{\ \ 2.2222222\ \ }$$

So 36°F is about **2°C** (nearest whole degree).

EXERCISES

Solve. Round answers to the nearest whole degree.

1. When the temperature reaches 90°F, Chris wears a bandanna soaked in cold water around her neck to play tennis. What is this temperature in degrees Celsius?

2. The highest surface temperature on the ocean is 30°C. How many degrees Fahrenheit is this?

3. In 1917, Death Valley, California experienced high temperatures of 49°C for 43 consecutive days. How many degrees above 100°F is this?

4. One year, the lowest temperature in Honolulu, Hawaii, was 53°F. Is this above or below 10°C? How many degrees above or below is it?

5. For every 550 feet above sea level, the boiling point of water at sea level (212°F) is lowered by 1°F. What is the boiling point of water in °C at 55,000 feet above sea level?

6. **a.** In the formula $C = \frac{5}{9}(F - 32)$, for what value of F will C = 0°?

   b. In the formula F = 1.8C + 32, for what value of C will F = 212°C?

PROJECT
   a. Record the daily high temperature in °F for one week.

Temp	Sun	Mon	Tues	Wed	Thur	Fri	Sat
°F	?	?	?	?	?	?	?
°C	?	?	?	?	?	?	?

   b. Use the formula to list the temperatures in °C (nearest degree).

   c. Make a double-line graph to show the data. There will be one line for each temperature scale.

Volume: PYRAMIDS

It takes three **pyramids** of popcorn to fill the box (rectangular prism). The pyramid and the rectangular prism have the same base and height.

1. Volume of the pyramid = __?__ × volume of the rectangular prism.

2. For a rectangular prism, $V = l \times w \times$ __?__ .

3. So for a pyramid, $V =$ __?__ $\times l \times w \times h$, or $\frac{l \times w \times h}{?}$.

EXAMPLE   The world's largest monument, the Pyramid of the Sun, near Mexico City has a square base. Each side is 467 yards long. The pyramid is 59 yards high. Find the volume.

4. *Complete:* $l =$ __?__ ; $w =$ __?__ ; $h =$ __?__

$V = \dfrac{l \times w \times h}{3}$

$V = \dfrac{467 \times 467 \times 59}{3}$

$V \approx 4{,}289{,}084 \text{ yd}^3$

The volume of the pyramid is about **4,289,084 yd³.**

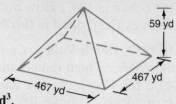

EXERCISES

Complete. Choose your answers from the box at the right.

1. A pyramid and a rectangular prism have equal bases and equal heights. So the volume of the prism is __?__ × the volume of the pyramid.

2. Volume is measured in __?__ units.

3. A rectangular prism has a volume of 111 cm³. The volume of a pyramid with the same base and height is __?__ cm³.

4. A pyramid has a volume of 360 in³. The volume of a rectangular prism with the same base and height is __?__ in³.

3
$\frac{1}{3}$
37
1080
square
cubic
333
120

Find the volume of each pyramid.

5.
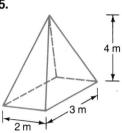
4 m
3 m
2 m

6.
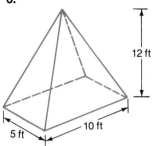
12 ft
10 ft
5 ft

7.
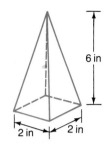
6 in
2 in 2 in

8.
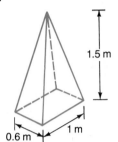
1.5 m
1 m
0.6 m

9.
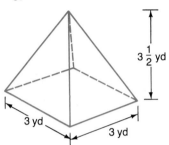
$3\frac{1}{2}$ yd
3 yd
3 yd

10.
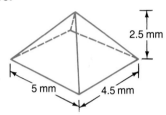
2.5 mm
5 mm 4.5 mm

	Length	Width	Height
11.	8 ft	12 ft	14 ft
12.	2 cm	3 cm	11 cm
13.	$2\frac{1}{2}$ in	$2\frac{1}{2}$ in	6 in

	Length	Width	Height
14.	100 yd	90 yd	100 yd
15.	3 m	2.4 m	3.6 m
16.	2.7 cm	1.5 cm	5.4 cm

17. The Great Pyramid of Egypt has a square base. Each side of the base is about 253 yards long. The height of the pyramid is about 162 yards.

a. Which has the greater volume, the Great Pyramid of Egypt or the Pyramid of the Sun (see the Example)?

b. How much greater?

18. The Transamerica Pyramid in San Francisco has a height of 260 meters and a square base with a perimeter of 140 meters. Find the volume to the nearest cubic meter.

MEASUREMENT: SURFACE AREA/VOLUME **309**

Math in Wallpapering

Estimate the cost of wallpapering a room that is $9\frac{1}{4}$ feet long and $11\frac{1}{2}$ feet wide. The ceiling is 9 feet high and the room has 2 doors and 5 windows.

CEILING HEIGHT	8 Feet	9 Feet	10 Feet
Size of Room in Feet	NUMBER OF SINGLE ROLLS		
8 × 10	9	10	11
10 × 10	10	11	13
10 × 12	11	12	14
10 × 14	12	14	15
12 × 12	12	14	15
12 × 14	13	15	16

NOTE: Subtract 1 roll for every door. Subtract 1 roll for every two windows.

READ   What are the facts?
Length: $9\frac{1}{4}$ ft
Width: $11\frac{1}{2}$ ft
Ceiling: 9 ft
Doors: 2
Windows: 5

PLAN   Use the table to find the number of rolls.

SOLVE

1. Round the length and width up to the next foot.

$9\frac{1}{4} \longrightarrow 10$ $11\frac{1}{2} \longrightarrow 12$

2. Read the table. Look in the column for a 9-foot ceiling.

$10 \times 12 \times 9 \longrightarrow 12$ rolls

3. Compute the allowance for doors and windows.

2 doors $\longrightarrow 2$ rolls
5 windows $\longrightarrow 2\frac{1}{2}$ rolls
Total Allowance: $4\frac{1}{2}$ rolls

4. Subtract: $12 - 4\frac{1}{2} = 7\frac{1}{2}$ Estimated number of rolls: **8**

CHECK   Is your answer reasonable?

1. In step 4 of the Example, why is $7\frac{1}{2}$ rounded up to 8?

EXERCISES   *Complete the table.*

Room	Length	Width	Ceiling	Doors	Windows	Number of Rolls
1. Kitchen	$7\frac{3}{4}$ ft	$9\frac{1}{2}$ ft	8 ft	2	1	?
2. Bedroom	$9\frac{1}{4}$ ft	$11\frac{1}{2}$ ft	9 ft	1	4	?

3. At a cost of $5.99 per roll, estimate the cost of wallpapering the bedroom in Exercise 2.

4. Estimate the cost of wallpapering the kitchen in Exercise 1 if the paper costs $7.10 per roll.

Chapter Summary

1. To change from larger to smaller metric units, you multiply by a number such as 10, 100, or 1000.

    1. Since 1 L = 1000 mL,
2.5 L = 2.5 × 1000,
or **2500 mL.**

2. To change from smaller to larger metric units, you divide by a number such as 10, 100, or 1000.

    2. Since 100 cm = 1 m,
42 cm = 42 ÷ 100,
or **0.42 m.**

3. To find the surface area of a rectangular prism, add the areas of the six faces.

    3.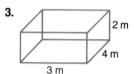

    Surface area: **52 m²**

4. To find the volume of a rectangular prism, find the product of the length, width, and height.

    $V = l \times w \times h$

    4.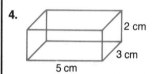

    Volume: **30 cm³**

5. To change degrees Celsius to degrees Fahrenheit, use the formula $F = 1.8C + 32$.

    5. $40°C = \underline{\ ?\ }\ °F$
$F = 1.8 \times 40 + 32$
$F = \mathbf{104°}$

6. To change degrees Fahrenheit to degrees Celsius, use the formula $C = \frac{5}{9}(F - 32)$.

    6. $68°F = \underline{\ ?\ }\ °C$
$C = \frac{5}{9}(68 - 32)$
$C = \mathbf{20°}$

7. To find the volume of a pyramid:
    ☐1 Multiply the area of the base and the height.
    ☐2 Then divide by 3.
    $V = \dfrac{l \times w \times h}{3}$

    7.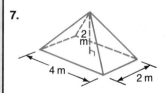

    Volume: About **5.3 m³**

Chapter Review

Part 1: VOCABULARY

For Exercises 1–8, choose from the box at the right the word(s) or number(s) that complete each statement.

1. The liter and the milliliter are commonly used metric units of ___?___ . (Page 294)

2. The milligram, gram, and kilogram are commonly used metric units of ___?___ . (Page 294)

3. To change from centimeters to millimeters, multiply by ___?___ . (Page 296)

4. To change from grams to kilograms, divide by ___?___ . (Page 296)

| 1000 |
| mass |
| 0 |
| 10 |
| $\frac{1}{3}$ |
| capacity |
| cubic feet |
| rectangular prism |

5. The six faces of a ___?___ are squares or rectangles. (Page 300)

6. A unit of volume in customary units is ___?___ . (Page 304)

7. Water freezes at ___?___ degrees Celsius. (Page 306)

8. A pyramid and rectangular prism have the same base and height. The volume of the pyramid is ___?___ of the volume of the rectangular prism. (Page 308)

Part 2: SKILLS

Choose the best estimate. Choose a, b, or c. (Pages 294–295)

9. Gas can   **a.** 4 L   **b.** 40 L   **c.** 40 mL

10. Water glass   **a.** 2.5 L   **b.** 25 L   **c.** 250 mL

11. Magazine   **a.** 0.5 mg   **b.** 0.5 g   **c.** 0.5 kg

12. Bowling ball   **a.** 8 mg   **b.** 8 g   **c.** 8 kg

Complete. (Pages 296–297)

13. 13 L = ___?___ ml

14. 25,000 g = ___?___ kg

15. 400 mL = ___?___ L

16. 59 mg = ___?___ g

17. 5 kg = ___?___ g

18. 28 m = ___?___ cm

19. 15 mm = ___?___ cm

20. 450 ml = ___?___ L

21. 875 kg = ___?___ g

22. 12 km = ___?___ m

23. 3721 mg = ___?___ g

24. 56 L = ___?___ mL

Add or subtract as indicated.   (Pages 298–299)

25. 5 gal 1 qt
        + 3 gal 7 qt

26. 8 ft 5 in
        − 3 ft 9 in

27. 1 T   900 lb
        −   1500 lb

28. 6 pt 2 c
        + 3 pt 2 c

Find the surface area of each rectangular prism.   (Pages 300–302)

29. Shoe Box

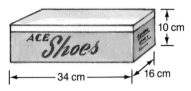

10 cm
34 cm
16 cm

30. Package

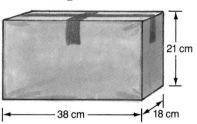

21 cm
38 cm
18 cm

31. Door

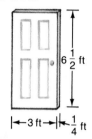

$6\frac{1}{2}$ ft
3 ft
$\frac{1}{4}$ ft

	Length	Width	Height		Length	Width	Height
32.	8 in	5 in	3 in	**34.**	9 m	6 m	2.3 m
33.	4.6 m	1.2 m	0.8 m	**35.**	21 yd	10 yd	9 yd

Find the volume of each rectangular prism.   (Pages 304–305)

	Length	Width	Height		Length	Width	Height
36.	5 ft	3 ft	6 ft	**38.**	6.7 cm	4.2 cm	5 cm
37.	4 in	4 in	4 in	**39.**	3.2 m	2.8 m	2 m

Find the volume of each pyramid.   (Pages 308–309)

40.
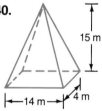
15 m
14 m
4 m

41.
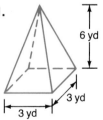
6 yd
3 yd
3 yd

42.
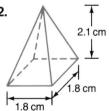
2.1 cm
1.8 cm
1.8 cm

43.
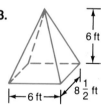
6 ft
6 ft
$8\frac{1}{2}$ ft

Part 3: APPLICATIONS

44. An airplane is carrying 104 passengers. The average mass of each passenger is 72 kilograms. The average mass of the luggage for each passenger is 36 kilograms. What was the total mass of the passengers and the luggage? (Pages 296–297)

45. A spiny anteater with a body temperature of 22°C has the lowest body temperature of any mammal. How many degrees Fahrenheit below normal human body temperature (98.6°F) is this? (Pages 306–307)

Chapter Test

Choose the best estimate. Choose a, b, or c.

1. Can of soup   **a.** 50 L   **b.** 500 mL   **c.** 50 mL

2. Tablespoon   **a.** 15 L   **b.** 150 mL   **c.** 15 mL

3. Can of tennis balls   **a.** 150 kg   **b.** 150 g   **c.** 150 mg

4. Pencil   **a.** 5 g   **b.** 5 mg   **c.** 5 kg

Complete.

5. 3.2 L = $\underline{\ ?\ }$ mL   6. 12.5 kg = $\underline{\ ?\ }$ g   7. 430 mL = $\underline{\ ?\ }$ L   8. 5010 mm = $\underline{\ ?\ }$ cm

Add or subtract as indicated.

9.
$$\begin{array}{r} 3\ \text{qt}\ 2\ \text{pt} \\ +\ 5\ \text{qt}\ 6\ \text{pt} \\ \hline \end{array}$$

10.
$$\begin{array}{r} 8\ \text{gal}\ 2\ \text{qt} \\ -\ 6\ \text{gal}\ 5\ \text{qt} \\ \hline \end{array}$$

11.
$$\begin{array}{r} 1\ \text{T}\ 300\ \text{lb} \\ -\ \ \ \ \ 600\ \text{lb} \\ \hline \end{array}$$

12.
$$\begin{array}{r} 2\ \text{mi}\ 4000\ \text{ft} \\ +\ 1\ \text{mi}\ 3250\ \text{ft} \\ \hline \end{array}$$

Find the surface area of each rectangular prism.

	Length	Width	Height
13.	6 cm	4 cm	2 cm
14.	1.8 m	9 m	4 m

	Length	Width	Height
15.	9 ft	$12\frac{1}{3}$ ft	2 ft
16.	17 yd	17 yd	17 yd

Find the volume of each rectangular prism.

	Length	Width	Height
17.	4 in	3 in	1 in
18.	8 cm	7 cm	3 cm

	Length	Width	Height
19.	2.4 m	10 m	3 m
20.	5 yd	$2\frac{1}{2}$ yd	1 yd

Find the volume of each pyramid.

21.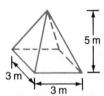

22.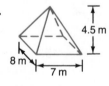

23.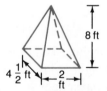

24. A chemist has 1.5 liters of water in a beaker. To this she adds 136 milliliters of salt solution and 245 milliliters of acid solution. How many liters is this in all?

25. The boiling point of water at the summit of Mt. Everest is 71°C. How many degrees Fahrenheit less than the boiling point of water at sea level (212°F) is this?

Choose the correct answer. Choose a, b, c, or d.

1. Add.

$$12 + 6 + 83 + 7$$

a. 108   **b.** 98   **c.** 118   **d.** 107

2. Round 809 to the nearest ten.

a. 800   **b.** 820
c. 900   **d.** 810

3. Add.

$$\frac{2}{3} + \frac{1}{5}$$

a. $\frac{3}{8}$ **b.** $\frac{3}{15}$ **c.** $\frac{2}{8}$ **d.** $\frac{13}{15}$

4. Divide.

$$54 \div \frac{1}{6}$$

a. 12   **b.** 9   **c.** 350   **d.** 324

5. How many square meters of carpet are needed to cover this rectangular-shaped floor?

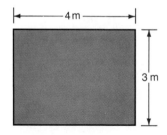

|←———4 m———→|

3 m

a. 14   **b.** 12   **c.** 28   **d.** 9

6. Divide.

$$130 \overline{)\, 41.6}$$

a. 0.32   **b.** 32
c. 3.2   **d.** 0.0032

7. Which numbers divide 54 evenly?

a. 6 and 7   **b.** 8 and 6
c. 2 and 9   **d.** 2 and 5

8. A softball team won 12 out of 16 games it played. What percent of its games did the team win?

a. 70   **b.** 80
c. 75   **d.** 85

9. Find the discount.

a. $4
b. $5
c. $12
d. $10

SALE
25% OFF
REGULARLY $16

10. Lorenzo paid $6.20 for lunch. A 15% tip is closest to:

a. 80¢   **b.** 90¢
c. $1.00   **d.** $1.20

11. What is $\frac{54}{81}$ in lowest terms?

a. $\frac{3}{4}$ **b.** $\frac{2}{3}$
c. $\frac{5}{9}$ **d.** $\frac{4}{9}$

12. Scores of seven students on a 10-point math quiz were 9, 8, 5, 4, 6, 5, and 10. What is the mode of the scores?

a. 6   **b.** 5
c. 4   **d.** 7

13. One side of a square measures 9 centimeters. What is the perimeter of the square?

a. 18 cm   b. 81 cm
c. 36 cm   d. 63 cm

14. Subtract.

$$53.08$$
$$-\ \ 8.59$$

a. 44.49   b. 45.39
c. 43.39   d. 47.09

15. Multiply.

$$26.5 \times 0.001$$

a. 0.265   b. 2.65
c. 0.0265   d. 26500

16. What is 20% of 50?

a. 20   b. 5
c. 40   d. 10

17. Which fraction has the greatest value?

a. $\frac{8}{9}$ b. $\frac{2}{3}$
c. $\frac{5}{7}$ d. $\frac{5}{6}$

18. The rate of sales tax in one state is 7%. What is the amount of tax on a car that costs $6500?

a. $445   b. $45.50
c. $455   d. $44.50

19. Richard gave the clerk $50 as a down payment for a trumpet. He made 10 monthly payments of $25 each. What was the total cost of the trumpet?

a. $300   b. $275
c. $250   d. $225

20. Maria bought a coat at 20% off the regular price of $120. How much did she pay for the coat?

a. $100   b. $24
c. $96   d. $90

21. Choose the best estimate.

$$711 \times 18$$

a. 20,000   b. 18,600
c. 16,000   d. 14,000

22. What percent of 48 is 36?

a. 75%   b. 60%
c. 80%   d. 40%

23. Each barrel below represents 50,000 liters. What is the total number of liters represented?

a. 500,000   b. 200,000
c. 20,000   d. 2,000,000

Circles and Applications

Holly's Budget

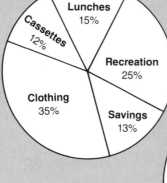

Lunches 15%

Cassettes 12%

Recreation 25%

Clothing 35%

Savings 13%

Raoul's Budget

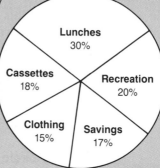

Lunches 30%

Cassettes 18%

Recreation 20%

Clothing 15%

Savings 17%

- Raoul spends a greater percent of his budget on cassettes than Holly. If they both earn $50 a month, and both follow their budget guidelines carefully, does this mean that Holly spends less on cassettes than Raoul?

- The budgets show that Raoul saves a greater percent of his earnings than Holly. You do not know how much each earns per month. Can you be sure that Raoul saves more money per month than Holly? Explain.

Meaning of Circumference

The **diameter** and **radius** of a circle are shown at the right.

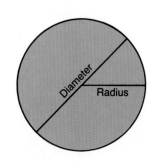

Diameter
Radius

> In a circle, the length of a diameter, *d,* is twice the length of a radius, *r.*
> $$d = 2 \times r$$

ACTIVITY Find three different circular objects. Mark a point along the outer edge of each object. Place the mark on the object at the point "0" on a ruler, yardstick, or meter stick. Roll the object along the ruler for one complete turn. The distance traveled in one complete turn is the *circumference, C.*

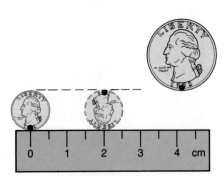

> The **circumference** of a circle is the distance around the circle.

1. Complete this table. Use a calculator to compute $C \div d$. Round to the nearest hundredth.

	Circumference, C	Diameter, d	C ÷ d
First object	?	?	?
Second object	?	?	?
Third object	?	?	?

For any circle, $C \div d$ is always the same number.

2. What number or numbers did you get for $C \div d$? The number $C \div d$ is approximately equal to 3.14 <u>or</u> $\frac{22}{7}$. Since $C \div d$ cannot be represented exactly by a fraction or decimal, the Greek letter π (called **pi**) is used. That is,
$$\pi = \frac{C}{d}.$$

3. For $\pi = 3.14$ and $d = 2$ centimeters, what does C equal?

4. For $\pi = \frac{22}{7}$ and $d = 21$ inches, what does C equal?

Circumference and Applications

You learned in the previous lesson that, for any circle,

$$\pi = \frac{C}{d} \quad \begin{array}{l} \textbf{C = circumference} \\ \textbf{d = diameter} \end{array}$$

Then $\pi \times d = \frac{C}{d} \times d$. ⟵ **Multiply each side by d.**

So $\pi d = C$, or $C = \pi d$. ⟵ **Formula**

Since $d = 2 \times r$, replace d with $2 \times r$.

$C = \pi \times d$
$C = \pi \times 2 \times r$, or
$C = 2\pi r$

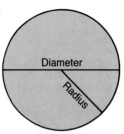
Diameter
Radius

> In any circle with **diameter,** d, and **radius,** r, you can use these formulas to find the **circumference,** C.
> $$C = \pi d \qquad \text{or} \qquad C = 2\pi r$$

EXAMPLE The radius of this bicycle wheel is 32 centimeters. How far will the bicycle travel when the wheel goes around once? Use 3.14 for π. Round your answer to the nearest whole number.

Method 1

$C = 2\pi r$ ⟵ $r = 32$ cm
$C \approx 2 \times 3.14 \times 32$
$C \approx 200.96$

Method 2

$C = \pi d$ ⟵ $d = 2 \times 32$, or 64 cm
$C \approx 3.14 \times 64$
$C \approx 200.96$

The wheel will travel about **201 centimeters.**

1. Which has the greater circumference, a circle with a radius of 32 centimeters or a circle with a radius of 40 centimeters?

EXERCISES

For Exercises 1–5, complete each sentence.

1. In any circle with diameter, d, and radius, r, $d = \underline{\ ?\ } \times r$.

2. The distance around a circle is called its $\underline{\ ?\ }$.

3. In any circle, $\frac{C}{d} = $ __?__ .

4. In any circle, $\frac{C}{2r} = $ __?__ .

5. In the formula $C = 2\pi r$, π is approximately equal to __?__ (decimal) or __?__ (fraction).

Customary Measures

Find the circumference. Use $\frac{22}{7}$ for π. Round each answer to the nearest whole number.

6. Bicycle Wheel

$r = 13$ in

7. Baseball

$r = 1\frac{5}{12}$ in

8. Clock

$d = 12\frac{1}{2}$ in

9. Moon

$d = 2250$ mi

Metric Measures

Find the circumference. Use 3.14 for π. Round each answer to the nearest whole number.

10. Basketball

$r = 12.4$ cm

11. Table Fan

$r = 11$ cm

12. Fountain Base

$d = 7.6$ m

13. Record

$d = 25.4$ cm

14. The top of a round table has a radius of 0.5 meter. Find the circumference of the table top.

15. The earth has a diameter of 7927 miles. Find the circumference of the earth. Use $\pi = \frac{22}{7}$.

16. Suppose you want to use a metal bar to make a basketball hoop with a radius of 9 inches. How long a metal bar will you need?

17. The length of a helicopter's propeller is 9.75 meters. What distance in meters does the tip of a blade travel in one revolution?

18. How much farther do you ride in one complete revolution of a merry-go-round by sitting in the outside lane, 21 feet from the center, than by sitting in the inside lane, 14 feet from the center?

19. A bicycle has wheels with diameters of 28 inches.

  a. How far does it travel when the wheels revolve once?

  b. How many times do the wheels revolve in a distance of 1 mile?

Angles and Triangles

An **angle** is formed by two **rays** that meet at a point called the vertex. A **degree** (symbol: °) is the unit of measure for angles. You use a **protractor** to measure angles. A protractor is one half of a circle, or a semicircle (180°).

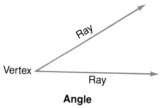

Angle

1. How many degrees are there in a circle?

This table shows how to use a protractor to measure an angle.

Angle	Measure	Name
Read the measure from this ray. This ray passes through 0. Read the outside scale. base	60°	**Acute.** An **acute angle** has a measure of less than 90°.
Place the center of the base at the vertex. The ray passes through 90° exactly. This indicates an angle of 90°.	90°	**Right.** A **right angle** has a measure of 90°. The sides of a right angle are **perpendicular** to each other.
The ray passes through 120° exactly. Read the outside scale.	120°	**Obtuse.** An **obtuse angle** has a measure of more than 90° but less than 180°.

When the sum of the measures of two angles is 90°, the angles are **complementary.**

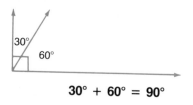

30° + 60° = 90°

The sum of the measures of two angles that form a line is 180°. The angles are **supplementary.**

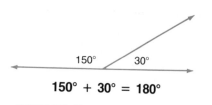

150° + 30° = 180°

EXERCISES   **a.** *Estimate the measure of each angle.*
b. *Classify each angle as acute, right, or obtuse.*

1.
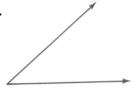

2.
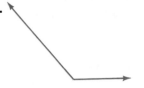

3.

Find the measure of angle x in each figure.

4.
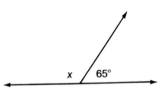

5.
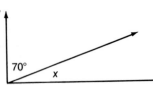

6.
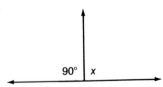

Measure each angle in each figure to the nearest degree. Then find the sum of the angles. Answer the questions.

Figure	Angle Measures	Sum of the Angle Measures	Questions
7. Acute Triangle	angle 1: ? angle 2: ? angle 3: ?	?	Are all the angles of an acute triangle less than 90°?
8. Right Triangle	angle 1: ? angle 2: ? angle 3: ?	?	What is the sum of the acute angles of the right triangle?
9. Obtuse Triangle	angle 1: ? angle 2: ? angle 3: ?	?	Are any of the angles of an obtuse triangle greater than 90°?

10. What is the sum of the measures of the angles in any triangle?

Find the missing angle for each triangle. Tell whether the triangle is acute, right, or obtuse.

11. 60°, 50°, ___?___

12. 40°, 40°, ___?___

13. 30°, 60°, ___?___

Circle Graphs

Circle graphs are used to compare parts to a whole. Circle graphs are often labeled in percents. The whole circle represents 100%.

1. How many degrees are there in a circle?

This table and graph show the percent of hours per day most students in Lynn's class spend on different activities.

Activity	Percent
School	25
Sleeping	$33\frac{1}{3}$
Eating	$8\frac{1}{3}$
Recreation	$16\frac{2}{3}$
Homework	$8\frac{1}{3}$
Other	$8\frac{1}{3}$

Typical Student Day

2. *Complete:* In the graph, school is __?__% of the circle.

Since the circle contains 360°, the school part of the circle contains 90°.

$$25\% \text{ of } 360° = \frac{1}{4} \times 360° = 90°$$ ◀ **Write a fraction or a decimal for the percent.**

3. Find the number of degrees for each of these parts.
   a. Sleeping   **b.** Eating   **c.** Recreation
   d. Homework   **e.** Other

To show each part of a circle graph, you draw angles. The vertex of each angle is the center of the circle.

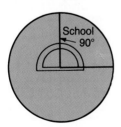

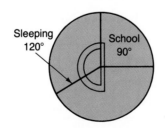

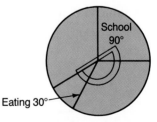

1. Place the protractor on a radius. Draw the angle for school.

2. Place the protractor on the "new" radius. Draw the angle for sleeping.

3. Place the protractor on the "newer" radius. Draw the angle for eating.

Each angle shown in the circle is called a **central angle**.

---

4. Draw a circle with a radius of one inch. Use a protractor to draw central angles of 90°, 120°, and 30°. Follow the steps shown on page 323.

5. Draw the next two central angles: **a.** Recreation: 60° **b.** Homework: 30°

6. How many degrees are left for "Other Activities"?

EXERCISES

1. What is the total percent of all the parts in a circle graph?

2. What is the total number of degrees in a circle graph?

·Brenda's take-home pay is $350 a week. The circle graph shows how she spends her earnings.

3. What percent does Brenda spend on food?

4. How much does Brenda spend on food per week? (HINT: 20% of $350 = __?__)

5. How much does Brenda save in four weeks?

6. How much does Brenda spend on rent in a year?

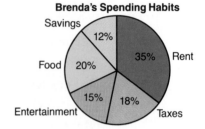

Brenda's Spending Habits

Savings 12%
Rent 35%
Food 20%
Entertainment 15%
Taxes 18%

For Exercises 7–8, draw a circle graph to show the data.

7.   **Immigration to U.S. Since 1820**

From	Percent
Asia	5
Europe	75
North and South America	18
Other	2

8.   **Vehicles Passing a Toll Booth**

Vehicle	Percent
Automobiles	47
Motorcycles	10
Trucks	25
Vans	18

9. A thousand students were asked to name their favorite sport.

a. Complete the chart.   **b.** Use the chart to draw a circle graph.

			Favorite Sports at Weldon High		
Sport	Number	Number Total		Percent of Total	Central Angle
Basketball	400	$\frac{400}{1000} = \frac{2}{5}$		40	144°
Football	250	?		?	?
Baseball	150	?		?	?
Soccer	50	?		?	?
Track	100	?		?	?
Other	50	?		?	?

Personal Budget

Claude has a part-time job at a supermarket. His earnings average $80 per week. After keeping a careful record of his spending for 6 weeks, Claude would like to plan a *budget*.

A **budget** is a plan for adjusting expenses to income.

Here is Claude's 6-week record of his expenses.

Item	Week 1	Week 2	Week 3	Week 4	Week 5	Week 6
Gas	$10.00	$12.00	$ 9.40	$10.00	$11.20	$ 9.80
Clothes	12.00	11.00	10.00	15.00	10.50	13.50
Entertainment	20.00	22.00	17.50	17.00	22.50	21.00
Food	12.40	16.00	15.00	18.00	15.60	19.00
Personal	9.60	9.00	8.50	10.10	10.20	10.20
Savings	12.00	5.00	20.00	11.00	10.50	13.50

1. What is the average amount Claude spends on food each week?

The Exercises will show you how Claude made a weekly budget.

EXERCISES

Use Claude's six-week record to find the average amount spent for each item.

1. Gasoline   **2.** Clothes   **3.** Entertainment

Complete this table to find what percent of average weekly income Claude spends on each item. Use the information to draw and complete the circle graph.

	Item	Average Amount Spent	Budget Percent
4.	Gas	$10.40	$\frac{10.40}{80} = \underline{\ ?\ } \%$
5.	Clothes	$12.00	?
6.	Entertainment	$20.00	?
7.	Food	?	?
8.	Personal	?	?
9.	Savings	?	?

Claude's Weekly Budget

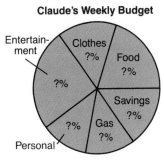

10. What is the total of the amounts in the "Average Amount Spent" column?

11. What is the total of the percents in the "Budget Percent" column?

Suppose that Claude's earnings increase to $120 per week. How much would he then budget for each of these items?
(HINT: 13% of $120 = Budget Amount for Gas)

12. Gas

13. Clothes

14. Entertainment

15. Savings

Maria works 22 hours a week and earns $5 an hour. She lives at home and plans to attend a local community college. The table shows Maria's weekly budget.

Maria's Budget	
Tuition for school	25%
Clothing	15%
Food	20%
Transportation	10%
Entertainment	10%
School supplies	8%
Miscellaneous expenses	12%

16. What is the sum of the percents in the table?

·17. Draw a circle graph to show how Maria plans to budget expenses.

Use the circle graph and Maria's weekly earnings to determine how much money is budgeted weekly for each of these items.

18. School tuition

19. Clothing

20. Food

21. Transportation

22. Entertainment

23. School supplies

24. How much more does Maria plan to spend for tuition than for entertainment?

25. How much less does Maria plan to spend for transportation than for food?

26. Why is a budget useful for good money management?

27. What are some advantages of a monthly budget over a weekly budget?

PROJECT

a. Prepare a weekly budget that might be useful to a high school student. Assume that the student has $50 a week in earnings.

b. Prepare a table that shows the items in the budget as well as the amount budgeted for each item, and what percent each amount is of the total earnings.

c. Make a circle graph to show the budget.

Mid-Chapter Review

Find the circumference. Round each answer to the nearest whole number. (Pages 319–320)

1. Tennis Ball

$r = 3.2$ cm; $\pi = 3.14$

2. Camera Lens

$r = 25$ mm; $\pi = 3.14$

3. Bicycle Gear

$r = 4\frac{2}{5}$ in; $\pi = \frac{22}{7}$

For Exercises 4–10, tell whether the given measure indicates an acute, right, or obtuse angle. (Pages 321–322)

4. $25°$ **5.** $88°$ **6.** $90°$ **7.** $142°$ **8.** $101°$ **9.** $10°$ **10.** $93°$

For Exercises 11–12, draw a circle graph to show the data. (Pages 323–324)

11. **Uses of Paper**

Use	Percent
Packaging	48
Writing Paper	29
Tissues	7
Other	16

12. **Garbage Recycled**

Year	Percent
1976	22
1977	23
1980	25
1985	30

13. John earns $300 per month working at a part-time job. He budgets 13% of his earnings for transportation. How much does he budget for transportation? (Pages 325–326)

14. Debra earns $380 per week. She budgets 12% of her earnings for clothing and 25% for food. How much more does she budget for food than on clothing? (Pages 325–326)

MAINTENANCE

Add or subtract. (Pages 140–141)

15. $\frac{1}{2} + \frac{3}{5}$ **16.** $\frac{7}{8} - \frac{1}{3}$ **17.** $2\frac{1}{5} + 3\frac{1}{4}$ **18.** $6\frac{5}{6} - 2\frac{1}{8}$

19. Use the table at the right to find the difference between the total combined areas of Lake Huron and Lake Ontario and the total area of Lake Superior. (Pages 2–3)

Lake	Area (square miles)
Superior	81,000
Huron	74,700
Ontario	34,850

Strategy: PATTERNS IN GEOMETRY

Sometimes you can solve a problem by **identifying a pattern.** Geometric figures can be used to make patterns. In a sequence of these figures, look for what changes and for what remains the same. This will help you to identify the pattern.

EXAMPLE 1 | Draw the next figure in the sequence at the right.

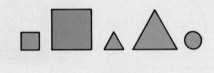

Think: The figures are in pairs. The pairs have the same shape. The second figure in each pair is larger.

1. Draw the next figure in this pattern.

Sometimes the figures are turned (rotated) to the right or to the left.

EXAMPLE 2 | Draw the next figure in this pattern.

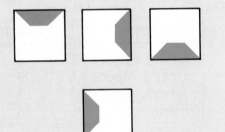

Think: The second square was turned 90° to the right. The third tile was turned an additional 90° to the right.

2. Draw the next figure in this pattern.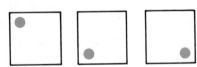

EXERCISES

Choose the figure that completes each pattern. Choose **a, b,** *or* **c.**

1.

a. **b.** **c.**

2.   **a.** **b.** **c.**

3.   **a.** **b.** **c.**

Draw the figure that completes each pattern.

4.

5.

6.

7.

8.

9.

For Exercises 10–11, choose a strategy from the box at the right that you can use to solve each problem.

a. *Name the strategy.*

b. *Solve the problem.*

Using a formula
Making a model
Choosing a computation method
Using patterns

10. Clara began jogging each day for exercise. She jogged 0.4 miles each day the first week, 0.6 miles each day the second week, 0.9 miles each day the third week. Following this plan, how far did she jog each day of the fourth week?

11. Don, Gladys, and Milo live on the same straight road. Don lives $2\frac{1}{4}$ miles from Gladys and $3\frac{1}{2}$ miles from Milo. Gladys lives farther from Don than from Milo. How far does Gladys live from Milo?

Area and Applications

The circle at the left below is divided into equal pie-shaped pieces.
Then the pieces are rearranged to form a parallelogram.

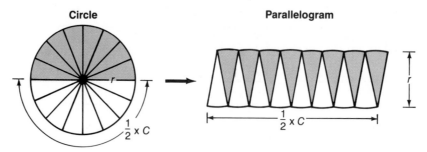

Circle   **Parallelogram**

$\frac{1}{2} \times C$ $\frac{1}{2} \times C$

1. What is the circumference of a circle with radius *r*?

2. What is $\frac{1}{2}$ the circumference of a circle with radius *r*?

3. What is the formula for the area of a parallelogram?

Area of a parallelogram: $A = \underbrace{\ \ b\ \ } \times h$

Area of a circle: $A = \frac{1}{2} \times \underbrace{\ C\ } \times r$

$A = \frac{1}{2} \times 2 \times \pi \times r \times r$

$A = \pi \times r \times r, \text{ or } A = \pi r^2$

For any circle with radius *r*,   **A(area) $= \pi r^2$.**

4. *Complete:* Area is measured in __?__ units.

EXAMPLE

A revolving sprinkler sprays a lawn for a distance of 6 yards. How
many square yards are sprayed when the sprinkler makes one
complete turn? Use $\frac{22}{7}$ for π. Round your answer to the nearest
whole number.

Think: $r = 6$ yd $A = $ __?__

$A = \pi r^2$

$A = \frac{22}{7} \times 6 \times 6$

$A = \frac{792}{7}, \text{ or } 113\frac{1}{7}$

The sprinkler sprays about **113 yd²** in one complete turn.

EXERCISES

For Exercises 1–4, complete each statement.

1. The area of a circle equals the product of __?__ and the radius squared.

2. The circumference of a circle equals the product of π and the __?__ , or the product of π and __?__ the radius.

3. The circumference of a circle with a radius of 1 centimeter and π = 3.14 is __?__ .

4. The area of a circle with a radius 1 centimeter and π = 3.14 is __?__ .

Customary Measures

> *Find the area. Use $\frac{22}{7}$ for π. Round each answer to the nearest whole number.*

5. Window: r = 10 ft   **6. Target:** r = $1\frac{1}{2}$ ft   **7. Plate:** r = 7 in   **8. Emblem:** r = $2\frac{1}{4}$ in

Metric Measures

> *Find the area. Use 3.14 for π. Round each answer to the nearest whole number.*

9. Garden: r = 5 m   **10. Tray:** r = 20 cm   **11. Mirror:** r = 23 cm   **12. Patio:** r = 3.5 m

   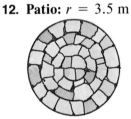

For Exercises 13–14, find the area of each shaded region.

13.

6 cm

8 cm

14.

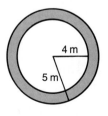

4 m

5 m

15. Which is greater, the area of a circle 6 inches in diameter or the area of a square 6 inches on a side? How much greater?

16. How many circles 2 inches in diameter can be cut out of a 6-inch square piece of paper? How much paper will be left over?

Volume and Applications: CYLINDERS

Finding the volume of a **cylinder** is similar to finding the volume of a rectangular prism.

1. What is the formula for the volume of a rectangular prism?

2. *Complete:* The base of a rectangular prism has the shape of a ?.

3. *Complete:* The area, *A*, of the base of a rectangular prism is the product of the ? and ?.

4. *Complete:* The base of a cylinder has the shape of a ?.

5. *Complete:* The area, *A*, of the base of a cylinder equals the product of π and the ?.

Rectangular Prism | **Cylinder**

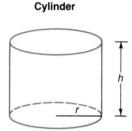

6. *Complete:* To find the volume of a rectangular prism or a cylinder, multiply the area of the base and the ?.

7. *Complete:* Volume is measured in ? units.

> The volume, *V*, of any cylinder with radius *r* and height *h* equals
> $$V = \pi \times r^2 \times h \quad \text{or} \quad V = \pi r^2 h.$$

EXAMPLE How many cubic feet of trash will this trash can hold?

Use $\pi = \frac{22}{7}$. Round your answer to the nearest whole number.

Think: $r = 1$ ft $\qquad h = 2\frac{1}{3}$ ft $\qquad V = \underline{\ ?\ }$

$V = \pi r^2 h$

$V = \frac{22}{7} \times 1 \times 1 \times \frac{7}{3}$

$V = \frac{22}{3}$, or $7\frac{1}{3}$ ft^3

It will hold about **7 ft³** of trash.

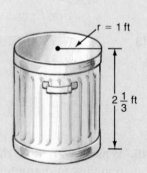

r = 1 ft

$2\frac{1}{3}$ ft

EXERCISES

For each of Exercises 1–6, choose the unit of measure from the box below.

inches	square inches	cubic inches

1. Distance around a circle
2. Area of a circle
3. Area of a square
4. Volume of a box
5. Perimeter of a rectangle
6. Volume of a cylinder

Customary Measures

Find the volume. Use $\frac{22}{7}$ for π. Round each answer to the nearest whole number.

7. Water Tower **8. Silo** **9. Fruit Juice Can** **10. Cookie Tin**

$r = 15$ ft; $r = 9$ ft; $r = 3$ in; $r = 4\frac{1}{2}$ in;
$h = 40$ ft $h = 60$ ft $h = 8$ in $h = 4$ in

Metric Measures

Find the volume. Use 3.14 for π. Round each answer to the nearest whole number.

11. Orange Juice **12. Mug** **13. Sauce Pan** **14. Planter**

$r = 2$ cm; $r = 4$ cm; $r = 6.5$ cm; $r = 0.9$ m;
$h = 8$ cm $h = 8.5$ cm $h = 5$ cm $h = 1$ m

15. The O'Leary's grain silo is shaped like a cylinder. The silo is 20 meters tall and has a radius of 7 meters. How many cubic meters of grain can the silo hold? Round your answer to the nearest whole number.

16. The tallest water tower in Union, New Jersey is 210 feet tall and has a radius of 7 feet. If one cubic foot equals $7\frac{1}{2}$ gallons, how many gallons of water will the tower hold (nearest gallon)? Use $\pi = \frac{22}{7}$.

Volume and Applications: CONES

At the Upside-Down V Drive-In, you can buy a Big V or a Little V soda. The Big V comes in a can (cylinder) and the Little V is served in a cone.

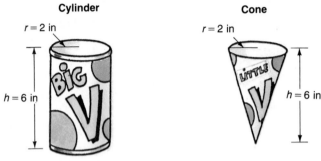

Cylinder

$r = 2$ in

$h = 6$ in

Cone

$r = 2$ in

$h = 6$ in

1. How many bases does a cone have?

2. What is the shape of the base of a cone?

The Big V and Little V containers have equal radii (plural of radius) and equal heights. So it takes **three** Little V's to fill one Big V.

3. What fraction of the volume of the Big V is the volume of the Little V?

4. *Complete:* The formula for the volume, *V*, of a cylinder with radius, *r*, and height, *h*, is $V = \underline{\ ?\ }$.

5. *Complete:* The formula for the volume, *V*, of a cone, with radius, *r*, and height, *h*, is $V = \underline{\ ?\ }$.

> The volume, *V*, of a cone with radius *r* and height, *h*, is
> $$V = \tfrac{1}{3} \times \pi \times r^2 \times h, \qquad \text{or} \qquad V = \tfrac{1}{3}\pi r^2 h.$$

6. *Complete:* Volume is expressed in $\underline{\ ?\ }$ units.

EXAMPLE	A cone-shaped pile of sand is 9 feet high and has a diameter of 6 feet. How many cubic feet of sand are there in the pile? Use $\frac{22}{7}$ for π. Round your answer to the nearest whole number.

Think: $h = 9$ ft $d = 6$ ft $V = \underline{\ ?\ }$

$V = \frac{1}{3}\pi r^2 h$

$V = \frac{1}{3} \times \frac{22}{7} \times 3 \times 3 \times 9$ ◀ $r = \frac{1}{2} \times 6 = 3$

$V = \frac{1782}{21} = 84\frac{18}{21}$ ft³

There are about **85 ft³** of sand in the pile.

EXERCISES

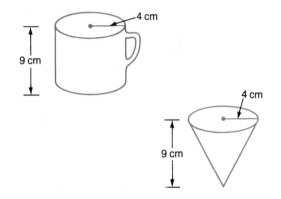

1. Find the volume of the cylinder-shaped cup. Round your answer to the nearest cubic centimeter.

2. Find the volume of the cup shaped like a cone. Round your answer to the nearest cubic centimeter.

3. Which cup has the greater volume? How much greater is it?

Customary Measures

Find the volume. Use $\frac{22}{7}$ for π. Round each answer to the nearest whole number.

4.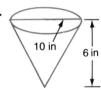

5.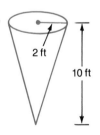

6.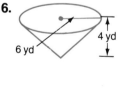

7.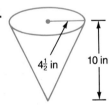

Metric Measures

Find the volume. Use 3.14 for π. Round each answer to the nearest whole number.

8.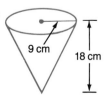

9.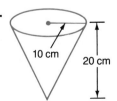

10.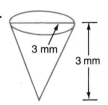

11.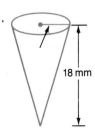

Volume of Spheres

Some storage tanks have the shape of a **sphere**. You can use this formula to find the volume of a sphere.

> The volume V, of a sphere with radius, r, is
> $$V = \tfrac{4}{3} \times \pi \times r \times r \times r$$
> or $V = \tfrac{4}{3}\pi r^3$. ⟵ $r^3 = r \times r \times r$

EXAMPLE

The radius of the storage tank shown above is 7.2 meters. Find the volume of the tank. Use 3.14 for π. Round your answer to the nearest whole number.

$$V = \tfrac{4}{3} \times \pi \times r \times r \times r \quad \longleftarrow \textbf{Replace r with 7.2.}$$

$$V = \tfrac{4}{3} \times 3.14 \times 7.2 \times 7.2 \times 7.2$$

$$\boxed{4}\ \boxed{\times}\ \boxed{3}\ \boxed{\cdot}\ \boxed{1}\ \boxed{4}\ \boxed{\times}\ \boxed{7}\ \boxed{\cdot}\ \boxed{2}\ \boxed{\times} \qquad \boxed{90.432}$$

$$\boxed{7}\ \boxed{\cdot}\ \boxed{2}\ \boxed{\times}\ \boxed{7}\ \boxed{\cdot}\ \boxed{2}\ \boxed{\div}\ \boxed{3}\ \boxed{=} \qquad \boxed{1562.6649}$$

The volume is about **1563 m³**. ◀ **Nearest whole number**

EXERCISES

Solve. Round each answer to the nearest whole number.

1. A spherical storage tank for natural gas has a radius of 36 feet. Find the volume of the tank. Use $\tfrac{22}{7}$ for π.

2. Find the volume of a basketball that has a radius of 12 centimeters. Use 3.14 for π.

3. Find the volume of a volleyball that has a radius of $4\tfrac{1}{5}$ inches. Use $\tfrac{22}{7}$ for π.

4. The kettle at the right has a radius of 29 centimeters. Find the volume of the bottom half of the kettle (half a sphere or a **hemisphere**). Use 3.14 for π.

Chapter Summary

1. To find the circumference, *C*, of a circle:
   a. Find the product of the radius and 2π.
   $$C = 2 \times \pi \times r, \text{ or } C = 2\pi r$$
   b. Find the product of the diameter and π.
   $$C = \pi \times d, \text{ or } C = \pi d$$

2. To use a protractor to measure an angle:
   1. Place the center of the base of the protractor at the vertex of the angle.
   2. Place the protractor so that one ray of the angle passes through "0" on the protractor.
   3. Read the measure of the angle from the second ray.

3. To draw a circle graph:
   1. Write a decimal for each percent.
   2. Multiply the decimal by $360°$. Round your answer to the nearest degree.
   3. Use a protractor to draw the graph.

4. To find the area of a circle, multiply π times the radius times the radius.
   $$A = \pi \times r \times r \quad \text{or} \quad A = \pi r^2$$

5. To find the volume, *V*, of a cylinder, multiply the area of the base and the height.
   $$V = \pi \times r \times r \times h \quad \text{or} \quad V = \pi r^2 h$$

6. To find the volume, *V*, of a cone, find the product of $\frac{1}{3}$ the area of the base and the height.
   $$V = \frac{1}{3} \times \pi \times r \times r \times h \quad \text{or} \quad V = \frac{1}{3} r^2 h$$

---

1. Find the circumference.
   a. $r = 5$ m
   $$C = 2 \times \pi \times r$$
   $$= 2 \times 3.14 \times 5$$
   $$= 31.4 \approx \textbf{31 m}$$
   b. $d = 3$ in
   $$C = \pi \times d$$
   $$= 3.14 \times 3$$
   $$= 9.42 \approx \textbf{9 in}$$

2. Find the measure of angle 1.

   Angle 1: **45°**

3. Body Tissue:
   Surface–13%;
   Other–87%
   $$0.13 \times 360° = 46.8 \approx \textbf{47°}$$
   $$0.87 \times 360° = 313.2 \approx \textbf{313°}$$
   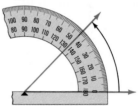

4. Find the area: $r = 8$ cm
   $$A = \pi \times r \times r$$
   $$= 3.14 \times 8 \times 8$$
   $$= 200.96 \approx \textbf{201 cm}^2$$

5. Find the volume.
   $$r = 2 \text{ ft}; h = 7 \text{ ft}$$
   $$V = \pi \times r \times r \times h$$
   $$= \frac{22}{7} \times 2 \times 2 \times 7$$
   $$= \textbf{88 ft}^3$$

6. Find the volume.
   $$r = 6 \text{ m}; h = 14 \text{ m}$$
   $$V = \frac{1}{3} \times \pi \times r \times r \times h$$
   $$= \frac{1}{3} \times 3.14 \times 6 \times 6 \times 14$$
   $$= 527.52 \approx \textbf{528 m}^3$$

Chapter Review

Part 1: VOCABULARY

For Exercises 1–6, choose from the box at the right the word(s) that complete(s) each statement.

1. In a circle, a __?__ is twice the length of a radius.   (Page 318)

2. An angle with a measure of less than 90° is an __?__ angle. (Page 321)

3. A plan for adjusting expenses to income is called a __?__ . (Page 325)

4. Area is measured in __?__ units.   (Page 330)

5. Volume is measured in __?__ units.   (Page 332)

6. The volume of a cone is __?__ of the volume of a cylinder with the same radius and height.   (Page 334)

> cubic
> budget
> diameter
> acute
> square
> obtuse
> $\frac{1}{3}$
> 3.14

Part 2: SKILLS

For Exercises 7–10, find the circumference. Round each answer to the nearest whole number.   (Pages 319–320)

7. Rim of a Dish

$r = 6$ cm

$\pi = 3.14$

8. Rim of a Pool

$d = 42$ in

$\pi = \frac{22}{7}$

9. Target

$r = 6\frac{1}{2}$ ft

$\pi = \frac{22}{7}$

10. Checker

$d = 3$ cm

$\pi = 3.14$

Use a protractor to find the measure of each angle to the nearest degree. Then identify the angle as acute, right, or obtuse. (Pages 321–322)

11.

12.

13.

14.
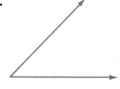

For Exercises 15–16, make a circle graph to show the data.
(Pages 323–324)

15.

Personal Budget	
Item	Percent
Clothing	4
Food	25
Housing	28
Other	43

16.

Books in a Library	
Type of Book	Percent
Fiction	45
History	25
Science	10
Reference	8
Others	12

Draw the next figure that completes each pattern. (Pages 328–329)

17.

18.

For Exercises 19–22, round each answer to the nearest whole number.

Find the area. (Pages 330–331)   *Find the volume.* (Pages 332–335)

19.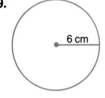
6 cm
$\pi = 3.14$

20.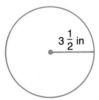
$3\frac{1}{2}$ in
$\pi = \frac{22}{7}$

21.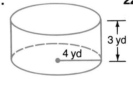
3 yd
4 yd
$\pi = \frac{22}{7}$

22.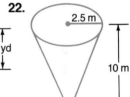
2.5 m
10 m
$\pi = 3.14$

Part 3: APPLICATIONS

Paul earns $420 per week. The table in Exercise 15 shows Paul's monthly budget. Use Paul's monthly earnings to determine how much money is budgeted monthly for each of these items.
(Pages 325–326)

23. Clothing   **24.** Food   **25.** Housing   **26.** Other

For Exercises 27–28, round each answer to the nearest whole number.

27. The sun has a radius of 432,500 miles. Find the circumference of the sun. Use $\frac{22}{7}$ for π. (Pages 319–320)

28. The top of a storage tank is circular. The radius of the top is 36 feet. Find the area of the top of the tank. Use $\frac{22}{7}$ for π. (Pages 330–331)

Chapter Test

For Exercises 1–4, round each answer to the nearest whole number.

Find the circumference.   *Find the area.*

1. Watch Face   **2.** Lamp Base   **3.** Bottom of Pan   **4.** Top of Scale

$d = 3$ cm $d = 6$ in $r = 7$ in $r = 10$ cm

$\pi = 3.14$ $\pi = \frac{22}{7}$ $\pi = \frac{22}{7}$ $\pi = 3.14$

Use a protractor to find the measure of each angle to the nearest degree. Identify each angle as acute, right, or obtuse.

5.   **6.**   **7.**   **8.**

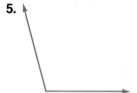      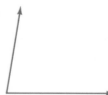   

Gabriella earns \$380 per week. The table at the right below shows her monthly budget.

9. Make a circle graph to show the data in the table.

10. How much money is budgeted monthly for food?

11. How much money is budgeted monthly for housing?

12. How much more is budgeted for housing than for clothing?

13. Draw the figure that completes the pattern at the right.

14. A circular garden has a radius of 4.2 meters. Find the area. Use 3.14 for π.

Personal Budget	
Item	**Percent**
Clothing	3
Food	26
Housing	30
Other	41

15. The radius of a jar of pickles is $1\frac{1}{2}$ inches. The jar has a height of 8 inches. Find the volume. Use $\frac{22}{7}$ for π.

Cumulative Maintenance Chapters 1–13

Choose the correct answer. Choose a, b, c, or d.

1. Which numbers divide evenly into 60?

  a. 2 and 9   **b.** 8 and 6
  c. 6 and 7   **d.** 2 and 5

2. Multiply.
$$35 \times 24$$

  a. 840   **b.** 740
  c. 210   **d.** 200

3. Round 809 to the nearest ten.

  a. 800   **b.** 820
  c. 900   **d.** 810

4. Divide.
$$8\overline{)2520}$$

  a. 315   **b.** 318
  c. 305   **d.** 316

5. What is the perimeter of a triangle whose sides are 8, 9, and 15?

  a. 25   **b.** 32
  c. 64   **d.** 16

6. Dale used 2 cups of flour to make 75 oatmeal cookies. How many cups of flour would he need to make 300 cookies?

  a. 2   **b.** 4
  c. 6   **d.** 8

7. What is the total value of eighteen $5-bills, six $10-bills, and seven $20-bills?

  a. $31   **b.** $236
  c. $218   **d.** $290

8. Divide.
$$\frac{3}{4} \div \frac{15}{16}$$

  a. $\frac{7}{31}$ **b.** $\frac{45}{64}$
  c. $\frac{5}{4}$ **d.** $\frac{4}{5}$

9. Subtract.
$$8\frac{3}{5} - 4\frac{4}{5}$$

  a. $4\frac{1}{5}$ **b.** $3\frac{1}{5}$
  c. $3\frac{4}{5}$ **d.** $4\frac{4}{5}$

10. Pam bought a shirt for $15.95. The sales tax rate was 4%. Find the total cost.

  a. $16.58   **b.** $63.80
  c. $22.33   **d.** $16.59

11. What is the volume in cubic centimeters of this box?

  a. 192
  b. 240
  c. 288
  d. 1536

BISCUIT MIX

16 cm

8 cm

12 cm

12. What number is 40% of 75?

    a. 187.5   **b.** 3000
    c. 45   **d.** 30

13. Solve the proportion for n.
$$\frac{12}{15} = \frac{n}{105}$$
    a. 105   **b.** 84
    c. 131   **d.** 91

14. Twenty-seven pounds of ground meat costs \$62.40. Estimate the cost of one pound.

    a. \$3   **b.** \$4
    c. \$2   **d.** \$6

15. Use $C = 2 \times \pi \times r$ to find the circumference of this circle. (Use $\pi = \frac{22}{7}$.)

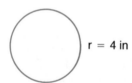
r = 4 in

    a. $12\frac{4}{7}$ in   **b.** $25\frac{1}{7}$ in
    c. 8 in   **d.** 24 in

16. Use $V = \pi \times r^2 \times h$ to find the volume of this tank. Round the answer to the nearest whole number. (Use $\pi = 3.14$.)

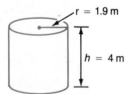
r = 1.9 m
h = 4 m

    a. 45 m^3 **b.** 46 m^3
    c. 452 m^3 **d.** 453 m^3

17. What percent of 35 is 1$\bar{4}$?

    a. 40%   **b.** 400%
    c. 60%   **d.** 600%

18. Solve for n.
$$n + 6 = 47$$
    a. 41   **b.** 40
    c. 53   **d.** 8

19. The number of home runs hit by six people are given below. Find the median.

    9   12   10   14   22   18

    a. 14   **b.** 13
    c. 12   **d.** 15

20. Find the surface area in square centimeters of this box.

    a. 400
    b. 780
    c. 800
    d. 520

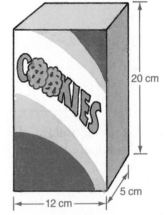
20 cm
5 cm
12 cm

21. On their vacation, the Cook family traveled 840 miles in 15 hours. Find the rate in miles per hour.

    a. 855   **b.** 56
    c. 825   **d.** 12,600

Probability

CHAPTER 14

January
```
S  M  T  W  T  F  S
   1  2  3  4  5  6  7
 8  9 10 11 12 13 14
15 16 17 18 19 20 21
22 23 24 25 26 27 28
29 30 31
```

February
```
S  M  T  W  T  F  S
          1  2  3  4
 5  6  7  8  9 10 11
12 13 14 15 16 17 18
19 20 21 22 23 24 25
26 27 28
```

March
```
S  M  T  W  T  F  S
          1  2  3  4
 5  6  7  8  9 10 11
12 13 14 15 16 17 18
19 20 21 22 23 24 25
26 27 28 29 30 31
```

April
```
S  M  T  W  T  F  S
                   1
 2  3  4  5  6  7  8
 9 10 11 12 13 14 15
16 17 18 19 20 21 22
23 24 25 26 27 28 29
```

May
```
S  M  T  W  T  F  S
    1  2  3  4  5  6
 7  8  9 10 11 12 13
14 15 16 17 18 19 20
21 22 23 24 25 26 27
28 29 30 31
```

June
```
S  M  T  W  T  F  S
             1  2  3
 4  5  6  7  8  9 10
11 12 13 14 15 16 17
18 19 20 21 22 23 24
25 26 27 28 29 30
```

July
```
S  M  T  W  T  F  S
                   1
 2  3  4  5  6  7  8
 9 10 11 12 13 14 15
16 17 18 19 20 21 22
23 24 25 26 27 28 29
30 31
```

August
```
S  M  T  W  T  F  S
       1  2  3  4  5
 6  7  8  9 10 11 12
13 14 15 16 17 18 19
20 21 22 23 24 25 26
27 28 29 30 31
```

September
```
S  M  T  W  T  F  S
                1  2
 3  4  5  6  7  8  9
10 11 12 13 14 15 16
17 18 19 20 21 22 23
24 25 26 27 28 29 30
```

October
```
S  M  T  W  T  F  S
 1  2  3  4  5  6  7
 8  9 10 11 12 13 14
15 16 17 18 19 20 21
22 23 24 25 26 27 28
29 30 31
```

November
```
S  M  T  W  T  F  S
          1  2  3  4
 5  6  7  8  9 10 11
12 13 14 15 16 17 18
19 20 21 22 23 24 25
26 27 28 29 30
```

December
```
S  M  T  W  T  F  S
                1  2
 3  4  5  6  7  8  9
10 11 12 13 14 15 16
17 18 19 20 21 22 23
24 25 26 27 28 29 30
31
```

Graph — Chances or Probability (vertical axis: 100%, 80%, 60%, 40%, 20%) vs. Number of People (horizontal axis: 0, 10, 20, 30, 40, 50, 60)

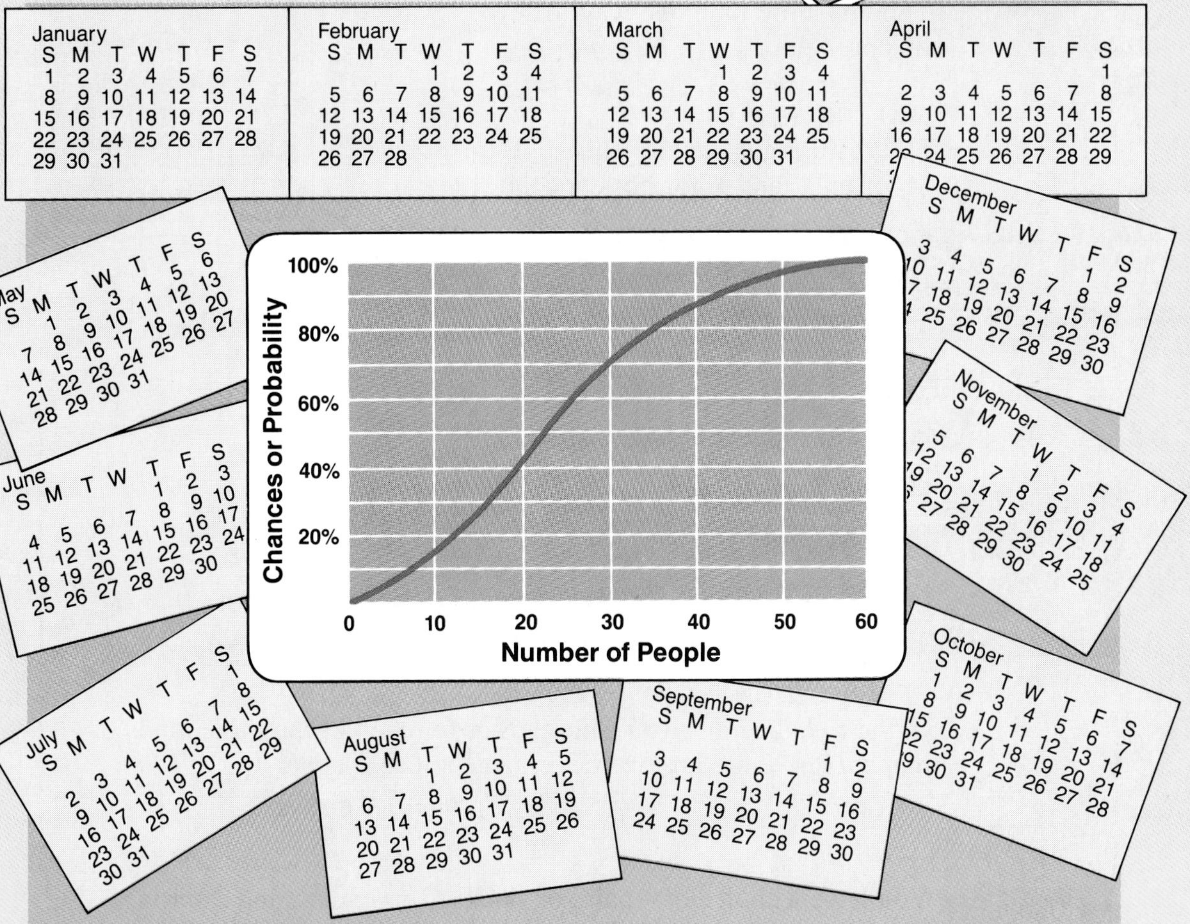

- There are 60 people in a room. According to the graph, what percent chance (probability) is there that at least two of them have the same birthday (month and day)?

- According to the graph, how many people would there have to be in the room for there to be a 50% chance that two of them have the same birthday?

- In a group of 23 people, the chances that at least two of them have the same birthday is 51%. What are the chances that all 23 people have birthdays on different days?

Counting and Tree Diagrams

It was "Burger Day" at the school cafeteria.
Look at the menu.

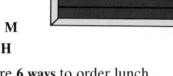

PROBLEM How many different ways can you order lunch?

There is more than one way to solve
this problem.

MAKING A LIST
Use **R** for regular, **C** for cheese, **P** for pizza,
M for milk, and **H** for hot chocolate.

1. R, M   2. R, H   3. C, M
4. C, H   5. P, M   6. P, H

This list shows, by counting, that there are **6 ways** to order lunch.

MAKING A MODEL: TREE DIAGRAM
1 Draw the "Hamburger" branches.
2 Draw the "Beverage" branches.
   There are two Beverage branches
   for each Hamburger branch.
3 Count. There are **6 ways.**

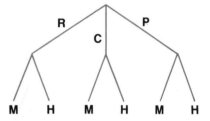

The tree diagram suggests a third way to solve the problem.

COMPUTATION
In the tree diagram, you can see that for <u>each</u> of the three kinds of
hamburgers there are two beverage choices. Thus,

$$3 \times 2 = 6 \qquad \textbf{There are 6 ways.}$$

EXAMPLE   You have a choice of 4 pairs of slacks, 2 sweaters, and 2 shirts.
How many different outfits can you wear?

READ   What are the facts?
4 pairs of slacks, 2 sweaters, 2 shirts

PLAN   Use the computation method.

SOLVE   **Think:** For <u>each</u> of the 4 pairs of slacks there are 2 sweater choices.
Thus, $4 \times 2 = 8.$

For <u>each</u> of the 8 slacks/sweater outfits there are 2 shirt choices.
Thus, $8 \times 2 = \textbf{16.}$

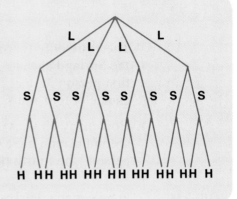
EXERCISES

The Bakery Burst sells white and pumpernickel breads in 12-ounce, 16-ounce, and 24-ounce loaves. The loaves are available in regular and thin slices.

Bread **Loaf Size** **Slices**

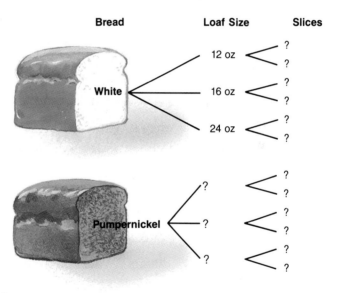

1. Draw and complete the tree diagram to find the total number of choices.

For Exercises 2–4, use the tree diagram you made in Exercise 1.

2. A customer wants to buy white bread. How many choices are there?

3. A customer wants to buy a 16-ounce loaf of pumpernickel bread. How many choices are there?

4. A customer wants a 24-ounce loaf of bread, sliced thin. How many choices are there?

The menu at Jack's Snacks offers 2 soups, 3 salads, 8 sandwiches, and 5 beverages. How many choices are there for each of these orders?

5. A soup and a salad

6. A sandwich and a beverage

7. Soup, sandwich, and a beverage

8. Soup, salad, and a beverage

Factorials

There are only four seats left for the Spring Pageant. You are the first person in line. So you have a choice of four seats.

1. How many choices does the second person have? the third person? the fourth person?

PROBLEM In how many different ways could these four people be seated?

There is more than one way to solve this problem.

USING A TREE DIAGRAM

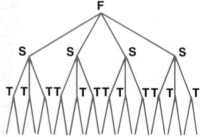

1. **F:** First person has 4 choices. Draw 4 branches.

2. **S:** Second person has 3 choices. Draw 3 branches.

3. **T:** Third person has 2 choices. Draw 2 branches.

4. **Count.** There are **24 ways.**

The tree diagram suggests a shorter way to solve the problem.

COMPUTATION: FACTORIALS

In the tree diagram, you can see that for <u>each</u> of the 4 seats you could choose, there are 3 choices left for the second person. For each of these 3 choices, there are 2 choices left for the third person.

Think:

$$\begin{pmatrix}\text{Number of}\\\text{choices for}\\\text{first person}\end{pmatrix} \times \begin{pmatrix}\text{Number of}\\\text{choices for}\\\text{second person}\end{pmatrix} \times \begin{pmatrix}\text{Number of}\\\text{choices for}\\\text{third person}\end{pmatrix} \times \begin{pmatrix}\text{Number of}\\\text{choices for}\\\text{fourth person}\end{pmatrix} = \begin{pmatrix}\text{Total}\\\text{number}\\\text{of choices}\end{pmatrix}$$

$$4 \quad \times \quad 3 \quad \times \quad 2 \quad \times \quad 1 \quad = \quad 24$$

A shortcut way to write this product is to use the **factorial** symbol.

$4 \times 3 \times 2 \times 1$ can be written as **4!** ◀ **Read 4! as 4 factorial.**

2. Use the factorial symbol to show in how many different ways 10 people could choose 10 seats.

3. Use a calculator to find 10!

EXERCISES

Complete.

1. 3! = _?_   **2.** 5! = _?_   **3.** 4! = _?_   **4.** $\frac{3!}{2!}$ = _?_   **5.** $\frac{5!}{3!}$ = _?_

You are taking a poll to find your classmates' preference for four recordings.

6. How many ways are there to choose first place?

7. How many ways are there to choose second place?

8. How many ways are there to choose third place?

9. How many ways are there to choose fourth place?

| Show Breeze |
| Purple Suns |
| Tumbleweeds |
| Zithers |

10. What is the total number of ways in which the choices can be made?

11. Three students line up to show their identification cards to a hall monitor. In how many ways can they line up?

12. Six parking spaces in a garage are to be assigned to the 6 executives of a large corporation. In how many ways can the parking spaces be assigned?

Three people get on a bus. There are only 4 seats available.

13. In how many ways can the first person choose a seat?

14. In how many ways can the second person choose a seat?

15. In how many ways can the third person choose a seat?

16. What is the total number of ways for 3 people to choose 4 seats?

Probability

How good are you at predicting the future? Try these. Write 1 if the event is certain to happen. Write 0 if you think it is impossible.

- The sun will rise tomorrow.
- It will snow in Miami on July 4.
- You will blink within 2 minutes.
- You will live to be 200 years old.

Most events do not fit into the 0 or 1 category. For example, there are two sides to a coin, a head and a tail.

1. What are the chances of tossing a head?

   The chance of tossing a head is **1 out of 2.**

   1 out of 2 means $\frac{1}{2}$.

2. Write a percent for $\frac{1}{2}$.

3. Will the chance of tossing a tail also be 1 out of 2?

   These 3 cards are shuffled (mixed up) and placed face down.

4. What is the chance or <u>probability</u> of picking the ace in one try? Write your answer as a fraction and as a percent.

   Let's add a four of clubs to the three cards.

5. What is the probability now of picking the ace? Write your answer as a fraction and as a percent.

> The **probability** of an event's happening is a number between 0 and 1. This number tells you how likely it is that the event will happen. The probability can also be written as a percent between 0 and 100.

---

6. On the first toss of a coin, you get a head. Does the probability $\frac{1}{2}$ mean that you will get a tail on the second toss?

EXERCISES

Write 1 if the event is certain to happen. Write 0 if you think it is impossible. Write P if you think it is possible for the event to happen.

1. It will rain tomorrow.

2. Man will walk on the planet Mars.

3. You will open a door today.

4. You will find a golden egg.

These three cards are placed face down. One card is drawn. Find each probability.

5. Drawing the 3

6. Drawing the ace

7. Drawing a red card

8. Drawing a black card

9. Drawing the five of hearts

10. Drawing a queen

These four cards are shuffled and placed face down. One card is drawn.

11. What is the probability of drawing the 10 of clubs?

12. What is the probability of drawing an ace?

13. What is the probability of drawing a black card?

14. What is the probability of drawing a red card?

15. Which is more likely to happen, drawing a red card or drawing a black card?

16. Which is more likely to happen, drawing a red card or drawing an ace?

There are 100 chances sold for a prize. Ellen bought 10 chances.

17. What is the probability that Ellen will win? Write your answer as a fraction and as a percent.

18. Which is more likely, that Ellen will win or that she will not win? Why?

19. Rosa tossed a coin. If it comes up heads, she will go to the movies. What is the probability she will not go?

20. There are 100 chances sold for a prize. Ken bought 5 chances. What is the probability he will win?

The Probability Ratio

In this game, 10 red marbles and 6 blue marbles were placed in a bowl. Without looking, you picked a blue marble. Were you lucky?

To find out, use the **probability ratio.** (You actually used this ratio in the previous lesson.)

$$\text{Probability ratio} = \frac{\text{Number of Successful Ways}}{\text{Number of Possible Ways}}$$

Use P(blue marble) to mean the "probability of choosing a blue marble."

P(blue marble) $= \dfrac{\text{Number of Blue Marbles}}{\text{Total Number of Marbles}}$ ← **Number of Successful Ways**
← **Number of Possible Ways**

$= \dfrac{6}{16} = \dfrac{3}{8}$

Since $\frac{3}{8} = 0.375$, or $37\frac{1}{2}\%$, P(blue marble) can also be given as **$37\frac{1}{2}\%$.**

1. Since the probability of picking a blue marble is less than 50%, were you lucky?

2. What is the probability of picking a red marble, or P(red marble)?

3. What is the sum of P(blue marble) and P(red marble)?

4. Knowing that the sum of the probabilities is 1 (or 100% if you use percents), how could you use subtraction to find the probability in Exercise 2?

There are 6 possible ways for a die (singular of dice) to turn up.

5. What is the probability of rolling a 1? a 2? a 5?

6. How many ways are there of rolling an even number?

7. What is the probability of rolling an even number?

This game involves a spinner. In playing this game, the player can choose the rule for winning from the following.

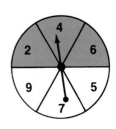

- You win if the arrow lands on the brown region.
- You win if the arrow lands on 2, 5, or 7.
- You win if the arrow lands on an odd number.

(NOTE: You lose if the arrow lands on any of the black lines no matter which rule you choose.)

8. Which rule would you choose? Why?

EXERCISES

The arrow is spun once.

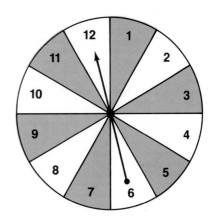

1. What is the probability of getting a 2?

2. You "win" if the arrow lands on an even number. What is your probability of winning?

3. How many numbers on the spinner are greater than 5?

4. What is the probability that the arrow lands on a number greater than 5?

5. You "win" if the arrow lands on 3, 5, 7, or 10. What is your probability of winning?

6. You "win" if the arrow lands on 8, 10, or 11. Is your probability of winning greater than, or less than, $\frac{1}{2}$?

You have 4 pennies, 3 nickels and 2 dimes in your pocket. One coin fell out.

7. What is the probability that the coin that fell out was a dime?

8. What is the probability that the coin that fell out was a penny?

These cards are placed in a box. One card is drawn without looking.

9. Which letter has the greatest probability of being drawn? Why?

10. Which letter has the least probability of being drawn?

11. Which two letters have the same probability of being drawn?

12. For which letter is the probability of being drawn greater than 40%? Why?

Mid-Chapter Review

You are buying a uniform. You can choose any of the options in the table. (Pages 344–345)

Sports Uniform		
Shorts	**Shirts**	**Shoes**
red white	red white striped	red white

1. How many choices are there in all?

2. If you want red shorts, how many choices for shirts and shoes do you have?

3. If you want red shoes, how many choices for shorts and shirts do you have?

4. Four cars line up in a row at the starting line. In how many ways can they line up? (Pages 346–347)

5. Five boat berths are to be assigned to 5 boat owners. In how many ways can the berths be assigned? (Pages 346–347)

These four cards are shuffled and placed face down. One card is drawn. (Pages 348–349)

6. What is the probability of drawing the four of diamonds?

7. What is the probability of drawing an ace?

These ten names are written on slips of paper and placed in a hat. One name is drawn. Find each probability. (Pages 350–351)

Alice	Jennifer	Ronald	Anthony	Joyce
Philip	Anna	Sylvia	Carlos	Michael

8. What is the probability that the name drawn begins with the letter "A"?

9. What is the probability of drawing a boy's name?

MAINTENANCE

Perform the indicated operations. (Pages 140–141 and 155–156)

10. $\frac{1}{2} + \frac{1}{3}$

11. $\frac{1}{4} + \frac{1}{7}$

12. $\frac{1}{5} + \frac{2}{3}$

13. $\frac{1}{3} + \frac{1}{6}$

14. $\frac{1}{2} \times \frac{1}{2}$

15. $\frac{1}{2} \times \frac{1}{3}$

16. $\frac{1}{4} \times \frac{2}{5}$

17. $\frac{1}{6} \times \frac{1}{6}$

Write the fractions in order from least to greatest.

18. $\frac{3}{4}, \frac{2}{5}, \frac{1}{8}, \frac{3}{10}$

19. $\frac{5}{6}, \frac{1}{2}, \frac{4}{7}, \frac{2}{3}$

20. $\frac{3}{7}, \frac{1}{5}, \frac{2}{9}, \frac{5}{8}$

Math in Advertising

How do manufacturers determine which colors to use on cereal boxes in order to attract the most customers? They do it by taking a *sample*. For example, they might ask people entering supermarkets in several cities in different parts of the country which of two or three cereal boxes they prefer. The designs and colors for new cereal boxes are based on the results of this sampling.

A **sample** is a group chosen to represent
a larger group called the **population.**

In order for the sample to represent the population, the sample must be <u>large enough</u> and it must be chosen <u>at random</u>; that is, completely by chance.

EXERCISES

1. This advertisement is based on a sample of Senator Wynn's friends. Is this a random sample? Explain.

**VOTE FOR
SENATOR
WYNN**
The person 9 out of 10 voters prefer!

2. This advertisement is based on a sample of four plumbers. Is the sample large enough? Explain.

3 out of 4 plumbers prefer **DRAIN OUT**

3. A supermarket manager asked the first six customers of the day which of two brands of soup they preferred. Is this a good sample? Explain.

4. A math workbook has 144 pages. An editor counts the number of problems on 75 different pages and divides by 75 to find the average number of problems per page. Is this a good sample? Explain.

PROJECT Estimate the number of left-handed students in your school by asking 25 students. Tell how you selected the students. Explain why you think the sample does, or does not, provide a good basis for the estimate.

Compound Events: AND

You have 8 sweaters. One is red, four are blue, and three are green.

1. What is the probability of picking a blue sweater without looking?

You also have 6 shirts: 3 white, 2 blue, and 1 green.

2. What is the probability of picking a blue shirt without looking?

PROBLEM   What is the probability of picking a blue sweater <u>and</u> picking a blue shirt?

There is more than one way to solve this problem.

MAKING A CHART

List all the possible ways of picking sweaters and shirts.
Let **B** = blue, **R** = red, **G** = green, and **W** = white.

Sweaters

	B	**B**	**B**	**B**	**R**	**G**	**G**	**G**
W	BW	BW	BW	BW	RW	GW	GW	GW
W	BW	BW	BW	BW	RW	GW	GW	GW
W	BW	BW	BW	BW	RW	GW	GW	GW
B	BB	BB	BB	BB	RB	GB	GB	GB
B	BB	BB	BB	BB	RB	GB	GB	GB
G	BG	BG	BG	BG	RG	GG	GG	GG

Shirts

3. Count the number of **BB**'s.

4. How many possible ways are there of picking sweaters and shirts?

5. Write the ratio.

$$P(BB) = \frac{\text{Number of BB's}}{\text{Number of Possible Ways}} = \frac{\square}{\square}$$

As a fraction in lowest terms, P(**BB**) = $\frac{1}{6}$.

COMPUTATION

This problem is an example of a compound "<u>and</u>" event. In this situation, you must pick a blue sweater <u>and</u> a blue shirt to be successful. You can make a chart to solve a compound "<u>and</u>" event or you can use the rule at the top of the next page.

> The probability of a compound "and" event is the product of the probabilities of each event.

6. What is the product of the probabilities in Exercises 1 and 2 on page 354?

7. Use the rule to find the probability of picking a red sweater and a white shirt. Check your answer with the chart.

EXERCISES

Use the information in the Problem on page 354 for Exercises 1–4.

1. What is the probability of picking a white shirt and a green sweater?

2. What is the probability of picking a green shirt and a green sweater?

3. What is the probability of picking a red sweater and a white shirt?

4. What is the probability of picking a blue shirt and a red sweater?

5. What sweater-shirt combination has the least probability of being chosen?

6. What sweater-shirt combination has the greatest probability of being chosen?

The pointer on each dial is spun once. The table lists all the possible ways in which the pointers can end up.

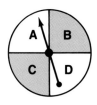

7. What is the probability that the first pointer stops on B?

8. What is the probability that the second pointer stops on 1?

9. What is the probability that the first pointer stops on B and the second pointer stops on 1?

Second Dial

	1	2	3	4
A	A1	A2	A3	A4
B	B1	B2	B3	B4
C	C1	C2	C3	C4
D	D1	D2	D3	D4

First Dial

10. What is the probability that one pointer stops on B and the other stops on an even number?

11. What is the probability that one pointer stops on C and one stops on an odd number?

12. What is the probability that one pointer stops on D and the other on a number less than 3?

13. What is the probability that one pointer stops on C and the other on a number greater than 0?

Compound Events: OR

Many games use two dice. In these games, the sum of the readings on the dice is the key.

In one game, you win if you roll a 7 or an 11 on the first roll.

PROBLEM   What is the probability of rolling a 7 or an 11?

There is more than one way to solve this problem.

MAKING A TABLE

This table shows all the possible ways for a pair of dice to land.

This table shows the <u>sum</u> for each of the ways shown in Table 1.

Second Die (Table 1)

First Die — a grid showing all possible pairs of dice faces.

Table 1

Second Die (Table 2)

+	1	2	3	4	5	6
1	2	3	4	5	6	(7)
2	3	4	5	6	(7)	8
3	4	5	6	(7)	8	9
4	5	6	(7)	8	9	10
5	6	(7)	8	9	10	(11)
6	(7)	8	9	10	(11)	12

First Die

Table 2

1. What is the total number of possible sums in Table 2?

2. Count the number of 7's and 11's.

3. Write the ratio.

$$P(7 \underline{\text{ or }} 11) = \frac{\text{Number of 7's and 11's}}{\text{Number of Possible Ways}} = \frac{\square}{\square}$$

As a fraction in lowest terms, $P(7 \underline{\text{ or }} 11) = \frac{2}{9}.$

COMPUTATION

This problem is an example of a compound "or" event. In this situation, you are successful if either a 7 or an 11 is rolled on the first try. You can make a table to solve a compound "or" event or you can use this rule.

> The probability of a compound "or" event is the sum of the probabilities of each event.

4. What is the probability of rolling a 7? an 11?

5. What is the sum of these probabilities?

EXERCISES

A pair of dice is tossed. Refer to Table 2 on page 356 to find each probability. Write a fraction in lowest terms for each probability.

1. Getting a sum of 5

2. Getting a sum of 3

3. Getting a sum of 5 or 3

4. Getting a sum of 8

5. Getting a sum of 12

6. Getting a sum of 8 or 12

7. Getting a sum of 2 or a sum greater than 9

8. Getting a sum of 13 or a sum less than 6

In this game, your "turn" consists of making two spins. Your score is the sum of the two spins. The table shows the possible sums.

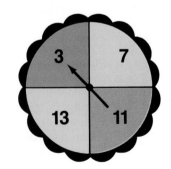

9. How many possible sums are there?

10. What is the probability of spinning a sum of 18? a sum of 25?

11. What is the probability of spinning a sum of 18 or a sum of 25?

12. What is the probability of spinning a sum of 14 or a sum of 22?

13. What is the probability of spinning a sum greater than 14 or a sum less than 14?

Second Spin

+	3	7	11	13
3	6	10	14	16
7	10	14	18	20
11	14	18	22	24
13	16	20	24	26

First Spin

Strategy: PREDICTING

Your job at the Spring Fling Concert is to manage the ice cream stand.

PROBLEM   Should you order the same amount of each flavor?

It is not possible to ask each of the 4000 students who will attend the concert what flavor each prefers. So you take a poll of 400 students.

Poll Question: *Which flavor do you prefer?*

Vanilla   *Chocolate*   *Vanilla Fudge*

The data was collected and organized in a table.

	Vanilla	Chocolate	Vanilla Fudge
Seniors	68	20	12
Juniors	69	21	10
Sophomores	63	24	13
Freshmen	50	35	15
Totals	**250**	**100**	**50**

1. Find the probability that a student will choose the following.

a. P(vanilla)   **b.** P(chocolate)   **c.** P(vanilla fudge)

As a percent:

P(vanilla) = **62.5%**   P(chocolate) = **25%**   P(vanilla fudge) = **12.5%**

You can use the results of your poll of 400 students to predict which ice cream flavors 4000 students will buy.

EXAMPLE   You plan to order 100 gallons of ice cream. About how many gallons of vanilla should you order?

Think: Since 62.5% of the students polled prefer vanilla, find 62.5% of 100 gallons.

62.5% = 0.625 and 0.625 × 100 = **62.5**

You should order about **63 gallons** of vanilla.

EXERCISES

Students at Littleton High surveyed 200 households to find the number of TV sets per household. The results are shown in the table at the right. Use this information for Exercises 1–12.

Number of TV sets	Number of Households
0	4
1	116
2	60
3 or more	20

For Exercises 1–6, find each probability for a household in Littleton. Write each probability as a fraction and as a percent.

1. The household has no TV sets.

2. The household has two TV sets.

3. The household has 3 or more TV sets.

4. The household has two or more TV sets.

5. The household has <u>at least</u> one TV set.

6. The household has <u>fewer than</u> 3 or <u>no</u> TV sets.

There are 1200 students at Littleton High.

7. About how many come from households that have 2 TV sets?

8. About how many came from homes that have 3 or more TV sets?

9. About how many come from households that have one or more TV sets?

10. About how many come from homes that have <u>fewer than</u> 3 or <u>no</u> TV sets?

11. About how many came from homes that have <u>at least</u> 2 TV sets?

12. About how many came from homes that have at least one TV set?

13. The probability of a person's winning a prize in a contest is $\frac{1}{1000}$. If 900,000 people enter the contest, predict how many can be expected to win a prize.

14. A poll shows that the probability that the District A voters will vote for a certain candidate is $\frac{3}{5}$. If 8700 people vote in District A, predict how many votes the candidate can expect to receive.

15. A company tests a sample of the radios it produces and finds that 1 out of every 100 is defective. Predict how many defective radios there will be in a shipment of 10,000.

16. In another sample, the company finds that 3 out of every 150 radios is defective. Predict how many defective radios there will be in a shipment of 7500.

What Are the Odds?

You've heard the expression "What are the odds?"

In a game of dice, you win if you roll a 7 on the first try. Recall from the table on page 356 that there are 6 ways of rolling a 7 out of 36 possible ways. This means there are 30 ways of <u>not</u> rolling a 7.

$$\textbf{Possible Ways} - \textbf{Successful Ways} = \textbf{Unsuccessful Ways}$$
$$36 \qquad - \qquad 6 \qquad = \qquad \textbf{30}$$

To find the *odds* of rolling a 7, you compare the number of successful ways to the number of unsuccessful ways.

$$\textbf{Odds in favor} = \frac{\textbf{Number of Successful Ways}}{\textbf{Number of Unsuccessful Ways}}$$

Odds in favor of rolling a $7 = \frac{6}{30} = \frac{1}{5}$ or

"1 to 5."

If the odds in favor are "1 to 5", then the **odds against** rolling a 7 are

"5 to 1."

1. From the table on page 356, how many ways are there of rolling a 7 <u>or</u> 11?

2. How many ways are there of <u>not</u> rolling a 7 <u>or</u> 11?

3. What are the odds in favor of rolling a 7 <u>or</u> 11?

4. What are the odds against rolling a 7 <u>or</u> 11?

EXERCISES

In a class of 24 girls and 18 boys, one name is drawn from a box to select a student council representative for the class.

1. What are the odds in favor of a girl's being chosen?

2. What are the odds in favor of a boy's being chosen?

3. Laura has two 1980 dimes, four 1982 dimes, and six 1985 dimes in her purse. She reaches in and picks one dime. What are the odds that she picks a 1982 dime?

4. Brian chooses a topic for a speech from slips of paper in a box. Of the ten topics in the box, Brian likes only four. What are the odds he chooses a topic he likes?

Chapter Summary

1. To find the number of different ways items can be chosen, multiply the number of choices for each item.

2. To find the number of different ways of arranging a certain number of items, write the factorial of the number of items and find the product.

3. To find the probability of an event, use this ratio:
$$P = \frac{\text{Number of Successful Ways}}{\text{Number of Possible Ways}}$$

4. To find the probability of a compound "and" event, multiply the probabilities of each event.

5. To find the probability of a compound "or" event, add the probabilities of each event.

6. To predict the number of times an event will occur, multiply the probability that the event will occur and the number of trials.

---

1. There are 4 different compact discs and 6 different tapes. How many different choices are there for a customer who wants to buy one compact disc and one tape?
$$4 \times 6 = \mathbf{24}$$

2. There are 5 students and 5 bicycles. In how many different ways could these 5 students choose a bicycle?
$$5! = 5 \times 4 \times 3 \times 2 \times 1 = \mathbf{120}$$

3. What is the probability of picking a red marble without looking from a group of 4 red marbles and 8 blue marbles?
$$P = \frac{4}{12}, \text{ or } \frac{1}{3}$$

4. $P(\text{white card}) = \frac{1}{4}$;
$P(\text{green card}) = \frac{2}{3}$
$P(\text{white card } \underline{\text{and}} \text{ green card}) = \frac{1}{4} \times \frac{2}{3} = \frac{1}{6}$

5. A pair of dice is tossed.
$P(4) = \frac{3}{36}, P(9) = \frac{4}{36}$
$P(4 \underline{\text{ or }} 9) = \frac{3}{36} + \frac{4}{36}$
$\qquad = \frac{7}{36}$

6. $P(\text{winning}) = \frac{4}{5}$
Number of trials: 80
Predicted wins:
$$\frac{4}{5} \times 80 = \frac{320}{5}$$
$$= \mathbf{64}$$

Chapter Review

Part 1: VOCABULARY

For Exercises 1–5, choose from the box at the right the word(s) or number(s) that complete(s) each statement.

1. To show all the possible ways that an event can happen, you can draw a __?__ .   (Page 344)

2. The product $5 \times 4 \times 3 \times 2 \times 1$ can be written as __?__ . (Page 346)

3. A number between 0 and 1 that tells how likely it is that a certain event will happen is called the __?__ of an event. (Page 348)

4. The probability of an event that cannot happen is __?__ . (Page 348)

5. The probability of an event that is certain to happen is __?__ . (Page 348)

> 1
> probability
> 0
> tree diagram
> 5!

Part 2: SKILLS AND APPLICATIONS

6. Class rings can be ordered in gold or white gold with red, blue, or amber colored stones. How many choices are there in all? (Pages 344–345)

7. For a certain bicycle there are 3 different frames, 3 different component groups, and 2 types of tires. You have chosen a frame. How many different choices for components and tires are there?   (Pages 344–345)

8. Five students line up for lunch. In how many different ways can they line up?   (Pages 346–347)

9. Ten chairs are assigned to ten remaining students. In how many ways can the chairs be assigned? (Pages 346–347)

The names of the seven days of the week are written on slips of paper and placed in a hat. One slip of paper is drawn. Find each probability. Write each answer as a fraction in lowest terms.   (Pages 348–351)

10. What is the probability of drawing "Sunday"?

11. What is the probability of drawing "Friday"?

12. What is the probability of drawing a day that begins with "T"?

13. What is the probability of drawing a day that begins with "M"?

14. What is the probability of drawing a day that begins with "H"?

15. What is the probability of drawing a day that begins with "S"?

Sue has six pairs of socks. One pair is red, three are navy blue, and two are brown. She also has five scarves. Two are red, one is blue, and two are green. She chooses a pair of socks and a scarf. Use this information for Exercises 16–21. Write a fraction in lowest terms for each answer. (Pages 354–355)

16. What is the probability of picking a pair of blue socks?

17. What is the probability of picking a blue scarf?

18. What is the probability of picking blue socks <u>and</u> a blue scarf?

19. What is the probability of picking a pair of red socks?

20. What is the probability of picking a green scarf?

21. What is the probability of picking red socks <u>and</u> a green scarf?

A bag contains four green marbles, six white marbles, and ten red marbles. One marble is picked from the bag. Find each probability. Write a fraction in lowest terms for each answer. (Pages 356–357)

22. What is the probability of picking a green marble?

23. What is the probability of picking a red marble?

24. What is the probability of picking a white marble?

25. What is the probability of picking a red <u>or</u> a white marble?

26. What is the probability of picking a white <u>or</u> a green marble?

27. What is the probability of picking a green <u>or</u> a red marble?

Solve. (Pages 358–359)

28. Charles is a member of a basketball team. Out of 10 free throws, he usually succeeds in making 7 baskets. Charles tried 30 free throws. Predict how many baskets he made.

29. Tina is on the bowling team. She usually bowls more than 170 in 2 out of 3 games. Predict the number of times she will bowl more than 170 in 90 games.

30. The Ace Game Company makes board games. About 4 out of every 100 games turn out to be defective. Predict how many of 1500 games produced will be defective.

31. During summer vacation in Smallville, there are afternoon rain showers about every 3 out of 7 days. Predict how many afternoon rain showers will occur during 70 days of summer vacation.

Chapter Test

1. You take a poll to find your class-mate's preference for six sports. What is the total number of ways in which choices can be made?

2. Herb is on a baseball team. He averages 3 hits in 12 times at bat. Predict the number of hits he will have after 160 times at bat.

The names of the twelve months of the year are written on slips of paper and placed in a hat. One name is drawn. Find each probability. Write each answer as a fraction in lowest terms.

3. What is the probability of drawing "January"?

4. What is the probability of drawing "August"?

5. What is the probability of drawing a month that begins with "J"?

6. What is the probability of drawing a month that begins with "A"?

Jane has four skirts. Two skirts are gray, one is blue, and one is red. She also has ten blouses. Four are white, two are red, and four are prints. On a certain day, Jane wants to wear one of her skirts and one of her blouses. Find each probability. Write each answer as a fraction in lowest terms <u>and</u> as a per cent.

7. Picking a gray skirt

8. Picking a red blouse

9. Picking a red skirt

10. Picking a white blouse

11. Picking a gray skirt <u>and</u> a white blouse

12. Picking a red skirt <u>and</u> a white blouse

13. Picking a blue skirt <u>and</u> a print blouse

14. Picking a gray skirt <u>and</u> a print blouse

A bag contains two green marbles, 4 red marbles and 4 blue marbles. One marble is picked. Find each probability. Write each answer as a fraction in lowest terms.

15. Picking a green marble

16. Picking a red marble

17. Picking a blue marble

18. Picking a green <u>or</u> a red marble

19. Picking a blue <u>or</u> a green marble

20. Picking a blue <u>or</u> a red marble

Cumulative Maintenance   Chapters 1–14

Choose the correct answer. Choose a, b, c, or d.

1. Write the numeral for ten thousand six.

   a. 10,000.6   b. 1006
   c. 10,600   d. 10,006

2. Which fraction has the smallest value?

   a. $\frac{1}{2}$ b. $\frac{1}{3}$
   c. $\frac{1}{4}$ d. $\frac{1}{5}$

3. Use the formula $A = \pi \times r \times r$ to find the area of a circle whose radius is 14. (Use $\pi = \frac{22}{7}$.)

   a. 308   b. 44
   c. 616   d. 88

4. What is the sale price of the grill? Use mental computation.

   Grill
   20% OFF
   Regular Price
   $80

   a. $16   b. $80
   c. $72   d. $64

5. A recipe that serves 4 people calls for 4 cups of flour. How many cups of flour are needed to make enough for 12 people?

   a. 36   b. 16
   c. 9   d. 12

6. Add:

   6.21
   13.5
   3.933
   +354.7

   a. 366.743   b. 367.843
   c. 377.734   d. 378.343

7. Use the formula $C = 2 \times \pi \times r$ to find the circumference of a circle whose radius is 10. (Use $\pi = 3.14$.)

   a. 62.8   b. 6.28
   c. 314   d. 31.4

8. Multiply: 306×250

   a. 765,000   b. 76,500
   c. 756,000   d. 75,600

9. This circle graph shows how one family spends its money. For which item do they spend the *most* money?

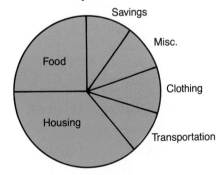

   a. Food   b. Savings
   c. Clothing   d. Housing

10. Kim earned $100 in tips during April, $185 in May, and $240 in June. Find the average monthly tips.

a. $150   b. $180
c. $165   d. $175

11. Find the median price of these five prices.

 $8.95 $12.50 $17.50
  $6.95 $11.25

a. $8.95   b. $9.95
c. $11.25   d. $12.50

12. What percent of 60 is 18?

a. 30   b. 45
c. 35   d. 42

13. A company tests a sample of the computers it produces and finds that 3 out of every 500 are defective. Predict how many defective computers there will be in 50,000 produced.

a. 500   b. 300
c. 1000   d. 1500

14. What is the perimeter in meters of this parking lot?

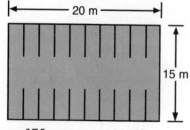

20 m
15 m

a. 375   b. 60
c. 70   d. 35

15. What is the total value of eighteen $5-bills, five $10-bills, and twelve $20-bills?

a. $380   b. $350
c. $370   d. $320

16. Jeff has a blue tie and a brown tie. He also has a white shirt, a brown shirt, and a plaid shirt. What is the probability that he will choose the blue tie <u>and</u> the plaid shirt?

a. 0   b. 1
c. $\frac{1}{5}$ d. $\frac{1}{6}$

17. Find the mode of these numbers.

15 12 15 25 18 16
18 24 19 21 16 15

a. 18   b. 24
c. 15   d. 21

18. Larry sold $2000 of boating equipment. He earns a 15% commission. How much does he earn?

a. $30   b. $300
c. $1700   d. $133

19. Divide.

$$\frac{3}{4} \div \frac{15}{16}$$

a. $\frac{7}{31}$ b. $\frac{45}{64}$
c. $\frac{5}{4}$ d. $\frac{4}{5}$

The distance, d, that an object will fall in t seconds is given by this formula:

$$d = 16t^2$$

- How many seconds will it take for the object to fall 16 ft? 64 ft?

- Will the distance fallen at the end of the third second be 3 times or 9 times the distance fallen at the end of the first second?

- Will the object be falling faster at the end of the fourth second than at the end of the third second? Explain.

Squares and Square Roots

1. Each of these figures is a square. Why?

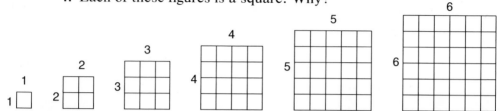

2. How many square units, □, are there in each square?

3. For which of these numbers can you draw a square?

    a. 40   **b.** 50   **c.** 49   **d.** 81

Numbers such as the following are called <u>perfect squares</u>.

$$1, 4, 9, 16, 25, 36, 49, \text{ and } 81$$

> A number is a **perfect square** if it has two equal factors.

You know that 49 is a perfect square because $49 = 7 \times 7$, or

$$49 = 7^2.$$ ◀ **Read as:** "7 squared"

The 2 in 7^2 is called an **exponent.**

4. What is the next perfect square after 81?

5. What are the two equal factors of 100?

> Finding one of the two equal factors of a number is called finding the **square root.** The symbol for square root is $\sqrt{}$.

6. *Complete:* Thus, the square root of 100, or $\sqrt{100} = \underline{\ ?\ }$.

EXERCISES

Complete. Choose your answers from the box at the right.

1. Because $16 = 4 \times 4$, the number 16 is a $\underline{\ ?\ }$.

2. The number 25 is a perfect square because it has $\underline{\ ?\ }$ equal factors.

3. The 2 in 8^2 is called an $\underline{\ ?\ }$.

4. Because $81 = 9 \times 9$, the number 9 is the $\underline{\ ?\ }$ of 81.

square
square root
perfect
   square
exponent
factor
two
one

Complete.

5. $2^2 = 4$ so $\sqrt{4} = \underline{\ ?\ }$. **6.** $3^2 = 9$ so $\sqrt{9} = \underline{\ ?\ }$. **7.** $5^2 = 25$ so $\sqrt{25} = \underline{\ ?\ }$.

8. $6^2 = \underline{\ ?\ }$ so $\sqrt{36} = 6$. **9.** $7^2 = 49$ so $\sqrt{?} = 7$. **10.** $\underline{\ ?\ }^2 = 1$ so $\sqrt{1} = 1$.

Find each square.

11. 8^2 **12.** 9^2 **13.** 11^2 **14.** 15^2 **15.** 10^2

16. 25^2 **17.** 30^2 **18.** 40^2 **19.** 21^2 **20.** 51^2

21. 19^2 **22.** 14^2 **23.** 60^2 **24.** 24^2 **25.** 200^2

Find each square root.

26. $\sqrt{4}$ **27.** $\sqrt{1}$ **28.** $\sqrt{64}$ **29.** $\sqrt{81}$ **30.** $\sqrt{36}$

31. $\sqrt{100}$ **32.** $\sqrt{225}$ **33.** $\sqrt{900}$ **34.** $\sqrt{289}$ **35.** $\sqrt{625}$

36. $\sqrt{441}$ **37.** $\sqrt{196}$ **38.** $\sqrt{1600}$ **39.** $\sqrt{3600}$ **40.** $\sqrt{4900}$

41. The area of a square is 169 square meters. Find the length of a side.

42. The area of a square is 144 square feet. Find the length of a side.

Find each answer.

43. 27^2 **44.** $\sqrt{400}$ **45.** $\left(\frac{1}{2}\right)^2$ **46.** $\left(\frac{2}{5}\right)^2$ **47.** $\sqrt{2500}$

48. $(1.2)^2$ **49.** $(1.3)^2$ **50.** 15^2 **51.** $\sqrt{0.04}$ **52.** $\sqrt{0.64}$

53. $\frac{\sqrt{1}}{\sqrt{25}}$ **54.** $\frac{\sqrt{4}}{\sqrt{81}}$ **55.** $\frac{\sqrt{25}}{\sqrt{81}}$ **56.** $\frac{\sqrt{16}}{\sqrt{25}}$ **57.** $\frac{\sqrt{4}}{\sqrt{144}}$

You can use the following formula to find the distance in feet that an object will fall in t seconds.

$$d = 16t^2 \quad \begin{aligned} d &= distance \\ t &= time \end{aligned}$$

58. How far will an object fall in 2 seconds?
(HINT: $d = 16 \times 2^2 = \underline{\ ?\ }$)

59. How far will an object fall in 5 seconds?

60. How far will an object fall in $\frac{1}{2}$ second?

61. How far will an object fall in $\frac{1}{4}$ second?

62. How long will it take an object to fall 64 feet?
(HINT: $64 = 16t^2$; $t = \underline{\ ?\ }$)

Skydivers fall freely at speeds of more than 160 kilometers per hour.

63. How long will it take an object to fall 256 feet?

Using a Table of Squares and Square Roots

You can find the square root of these perfect squares without using paper and pencil.

$$1, 4, 9, 16, 25, 36, 49, 81, 100$$

You can also find the square root of numbers that are <u>not</u> perfect squares without using paper and pencil.

EXAMPLE 1 $\sqrt{23} = \underline{\quad ?\quad}$

1 Find 23 in the NUMBER column in the table on page 371.

2 Look directly to the right. Read the square root of 23 in the SQUARE ROOT column.

$\sqrt{23}$ is about **4.796**

Number	Square	Square Root
21	441	4.583
22	484	4.690
23	529	4.796
24	576	4.899
25	625	5.000

3 **Check:** Is 4.796^2 about equal to 23?

$\boxed{4}\;\boxed{\cdot}\;\boxed{7}\;\boxed{9}\;\boxed{6}\;\boxed{\times}\;\boxed{4}\;\boxed{\cdot}\;\boxed{7}\;\boxed{9}\;\boxed{6}\;\boxed{=}\qquad \boxed{\text{23.001616}}$

As the check shows, you can use the calculator to square a number. You can also use the table to square numbers. To find 21^2, first find 21 in the NUMBER column. Look directly to the right. Read the square of 21 in the SQUARE column.

1. What is 21^2?

You can use the table to find the square roots of some numbers greater than 150. First, find the number in the SQUARE column. Then read the square root in the NUMBER column.

EXAMPLE 2 $\sqrt{4761} = \underline{\quad ?\quad}$

1 Find 4761 in the SQUARE column.

2 Look directly to the left. Read the number. ⟶ **69**

$\sqrt{4761} = $ **69**

Number	Square
66	4356
67	4489
68	4624
69	4761
70	4900

Table of Squares and Square Roots

Number	Square	Square Root	Number	Square	Square Root	Number	Square	Square Root
1	1	1.000	51	2601	7.141	101	10,201	10.050
2	4	1.414	52	2704	7.211	102	10,404	10.100
3	9	1.732	53	2809	7.280	103	10,609	10.149
4	16	2.000	54	2916	7.348	104	10,816	10.198
5	25	2.236	55	3025	7.416	105	11,025	10.247
6	36	2.449	56	3136	7.483	106	11,236	10.296
7	49	2.646	57	3249	7.550	107	11,449	10.344
8	64	2.828	58	3364	7.616	108	11,664	10.392
9	81	3.000	59	3481	7.681	109	11,881	10.440
10	100	3.162	60	3600	7.746	110	12,100	10.448
11	121	3.317	61	3721	7.810	111	12,321	10.536
12	144	3.464	62	3844	7.874	112	12,544	10.583
13	169	3.606	63	3969	7.937	113	12,769	10.630
14	196	3.742	64	4096	8.000	114	12,996	10.677
15	225	3.873	65	4225	8.062	115	13,225	10.724
16	256	4.000	66	4356	8.124	116	13,456	10.770
17	289	4.123	67	4489	8.185	117	13,689	10.817
18	324	4.243	68	4624	8.246	118	13,924	10.863
19	361	4.359	69	4761	8.307	119	14,161	10.909
20	400	4.472	70	4900	8.367	120	14,400	10.954
21	441	4.583	71	5041	8.426	121	14,641	11.000
22	484	4.690	72	5184	8.485	122	14,884	11.045
23	529	4.796	73	5329	8.544	123	15,129	11.091
24	576	4.899	74	5476	8.602	124	15,376	11.136
25	625	5.000	75	5625	8.660	125	15,625	11.180
26	676	5.099	76	5776	8.718	126	15,876	11.225
27	729	5.196	77	5929	8.775	127	16,129	11.269
28	784	5.292	78	6084	8.832	128	16,384	11.314
29	841	5.385	79	6241	8.888	129	16,641	11.358
30	900	5.477	80	6400	8.944	130	16,900	11.402
31	961	5.568	81	6561	9.000	131	17,161	11.446
32	1024	5.657	82	6724	9.055	132	17,424	11.489
33	1089	5.745	83	6889	9.110	133	17,689	11.533
34	1156	5.831	84	7056	9.165	134	17,956	11.576
35	1225	5.916	85	7225	9.220	135	18,225	11.619
36	1296	6.000	86	7396	9.274	136	18,496	11.662
37	1369	6.083	87	7569	9.327	137	18,769	11.705
38	1444	6.164	88	7744	9.381	138	19,044	11.747
39	1521	6.245	89	7921	9.434	139	19,321	11.790
40	1600	6.325	90	8100	9.487	140	19,600	11.832
41	1681	6.403	91	8281	9.539	141	19,881	11.874
42	1764	6.481	92	8464	9.592	142	20,164	11.916
43	1849	6.557	93	8649	9.644	143	20,449	11.958
44	1936	6.633	94	8836	9.695	144	20,736	12.000
45	2025	6.708	95	9025	9.747	145	21,025	12.042
46	2116	6.782	96	9216	9.798	146	21,316	12.083
47	2209	6.856	97	9409	9.849	147	21,609	12.124
48	2304	6.928	98	9604	9.899	148	21,904	12.166
49	2401	7.000	99	9801	9.950	149	22,201	12.207
50	2500	7.071	100	10,000	10.000	150	22,500	12.247

EXERCISES

Copy the table at the right. For each square or square root in Exercises 1–6, find the number in your table. Then draw an arrow pointing to the number's square or square root. Write the answer.

Number	Square	Square Root
86	7396	9.274
87	7569	9.327
88	7744	9.381
89	7921	9.434
90	8100	9.487

1. 87^2 **2.** 90^2 **3.** $\sqrt{86}$

4. $\sqrt{88}$ **5.** $\sqrt{7921}$ **6.** $\sqrt{7396}$

Use the table on page 371 to find each square or square root.

7. 45^2 **8.** 73^2 **9.** 80^2 **10.** 115^2 **11.** 121^2 **12.** 62^2

13. 125^2 **14.** 110^2 **15.** 148^2 **16.** 135^2 **17.** 66^2 **18.** 91^2

19. $\sqrt{70}$ **20.** $\sqrt{18}$ **21.** $\sqrt{141}$ **22.** $\sqrt{118}$ **23.** $\sqrt{105}$ **24.** $\sqrt{81}$

25. $\sqrt{63}$ **26.** $\sqrt{40}$ **27.** $\sqrt{126}$ **28.** $\sqrt{148}$ **29.** $\sqrt{90}$ **30.** $\sqrt{5}$

31. $\sqrt{3}$ **32.** $\sqrt{11}$ **33.** $\sqrt{49}$ **34.** $\sqrt{57}$ **35.** $\sqrt{84}$ **36.** $\sqrt{98}$

37. $\sqrt{324}$ **38.** $\sqrt{900}$ **39.** $\sqrt{676}$ **40.** $\sqrt{529}$ **41.** $\sqrt{1600}$ **42.** $\sqrt{196}$

43. $\sqrt{484}$ **44.** $\sqrt{729}$ **45.** $\sqrt{9801}$ **46.** $\sqrt{9216}$ **47.** $\sqrt{19,881}$ **48.** $\sqrt{14,884}$

49. $\sqrt{841}$ **50.** $\sqrt{2209}$ **51.** $\sqrt{1681}$ **52.** $\sqrt{5776}$ **53.** $\sqrt{13,689}$ **54.** $\sqrt{10,816}$

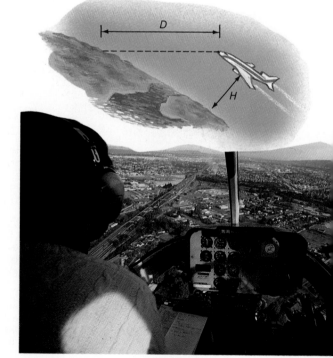

You can use the following formula to find the distance in miles from a viewer to the horizon.

$$D = 1.2 \times \sqrt{H}$$

H: number of feet from the ground

Use this formula for Exercises 55–57. Round each answer to the nearest whole number.

55. Find the distance to the horizon from a plane. The plane is 3025 feet above the ground.

56. Find the distance to the horizon from a plane flying at an altitude (height) of 22,500 feet.

57. The height of a plane above the ground is 20,736 feet. Find the distance to the horizon.

The Rule of Pythagoras

A **right triangle** is a triangle with one right angle (90°). The ⌐ in the triangle stands for a right angle. The side opposite the right angle is the **hypotenuse** and it is the longest side. The other sides are **legs**.

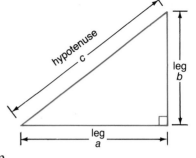

1. Measure each side of the triangle at the right to the nearest centimeter.

   a. $a = $ _?_ cm   **b.** $b = $ _?_ cm   **c.** $c = $ _?_ cm

   2. Now square each measure.

      a. $a^2 = $ _?_   **b.** $b^2 = $ _?_   **c.** $c^2 = $ _?_

   3. Add a^2 and b^2. Compare this sum with c^2.

      This comparison suggests the Rule of Pythagoras.

   > Rule of Pythagoras: $(\text{Hypotenuse})^2 = (\text{leg})^2 + (\text{leg})^2$, or
   > $$c^2 = a^2 + b^2$$

      You can use this rule to determine if a triangle is a right triangle.

   4. *Complete:* If triangle *ABC* below is a right triangle, then $6^2 = $ _?_ $+$ _?_.

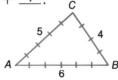
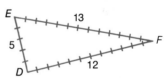

   5. *Complete:* If triangle *DEF* above is a right triangle, then $13^2 = $ _?_ $+$ _?_.

EXAMPLE

a. Is triangle *ABC* a right triangle?

$c^2 = a^2 + b^2$ ⟵ **Write the rule.**

$6^2 \overset{?}{=} 4^2 + 5^2$ ⟵ **Replace a, b, and c.**

$36 \overset{?}{=} 16 + 25$

$36 \neq 41$ ◀ **The ≠ means "is not equal to."**

Triangle *ABC* is **not** a right triangle.

b. Is triangle *DEF* a right triangle?

⟶ $c^2 = a^2 + b^2$

⟶ $13^2 \overset{?}{=} 5^2 + 12^2$

$169 \overset{?}{=} 25 + 144$

$169 = 169$

Triangle *DEF* is a right triangle.

EXERCISES

*Complete. Choose your answers from the box at
the right.*

1. The measure of a right angle is __?__ .

2. In a right triangle, the side opposite the right angle is the __?__ .

3. The longest side of a right triangle is the __?__ .

*For Exercises 4–6, each triangle is a right triangle. What is the length
of the hypotenuse?*

4.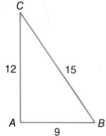

5.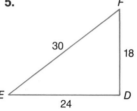

6.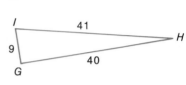

*Use the Rule of Pythagoras to determine which triangles are right
triangles. Write Yes or No.*

7.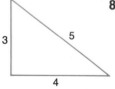

8.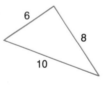

9.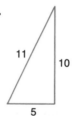

10.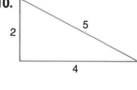

11.

12.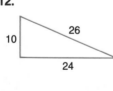

13.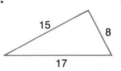

14.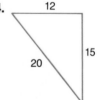

15. Martha made a desk. She attached a
board to the wall and supported the
board with a brace 5 feet long. The
brace meets the wall 3 feet below
the table and meets the table 2 feet
out from the wall. Does the board
make a right angle with the wall?

16. The lengths of the sides of a
triangular jogging path are 20
meters, 21 meters, and 29 meters.
Does the path form a right triangle?

Strategy: USING A DRAWING AND A FORMULA

When you know the lengths of two sides of a right triangle, you can use the Rule of Pythagoras to find the third side.

EXAMPLE 1
Find the length of the hypotenuse of this sail. Round your answer to the nearest tenth.

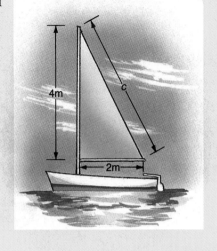

READ What are the facts?

$a = 2$ m and $b = 4$ m

PLAN Use the Rule of Pythagoras to find c, the hypotenuse.

SOLVE $c^2 = a^2 + b^2$ **Replace a with 2 and b with 4.**

$c^2 = 2^2 + 4^2$

$c^2 = 4 + 16$

$c^2 = 20$ **To find c, find the square root of 20.**

$c = \sqrt{20}$

$c = 4.472$ **From the table** or [2] [0] [√]   4.4721359

CHECK Did you use the facts correctly in the problem?

To the nearest tenth, the hypotenuse is about **4.5 meters.**

It is helpful to draw a right triangle and label the sides.

EXAMPLE 2
The diagonal of a television screen is 16 inches. One of the sides is 12 inches long. What is the length of the other side? Round your answer to the nearest tenth.

Diagonal

16 in–TV Screen

READ What are the facts?

Diagonal: 16 in   One side: 12 in

PLAN Draw a right triangle and label the sides. Use the Rule of Pythagoras to find the unknown side.

SOLVE

$$c^2 = a^2 + b^2$$ ◀ Replace c with 16 and a with 12.

$$16^2 = 12^2 + b^2$$

$$256 = 144 + b^2$$ ◀ Subtract 144 from each side.

$$256 - 144 = 144 + b^2 - 144$$

$$112 = b^2$$

$$\sqrt{112} = b$$

$$10.583 = b \quad \text{or} \quad \boxed{1}\boxed{1}\boxed{2}\boxed{\sqrt{}} \quad \boxed{10.583005}$$

CHECK Did you use the facts correctly in the solution?

The length of the other side is about **10.9 inches.**

EXERCISES

In Exercises 1–7, find the length of the hypotenuse of each right triangle. Use the Table of Squares and Square Roots on page 371.

1.
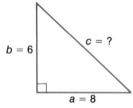
b = 6, c = ?, a = 8

2.
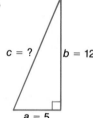
c = ?, b = 12, a = 5

3.
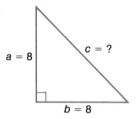
a = 8, c = ?, b = 8

4. $a = 4$ yd
$b = 9$ yd
$c = \underline{\ ?\ }$

5. $a = 5$ ft
$b = 6$ ft
$c = \underline{\ ?\ }$

6. $a = 2$ mm
$b = 2$ mm
$c = \underline{\ ?\ }$

7. $a = 3$ m
$b = 9$ m
$c = \underline{\ ?\ }$

Solve. Round answers to the nearest tenth.

8.
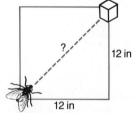

?, 12 in, 12 in

How far is it from the fly to the sugar?

9.
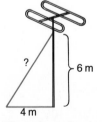
?, 6 m, 4 m

How long is the wire holding the antenna?

10.
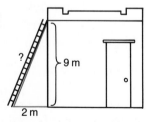
?, 9 m, 2 m

How long is the ladder?

In Exercises 11–13, find the length of the unknown side of each triangle.
Use the Table of Squares and Square Roots on page 371.

11.

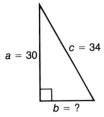

$a = 30$
$c = 34$
$b = ?$

12.

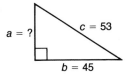

$a = ?$
$c = 53$
$b = 45$

13.

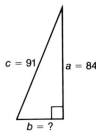

$c = 91$
$a = 84$
$b = ?$

Solve. Find the answers to the nearest tenth.

14.

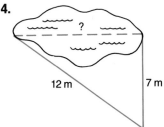

12 m   7 m
?

How far is it
across the pond?

15.

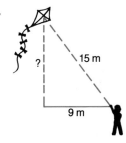

?   15 m
9 m

About how high
is the kite?

16.

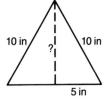

10 in   10 in
?
5 in

What is the height
of the triangle?

17. A ladder is placed against the side
of the house as shown in the figure
below. Find the length of the
ladder. Round your answer to the
nearest tenth.

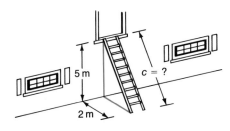

5 m
2 m
$c = ?$

18. The front wall of an A-frame house
is as shown in the figure below.
How tall is the front of the house?
Round your answer to the nearest
tenth.

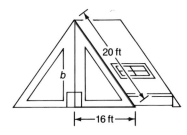

20 ft
b
16 ft

19. Tama's kite is 30 yards above the
ground. A tree is directly under the
kite. Tama is standing 16 yards away
from the tree. How long is the kite
string? (HINT: Draw a figure.)

20. A square-shaped park has an area of
49 square kilometers. Doug took a
diagonal path through the park. To
the nearest kilometer, how far did
Doug walk? (HINT: Draw a figure.)

Mid-Chapter Review

Find each square or square root.   (Pages 368–369)

1. 1^2 **2.** 5^2 **3.** 17^2 **4.** 37^2 **5.** $(\frac{1}{4})^2$ **6.** $(0.5)^2$

7. $\sqrt{1}$ **8.** $\sqrt{25}$ **9.** $\sqrt{121}$ **10.** $\sqrt{0.25}$ **11.** $\sqrt{0.36}$ **12.** $\sqrt{6400}$

Use the Table of Squares and Square Roots on page 371 to find each square or square root.   (Pages 370–372)

13. 71^2 **14.** 126^2 **15.** 139^2 **16.** $\sqrt{28}$ **17.** $\sqrt{143}$ **18.** $\sqrt{13,924}$

Three sides of a triangle are given. Use the Rule of Pythagoras to determine which triangles are right triangles. Write Yes or No.
(Pages 373–374)

19. 17 cm, 20 cm, 25 cm

20. 16 ft, 30 ft, 34 ft

21. 11 cm, 60 cm, 61 cm

22. 8 in, 11 in, 15 in

In Exercises 23–26, two sides of a right triangle are given. Find the length of the missing side. Use the Table of Squares and Square Roots on Page 371.   (Pages 375–377)

23. $a = 7$ m; $b = 25$ m; $c = \underline{\ ?\ }$

24. $a = 11$ cm; $b = \underline{\ ?\ }$; $c = 16$ cm

25. $a = \underline{\ ?\ }$; $b = 15$ ft; $c = 39$ ft

26. $a = 12$ in; $b = 16$ in; $c = \underline{\ ?\ }$

Solve.   (Pages 375–377)

27. The top of a 6-meter ladder reaches a window ledge. The ledge is 5 meters above the ground. How far from the house is the foot of the ladder? Round your answer to the nearest tenth.

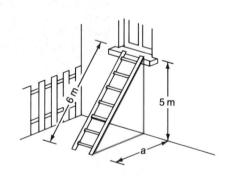

MAINTENANCE

Solve.   (Pages 279–280)

28. 4% of 86 = $\underline{\ ?\ }$ **29.** 30% of 240 = $\underline{\ ?\ }$ **30.** 12 = $\underline{\ ?\ }$% of 150

31. The volume of a rectangular prism is 480 cubic inches. The base has a length of 5 inches and a width of 8 inches. What is the height of the prism?   (Pages 304–305)

32. Which is greater, the area of a circle 8 centimeters in diameter or the area of a square 8 centimeters on a side? How much greater? (Pages 191–193 and 330–331)

Triangles: ANGLES AND SIDES

ACTIVITY **1** Cut a square sheet of paper along the diagonal. Label the two resulting triangles as shown in Figures 1 and 2.

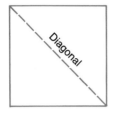 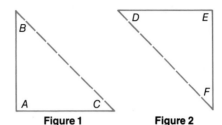 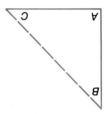

Figure 1   Figure 2   Figure 3

2 Place triangle *ABC* on top of triangle *DEF* so that the two triangles "fit." To do this, turn triangle *ABC* as shown in Figure 3. Because the two triangles fit, they are <u>congruent</u>.

> **Congruent triangles** have the same size and shape.

Since ∠*C* fits into ∠*D*, they have the same measure.

Thus, ∠*C* and ∠*D* are **corresponding angles.**

1. Name the two other pairs of corresponding angles in Figures 2 and 3. Since side *AB* fits onto side *EF*, they have the same measure.

Thus, side *AB* and side *EF* are **corresponding sides.**

NOTE: Side *AB* is opposite ∠*C*, and side *EF* is opposite ∠*D*.

This suggests the following.

> In congruent triangles, corresponding sides are opposite corresponding angles.

·2. Since ∠*A* and ∠*E* are corresponding angles, side __?__ and side __?__ are corresponding sides.

3. Since ∠*B* and ∠*F* are corresponding angles, side __?__ and side __?__ are corresponding sides.

Similar Triangles

1. Are these two triangles congruent? Why or why not?

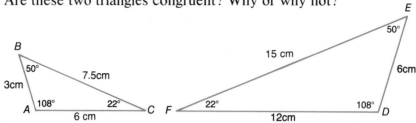

Two triangles have the same shape if the angles of one triangle have the same measure as the angles of the second triangle.

Angle A has the same measure as angle D. Thus, $\angle A$ and $\angle D$ are **corresponding angles.**

2. Which angle in triangle *DEF* corresponds to
   a. angle *B?*   **b.** angle *C?*

Therefore, triangle *ABC* and triangle *DEF* have the same shape.

Triangles that have the same shape are **similar triangles.**

In similar triangles, **corresponding sides** are opposite corresponding angles. Since $\angle B$ and $\angle E$ are corresponding angles, side *AC* and side *DF* are corresponding sides. You can use a ratio to compare corresponding sides.

$$\text{Triangle 1} \longrightarrow \frac{AC}{DF} = \frac{6}{12} = \frac{1}{2} \longleftarrow \text{Triangle 2}$$

3. What are the corresponding sides for
   a. $\angle A$ and $\angle D$?   **b.** $\angle C$ and $\angle F$?

4. Write a ratio that compares
   a. *BC* to *EF*   **b.** *AB* to *DE*

5. How do these ratios compare to the ratio of *AC* to *DF*?

This suggests the following.

In similar triangles, corresponding sides have equivalent ratios.

| EXAMPLE | Norman is 1.8 meters tall. To find the height of the school, he had Patty measure his shadow and the school's shadow at the same time of day. Find the height of the school to the nearest tenth. |

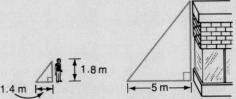

1.8 m

1.4 m

5 m

1 Write a ratio that compares the school's unknown height to the school's shadow.

$\dfrac{h}{5}$ ← School's height ← School's shadow

2 Write a ratio that compares Norman's height to Norman's shadow.

$\dfrac{1.8}{1.4}$ ← Norman's height ← Norman's shadow

3 Use the ratios to write a proportion.

$\dfrac{h}{5} = \dfrac{1.8}{1.4}$

4 Solve the proportion.

$$1.4 \times h = 5 \times 1.8$$
$$1.4h = 9$$
$$\dfrac{1.4h}{1.4} = \dfrac{9}{1.4}$$
$$h = \mathbf{6.43}$$

Find 9 ÷ 1.4 (two decimal places).

To the nearest tenth, the school's height is about **6.4 meters.**

EXERCISES

The triangles at the right are similar. Name the corresponding sides and angles. Find the ratio of the corresponding sides.

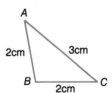

1. $\angle A$ corresponds to __?__.

$\angle B$ corresponds to __?__.

$\angle C$ corresponds to __?__.

2. Side AB corresponds to side __?__. Ratio: __?__

Side BC corresponds to side __?__. Ratio: __?__

Side AC corresponds to side __?__. Ratio: __?__

Each pair of triangles is similar. Complete the proportion.

3.

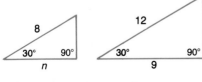

$$\dfrac{n}{8} = \dfrac{?}{12}$$

4.

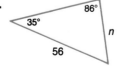

$$\dfrac{n}{56} = \dfrac{20}{?}$$

5.

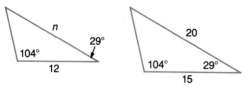

$$\frac{n}{12} = \frac{20}{?}$$

6.

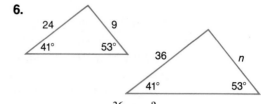

$$\frac{36}{n} = \frac{?}{9}$$

7. Sue is 1.6 meters tall. On a sunny day, her shadow is 2 meters long at the same time as the shadow of a TV tower is 70 meters long. Find the height of the tower.

8. On the same sunny day, the shadow of a tree was 14 yards long. Use the similar triangles below to find the height of the tree. Round your answer to the nearest tenth of a yard.

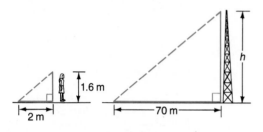

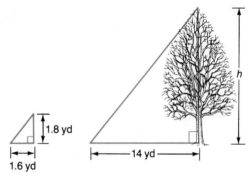

9. James found the height of a street lamp by placing a mirror on the sidewalk. Then he walked backwards until he saw the top of the streetlight in the mirror. The figure below shows how he formed similar triangles. Find the height of the street lamp. Round your answer to the nearest tenth of a meter.

10. Find the distance across this lake. Round your answer to the nearest tenth of a yard.

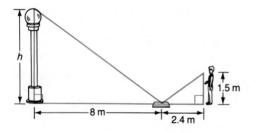

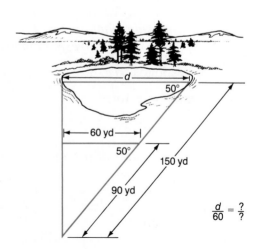

$$\frac{d}{60} = \frac{?}{?}$$

Tangent Ratios and Applications

If you know the measure of an acute angle in a right triangle, you can find the length of a leg without using similar triangles. To do this, you use the **tangent ratio.**

For angle A in the figure at the right, the **"tangent of angle A"** or simply **"tan A"** is defined as follows.

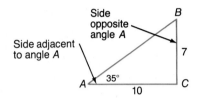

$$\text{tan } A = \frac{\text{length of side opposite angle } A}{\text{length of side adjacent to angle } A}$$

For the triangle above, $\tan 35° = \frac{7}{10} = 0.7$.

·1. Find tan A for the similar triangles shown at the right.

2. Does tan A = 0.7 for each triangle?

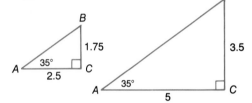

This suggests the following.

The tangent of a given angle is the same even if the lengths of the legs are not the same.

You can use this table to find tan A.

Table of Tangents			
Angles	Tangents	Angles	Tangents
5°	0.087	50°	1.19
10°	0.176	55°	1.43
15°	0.268	60°	1.73
20°	0.364	65°	2.14
25°	0.466	70°	2.75
30°	0.577	75°	3.73
35°	0.700	80°	5.67
40°	0.839	85°	11.43
45°	1.00		

tan 65° = 2.14

tan 30° = 0.577

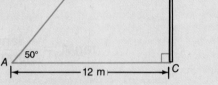

EXAMPLE Find the height of this flagpole. Round your answer to the nearest meter.

READ What are the facts?

Angle *A:* 50°   Adjacent side: 12 m

PLAN Use the tangent ratio and the table.

SOLVE
$$\tan A = \frac{\text{length of side opposite angle } A}{\text{length of side adjacent to angle } A}$$

$$\tan 50° = \frac{h}{12}$$ ◀ **Find tan 50° in the Tangent Table.**

$$1.19 = \frac{h}{12}$$ ◀ **Multiply each side by 12.**

$$1.19 \times 12 = \frac{h}{12} \times \frac{12}{1}$$

$$14.28 = h$$

CHECK Did you use the facts correctly in the problem?

The flagpole is about **14 meters high.**

EXERCISES

In Exercises 1–8, find tan A. Write a two-place decimal for your answer.

1.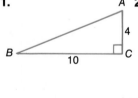

2.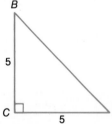

3.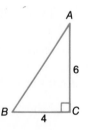

4.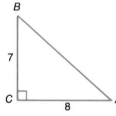

5.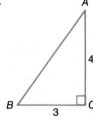

6.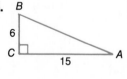

7.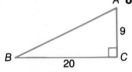

8.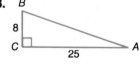

In Exercises 9–13, use the Table of Tangents to find each of the following.

9. tan 10°   **10.** tan 85°   **11.** tan 25°   **12.** tan 60°   **13.** tan 45°

In Exercises 14–17, use the Table of Tangents to solve each problem. Round each answer to the nearest whole number.

14. $\tan 10° = \frac{n}{20}$ **15.** $\tan 35° = \frac{n}{9}$ **16.** $\tan 20° = \frac{n}{12}$ **17.** $\tan 85° = \frac{n}{5}$

In Exercises 18–20, use the Table of Tangents to find the lengths of the sides. Round each answer to the nearest whole number.

18.

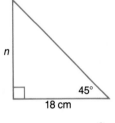

$n = \underline{\ ?\ }$

19.

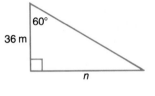

$n = \underline{\ ?\ }$

20.

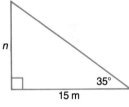

$n = \underline{\ ?\ }$

In Exercises 21–24, use the tangent ratio and the Table of Tangents to solve each problem. Round each answer to the nearest whole number.

21. Find the height, *h*, of the building shown below.

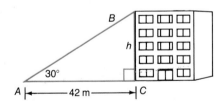

22. Find the height, *h*, of the mountain shown below.

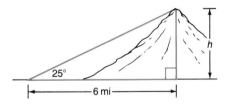

23. Find the distance, *d*, across the widest point of Sweet Moss Lake below.

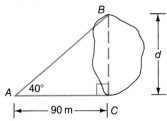

24. Find the width, *w*, of Lake Unknown.

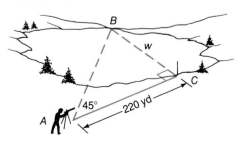

Finding Reaction Time

> **Reaction distance** is the distance a car travels during the time it takes the driver to begin braking. The time it takes the driver to begin braking is called **reaction time.**

The faster the reaction time, the shorter the reaction distance. This activity will give you the opportunity to estimate how fast a reaction time you have. You and a partner will need a yard stick and a calculator.

ACTIVITY

1 The first person holds the yard stick so that the end of the yard stick is even with the top of the desk. The second person places one hand on the edge of the desk.

2 The first person releases the yard stick. The second person then grabs the yard stick with a free hand <u>without</u> moving the other hand from the desk.

3 Record the distance that the yard stick dropped before the second person was able to grab it.

4 Use this formula and a calculator to find the reaction time to the nearest hundredth of a second.

Reaction time in seconds ▶ $t = \frac{\sqrt{d}}{13.86}$ ◀ **Distance the yard stick fell in inches**

5 Perform this activity three times. Then find the average.

6 Now change positions with your partner and repeat steps **1** through **5**.

EXERCISES

1. Make a list of the average reaction times for each person in the class. Write the list in order beginning with the fastest reaction time.

2. Find the mean, median, and mode for the class.

Chapter Summary

1. To square a number, multiply the number by itself.

2. To find the square root of a number, find one of the two equal factors of the number.

3. To use a square and square root table:
   1. Find the number in the NUMBER column.
   2. Look directly to the right.
      a. Read the square of the number in the SQUARE column.
      b. Read the square root of the number in the SQUARE ROOT column.

4. To find the square root of some numbers greater than 150:
   1. Find the number in the SQUARE column.
   2. Look directly to the left. Read the number.

5. To find the measure of a side of a right triangle:
   1. Draw a right triangle. Label the sides with the given measures.
   2. Write the rule of Pythagoras. Replace two letters in the rule by the corresponding given measures.
   3. Solve for the third measure.

6. To find the length of an unknown side in a pair of similar triangles:
   1. Use the unknown side and a known side in one triangle to write a ratio.
   2. Use sides that correspond in the second triangle to write another ratio.
   3. Use the two ratios to write a proportion.
   4. Solve the proportion.

7. To use the tangent ratio to find a leg of a right triangle:
   1. Draw and label a right triangle.
   2. Use the known angle, the known leg, and the unknown leg to write the tangent ratio.
   3. Solve for the unknown leg.

---

1. $6^2 = 6 \times 6 = \mathbf{36}$

2. Since $121 = 11 \times 11$,
   $\sqrt{121} = \mathbf{11}$.

3. $52^2 = \mathbf{2704}$
   $\sqrt{54} = \mathbf{7.348}$

Number	Square	Square Root
51	2601	7.141
52 →	2704	7.211
53	2809	7.280
54 —	2916 →	7.348

4. $\sqrt{10,609} = \mathbf{103}$

Number	Square	Square Root
101	10,201	10.050
102	10,404	10.100
103 ←	10,609	10.149

5. In a right triangle, $a = 7$ m and $b = 24$ m. Find c.
   $$c^2 = a^2 + b^2$$
   $$c^2 = 7^2 + 24^2$$
   $$c^2 = 49 + 576$$
   $$c^2 = 625$$
   $$c = \sqrt{625} = \mathbf{25 \text{ m}}$$

6. Find h.
   $$\frac{h}{6} = \frac{3}{2}$$
   $$h \times 2 = 6 \times 3$$
   $$2h = 18$$
   $$\frac{2h}{2} = \frac{18}{2}$$
   $$h = \mathbf{9}$$

7. Find h.
   $$\tan 40° = \frac{h}{20}$$
   $$0.839 = \frac{h}{20}$$
   $$0.839 \times 20 = \frac{h}{20} \times \frac{20}{1}$$
   $$\mathbf{16.78} = h$$

Chapter Review

Part 1: VOCABULARY

For Exercises 1–6, choose from the box at the right the word(s) that complete(s) each statement.

tangent
square root
hypotenuse
square
identical
right
similar

1. The product of a number and itself is the _?_ of the number. (Page 368)

2. One of the two equal factors of a number is the _?_ of the number. (Page 368)

3. A triangle with one right angle is called a _?_ triangle. (Page 373)

4. In a right triangle, the side opposite the right angle is the _?_ . (Page 373)

5. Triangles that have the same shape are _?_ triangles. (Page 380)

6. The length of the side opposite an angle to the length of the side adjacent to the angle is the _?_ ratio. (Page 383)

Part 2: SKILLS

Find each square or square root. (Pages 368–369)

7. 6^2 8. 9^2 9. 15^2 10. 21^2 11. 13^2 12. 20^2

13. $\sqrt{9}$ 14. $\sqrt{25}$ 15. $\sqrt{81}$ 16. $\sqrt{100}$ 17. $\sqrt{225}$ 18. $\sqrt{144}$

For Exercises 19–42, find each answer. Use the Table of Squares and Square Roots on page 371. (Pages 370–372)

19. 27^2 20. 56^2 21. 89^2 22. 43^2 23. 112^2 24. 131^2

25. 18^2 26. 37^2 27. 74^2 28. 92^2 29. 134^2 30. 147^2

31. $\sqrt{75}$ 32. $\sqrt{103}$ 33. $\sqrt{114}$ 34. $\sqrt{62}$ 35. $\sqrt{95}$ 36. $\sqrt{28}$

37. $\sqrt{3481}$ 38. $\sqrt{961}$ 39. $\sqrt{9025}$ 40. $\sqrt{19,044}$ 41. $\sqrt{11,236}$ 42. $\sqrt{729}$

Three sides of a triangle are given. Determine which triangles are right triangles. Write Yes or No. (Pages 373–374)

43. 11 cm, 50 cm, 54 cm

44. 16 in, 30 in, 33 in

45. 15 m, 20 m, 25 m

46. 5 ft, 12 ft, 13 ft

For Exercises 47–50, find the length of the unknown side of the triangle.
Round each answer to the nearest tenth. (Pages 375–377)

47.
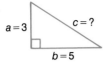

48.
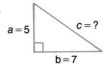

49.

50.
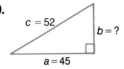

For Exercise 51, the pair of triangles is similar. Complete the proportion. Find the unknown side. (Pages 380–382)

For Exercise 52, find tan A. Write a two-place decimal for your answer. (Pages 383–384)

51.

$$\frac{n}{12} = \frac{?}{6}$$

52.
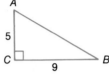

Part 3: APPLICATIONS

53. The sail on the boat below is 6 meters high and 2 meters wide. Find the length of the third side. Round your answer to the nearest tenth. (Pages 375–377)

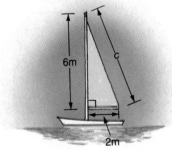

54. A ladder is placed against the side of a house as shown below. How far is the foot of the ladder from the base of the house? Round your answer to the nearest tenth. (Pages 375–377)

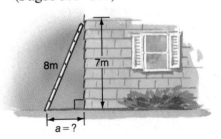

55. On a sunny day, Sue's shadow is 2 meters long when the shadow of a tower is 70 meters long. Find the height of the tower. (Pages 380–382)

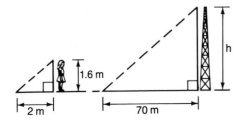

56. Find the height, h, of the plane above the ground. (HINT: $\tan 15° = \frac{h}{?}$) (Pages 383–384)

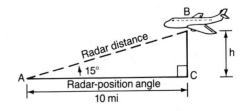

Chapter Test

Find each square or square root.

1. 3^2 **2.** 16^2 **3.** $\sqrt{16}$ **4.** $\sqrt{36}$ **5.** $\sqrt{144}$

For Exercises 6–10, use the Table of Squares and Square Roots on page 371.

6. 45^2 **7.** 37^2 **8.** $\sqrt{96}$ **9.** $\sqrt{104}$ **10.** $\sqrt{1521}$

Three sides of a triangle are given. Determine which triangles are right triangles. Write Yes or No.

11. 20 yd, 22 yd, 29 yd   **12.** 24 m, 32 m, 40 m

For Exercises 13–16, use the Table of Squares and Square Roots on page 371 to help you find the unknown side of each right triangle. Round each answer to the nearest tenth.

13.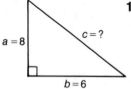

14.

15.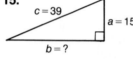

16.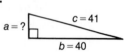

For Exercises 17–18, each pair of triangles is similar. Complete the proportion.

17.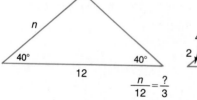

$$\frac{n}{12} = \frac{?}{3}$$

18.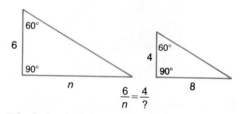

$$\frac{6}{n} = \frac{4}{?}$$

19. A ramp to a bridge is 24 meters long and 7 meters high as shown below. Find the length of the ramp. Use the Table of Squares and Square Roots on page 371.

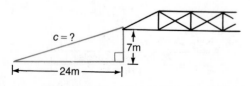

20. Find the height of the building shown below. (HINT: $\tan 28° = \frac{?}{45} = 0.531$)

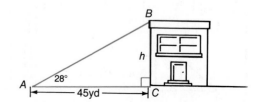

Cumulative Maintenance   Chapters 1–15

Choose the correct answer. Choose a, b, c, or d.

1. Subtract.

$$53.17 - 8.59$$

  a. 44.58   **b.** 45.42
  c. 44.68   **d.** 45.68

2. Choose the best estimate.

$$8.1 \times 11.8$$

  a. 96   **b.** 90
  c. 88   **d.** 99

3. Divide.

$$4.06 \overline{)\, 0.02436}$$

  a. 0.006   **b.** 6000
  c. 0.0600   **d.** 0.6

4. *Complete:* 12 kg = $\underline{\ ?\ }$ g

  a. 1.2   **b.** 12000
  c. 1200   **d.** 120

5. Add.

$$\frac{1}{3} + \frac{1}{4}$$

  a. $\frac{1}{7}$ **b.** $\frac{7}{12}$
  c. $\frac{2}{12}$ **d.** $\frac{1}{6}$

6. Find the missing terms in the pattern.

$$26, 23, \underline{\ ?\ }, \underline{\ ?\ }, 14, 11$$

  a. 21, 19   **b.** 19, 15
  c. 20, 17   **d.** 20, 18

7. Choose the best estimate.

$$20.1 \div 3.9$$

  a. 3   **b.** 5
  c. 4   **d.** 6

8. Multiply.

$$26.5 \times 0.001$$

  a. 0.265   **b.** 2.65
  c. 0.0265   **d.** 26500

9. Find the mode of these numbers.

  15   12   18   25   18   16   15
  18   24   19   21   16   15   18

  a. 18   **b.** 24
  c. 15   **d.** 21

10. What is the probability of getting a 3 on one spin of the game below?

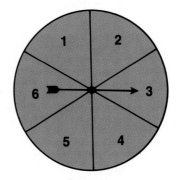

  a. $\frac{1}{6}$ **b.** $\frac{1}{3}$
  c. $\frac{1}{5}$ **d.** 1

11. A bag contains four yellow marbles and seven blue marbles. One marble is drawn. What is the probability that it is yellow?

   a. $\frac{7}{11}$ **b.** $\frac{7}{4}$

   c. $\frac{4}{7}$ **d.** $\frac{4}{11}$

---

12. What is $\frac{7}{8}$ of 56?

   a. 49   **b.** 64

   c. 50   **d.** 48

---

13. Add.

$$4\tfrac{2}{15} + 3\tfrac{8}{15}$$

   a. $6\frac{9}{15}$ **b.** $7\frac{2}{3}$

   c. $7\frac{6}{15}$ **d.** $8\frac{1}{15}$

---

14. Use the Rule of Pythagoras,

$$c^2 = a^2 + b^2$$

to find c when $a = 6$ and $b = 8$.

   a. 100   **b.** 7

   c. 10   **d.** 14

---

15. Use the table below to find $\sqrt{113}$.

Table of Squares and Square Roots		
Number	Square	Square Root
110	12,100	10.488
111	12,321	10.536
112	12,544	10.583
113	12,769	10.630
114	12,996	10.677

   a. 10.677   **b.** 12,769

   c. 10.583   **d.** 10.630

---

16. Use the formula $C = 2 \times \pi \times r$ to find the circumference of a circle whose radius is 5. (Use $\pi = 3.14$.)

   a. 3.14   **b.** 31.4

   c. 0.0314   **d.** 15.70

---

17. Larry earns \$6 per hour. How much more will he earn by working 8 hours than by working 6 hours?

   a. \$24   **b.** \$48

   c. \$36   **d.** \$12

---

18. Use the formula $A = \pi \times r \times r$ to find the area of a circle whose radius is 7. (Use $\pi = \frac{22}{7}$.)

   a. 154   **b.** 22

   c. 144   **d.** 75

---

19. Candy jogged for an hour in 37°C-temperatures. Use the formula $F = 1.8C + 32$ to determine how many degrees Fahrenheit this is. Round your answer to the nearest whole degree.

   a. 98°   **b.** 124°

   c. 61°   **d.** 99°

---

20. What is the volume in cubic centimeters of this box?

   a. 850

   b. 1000

   c. 900

   d. 1200

15 cm

8 cm

10 cm

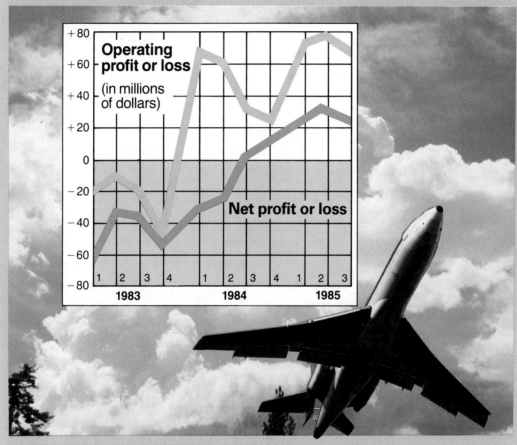

This graph shows the operating profit and loss and the net profit or loss for a large airline company in a recent year.

- How can you tell sharp increases or decreases in profit just by looking at the graph?

- The graph for net profit or loss in the fourth quarter of 1983 is at −55. What does this mean?

- In what year was there the least amount of change in the net profit?

- Use the graph to predict the operating profits in 1986. Explain your answer.

Integers

The temperature at which water will freeze is 32° **below** 0° Fahrenheit. This is the number description.

−32 read as: **negative thirty-two**

The temperature at which water will boil is 212° Fahrenheit. The number description is a positive number.

212

1. What kind of number, positive or negative, does each word represent?

   a. Increase   **b.** Withdrawal   **c.** Drop   **d.** Gain

EXAMPLE Write a positive or negative number to represent each word description. Name the word that is the clue.

Word Description	Number	Clue
a. An increase in temperature of 5°F	5	**Increase**
b. A mini-sub 25 meters below sea level	−25	**Below**
c. A deposit of $55	55	**Deposit**
d. A loss of 5 yards	−5	**Loss**

In the Example, numbers such as the positive and negative numbers and zero are called **integers.**

If you lay a thermometer on its side, the positive numbers are to the right of zero and the negative numbers are to the left of zero.

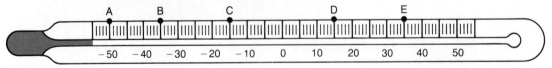

2. What integer is represented by each letter on the thermometer?

EXERCISES

Write a positive or negative number to represent each word description.

1. 120 meters above sea level

2. A business profit of $10,000

3. Five years ago

4. Ten seconds after rocket lift-off

5. A loss of 500 feet in altitude

6. A pay raise of $0.30 per hour

7. Ten years from now

8. A team penalty of 5 yards

9. A weight loss of 3 pounds

10. A weight gain of 2 kilograms

11. A rise of 6° in temperature

12. A business loss of $6000

13. A rainfall 4 inches below normal

14. A rainfall 2 inches above normal

15. A deposit of $80 in a savings account

16. A forward speed of 62 kilometers per hour

For Exercises 17–22, write the integer represented by each letter.

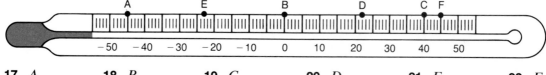

17. *A*   18. *B*   19. *C*   20. *D*   21. *E*   22. *F*

For Exercises 23–28, write the integer represented by each letter.

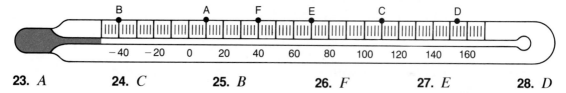

23. *A*   24. *C*   25. *B*   26. *F*   27. *E*   28. *D*

Write an integer to represent each height or depth. Write a positive integer, a negative integer, or zero.

29. The highest city in the United States is Leadville, Colorado. The city is 10,200 feet above sea level.

30. The lowest town in the United States is Calipatria, California. The town is 184 feet below sea level.

31. The lowest point in North America is Death Valley, California, which is 282 feet below sea level.

32. The highest point in Texas is Guadalupe Peak. This point is 8749 feet above sea level.

33. The highest point in North America is in Alaska at the top of Mt. McKinley. This point is 20,320 feet above sea level.

34. The Gulf of Mexico is at sea level.

Guadalupe Peak

Comparing Integers

1. How many units is 20 from 0?

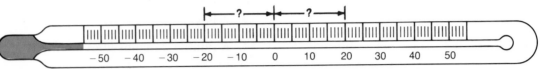

2. How many units is −20 from 0?

The integers 20 and −20 are the same distance from 0. They are on
opposite sides of 0. Thus, 20 and −20 are **opposites.** The number 0 is
its own opposite.

3. What is the opposite of each of these integers?

    a. −8   **b.** 35   **c.** 24   **d.** −12

4. As you read <u>from left to right</u> on the thermometer, do the numbers
become larger or smaller?

The **number line** is a model of the thermometer. You can use the
number line and the symbols < (less than) and > (greater than) to
compare integers.

EXAMPLE   Replace each ● with < or >. Refer to this number line.

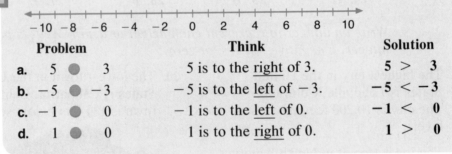

	Problem		Think	Solution
a.	5 ●	3	5 is to the <u>right</u> of 3.	5 > 3
b.	−5 ●	−3	−5 is to the <u>left</u> of −3.	−5 < −3
c.	−1 ●	0	−1 is to the <u>left</u> of 0.	−1 < 0
d.	1 ●	0	1 is to the <u>right</u> of 0.	1 > 0

EXERCISES

For Exercises 1–5, refer to the thermometer above.

1. The integer −30 is __?__ units to the __?__ of 0.

2. The integer 30 is __?__ units to the __?__ of 0.

3. The integers −30 and 30 are the __?__ distance from 0.

4. The integers −30 and 30 are on __?__ sides of 0.

5. Integers that are the same distance from 0 but are on opposite sides of
0 are called __?__ .

Write the opposite of each integer.

6. 8   **7.** −10   **8.** 75   **9.** 0   **10.** −56   **11.** 87   **12.** 120

13. −32   **14.** −100   **15.** 311   **16.** −212   **17.** −5000   **18.** 7500   **19.** 9125

Write the opposite of each word description.

20. 55 meters above sea level

21. 20 miles south

22. 18 revolutions faster

23. 3 hours before

24. Spend 14 dollars.

25. Lose 5 pounds.

Replace each ● with < or >. Refer to the number line below.

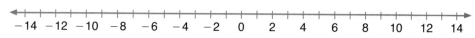

26. 0 ● 6

27. 0 ● −6

28. −3 ● 3

29. 3 ● −3

30. −1 ● −2

31. 1 ● −2

32. −2 ● −1

33. 5 ● −2

34. −5 ● 2

35. −10 ● 9

36. 12 ● 7

37. −8 ● −13

Write in order from least to greatest.

38. 3, −4, 0, −6, 6

39. −8, −3, −1, 7, −7

40. 6, 4, 0, −5, −7

41. −25, 16, 0, −8, 5

Replace each ● with < or >. Temperatures are given in degrees Fahrenheit.

42. The lowest temperature ever recorded in Wisconsin was −54°. The lowest temperature ever recorded in Tennessee was −32°.

Complete: −54° ● −32°

43. The lowest temperature ever recorded in New Hampshire was −46°. The lowest temperature ever recorded in Illinois was −35°.

Complete: −35° ● −46°

44. The lowest temperature ever recorded in Louisiana was −16°. The lowest temperature ever recorded in Georgia was −15°.

Complete: −15° ● −16°

Meaning of Addition: LIKE SIGNS

ACTIVITY   1. Draw and label two number lines as in Figure 1 below.
    2. Cut out the two number lines and place them flat on your desk, with line **A** directly over line **B** as shown below.

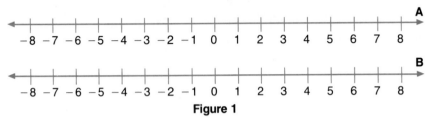

Figure 1

You are now ready to add integers.

Add: $-2 + (-5) = \underline{\ ?\ }$

Step 1: The first addend, -2, is a negative number. Therefore, "slide" line **B** to the left by 2 units. The 0 on line **B** should now be directly under -2 on line **A**.

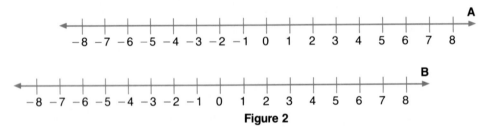

Figure 2

Step 2: Find the second addend, -5, on line **B**. It should be directly under -7 on line **A**, which is the sum. Thus,
$$-2 + (-5) = \mathbf{-7.}$$

With lines **A** and **B** in this same position (Figure 2), you can read the sum for other addition problems whose first addend is -2.

EXERCISES

Use the number lines to find these sums.

1. $-2 + (-4) = \underline{\ ?\ }$ 2. $-2 + (-2) = \underline{\ ?\ }$ 3. $-2 + (-3) = \underline{\ ?\ }$
4. $-4 + (-2) = \underline{\ ?\ }$ 5. $-5 + (-2) = \underline{\ ?\ }$ 6. $-3 + (-3) = \underline{\ ?\ }$

7. In Exercises 8–10, the first addend is a positive number. In which direction (left or right) should you slide line **B**?

Use the number lines to find these sums.

8. $5 + 3 = \underline{\ ?\ }$ 9. $4 + 4 = \underline{\ ?\ }$ 10. $2 + 3 = \underline{\ ?\ }$

Adding Integers: LIKE SIGNS

The temperature at 7 P.M. was $-2°C$. Three hours later, the temperature dropped 3 degrees.

1. Is it colder at 10 P.M. than at 7 P.M.?

2. Is the temperature at 10 P.M. less than $-2°$ or greater than $-2°$?

3. How could you find the temperature at 10 P.M.?

EXAMPLE 1

Add: $-2 + (-3) = $ __?__

Draw a number line.

First addend: Draw a dot at -2.

Second addend: Starting at -2, draw an arrow that is 3 units long. Since the second addend is a <u>negative number</u>, the arrow is drawn to the <u>left</u> of -2.

The tip of the arrow is at -5, the answer.

3 units

$$-8\ -7\ -6\ -5\ -4\ -3\ -2\ -1\ \ 0\ \ 1\ \ 2\ \ 3\ \ 4\ \ 5\ \ 6\ \ 7\ \ 8\ \ 9\ \ 10$$

$$-2 + (-3) = -5$$

EXAMPLE 2

Add: $5 + 3 = $ __?__

3 units

The tip of the arrow is at 8, the answer.

$$-8\ -7\ -6\ -5\ -4\ -3\ -2\ -1\ \ 0\ \ 1\ \ 2\ \ 3\ \ 4\ \ 5\ \ 6\ \ 7\ \ 8\ \ 9\ \ 10$$

$$5 + 3 = 8$$

4. Why does the arrow for the second addend, 3, point to the right?

EXERCISES

*Match each of Exercises 1–3 with a number line from **a–d**.*

1. $-2 + (-3) = -5$

2. $-3 + (-4) = -7$

3. $3 + 4 = 7$

a.
4 units

b.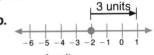
3 units

c.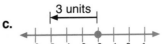
3 units

d.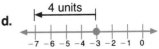
4 units

INTEGERS: ADDITION/SUBTRACTION

Use a number line to add.

4. $1 + 3$ **5.** $4 + 4$ **6.** $8 + 2$ **7.** $0 + 4$

8. $9 + 3$ **9.** $4 + 2$ **10.** $10 + 0$ **11.** $7 + 3$

12. $-1 + (-1)$ **13.** $-3 + (-4)$ **14.** $-4 + (-3)$ **15.** $-3 + (-2)$

16. $-2 + (-2)$ **17.** $-1 + 0$ **18.** $0 + (-7)$ **19.** $-5 + (-2)$

20. $0 + (-4)$ **21.** $-7 + (-1)$ **22.** $-6 + (-3)$ **23.** $-1 + (-5)$

24. $-5 + (-3)$ **25.** $-4 + (-5)$ **26.** $-2 + (-5)$ **27.** $-8 + 0$

28. $4 + 7$ **29.** $-4 + (-7)$ **30.** $-4 + (-6)$ **31.** $4 + 6$

32. $-1 + 0$ **33.** $0 + (-1)$ **34.** $5 + 5$ **35.** $-5 + (-5)$

36. $4 + 4$ **37.** $11 + 1$ **38.** $-6 + (-1)$ **39.** $3 + 4$

40. $-3 + (-6)$ **41.** $6 + 3$ **42.** $0 + 12$ **43.** $0 + (-12)$

44. The sum of two positive integers is a __?__ integer.

45. The sum of two negative integers is a __?__ integer.

Use a number line to find the missing number.

46. $4 + \underline{\ ?\ } = 11$ **47.** $3 + \underline{\ ?\ } = 8$ **48.** $\underline{\ ?\ } + 7 = 11$

49. $\underline{\ ?\ } + 5 = 14$ **50.** $-3 + \underline{\ ?\ } = -8$ **51.** $-5 + \underline{\ ?\ } = -6$

52. $\underline{\ ?\ } + (-8) = -8$ **53.** $12 + \underline{\ ?\ } = 12$ **54.** $\underline{\ ?\ } + (-10) = -12$

55. $-4 + (-5) + (-6) = \underline{\ ?\ }$ **56.** $-3 + (-8) + (-5) = \underline{\ ?\ }$

For Exercises 57–60:

  a. *Represent each problem by the sum of two integers.*

  b. *Then use a number line to find the sum.*

57. The temperature at 7 P.M. was 0°C. Two hours later, the temperature was 3 degrees below zero. How much did the temperature fall?

58. Albert's team lost 3 penalty points the first quarter. The team lost 2 penalty points the next quarter. Find the total number of points lost.

59. Raoul had to write checks for $4 and $7 to pay some bills. By how much did he decrease his checking account?

60. Susan lost 8 pounds last month. This month she lost 7 pounds. How much did she lose in all?

Meaning of Addition: UNLIKE SIGNS

You can use the "sliding" number lines to add integers with unlike signs. The steps are the same as for adding like signs.

Like Signs	Unlike Signs
$8 + 4$	$7 + (-2)$
$-4 + (-6)$	$-9 + 5$

ACTIVITY   Add: $3 + (-2) = \underline{\ ?\ }$

⊡1 The first addend, 3, is a <u>positive</u> integer. Therefore, "slide" line **B** to the <u>right</u> by 3 units.

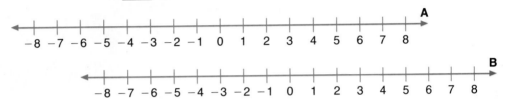

⊡2 Find the second addend, -2, on line **B.** It should be directly under 1 on line **A,** which is the sum. Thus,

$$3 + (-2) = 1$$

With lines **A** and **B** in this same position, you can read the sum for other addition problems whose first addend is 3.

EXERCISES

Use the number lines to find these sums.

1. $3 + (-5) = \underline{\ ?\ }$ **2.** $3 + (-8) = \underline{\ ?\ }$ **3.** $3 + (-3) = \underline{\ ?\ }$

4. $4 + (-2) = \underline{\ ?\ }$ **5.** $2 + (-7) = \underline{\ ?\ }$ **6.** $6 + (-8) = \underline{\ ?\ }$

7. In Exercises 8–10, the first addend is a negative number. In which direction (left or right) should you slide line **B?**

Use the number lines to find these sums.

8. $-7 + 6 = \underline{\ ?\ }$ **9.** $-4 + 6 = \underline{\ ?\ }$ **10.** $-5 + 5 = \underline{\ ?\ }$

11. In Exercise 3, the sum is 0. What are 3 and -3 called?

12. In Exercise 10, the sum is 0. What are -5 and 5 called?

13. Write a statement about the sum of opposites.

Adding Integers: UNLIKE SIGNS

When Alex faded back to pass, he was tackled for a 3-yard loss. On the next play, he ran for a gain of 8 yards.

1. What integer represents each of these?

   a. A 3-yard loss

   b. A gain of 8 yards

2. How would you use these two integers to find how much yardage Alex gained or lost after the two plays?

   You can use a number line to find this sum.

EXAMPLE 1 $-3 + 8 = \underline{\ ?\ }$

Draw a number line.

First addend: Draw a dot at -3.

Second addend: Starting at -3, draw an arrow 8 units long. Since the second addend is a <u>positive</u> number, the arrow is drawn to the <u>right</u> of -3.

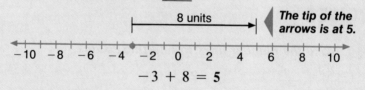

$-3 + 8 = 5$

---

3. Would the answer be the same if he gained 8 yards on the first play and lost 3 yards on the second play? Let's make sure.

EXAMPLE 2 $8 + (-3) = \underline{\ ?\ }$

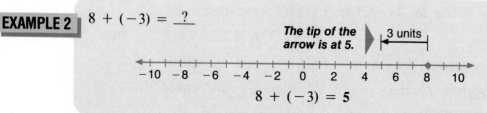

$8 + (-3) = 5$

EXERCISES

Match each of Exercises 1–4 with a number line from **a–d.**

1. $-4 + 5$

2. $-5 + 4$

3. $3 + (-4)$

4. $2 + (-5)$

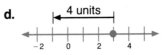

a. 4 units

b. 5 units

c. 5 units

d. 4 units

Use a number line to add.

5. $-8 + 4$ **6.** $-9 + 6$ **7.** $-3 + 5$ **8.** $-2 + 10$

9. $-7 + 8$ **10.** $-11 + 4$ **11.** $-6 + 10$ **12.** $-4 + 8$

13. $5 + (-3)$ **14.** $6 + (-12)$ **15.** $2 + (-1)$ **16.** $6 + (-6)$

17. $9 + (-9)$ **18.** $7 + (-4)$ **19.** $1 + (-6)$ **20.** $4 + (-8)$

21. $-3 + 11$ **22.** $-9 + 11$ **23.** $11 + (-9)$ **24.** $11 + (-3)$

25. $10 + (-8)$ **26.** $-8 + 10$ **27.** $-8 + 8$ **28.** $8 + (-8)$

29. $2 + (-7)$ **30.** $-8 + 6$ **31.** $-6 + 12$ **32.** $10 + (-15)$

Use a number line to find the missing number.

33. $3 + \underline{\ ?\ } = -10$ **34.** $-6 + \underline{\ ?\ } = 11$ **35.** $-2 + \underline{\ ?\ } = 7$

36. $-1 + \underline{\ ?\ } = 8$ **37.** $\underline{\ ?\ } + (-1) = 5$ **38.** $\underline{\ ?\ } + (-5) = -4$

39. $\underline{\ ?\ } + (-3) = 3$ **40.** $8 + \underline{\ ?\ } = 2$ **41.** $-7 + \underline{\ ?\ } = -2$

Represent each change in temperature by the sum of two integers. Then use the thermometer scale at the right to find the sum.

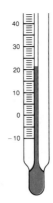

42. The temperature was 21°C and rose 9 degrees. Find the new temperature.

43. The temperature was 0°C and dropped 3 degrees. Find the new temperature.

44. The temperature was 0°C and rose 10 degrees. Find the new temperature.

45. The temperature was −4°C and rose 4 degrees. Find the new temperature.

46. The temperature was −5°C and rose 7 degrees. Find the new temperature.

Mid-Chapter Review

Write a positive or negative number to represent each word description.
(Pages 394–395)

1. 175 feet below sea level

2. A 3°F rise in temperature

3. A profit of $2000

4. A weight loss of 2 pounds

Write the opposite of each integer. (Pages 396–397)

5. −6   6. 2   7. 15   8. −40   9. 0   10. −102   11. 78

Replace each ● with < or >. (Pages 396–397)

12. 0 ● −4   13. 1 ● 3   14. −5 ● 5   15. −1 ● −6

6. 7 ● 3   17. −5 ● −2   18. 10 ● −17   19. −21 ● −45

Use a number line to add. (Pages 399–400)

20. 2 + 4   21. 3 + 7   22. 4 + 1   23. 0 + 8

24. −1 + (−3)   25. −5 + (−2)   26. −7 + (−4)   27. −6 + (−1)

(Pages 402–403)

28. −1 + 5   29. 3 + (−2)   30. 7 + (−1)   31. −3 + 4

32. 6 + (−5)   33. −5 + 8   34. −9 + 6   35. 8 + (−4)

For Exercises 36–37, represent each change in temperature by the sum of two integers. Then use a number line to find the sum. (Pages 402–403)

36. The temperature was 24°C and rose 5°. Find the new temperature.

37. From 0°C, the temperature dropped 4°. Find the new temperature.

MAINTENANCE

38. Sam bought two writing pads for 85¢ each and three pens for 79¢ each. How much did he pay in all? (Pages 41–42)

39. A 10-ounce can of fruit costs 69¢. A 14-ounce can of the same fruit costs 98¢. Find the better buy. (Pages 66–67)

40. The names of the seven days of the week are written on slips of paper and placed in a hat. One slip of paper is drawn. What is the probability of drawing a day that begins with "T"? (Pages 350–351)

41. In a cupboard there are 4 cans of tomato soup, 2 cans of vegetable soup, and 6 cans of clam chowder. If you choose a can without looking, what is the probability that it will be tomato soup? (Pages 350–351)

Math and Automobiles

Some motor oils protect an engine over a wide range of temperatures. These oils have code names such as **5W–20** or **10W–40.**

The chart below shows the temperature ranges for which four different oils provide adequate protection to an engine.

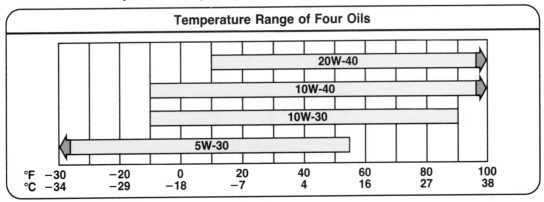

Temperature Range of Four Oils

EXAMPLE	Which oils can be used for temperatures between 0°F and 70°F?
READ	What are the facts? Temperature range: 0°F to 70°F
PLAN	Read the chart. Find the bars that extend between 0°F and 70°F.
SOLVE	The bars representing the oils **10W–40** and **10W–30** extend between 0°F and 70°F.
CHECK	Did you use all the facts correctly in the solution?

EXERCISES

1. Which oil can be used only with temperatures above 10°F?

2. Which oil can be used only in winter temperatures?

3. Which oil can be used with temperatures below −29°C?

4. Which oil can be used in temperatures from −10°F to over 100°F?

5. How many degrees Fahrenheit are there between the highest and lowest temperatures in which Oil 10W–30 can be used?

6. How many degrees Celsius are there between the highest and lowest temperatures in which Oil 10W–30 can be used?

Subtracting Integers

Compare this subtraction problem with the addition problem.

a. $7 - 3 = 4$

b. $7 + (-3) =$ ___?___ 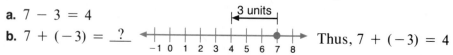 Thus, $7 + (-3) = 4$

1. What are 3 and -3 called?

In **a**, you <u>subtracted 3</u>. In **b**, you <u>added the opposite of 3</u> and got the same answer as in **a**.

2. *Complete.*

Subtracting an integer is the same as adding its ___?___ .

EXAMPLE 1   Write an addition problem for each subtraction problem.

Subtraction Problem	Read	Think	Addition Problem
a. $8 - 4$	8 minus 4	Opposite of 4: -4	$8 + (-4)$
b. $7 - (-6)$	7 minus -6	Opposite of -6: 6	$7 + 6$
c. $-4 - 2$	-4 minus 2	Opposite of 2: -2	$-4 + (-2)$
d. $-9 - (-1)$	-9 minus -1	Opposite of -1: 1	$-9 + 1$

To solve a subtraction problem, you solve the related addition problem.

EXAMPLE 2   **a.** $4 - 7 =$ ___?___   **b.** $-1 - (-5) =$ ___?___

First write an addition problem for the subtraction problem.

a. $4 - 7 = 4 + (-7)$ **b.** $-1 - (-5) = -1 + 5$

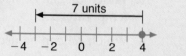   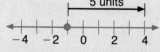

The tip of the arrow is at -3.   The tip of the arrow is at 4.

So, $4 - 7 = -3$.   So, $-1 - (-5) = 4$.

Complete.

3. $5 - 6 = 5 +$ ___?___

4. $-3 - (-1) = -3 +$ ___?___

EXERCISES

Complete the table.

	Subtraction Problem	Think	Addition Problem
1.	6 − 2	The opposite of 2 is __?__ .	6 + __?__
2.	14 − 3	The opposite of 3 is __?__ .	14 + __?__
3.	−6 − 1	The opposite of 1 is __?__ .	−6 + __?__
4.	8 − (−5)	The opposite of −5 is __?__ .	8 + __?__
5.	−4 − (−6)	The opposite of −6 is __?__ .	−4 + __?__

Use a number line to subtract.

6. 3 − 5   **7.** 4 − 5   **8.** −1 − (−2)   **9.** −3 − (−6)

10. −5 − 1   **11.** −10 − 9   **12.** 12 − 8   **13.** 8 − 12

14. −3 − 7   **15.** −7 − 3   **16.** −14 − (−8)   **17.** −8 − (−14)

18. 0 − 8   **19.** 0 − 7   **20.** 0 − (−5)   **21.** 0 − (−12)

22. −3 − (−7)   **23.** −2 − (−11)   **24.** 6 − (−2)   **25.** 7 − (−11)

26. 13 − 8   **27.** 13 − 11   **28.** 10 − (−8)   **29.** 9 − (−4)

30. −6 − 10   **31.** −9 − 5   **32.** −1 − (−8)   **33.** −4 − (−1)

34. −3 − 0   **35.** −9 − 0   **36.** −7 − (−5)   **37.** −5 − (−7)

Add or subtract. Use a number line.

38. 5 + 4   **39.** 3 − 2   **40.** 3 + (−6)   **41.** −4 − 10

42. −6 + 6   **43.** −7 + 0   **44.** 1 − 9   **45.** −2 − 5

46. 8 + (−7)   **47.** −7 + (−8)   **48.** 5 − 8   **49.** 8 − (−5)

50. 8 − 5   **51.** 10 − 3   **52.** 3 + (−10)   **53.** −10 + (−3)

Write an addition problem for each subtraction problem. Then find each answer without using a number line.

54. 6 − 5   **55.** 9 − 8   **56.** −3 − (−1)   **57.** −4 − (−2)

58. −1 − (−6)   **59.** −3 − (−9)   **60.** −12 − (−3)   **61.** −11 − (−9)

62. −8 − (−1)   **63.** −6 − (−5)   **64.** −15 − 5   **65.** −16 − 10

Strategy: USING A CHART

A combination of cold temperatures and high wind can make a person feel colder than the actual temperature. This is called **wind–chill.** Meteorologists use charts like this one to determine the windchill.

Winds in mph	Temperatures in °F																
	35	30	25	20	15	10	5	0	−5	−10	−15	−20	−25	−30	−35	−40	−45
5	33	27	21	19	12	7	0	−5	−10	−15	−21	−26	−31	−36	−42	−47	−52
10	22	16	10	3	−3	−9	−15	−22	−27	−34	−40	−46	−52	−58	−64	−71	−77
15	16	9	2	−5	−11	−18	−25	−31	−38	−45	−51	−58	−65	−72	−78	−85	−92
20	12	4	−3	−10	−17	−24	−31	−39	−46	−53	−60	−67	−74	−81	−88	−95	−103
25	8	1	−7	−15	−22	−29	−36	−44	−51	−59	−66	−74	−81	−88	−96	−103	−110
30	6	−2	−10	−18	−25	−33	−41	−49	−56	−64	−71	−79	−86	−93	−101	−109	−116
35	4	−4	−12	−20	−27	−35	−43	−52	−58	−67	−74	−82	−89	−97	−105	−113	−120
40	3	−5	−13	−21	−29	−37	−45	−53	−60	−69	−76	−84	−92	−100	−107	−115	−123
45	2	−6	−14	−22	−30	−38	−46	−54	−62	−70	−78	−85	−93	−102	−109	−117	−125

EXAMPLE The wind is blowing at 20 miles per hour and the temperature is 15°F. How much colder than 15°F does it feel?

READ What are the facts?

Wind: 20 miles per hour   Temperature: 15°F

PLAN To find how much colder it is, first answer this "hidden question."

What is the windchill?

To answer this question, you have to use the chart.

SOLVE

1. Find 20 in the "mph" column. Look directly right to the number under the "15" column.

2. Read the number: −17   The windchill is **−17°F.** This answer means that the temperature of 15°F feels like −17°F.

3. Now you can find how much colder it feels. This means that you want to find the difference between the actual temperature and the windchill. So subtract −17°F from 15°F.

$$15 - (-17) = \underline{\ ?\ }$$ ◀ *Rewrite as addition.*

$$15 + 17 = 32°$$

The windchill makes it feel **32° colder.**

CHECK Did you use all the facts correctly in solving the problem?

EXERCISES

For Exercises 1–4, find the windchill. Use the table on page 408.

Winds in mph	Temperature	Wind–Chill
1. 10	0°F	?
2. 20	20°F	?

Winds in mph	Temperature	Wind–Chill
3. 40	25°F	?
4. 20	10°F	?

5. On a certain day in Portland, Maine, the wind is blowing at 30 miles per hour and the temperature is 25°F. How much colder than 25°F does it feel?

6. On a certain day in Anchorage, Alaska, the wind is blowing at 40 miles per hour and the temperature is 5°F. How much colder than 5°F does it feel?

7. At 1:00 p.m. the windchill temperature was −3°F. At 7:00 p.m. the wind was blowing at 20 miles per hour and the actual temperature was 10°F. How much colder did it feel at 7:00 p.m. than at 1:00 p.m.?

8. At an Antarctic outpost, the windchill temperature was −39°F. An hour later, the wind was blowing at 45 miles per hour and the actual temperature was −10°F. How much colder did it feel an hour later?

9. Sarah is listening to a weather report. She learns that the windchill temperature is −49°F and the actual temperature is 0°F. What is the speed of the wind?

10. A meteorologist says the windchill temperature is −25°F and the wind is blowing at 30 miles per hour. What is the actual temperature?

For Exercises 11–12, choose a strategy from the box at the right that you can use to solve each problem.
a. *Name the strategy.*
b. *Solve the problem.*

> **Using a pattern**
> **Using more than one step**
> **Using a formula**

11. The surface of the planet Venus has a temperature of 454°C How many degrees Fahrenheit is this?

12. Use the table below to find the temperature of a chemical solution 10 minutes after the start of the experiment.

Time (min)	2	4	6	8	10
Temperature	76°	79°	83°	88°	?

Solving Addition Equations

Recall that to <u>solve an equation for *n*</u> means to <u>get *n* alone on one side of the equation</u>. To solve an equation, you use the idea of <u>opposites</u>.

> The sum of an integer and its opposite equals 0.
>
> $5 + (-5) = 0$ $\qquad\qquad$ $-3 + 3 = 0$

Drawing a model of the equation may help you understand the method.

EXAMPLE 1 Solve and check: $n + 6 = -4$

1 Use ▌ to represent *n*, ■ to represent 1, and ■ to represent -1.

$$n + \quad 6 \quad = \quad -4$$

▌ + ■■■■■■ = ■■■■

2 To get *n* alone, add the opposite of 6 to the left side of the equation. To keep the equation in balance, also add the opposite of 6 to the right side.

$$n + \; 6 + (-6) \; = \; -4 + (-6)$$

▌ + $\dfrac{■■■■■■}{■■■■■■}$ = $\dfrac{■■■■}{■■■■■■}$

3 Since ■■ represents $1 + (-1)$, ■■ equals 0. Remove each pair of ■■ from the equation.

▌ = $\dfrac{■■■■}{■■■■■■}$

$$n = -10$$

4 **Check:** Replace *n* in the original equation by -10.

Does $-10 + 6 = -4$?

$$-10 \quad + \quad 6 \quad = \quad -4$$

$\dfrac{■■■■■}{■■■■■}$ + $\dfrac{■■■}{■■■}$ = ■■■■

Remove pairs of ■■.

$$-4 \quad = \quad -4$$

■■■■ = ■■■■ ◄——— *It checks.*

Now let's solve an equation without drawing a model.

EXAMPLE 2

Solve and check: $n + 18 = -7$

$$n + 18 = -7$$ ◀ **Add the opposite of 18 to each side.**

$$n + 18 + (-18) = -7 + (-18)$$
$$n = -25$$

Check: $n + 18 = -7$ ◀ **Replace n with −25.**
$$-25 + 18 \stackrel{?}{=} -7$$
$$-7 = -7 \quad ✓$$

EXERCISES

For Exercises 1–4, a model is used to represent an equation.
■ *represents n,* ■ *represents 1, and* ■ *represents −1.*
Copy the model. Then draw ■*'s or* ■*'s to show what number you would add to each side of the equation in order to solve for n.*

1. ■ + ■ = ■■■■

2. ■ + ■■■■■ = ■

3. ■ + ■■ = ■■■■

4. ■ + ■■■ = ■■

Write the number you would add to each side of the equation in order to solve for n.

5. $n + 2 = -5$ **6.** $n + 1 = 4$ **7.** $n + 8 = 6$ **8.** $n + 3 = -8$

9. $n + 4 = -1$ **10.** $n + 5 = 7$ **11.** $n + 6 = -5$ **12.** $n + 7 = 2$

Solve for n by drawing a model. Let ■ = *n,* ■ = *1, and* ■ = *−1.*
Check each answer.

13. $n + 1 = -4$ **14.** $n + 3 = -6$ **15.** $n + 7 = -1$ **16.** $n + 6 = -5$

17. $n + 3 = -7$ **18.** $n + 10 = -4$ **19.** $n + 7 = -3$ **20.** $n + 9 = 2$

21. $n + 6 = 4$ **22.** $n + 1 = 0$ **23.** $n + 8 = 10$ **24.** $n + 7 = 8$

25. $n + 5 = 9$ **26.** $n + 1 = -1$ **27.** $n + 3 = -3$ **28.** $n + 8 = -5$

Solve and check each equation.

29. $n + 11 = -8$ **30.** $n + 9 = -6$ **31.** $n + 5 = 10$ **32.** $n + 12 = 19$

33. $n + 13 = 7$ **34.** $n + 10 = 4$ **35.** $n + 18 = -22$ **36.** $n + 15 = -35$

37. $n + 14 = 32$ **38.** $n + 21 = 47$ **39.** $n + 12 = -7$ **40.** $n + 16 = -11$

41. $n + 8 = 20$ **42.** $n + 11 = 29$ **43.** $n + 27 = -36$ **44.** $n + 20 = -21$

Solving Subtraction Equations

To solve a subtraction equation, first rewrite the left side of the equation as an addition equation.

Left side	Read	Think	Answer
$x - 4$	x minus 4	Opposite of 4: -4	$x + (-4)$
$n - (-2)$	n minus -2	Opposite of -2: 2	$n + 2$

EXAMPLE 1

Solve and check: $x - 4 = -7$

1. Rewrite as an addition equation: $x + (-4) = -7$

2. Proceed as in Example 1 on page 410.

   Use ▌ to represent x, ■ to represent 1, and ▪ to represent -1.

   $$x + \quad (-4) \quad = \quad -7$$

3. Add the opposite of -4 to each side of the equation.

   $$x + (-4) + 4 = \quad -7 + 4$$

4. Remove each pair of □■.

   $$x = -3$$

5. Check: Replace x in the original equation by -3.

   Does $-3 - 4 = -7$?

   Rewrite this as an addition problem.

   $$-3 + (-4) = -7$$

   Now model the check.

   $$-3 + (-4) = \quad -7$$

   ■■■ + ■■■■ = ▪▪▪▪▪▪▪ ← *It checks.*

Now let's try one without drawing a model.

EXAMPLE 2   Solve and check: $x - 6 = -9$
Rewrite as an addition equation.

$x + (-6) = -9$ ◀ **Add the opposite of -6 to each side.**   Check: $x - 6 = -9$

$x + (-6) + 6 = -9 + 6$

$x = -3$

$-3 - 6 \stackrel{?}{=} -9$ ◀ **Replace x with -3.**

$-3 + (-6) \stackrel{?}{=} -9$ ◀ **Rewrite as addition.**

$-9 = -9$ ✓

EXERCISES

For Exercises 1–8, rewrite each equation as an addition equation.

1. $x - 3 = 1$ **2.** $x - 4 = 3$ **3.** $x - 1 = -6$ **4.** $x - 5 = -2$

5. $x - (-2) = 4$ **6.** $x - (-1) = 1$ **7.** $x - (-7) = -4$ **8.** $x - (-3) = -8$

Write the number you would add to each side of the equation in order to solve for x.

9. $x - 2 = 3$ **10.** $x - 4 = 6$ **11.** $x - 9 = -5$

12. $x - 7 = -16$ **13.** $x - (-5) = 9$ **14.** $x - (-10) = 7$

15. $x - (-8) = -11$ **16.** $x - (-12) = -9$ **17.** $x - (-1) = 5$

Solve for x by drawing a model. Let ▌ $= x$, ■ $= 1$, *and* ▨ $= -1$.

18. $x - 2 = -6$ **19.** $x - 3 = -7$ **20.** $x - 5 = 8$ **21.** $x - 7 = 1$

22. $x - (-5) = 2$ **23.** $x - (-2) = 6$ **24.** $x - 6 = -1$ **25.** $x - 4 = -5$

26. $x - 9 = 4$ **27.** $x - 8 = 3$ **28.** $x - (-7) = 2$ **29.** $x - (-4) = -10$

Solve and check each equation.

30. $x - 4 = -3$ **31.** $x - 2 = -1$ **32.** $x - 8 = -8$

33. $x - 5 = -5$ **34.** $x - 9 = 3$ **35.** $x - 11 = 5$

36. $x - 12 = 21$ **37.** $x - 10 = 18$ **38.** $x - (-9) = 12$

39. $x - (-11) = -15$ **40.** $x - (-8) = -17$ **41.** $x - (-20) = 8$

42. $x - 12 = -31$ **43.** $x - 16 = 24$ **44.** $x - (-2) = 13$

45. $x - (-17) = -10$ **46.** $x - 32 = 14$ **47.** $x - 15 = 0$

48. $x - 30 = -22$ **49.** $x - (-21) = 12$ **50.** $x - (-9) = -11$

51. $x - 29 = -30$ **52.** $x - 16 = -13$ **53.** $x - 42 = 18$

The Slippery Cricket

Christopher, the slippery cricket, was hopping near a well one evening. He miscalculated the distance of a hop and fell 45 feet to the bottom of a well. He was so tired and discouraged that he waited until morning to start the climb to the top of the well.

Christopher climbed 3 feet each day. However, Christopher slipped back 2 feet at night because the walls of the well were slippery.

PROBLEM How many days did it take Christopher to reach the top of the well?

Think End of first day: 3 feet above the bottom
Second morning: 1 foot above the bottom
End of second day: 4 feet above the bottom

There is more than one way to solve this problem.

Method 1 MAKING A DRAWING

Make a drawing for the third morning, the end of the third day, the fourth morning, and so on until you reach 45 feet.

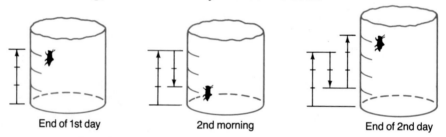

End of 1st day   2nd morning   End of 2nd day

Method 2 MAKING A TABLE

Continue this table until you reach a height of 45 feet.

The number of the day.	1	2	3	4	5	6	7
The height in feet at end of day.	3	4	?	?	?	?	?

Method 3 FINDING A PATTERN

Only complete as much of the table as you need to see a pattern that you can use to predict the answer to the problem.

EXERCISES

1. Solve the problem using the method that you prefer.

2. If you can see the pattern between "the number of the day" and the "height in feet at the end of the day," write a formula that describes the pattern.

Chapter Summary

1. To identify an integer on a number line:
   1. Start on 0. Numbers to the right of 0 are positive. Numbers to the left of 0 are negative.
   2. Read the number.

2. To write the opposite of a number, identify the number that is the same distance from 0 on the number line <u>and</u> in the opposite direction from 0.

3. To compare two integers, use a number line to determine which integer is to the right of the other. The integer that is farthest to the right is the greater integer.

4. To add two integers on the number line:
   1. Start at the point that represents the first integer.
   2. Draw an arrow to represent the second integer.
      a. Move to the right if the second integer is positive.
      b. Move to the left if the second integer is negative.
   3. Read the answer at the tip of the arrow.

5. To write an addition problem for a subtraction problem, add the opposite of the number you are subtracting.

6. To subtract two integers:
   1. Write an addition problem for the subtraction problem.
   2. Add.

7. To solve an addition equation such as $n + 5 = -3$, add -5 to each side of the equation.

8. To solve a subtraction equation such as $x - 3 = -9$, add 3 to each side of the equation.

1.

   $A: -3 \qquad B: -1 \qquad C: 3$

2. Integer: 1   0   3 -5
   Opposite: -1 0 -3 5

3.
   Replace ● with $<$ or $>$.
   $-1 \; ● \; -4$
   $-1 > -4$ ◀ -1 **is to the right of** -4.

4. Add: $1 + 3 = \underline{\;?\;}$

            3 units

   $1 + 3 = 4$

   Add: $-1 + 4 = \underline{\;?\;}$

        4 units

   $-1 + 4 = 3$

5. Subtraction   Addition
   $10 - 6 \longrightarrow$ **$10 + (-6)$**
   $8 - (-4) \longrightarrow$ **$8 + 4$**

6. Subtract: $2 - 4 = \underline{\;?\;}$
   $2 - 4 = 2 + (-4)$

        4 units

   $2 - 4 = -2$

7. Solve:
   $$n + 5 = -3$$
   $$n + 5 + (-5) = -3 + (-5)$$
   $$n = -8$$

8. Solve: $x - 3 = -9$
   $$x - 3 + 3 = -9 + 3$$
   $$x = -6$$

Chapter Review

Part 1: VOCABULARY

For Exercises 1–3, choose from the box at the right the word(s) that complete(s) each statement.

1. Numbers to the right of 0 on a number line are ___?___ numbers. (Page 394)

2. Numbers to the left of 0 on a number line are ___?___ numbers. (Page 394)

3. The sum of a number and its ___?___ is zero.   (Page 401)

> opposite
> negative
> sign
> positive

Part 2: SKILLS

For Exercises 4–9, use the number line to write the integer represented by each letter.   (Pages 394–395)

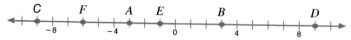

4. *A*   5. *B*   6. *C*   7. *D*   8. *E*   9. *F*

Write the opposite of each integer.   (Pages 396–397)

10. 8   11. 200   12. -56 13. -81 14. 72   15. -10 16. 325

Replace each ● *with* < *or* >.   (Pages 396–397)

17. 6 ● -3 18. -4 ● 4   19. -10 ● -8 20. -2 ● -4

21. -12 ● 4   22. -15 ● -9 23. 5 ● -16 24. 0 ● -17

Use a number line to add.   (Pages 399–400)

25. $5 + 9$ 26. $-4 + (-7)$ 27. $0 + (-4)$ 28. $7 + 10$

29. $21 + 4$ 30. $-12 + (-15)$ 31. $-2 + (-11)$ 32. $13 + 9$

(Pages 402–403)

33. $3 + (-7)$ 34. $-4 + 8$ 35. $10 + (-4)$ 36. $-6 + 2$

37. $-10 + 6$ 38. $4 + (-9)$ 39. $8 + (-10)$ 40. $-9 + 3$

Use a number line to subtract.   (Pages 406–407)

41. $7 - 8$ 42. $4 - 7$ 43. $2 - (-6)$ 44. $4 - (-5)$

45. $-6 - 3$ 46. $-5 - 10$ 47. $-5 - (-12)$ 48. $-3 - (-10)$

Solve and check.   (Pages 410–411)

49. $n + 5 = -12$ **50.** $n + 6 = -1$ **51.** $n + 10 = -7$ **52.** $n + 12 = -9$

53. $n + 10 = 5$ **54.** $n + 15 = 8$ **55.** $n + 11 = 4$ **56.** $n + 20 = 13$

Solve and check.   (Pages 412–413)

57. $x - 4 = 4$ **58.** $x - 47 = -1$ **59.** $x - 8 = 20$ **60.** $x - 30 = 22$

61. $x - 8 = -40$ **62.** $x - 14 = -7$ **63.** $x - (-1) = 10$ **64.** $x - (-4) = 15$

65. $x - 21 = -12$ **66.** $x - 17 = 9$ **67.** $x - (-8) = 2$ **68.** $x - 3 = -9$

Part 3: APPLICATIONS

For Exercises 69–70, write an integer to represent each height or depth. Write a positive integer, a negative integer or zero.   (Pages 394–395)

69. The lowest point in Vermont is Lake Champlain which is 29 meters above sea level.

70. The greatest known depth of the Atlantic Ocean is in the Puerto Rico Trench. This point is 8648 meters below the surface.

For Exercises 71–72, replace each ● with < or >.   (Pages 396–397)

71. The lowest temperature ever recorded in Nebraska was $-47°F$. The lowest temperature ever recorded in Nevada was $-50°F$.
Complete: $-47°$ ● $-50°$

72. The lowest temperature ever recorded in West Virginia was $-37°F$. The lowest temperature recorded in Wyoming was $-63°F$.
Complete: $-63°$ ● $-37°$

For Exercises 73–75, represent each problem by the sum of two integers. Then use a number line to find the sum.   (Pages 399–400 and 402–403)

73. The temperature was $7°C$ and rose $3°$. Find the new temperature.

74. The temperature was $-2°C$ and rose $8°$. Find the new temperature.

75. Ron lost 3 pounds last month. This month he lost 2 pounds. How much did he lose in all?

76. The wind is blowing at 35 miles per hour and the temperature is $30°F$. How much colder does the wind chill make it feel? Use the table on page 408.   (Pages 408–409)

Chapter Test

For Exercises 1–6, use this number line to write the integer represented by each letter.

1. *A*   **2.** *B*   **3.** *C*   **4.** *D*   **5.** *E*   **6.** *F*

Replace each ● *with < or >.*

7. -8 ● 7 **8.** -5 ● -2 **9.** -12 ● -5 **10.** 3 ● -8

Use a number line to add.

11. $4 + 7$ **12.** $-3 + (-4)$ **13.** $-2 + (-1)$ **14.** $7 + 8$

15. $-6 + 3$ **16.** $-7 + 8$ **17.** $5 + (-9)$ **18.** $10 + (-4)$

Use a number line to subtract.

19. $2 - 6$ **20.** $4 - 10$ **21.** $-3 - 9$ **22.** $-7 - 3$

23. $3 - (-5)$ **24.** $6 - (-10)$ **25.** $-4 - (-2)$ **26.** $-5 - (-8)$

Solve and check.

27. $n + 4 = -8$ **28.** $n + 7 = 2$ **29.** $x - 5 = 16$ **30.** $x - 6 = -5$

31. $x - 2 = -9$ **32.** $n + 16 = 12$ **33.** $n + 12 = -10$ **34.** $x - 7 = 13$

35. $n + 18 = -20$ **36.** $x - 17 = -11$ **37.** $x - (-3) = 6$ **38.** $x - (-8) = -14$

39. $n + 13 = 6$ **40.** $x - (-10) = -2$ **41.** $n + 8 = -15$ **42.** $n + 32 = 15$

For Exercises 43–44, represent each change in temperature by the sum of two integers. Then use a number line to find the sum.

43. The temperature was $3°C$ and rose 4 degrees. Find the new temperature.

44. The temperature was $6°C$ and fell $10°$. Find the new temperature.

For Exercise 45, use the table at the right.

45. At 1:00 P.M. the wind chill temperature was $-3°F$. At 8:00 P.M. the wind was blowing at 15 miles per hour and the temperature was $10°F$. How much colder did it feel at 8:00 P.M. than at 1:00 P.M.?

Winds in mph	Temperature in °F							
	35	30	25	20	15	10	5	0
5	33	27	21	19	12	7	0	5
10	22	16	10	3	-9	-9	-15	-22
15	16	9	2	-5	-11	-18	-25	-31
20	12	4	3	-10	-17	-24	-31	-39
25	8	1	7	-15	-22	-29	-36	-44

Cumulative Maintenance Chapters 1–16

Choose the correct answer. Choose a, b, c, or d.

1. Subtract: 5347 − 2696

    a. 2651   **b.** 3751

    c. 3651   **d.** 2751

---

2. Multiply: $\frac{12}{25} \times \frac{10}{27}$

    a. $\frac{6}{14}$ **b.** $\frac{8}{35}$

    c. $\frac{8}{45}$ **d.** $\frac{1}{5}$

---

3. Choose the figure that completes the pattern.

a. **b.** **c.** **d.**

---

4. Use the graph below to find how much the price of a certain stock rose from 1986 to 1987.

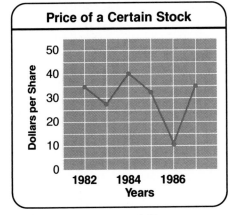

    a. $35   **b.** $50

    c. $25   **d.** $30

---

5. Choose the word that best describes this angle.

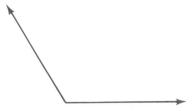

    a. acute   **b.** obtuse

    c. right   **d.** none of these

---

6. Compute:
$$18 \div 3 + 7 \times 2$$

    a. 23   **b.** 26

    c. 3.6   **d.** 20

---

7. Subtract:
$$-27 - (-13)$$

    a. −14   **b.** 14

    c. −40   **d.** 40

---

8. A boat cost $1800 in 1977. Four years later it cost $2400. Find the percent of increase.

    a. 40%   **b.** 25%

    c. 75%   **d.** $33\frac{1}{3}$%

---

9. Find the interest on $600 at 12% for 6 months. Use the formula:
$$i = p \times r \times t \ (t \text{ is in years})$$

    a. $45   **b.** $45

    c. $28   **d.** $36

10. There are 7 purple jelly beans and 3 green jelly beans in a bag. One jelly bean is drawn. What is the probability that it is green?

a. $\frac{7}{10}$ **b.** $\frac{3}{10}$

c. $\frac{3}{7}$ **d.** $\frac{7}{3}$

11. Carl deposits checks for $136.50 and $87.63. He receives 2 twenty-dollar bills and 3 five-dollar bills. What is Carl's net deposit?

a. $199.13   **b.** $169.13
c. $224.13   **d.** $231.13

12. Find the mean of the following five numbers.

12   17   24   15   12

a. 24   **b.** 12
c. 18   **d.** 16

13. Find the volume in cubic meters of this tool box.

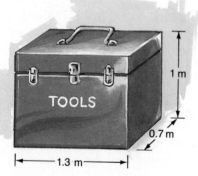

1 m

0.7 m

1.3 m

a. 0.91   **b.** 9.1
c. 0.091   **d.** 91

14. In the circle graph below, how many degrees represent the section titled "Soup"?

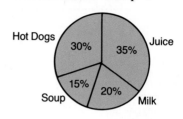

Hot Dogs   30%   35%   Juice
15%   20%
Soup   Milk

a. 75°   **b.** 54°
c. 108°   **d.** 70°

15. Find the height in feet of this flagpole. (Hint: tan 32° = .625)

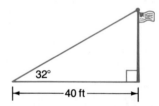

32°

40 ft

a. 25   **b.** 250
c. 0.016   **d.** 16

16. Which number has the *smallest* value?

a. −7   **b.** 7
c. 3   **d.** 0

17. How many meters long is the ramp? Use $c^2 = a^2 + b^2$.

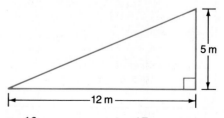

5 m

12 m

a. 13   **b.** 17
c. 10   **d.** 15

Integers: Multiplication/Division

CHAPTER 17

Caribou
−23°, 89°

Seattle
15°, 87°

Great Falls
−30°, 87°

Duluth
−24°, 95°

Chicago
−8°, 102°

New York
−1°, 102°

Salt Lake City
−4°, 101°

Washington, D.C.
12°, 103°

San Francisco
31°, 97°

Denver
5°, 100°

Kansas City
−8°, 106°

Phoenix
35°, 115°

Charleston
15°, 100°

Dallas
15°, 113°

New Orleans
25°, 101°

Miami
32°, 95°

The map shows the high and low temperatures for sixteen cities in a recent year. Temperatures are given in degrees Fahrenheit.

- Will the mean of the high temperatures for the sixteen cities be a positive number or a negative number? Explain.

- Will the mean of the low temperatures for Great Falls, Kansas City, Duluth, Chicago, and Caribou be a positive or a negative number? Explain.

- Will the range of the high and low temperatures for each city be a positive or a negative number? Explain.

Multiplication: PATTERNS AND UNLIKE SIGNS

Look at the pattern of the following multiplication problems.

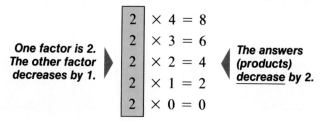

One factor is 2.
The other factor
decreases by 1.

$2 \times 4 = 8$
$2 \times 3 = 6$
$2 \times 2 = 4$
$2 \times 1 = 2$
$2 \times 0 = 0$

The answers
(products)
decrease by 2.

1. The next problem would be $2 \times (-1)$. If the pattern continues,
$$2 \times (-1) = \underline{\ ?\ }$$

2. Use the pattern to find these products.

   a. $2 \times (-2) = \underline{\ ?\ }$ b. $2 \times (-3) = \underline{\ ?\ }$ c. $2 \times (-4) = \underline{\ ?\ }$

3. What is the pattern in Column A? Use the pattern to find the missing products.

4. What is the pattern in Column B? Use the pattern to find the missing products.

Column A

$3 \times 3 = 9$
$3 \times 2 = 6$
$3 \times 1 = 3$
$3 \times 0 = 0$
$3 \times (-1) = \underline{\ ?\ }$
$3 \times (-2) = \underline{\ ?\ }$
$3 \times (-3) = \underline{\ ?\ }$

Column B

$3 \times 5 = 15$
$2 \times 5 = 10$
$1 \times 5 = 5$
$0 \times 5 = 0$
$-1 \times 5 = \underline{\ ?\ }$
$-2 \times 5 = \underline{\ ?\ }$
$-3 \times 5 = \underline{\ ?\ }$

Complete.

5. The product of a positive integer and a negative integer is a __?__ integer.

6. The product of a positive integer or a negative integer and zero is __?__ .

EXERCISES

In Exercises 1–9, study each pattern. Then write the missing product(s).

1. $8 \times 1 = 8$
   $8 \times 0 = 0$
   $8 \times (-1) = -8$
   $8 \times (-2) = \underline{\ ?\ }$

2. $2 \times 4 = 8$
   $1 \times 4 = 4$
   $0 \times 4 = 0$
   $-1 \times 4 = \underline{\ ?\ }$

3. $0 \times 6 = 0$
   $-1 \times 6 = -6$
   $-2 \times 6 = -12$
   $-3 \times 6 = \underline{\ ?\ }$

4. $7 \times 1 = 7$
$7 \times 0 = 0$
$7 \times (-1) = -7$
$7 \times (-2) = \underline{?}$

5. $11 \times 4 = 11$
$11 \times 0 = 0$
$11 \times (-1) = \underline{?}$
$11 \times (-2) = \underline{?}$

6. $-1 \times 10 = -10$
$-2 \times 10 = -20$
$-3 \times 10 = \underline{?}$
$-4 \times 10 = \underline{?}$

7. $-2 \times 5 = -10$
$-3 \times 5 = -15$
$-4 \times 5 = \underline{?}$
$-5 \times 5 = \underline{?}$

8. $14 \times 1 = 14$
$14 \times 0 = 0$
$14 \times (-1) = \underline{?}$
$14 \times (-2) = \underline{?}$

9. $1 \times 7 = 7$
$0 \times 7 = 0$
$-1 \times 7 = \underline{?}$
$-2 \times 7 = \underline{?}$

Multiply.

10. $4 \times (-2)$

11. $6 \times (-1)$

12. -3×5

13. -10×10

14. -5×0

15. 0×7

16. -8×11

17. -9×7

18. $11 \times (-4)$

19. $11 \times (-9)$

20. -7×18

21. -8×15

22. -4×3

23. -12×13

24. $0 \times (-20)$

25. -100×0

26. $6 \times (-6)$

27. $4 \times (-21)$

28. $8 \times (-10)$

29. $20 \times (-8)$

30. -15×3

31. -18×5

32. -200×7

33. -1000×5

34. $8 \times (-5)$

35. $7 \times (-17)$

36. $41 \times (-9)$

37. $83 \times (-7)$

38. -4×90

39. $100 \times (-50)$

Use positive and negative integers to represent each situation by a multiplication problem. Then find the product.

40. Twice a loss of 300 feet in altitude

41. Four penalties of 15 yards each

42. Two debts of $45 each

43. Three times a temperature drop of $6°F$

44. Five times a debt of $100

45. Seven times a depth of 3 meters below sea level

46. Twice a drop in temperature of $1°C$

47. Six times a loss of 4 yards in rushing

48. Two drops in the water level of 1.5 centimeters each

49. Three weight losses of 2 pounds each

Multiplication: PATTERNS AND LIKE SIGNS

Look at the pattern of the following multiplication problems.

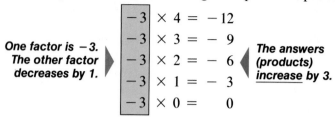

One factor is −3. The other factor decreases by 1. ➤

$-3 \times 4 = -12$
$-3 \times 3 = -9$
$-3 \times 2 = -6$
$-3 \times 1 = -3$
$-3 \times 0 = 0$

◀ The answers (products) <u>increase</u> by 3.

1. The next problem would be $-3 \times (-1)$. If the pattern continues,
$$-3 \times (-1) = \underline{\ ?\ }$$

2. Use the pattern to find these products.
   a. $-3 \times (-2) = \underline{\ ?\ }$ **b.** $-3 \times (-3) = \underline{\ ?\ }$ **c.** $-3 \times (-4) = \underline{\ ?\ }$

3. What is the pattern in Column A? Use the pattern to find the missing products.

Column A	Column B
$-4 \times 3 = -12$	$3 \times -5 = -15$
$-4 \times 2 = -8$	$2 \times -5 = -10$
$-4 \times 1 = -4$	$1 \times -5 = -5$
$-4 \times 0 = 0$	$0 \times -5 = -0$
$-4 \times (-1) = \underline{\ ?\ }$	$-1 \times -5 = \underline{\ ?\ }$
$-4 \times (-2) = \underline{\ ?\ }$	$-2 \times -5 = \underline{\ ?\ }$
$-4 \times (-3) = \underline{\ ?\ }$	$-3 \times -5 = \underline{\ ?\ }$

4. What is the pattern in Column B? Use the pattern to find the missing products.

5. *Complete.*

   The product of two negative integers is a $\underline{\ ?\ }$ integer.

EXERCISES

In Exercises 1–9, study each pattern. Then write the missing product(s).

1.
$2 \times (-5) = -10$
$1 \times (-5) = -5$
$0 \times (-5) = 0$
$-1 \times (-5) = 5$
$-2 \times (-5) = \underline{\ ?\ }$

2.
$-6 \times 2 = -12$
$-6 \times 1 = -6$
$-6 \times 0 = 0$
$-6 \times (-1) = 6$
$-6 \times (-2) = \underline{\ ?\ }$

3.
$-3 \times 2 = -6$
$-3 \times 1 = -3$
$-3 \times 0 = 0$
$-3 \times (-1) = 3$
$-3 \times (-2) = \underline{\ ?\ }$

4.
$2 \times (-2) = -4$
$1 \times (-2) = -2$
$0 \times (-2) = 0$
$-1 \times (-2) = 2$
$-2 \times (-2) = \underline{\ ?\ }$

5.
$-8 \times 2 = -16$
$-8 \times 1 = -8$
$-8 \times 0 = 0$
$-8 \times (-1) = \underline{\ ?\ }$
$-8 \times (-2) = \underline{\ ?\ }$

6.
$2 \times (-9) = -18$
$1 \times (-9) = -9$
$0 \times (-9) = 0$
$-1 \times (-9) = \underline{\ ?\ }$
$-2 \times (-9) = \underline{\ ?\ }$

7.
$2 \times (-6) = -12$
$1 \times (-6) = -6$
$0 \times (-6) = 0$
$-1 \times (-6) = \underline{\ ?\ }$
$-2 \times (-6) = \underline{\ ?\ }$

8.
$-11 \times 2 = -22$
$-11 \times 1 = -11$
$-11 \times 0 = 0$
$-11 \times (-1) = \underline{\ ?\ }$
$-11 \times (-2) = \underline{\ ?\ }$

9.
$-12 \times 2 = -24$
$-12 \times 1 = -12$
$-12 \times 0 = 0$
$-12 \times (-1) = \underline{\ ?\ }$
$-12 \times (-2) = \underline{\ ?\ }$

Multiply.

10. $-5 \times (-4)$

11. 5×5

12. $-8 \times (-2)$

13. 6×2

14. $-8 \times (-7)$

15. 8×10

16. $-8 \times (-11)$

17. $-6 \times (-7)$

18. $-7 \times (-13)$

19. $-8 \times (-16)$

20. $-12 \times (-5)$

21. $-17 \times (-7)$

22. $-9 \times (-7)$

23. $-5 \times (-12)$

24. $-11 \times (-11)$

25. $-2 \times (-21)$

26. $-8 \times (-2)$

27. $-3 \times (-10)$

28. $-10 \times (-22)$

29. $-4 \times (-32)$

30. $-43 \times (-9)$

31. $-6 \times (-7)$

32. $-4 \times (-9)$

33. $-4 \times (-12)$

34. $-18 \times (-9)$

35. $-17 \times (-74)$

36. $-32 \times (-41)$

37. $-63 \times (-27)$

38. $-35 \times (-16)$

39. $-32 \times (-100)$

40. $-10 \times (-72)$

41. $-45 \times (-22)$

42. $-8 \times (-45)$

43. $-14 \times (-11)$

44. 18×36

45. 40×16

46. $-16 \times (-200)$

47. $-24 \times (-100)$

48. $-13 \times (-27)$

49. $-202 \times (-68)$

50. 47×63

51. -51×35

52. $75 \times (-25)$

53. $-41 \times (-10)$

54. $15 \times (-42)$

55. 21×37

56. -55×81

57. $-16 \times (-79)$

58. A glacier moves at a rate of 5 centimeters per year. How far will it move in 100 years?

59. During a storm the temperature fell 2° an hour for 4 hours. How much did the temperature fall during the storm?

60. A certain stock loses 3 points on each of 3 days. Write the total loss as an integer.

61. The product of two integers is 9. The sum of the integers is −6. Find the integers.

Dividing Integers

For each multiplication problem, you can write two division problems.

Multiplication	Related Division
$3 \times 9 = 27$	$27 \div 3 = 9$ and $27 \div 9 = 3$

You can use this fact to determine whether the quotient is negative or positive. The quotients in the division problems below are negative.

Multiplication	Related Division
$-7 \times (-3) = 21$	$21 \div (-7) = -3$ and $21 \div (-3) = -7$
$-8 \times (-4) = 32$	$32 \div (-8) = -4$ and $32 \div (-4) = -8$

1. Name the divisor and dividend in each of the following.

   a. $21 \div (-7) = -3$ b. $21 \div (-3) = -7$

   c. $32 \div (-8) = -4$ d. $32 \div (-4) = -8$

2. In Exercise 1, what kind of integer is each divisor?

3. In Exercise 1, what kind of integer is each dividend?

4. *Complete.*
   A positive integer divided by a negative integer is a __?__ integer.

Each quotient in Column 1 is negative.

	Column 1	Column 2
$-6 \times 2 = -12$	$-12 \div 2 = -6$	$-12 \div (-6) = 2$
$-5 \times 9 = -45$	$-45 \div 9 = -5$	$-45 \div (-5) = 9$

5. In Column 1, what kind of integer is each divisor?

6. In Column 1, what kind of integer is each dividend?

7. *Complete.*
   A negative integer divided by a positive integer is a __?__ integer.

Each quotient in Column 2 is positive.

8. In Column 2, what kind of integer is each divisor?

9. In Column 2, what kind of integer is each dividend?

10. *Complete.*
    A negative integer divided by a negative integer is a __?__ integer.

EXERCISES

Complete.

1. $-6 \times (-9) = 54$
$54 \div (-6) = \underline{\ ?\ }$
$54 \div (-9) = \underline{\ ?\ }$

2. $-15 \times (-4) = 60$
$60 \div (-15) = \underline{\ ?\ }$
$60 \div (-4) = \underline{\ ?\ }$

3. $-11 \times 12 = -132$
$-132 \div (-11) = \underline{\ ?\ }$
$-132 \div 12 = \underline{\ ?\ }$

4. $-10 \times 8 = -80$
$-80 \div (-10) = \underline{\ ?\ }$
$-80 \div 8 = \underline{\ ?\ }$

5. $16 \times 12 = 192$
$192 \div 16 = \underline{\ ?\ }$
$192 \div 12 = \underline{\ ?\ }$

6. $-9 \times (-18) = 162$
$162 \div (-9) = \underline{\ ?\ }$
$162 \div (-18) = \underline{\ ?\ }$

Divide.

7. $-30 \div (-6)$

8. $-63 \div (-9)$

9. $-36 \div (-12)$

10. $-35 \div (-5)$

11. $18 \div 2$

12. $16 \div 8$

13. $15 \div 5$

14. $156 \div 12$

15. $81 \div (-9)$

16. $21 \div (-7)$

17. $49 \div (-7)$

18. $36 \div (-4)$

19. $-39 \div 13$

20. $-45 \div 15$

21. $-138 \div 6$

22. $-154 \div 77$

23. $-100 \div 5$

24. $-200 \div 40$

25. $744 \div 8$

26. $756 \div 9$

27. $-300 \div (-75)$

28. $-200 \div (-8)$

29. $-756 \div (-28)$

30. $-442 \div (-17)$

31. $72 \div (-12)$

32. $55 \div (-11)$

33. $-98 \div 14$

34. $-133 \div 19$

35. $215 \div 5$

36. $567 \div 7$

37. $504 \div (-14)$

38. $-345 \div (-15)$

Perform the indicated operation.

39. $6 + (-11)$

40. $-8 + 2$

41. $5 - 9$

42. $-4 \times (-9)$

43. -6×8

44. $72 \div (-8)$

45. $13 + (-42)$

46. $-126 \div (-14)$

47. $-40 \times (-97)$

48. $-7 - (-4)$

49. $-312 \div 8$

50. $-60 + (-24)$

51. Temperatures on the moon reach 100°C in the daytime. At night, temperatures drop to -150°C. What is the mean temperature?

52. The temperature on Mars reaches 27°C during the day and -125°C at night. What is the mean temperature?

53. The low temperatures in a northern city are -2°C, -3°C, 0°C, and -7°C on four successive days. What is the mean temperature for the four days?

Rational Numbers

Numbers such as these are **positive rational numbers.**

$$\frac{1}{4} \qquad 1 \qquad \frac{7}{3} \qquad 3\frac{2}{3} \qquad 15$$

1. Name a fraction for : **a.** $3\frac{2}{3}$ **b.** 15

Numbers such as these are **negative rational numbers.**

$$-\frac{1}{5} \qquad -1 \qquad -\frac{9}{5} \qquad -2\frac{1}{5} \qquad -27$$

You write a fraction for negative rational numbers in the same way as you do for positive rational numbers.

$$-2\frac{1}{5} = -\frac{11}{5} \qquad\qquad -27 = -\frac{27}{1}$$

> Any rational number can be written as a fraction.

Zero is neither negative nor positive. You can write a fraction for 0.

$$0 = \frac{0}{-5}$$

2. Name a fraction for 0 that has a different denominator.

3. Why can't the denominator be 0?

You can show the rational numbers on a number line.

Write the rational number represented by each letter.

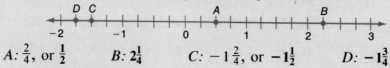

$A: \frac{2}{4}$, or $\frac{1}{2}$ $B: 2\frac{1}{4}$ $C: -1\frac{2}{4}$, or $-1\frac{1}{2}$ $D: -1\frac{3}{4}$

Recall that numbers become larger as you move from left to right on the number line. Thus,

$$-1\frac{1}{2} > -1\frac{3}{4}$$

4. Arrange the rational numbers in the Example in order from least to greatest.

EXERCISES

Write a fraction for each rational number.

1. 5   **2.** 3   **3.** -10 **4.** -9 **5.** $2\frac{1}{4}$ **6.** $1\frac{1}{3}$

7. $-3\frac{1}{4}$ **8.** $-7\frac{1}{6}$ **9.** -7 **10.** -5 **11.** $-3\frac{1}{3}$ **12.** $-1\frac{1}{12}$

13. 17   **14.** 52   **15.** 0   **16.** 26   **17.** $-1\frac{4}{5}$ **18.** $-3\frac{1}{8}$

19. -6 **20.** 1   **21.** $5\frac{1}{10}$ **22.** $9\frac{5}{12}$ **23.** $-7\frac{1}{2}$ **24.** $-2\frac{4}{5}$

Write the rational number represented by each letter.

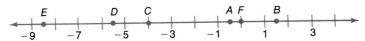

25. A **26.** B **27.** C **28.** D **29.** E **30.** F

31. Arrange the rational numbers in Exercises 25–30 in order from least to greatest.

Write the rational number represented by each letter.

32. G **33.** P **34.** K **35.** M **36.** R **37.** H

38. Arrange the rational numbers in Exercises 32–37 in order from least to greatest.

Replace each ● *with* $<$, $>$, *or* $=$.

39. $-\frac{4}{5}$ ● $-\frac{8}{10}$ **40.** $-\frac{4}{5}$ ● $\frac{4}{5}$ **41.** $-\frac{3}{4}$ ● -0.75 **42.** 0.75 ● $-\frac{3}{4}$

43. $-\frac{3}{4}$ ● $\frac{9}{12}$ **44.** $\frac{2}{3}$ ● $\frac{10}{15}$ **45.** $-\frac{2}{3}$ ● $-\frac{6}{9}$ **46.** -0.25 ● $-\frac{1}{4}$

47. $\frac{1}{4}$ ● $-\frac{2}{8}$ **48.** $-\frac{3}{8}$ ● 0.375 **49.** $\frac{3}{8}$ ● $\frac{12}{32}$ **50.** $-\frac{2}{5}$ ● -0.4

For Exercises 51–58, arrange the rational numbers in order from least to greatest.

51. 1; $\frac{1}{4}$; $-2\frac{1}{5}$; $3\frac{1}{3}$; $-5\frac{1}{5}$; $-\frac{1}{4}$

52. -8; 0; $4\frac{1}{3}$; $-7\frac{1}{2}$; $-\frac{2}{5}$; 2

53. 2; $-\frac{3}{4}$; 1; $-\frac{2}{3}$; $\frac{4}{5}$; $-1\frac{1}{2}$

54. 0; $-\frac{2}{3}$; $1\frac{1}{2}$; $-2\frac{3}{4}$; $\frac{2}{3}$; $-3\frac{1}{5}$

55. $2\frac{1}{4}$; $-1\frac{5}{6}$; $-1\frac{7}{8}$; $1\frac{1}{2}$; 0; $-\frac{11}{16}$

56. $-\frac{7}{8}$; 1; -2; $\frac{15}{16}$; -1; $\frac{8}{9}$

57. $-1\frac{1}{8}$; -3; $-2\frac{1}{2}$; -1; $-\frac{1}{4}$; $-2\frac{7}{8}$

58. $\frac{1}{5}$; 0; $-\frac{1}{4}$; $1\frac{1}{2}$; $-2\frac{7}{10}$; -5

Give an example to show that each sentence is true.

59. Every integer is a rational number.

60. Some rational numbers are negative integers.

61. Zero is a rational number.

62. Positive rational numbers are greater than negative rational numbers.

Mid-Chapter Review

Multiply.   (Pages 422–423)

1. $6 \times (-2)$ **2.** $8 \times (-1)$ **3.** -9×1 **4.** -12×5

5. -21×8 **6.** $7 \times (-16)$ **7.** $10 \times (-100)$ **8.** -120×0

(Pages 424–425)

9. 4×7 **10.** 8×12 **11.** $-6 \times (-1)$ **12.** $-3 \times (-9)$

13. 15×2 **14.** $-10 \times (-10)$ **15.** 41×28 **16.** $-12 \times (-31)$

Divide.   (Pages 426–427)

17. $36 \div 4$ **18.** $-42 \div 6$ **19.** $81 \div (-9)$ **20.** $-12 \div (-3)$

21. $72 \div 9$ **22.** $-28 \div 4$ **23.** $-54 \div (-6)$ **24.** $100 \div (-25)$

Write a fraction for each rational number.   (Pages 428–429)

25. 7 **26.** 15 **27.** -8 **28.** $1\frac{2}{3}$ **29.** $-4\frac{1}{5}$ **30.** 4

31. -10 **32.** $-3\frac{1}{5}$ **33.** $4\frac{1}{4}$ **34.** $6\frac{5}{8}$ **35.** -17 **36.** 0

Write the rational number represented by each letter.

37. A **38.** B **39.** C **40.** D **41.** E **42.** F

43. Arrange the rational numbers in Exercises 37–42 in order from least to greatest.   (Pages 428–429)

MAINTENANCE

44. Steven sold 52 magazine subscriptions. Stella sold 63. They were paid \$2 per subscription sold. How much more than Steven did Stella earn?   (Pages 32–33)

45. Stuart wants to make enough punch to fill 50 cups. Each cup holds 90 milliliters. How many liters of punch should he make? (1 liter = 1000 milliliters)   (Pages 296–297)

46. Rhoda ordered a sandwich for \$2.75 and a glass of fruit juice for \$0.60. The sales tax was 5%. Find her total bill.   (Pages 254–255)

47. A drawer contains 2 blue pens, 3 red pens, and 5 green pens. If you choose a pen without looking, what is the probability that it will be a green pen?   (Pages 350–351)

Math and Saving Energy

In warm weather, the temperature outside is higher than the temperature inside a house. The following formula and this table can be used to estimate the rate at which heat passes from the outside through the walls or windows. This is called **heat transfer.** It is measured in Btu's per hour.

Surface	U
Concrete, 6 inches thick	0.58
Glass, single pane	1.13
Brick, 8 inches thick	0.41
Wood, 2 inches thick	0.43

$$\text{Heat transfer} = a \times U(i - o)$$

a: surface area in feet
U: heat transfer number from the table
i: inside temperature in °F
o: outside temperature in °F

EXAMPLE

Estimate the rate at which heat is passing through a concrete wall (6 inches thick) when the inside temperature is $72°F$ and the outside temperature is $90°F$. The area of the wall is 200 square feet.

READ

What are the facts?

Surface area: $a = 200$ Inside temperature: $72°F$
$U = 0.58$ (From the table)   Outside temperature: $90°F$

PLAN

Replace the variables in the formula with the facts.

SOLVE

Heat transfer $= a \times U(i - o)$

Heat transfer $= 200 \times 0.58(72 - 90)$ ← *Do the work in parentheses first.*

Heat transfer $= 200 \times 0.58(-18)$ ← *Use paper and pencil or use a calculator.*

Heat transfer $= 116 \times (-18)$

Heat transfer $= -2088$ ← *The negative number means that heat is lost from the outside.*

CHECK

Did you use all the facts correctly in the formula?

Thus, heat is being lost at the rate of **about 2088 Btu's per hour.**

EXERCISES

Find the heat transfer for these surfaces. Assume the same area and temperatures in the Example.

1. Glass, single pane   2. Brick, 8 inches thick   3. Wood, 2 inches thick

4. Arrange your answers in Exercises 1, 2, and 3 and the answer in the Example in order from least to greatest. Write the name of the surface with each value.

5. Which surface allows the least amount of heat transfer?

INTEGERS: MULTIPLICATION/DIVISION   **431**

Solving Multiplication Equations

The following are examples of multiplication equations.

$2n = -6$ ◄——— **2n means 2 × n** $-4y = 36$ ◄——— **−4y means −4 × y**

Drawing a model may help you understand how to solve such equations.

ACTIVITY   Solve and check: $2n = -6$

☐1 Use ▮ to represent n and ▨ to represent -1.

$$2n = \quad -6$$

▮▮ = ▨▨▨▨▨▨

☐2 You know what 2 ▮'s equal. To find what one ▮ equals, divide each side into **two** equal groups.

▮ ▮ = ▨▨▨, ▨▨▨

Groups ——→ 1   2 = 1   2 ◄—— **Groups**

☐3 Thus, each ▮ on the left side can be paired with three ▨'s on the right side.

▮ = ▨▨▨

$$n = \ -3$$

☐4 **Check.** Replace n in the equation with -3.

$$2n = -6$$
$$2 \times (-3) \overset{?}{=} -6$$
$$-6 = -6 \quad \text{It checks!}$$

In $2n = -6$, 2 is called the **coefficient** of n.

1. Into how many groups would you divide each side of these equations?

    a. $3y = -18$ **b.** $4x = 32$ **c.** $6s = 42$ **d.** $9n = -54$

2. What is the coefficient in a? b? c? d?

3. *Complete.*
    To solve a multiplication equation, divide each side by the __?__ of the unknown.

In the example, you solve an equation without drawing a model.

Solve and check: $-6s = 48$

$-6s = 48$ ◀ *Divide each side by -6.*

$\dfrac{-6s}{-6} = \dfrac{48}{-6}$

$s = -8$ ◀ *The variable s is alone.*

Check: $-6s = 48$ ◀ *Replace s with -8.*

$-6 \times (-8) \overset{?}{=} 48$

$48 = 48$ It checks!

EXERCISES

Write the number by which you would divide each side of the equation in order to solve for the unknown.

1. $2n = -10$
2. $4x = 36$
3. $-2y = 8$
4. $-5p = 15$

5. $-10t = 80$
6. $-7n = -42$
7. $6c = -24$
8. $9m = -45$

Solve for n by drawing a model. Let ▮ $= n$, □ $= 1$, *and* ■ $= -1$.
Check each answer.

9. $3n = 6$
10. $2n = -10$
11. $4n = 8$
12. $5n = 10$

13. $6n = 6$
14. $3n = -9$
15. $2n = -2$
16. $5n = -10$

17. $2n = -12$
18. $3n = 12$
19. $4n = -16$
20. $2n = -4$

Solve and check each equation.

21. $-4n = 32$
22. $-6x = 36$
23. $-7a = 21$
24. $-8x = 32$

25. $-9m = 81$
26. $17d = -34$
27. $18t = -54$
28. $10n = -90$

29. $15x = -75$
30. $6y = -66$
31. $-2p = -6$
32. $-8x = -64$

33. $-18t = -288$
34. $-25n = -625$
35. $-10c = -80$
36. $-4y = 144$

37. $-12x = 132$
38. $17e = -85$
39. $-20f = -800$
40. $11k = -143$

41. $72n = -360$
42. $18y = 0$
43. $-18r = 90$
44. $16s = -80$

45. $n + 11 = -8$
46. $9n = 72$
47. $n - (-3) = 15$
48. $-7r = 70$

49. $x - 5 = -33$
50. $m + 18 = -68$
51. $-8p = -96$
52. $16y = 96$

53. $t + (-14) = 87$
54. $12a = 60$
55. $n - (-13) = 2$
56. $5c = 205$

57. $10d = -130$
58. $x + 12 = -47$
59. $5f = -195$
60. $y - 6 = -10$

61. $p + (-1) = -1$
62. $-13a = -26$
63. $n - 17 = -90$
64. $18x = 0$

Solving Division Equations

The following are examples of **division equations.**

$\frac{n}{4} = -9$ ◄——— $\frac{n}{4}$ means $\frac{1}{4}n$. $\frac{x}{-5} = 12$ ◄——— $\frac{x}{-5}$ means $-\frac{1}{5}x$.

To get the unknown alone in a division equation, you use reciprocals.

> ### The product of a number and its reciprocal is 1.
> $$4 \times \tfrac{1}{4} = 1 \qquad\qquad \tfrac{3}{5} \times \tfrac{5}{3} = 1$$

The same is true for negative numbers.

$$-4 \times (-\tfrac{1}{4}) = 1 \qquad\qquad -\tfrac{3}{5} \times (-\tfrac{5}{3}) = 1$$

1. What is the reciprocal of each of the following?

   a. -2 **b.** -1 **c.** $-\frac{1}{6}$ **d.** $-\frac{4}{5}$

EXAMPLE Solve and check: $\frac{n}{-5} = 12$

$\frac{n}{-5} = 12$ ◄ **Since $\frac{n}{-5}$ means $-\frac{1}{5}n$, multiply by -5.**   Check: $\frac{n}{-5} = 12$ ◄ **Replace n with -60.**

$\frac{n}{-5} \times \frac{-5}{1} = 12 \times (-5)$ ◄ $\frac{\overset{1}{-5}}{1} \times \frac{n}{\underset{1}{-5}} = n$ $\frac{-60}{-5} \overset{?}{=} 12$

$n = -60$ $12 = 12$ It checks!

EXERCISES

Write the number by which you would multiply each side of the equation in order to solve for the unknown.

1. $\frac{n}{-5} = 10$ **2.** $\frac{x}{-8} = 9$ **3.** $\frac{n}{4} = 35$ **4.** $\frac{a}{6} = -42$

5. $\frac{c}{3} = 18$ **6.** $\frac{n}{-2} = -67$ **7.** $\frac{y}{-8} = -19$ **8.** $\frac{n}{7} = -2$

Solve and check each equation.

9. $\frac{n}{-3} = -2$ **10.** $\frac{m}{-2} = -9$ **11.** $\frac{a}{-5} = 1$ **12.** $\frac{n}{-7} = 1$

13. $\frac{t}{-5} = -20$ **14.** $\frac{n}{-3} = -8$ **15.** $\frac{n}{6} = -12$ **16.** $\frac{x}{4} = -9$

17. $\frac{n}{-5} = 0$ **18.** $\frac{y}{-6} = -1$ **19.** $\frac{d}{6} = -30$ **20.** $\frac{n}{19} = -2$

21. $\frac{e}{-12} = 50$ **22.** $\frac{n}{-24} = 32$ **23.** $\frac{f}{-9} = -7$ **24.** $\frac{n}{-15} = -10$

25. $\frac{n}{-2} = -24$ **26.** $\frac{r}{-6} = -120$ **27.** $\frac{n}{10} = -9$ **28.** $\frac{w}{90} = -1$

Solving Equations: TWO OPERATIONS

RULES

To solve an equation that requires more than one operation:
1. Add the same positive or negative number to each side.
2. Multiply or divide each side by the same number.

EXAMPLE 1

Solve and check: $3y - 10 = -19$

$3y - 10 = -19$ ◀ **Add 10 to each side.**

1. $3y - 10 + 10 = -19 + 10$

2. $3y = -9$ ◀ **Divide each side by 3.**

$\frac{3y}{3} = \frac{-9}{3}$

$y = -3$

Check: $3y - 10 = -19$ ◀ **Replace y with −3.**

$3 \times (-3) - 10 \stackrel{?}{=} -19$

$-9 - 10 \stackrel{?}{=} -19$

$-19 = -19$ It checks!

EXAMPLE 2

Solve and check: $\frac{n}{-2} + 1 = -3$

$\frac{n}{-2} + 1 = -3$ ◀ **Add −1 to each side.**

1. $\frac{n}{-2} + 1 + (-1) = -3 + (-1)$

$\frac{n}{-2} = -4$ ◀ **Multiply each side by −2.**

2. $\frac{n}{-2} \times \frac{-2}{1} = -4 \times (-2)$

$n = 8$

Check: $\frac{n}{-2} + 1 = -3$ ◀ **Replace n with 8.**

$\frac{8}{-2} + 1 \stackrel{?}{=} -3$

$-4 + 1 \stackrel{?}{=} -3$

$-3 = -3$ It checks!

EXERCISES

Write the equation that results after you do step **1**.

1. $3k + 2 = 5$ **2.** $7n - 1 = 15$ **3.** $7n + 20 = 13$ **4.** $4x + 9 = -15$

5. $\frac{r}{-4} - 13 = 23$ **6.** $\frac{w}{3} - 6 = -14$ **7.** $\frac{n}{-9} + 14 = 10$ **8.** $\frac{n}{-6} + 9 = -11$

Solve and check each equation.

9. $2y + 6 = 22$ **10.** $8t - 15 = 17$ **11.** $-5m - 12 = 3$

12. $11s + 10 = 120$ **13.** $-5w - 9 = -19$ **14.** $-3t + 17 = 5$

15. $4q + 17 = 53$ **16.** $-3p - 4 = 17$ **17.** $\frac{w}{8} + 1 = 9$

18. $\frac{m}{6} - 3 = 11$ **19.** $\frac{z}{-2} + 7 = 8$ **20.** $\frac{y}{-4} - 9 = 2$

Strategy: WRITING AN EQUATION

The key step in writing an equation to solve a word problem is **translating the facts into an equation.** Watch for key words.

EXAMPLE 1

The number of points scored last season <u>minus</u> 12 is 236. How many points were scored last season?

READ

Use the information to choose the variable for the unkown.

Let p = the number of points scored last season.

PLAN

Write an equation for the problem.

Think: The number of points scored last season minus 12 is 236.

$$p \qquad - \quad 12 = 236.$$

SOLVE

$$p - 12 = 236 \quad \blacktriangleleft \text{ } \textit{To get p alone, add 12 to each side.}$$
$$p - 12 + 12 = 236 + 12$$
$$p = 248$$

CHECK

Does $248 - 12 = 236$?   Yes. ✓

There were **248 points** scored last season.

EXAMPLE 2

The <u>quotient</u> of the total cost <u>and</u> 12 payments is $25. Find the total cost.

READ

Use the information to choose a variable for the unknown.

Let c = the total cost.

PLAN

Write an equation for the problem.

Think: The quotient of the total cost and 12 equals 25.

Translate: $\dfrac{c}{12}$ $=$ 25

SOLVE

$$\frac{c}{12} = 25 \quad \blacktriangleleft \text{ } \textit{To get c alone, multiply each side by 12.}$$
$$\frac{\overset{1}{\cancel{12}}}{1} \times \frac{c}{\underset{1}{\cancel{12}}} = 25 \times 12$$
$$c = 300$$

CHECK

Does $300 \div 12 = 25$?   Yes. ✓

The total cost is **$300.**

EXERCISES

Write an equation for each word sentence. Do not solve the equation.

1. Twice the number of millimeters of rain, *r*, is 82.

2. The number of bushels of apples, *a*, minus 18 is 59.

3. The number of desks, *d*, decreased by 40 is 98.

4. The sum of the number of tickets sold, *t*, and 48 is 261.

5. The number of points, *p*, divided by 10 questions is 12.

6. The product of the number of hours, *h*, and $5.50 is $192.50.

7. The length of a room, *l*, minus 3 meters is 12 meters.

8. Twelve dollars more than the cost of a jacket, *j*, is $80.

9. The quotient of the total cost, *c*, and 24 is $30.

10. The temperature, *t*, plus 5° equals 72°.

11. The difference between Nina's score, *s*, and 45 is 50.

12. Four times the number of quarters, *q*, is 28.

Solve each problem.

13. Thirty-one less than the number of marathon runners is 52. How many marathon runners are there?

14. Five times the distance Sally traveled is 890 miles. How far did Sally travel?

15. The number of cars washed divided by 6 is 125. How many cars were washed?

16. The number of passengers increased by a flight crew of 9 is 340. Find the number of passengers.

17. The number of quarts of oil multiplied by $1.10 is $7.70. Find the number of quarts of oil.

18. The cost of the meal plus a tip of $5.45 is $32.70. How much did the meal cost?

19. The difference between the number of $20-bills and 6 is 18. How many $20-bills are there?

20. Five times the number of books of fiction is 675. How many books of fiction are there?

21. The quotient of the number of names on one page of a telephone book and 6 is 32. Find the number of names.

22. Ninety-five calories plus the number of calories in a salad totals 164. How many calories are there in the salad?

23. Rico's number of hits divided by 30 equals 0.400. Find Rico's number of hits.

24. A painter's weekly salary minus $65 is $312. What is the painter's weekly salary?

Equations and Patterns

If there is a pattern between two sets of numbers, then there is an equation that describes the pattern.

Is there a pattern in this table?

x	y
1	7
2	9
3	11
4	13
5	15
6	17

1 Compare what happens to x to what happens to y.

Each time x increases by 1, y increases by 2. Therefore y increases twice as fast as x.

Equation: $y = 2x$

2 **Check:** Replace x in $y = 2x$ with 1, 2, 3, 4, 5, and 6.

When $x = 1$:
$$y = 2 \times 1 = 2$$

In the table, $y = 7$ (not 2) when $x = 1$. You need to add 5 to $y = 2x$ so that $y = 7$ when $x = 1$.

Equation: $y = 2x + 5$

3 **Check:** Replace x in $y = 2x + 5$ with 1, 2, 3, 4, 5, and 6.

When $x = 1$:
$$y = 2 \times 1 + 5 \quad \blacktriangleleft \text{ Multiply first. Then add.}$$
$$y = 2 + 5 = 7 \quad \text{It checks! } \checkmark$$

When $x = 2$:
$$y = 2 \times 2 + 5$$
$$y = 4 + 5 = 9 \quad \text{It checks! } \checkmark$$

Complete the check in step **3** to show that $y = 2x + 5$ is the equation.

EXERCISES

Find the equation.

1.

x	y
2	1
3	3
4	5
5	7
6	9

2.

x	y
1	3
2	6
3	9
4	12
5	15

3.

x	y
1	10
2	13
3	16
4	19
5	22

4.

x	y
1	14
2	15
3	16
4	17
5	18

5.

x	y
1	7
2	11
3	15
4	19
5	23

6.

x	y
1	1
2	6
3	11
4	16
5	21

7.

x	y
2	1
4	2
6	3
8	4
10	5

8.

x	y
1	9
2	8
3	7
4	6
5	5

Chapter Summary

1. To multiply two integers with <u>unlike</u> signs:
   - 1 Multiply as with whole numbers.
   - 2 Insert a negative sign before the product.

2. To multiply two integers with like signs, multiply as with whole numbers. The product is a positive integer.

3. To divide two integers:
   - 1 Divide as with whole numbers.
   - 2 **a.** When the two integers have like signs, the quotient is positive.
       b. When the two integers have unlike signs, the quotient is negative.

4. To solve a multiplication equation such as $-6x = 48$, divide each side of the equation by -6.

5. To solve a division equation such as $\frac{n}{-2} = -5$, multiply each side of the equation by -2.

6. To solve an equation:
   - 1 Add the same positive or negative number to each side.
   - 2 Multiply or divide each side by the same number.

7. To use an equation to solve a word problem:

   - 1 Choose a variable to represent the unknown.

   - 2 Write an equation for the problem.
   - 3 Solve the equation.
   - 4 Check your answer with the statements in the problem. Answer the question.

---

1. Multiply.
$$-8 \times 7 = \mathbf{-56}$$
$$9 \times (-5) = \mathbf{-45}$$

2. Multiply.
$$-4 \times (-5) = \mathbf{20}$$
$$14 \times 9 = \mathbf{126}$$

3. Divide.
$$-36 \div (-4) = \mathbf{9}$$
$$12 \div (-2) = \mathbf{-6}$$

4. Solve: $-6x = 48$
$$\frac{-6x}{-6} = \frac{48}{-6}$$
$$x = \mathbf{-8}$$

5. Solve: $\frac{n}{-2} = -5$
$$\frac{n}{-2} \times \frac{\mathbf{-2}}{\mathbf{1}} = -5 \times \mathbf{-2}$$
$$n = \mathbf{10}$$

6. Solve: $4x - 5 = -25$
$$4x - 5 + \mathbf{5} = -25 + \mathbf{5}$$
$$4x = -20$$
$$\frac{4x}{4} = \frac{-20}{4}$$
$$x = \mathbf{-5}$$

7. Three times the number of dimes, *d,* is 27. How many dimes are there?

   Let d = the number of dimes.
   $$3d = 27$$
   $$\frac{3d}{3} = \frac{27}{3}; d = 9$$
   Does $3 \times 9 = 27$? Yes ✓
   There are **9 dimes.**

Chapter Review

Part 1: VOCABULARY

For Exercises, 1–5, choose from the box at the right the word(s) that complete(s) each statement.

1. The product of two integers having unlike signs is __?__. (Page 422)

2. When you multiply by zero, the product is __?__. (Page 422)

3. The quotient of two negative integers is __?__. (Page 426)

4. Any rational number can be written as a __?__. (Page 428)

5. In $3n = -9$, 3 is called the __?__ of n. (Page 432)

> zero
> coefficient
> negative
> fraction
> positive

Part 2: SKILLS

Multiply. (Pages 422–423)

6. $6 \times (-15)$ 7. -9×7 8. -14×3 9. $21 \times (-4)$

10. -5×16 11. -8×25 12. $5 \times (-21)$ 13. $8 \times (-17)$

(Pages 424–425)

14. 8×6 15. 9×5 16. 12×6 17. 7×5

18. $-3 \times (-7)$ 19. $-5 \times (-4)$ 20. $-3 \times (-6)$ 21. $-2 \times (-9)$

Divide. (Pages 426–427)

22. $-25 \div 5$ 23. $36 \div (-9)$ 24. $-60 \div 12$ 25. $320 \div (-10)$

26. $-400 \div (-20)$ 27. $65 \div (-13)$ 28. $-600 \div (-25)$ 29. $624 \div (-12)$

Write a fraction for each rational number. (Pages 428–429)

30. 4 31. -12 32. $3\frac{5}{6}$ 33. $-7\frac{1}{4}$ 34. $-3\frac{1}{2}$ 35. $4\frac{1}{6}$

Write the rational number represented by each letter. (Pages 428–429)

36. A 37. B 38. C 39. D 40. E 41. F

42. Arrange the rational numbers in Exercises 36–41 in order from least to greatest.

For Exercises 43–46, arrange the numbers in order from least to greatest.   (Pages 428–429)

43. $6\frac{1}{2}$; $-3\frac{1}{4}$; $3\frac{1}{5}$; 1; $-2\frac{1}{2}$; 0

44. $\frac{1}{2}$; $-\frac{1}{2}$; 1; -2; $\frac{1}{4}$; $-\frac{1}{4}$

45. -3; $-4\frac{1}{2}$; 2; $4\frac{1}{4}$; 0; -1

46. $\frac{6}{7}$; $-\frac{5}{7}$; $\frac{5}{6}$; $-\frac{5}{6}$; $-\frac{6}{7}$; $\frac{5}{7}$

Solve and check each equation.   (Pages 432–433)

47. $-5x = -30$

48. $6y = -48$

49. $-4c = -20$

50. $-9t = -27$

51. $2n = 18$

52. $3a = -45$

53. $-8y = 128$

54. $-7x = -70$

(Page 434)

55. $\frac{n}{-6} = -7$

56. $\frac{b}{5} = -6$

57. $\frac{w}{-2} = -9$

58. $\frac{t}{-3} = 7$

59. $\frac{x}{4} = 10$

60. $\frac{y}{-3} = -9$

61. $\frac{n}{10} = -12$

62. $\frac{c}{-9} = -15$

(Page 435)

63. $-3x + 5 = -4$

64. $6a - 10 = -16$

65. $-5d - 3 = -13$

66. $8f + 11 = 35$

67. $4y + 16 = 8$

68. $-11b - 9 = 2$

69. $3t + 4 = 16$

70. $-9s - 16 = -34$

71. $-3q - 12 = 6$

Part 3: APPLICATIONS

Use positive and negative integers to represent each situation by a multiplication problem. Then find the product.   (Pages 422–423)

72. Two penalties of 10 yards each

73. Four times a loss of 6 pounds

74. Three times a depth of 200 feet below sea level

75. Seven times a loss of 3 yards rushing

Write an equation for each word sentence. Do _not_ solve the equation. (Pages 436–437)

76. Tom's age, a, increased by 9 is 25.

77. One foot less than Sue's height, h, is 4.

78. The length of a field, f, times 5 is 620 yards.

79. A telephone bill, b, divided by 5 is $12.

Write an equation for each problem. Then solve the equation. (Pages 436–437)

80. Three times the age of Mr. Ellis is 60. How old is Mr. Ellis?

81. Three feet less than a tree's height is 4. How tall is the tree?

82. The sale price increased by $5.50 is $20.00. What is the sale price?

83. The electric bill divided by 3 is $30.00. How much is the bill?

Chapter Test

Multiply.

1. 2×9 **2.** $-5 \times (-10)$ **3.** $6 \times (-4)$ **4.** -8×9 **5.** $-12 \times (-7)$

6. 10×15 **7.** -13×21 **8.** $14 \times (-17)$ **9.** -16×21 $11 \times (-13)$

Divide.

11. $-48 \div (-6)$ **12.** $-96 \div 6$ **13.** $80 \div (-16)$ **14.** $-24 \div (-12)$

15. $-735 \div 15$ **16.** $-351 \div (-13)$ **17.** $391 \div (-23)$ **18.** $510 \div (-17)$

Write a fraction for each rational number.

19. -3 **20.** $4\frac{1}{3}$ **21.** $5\frac{1}{2}$ **22.** $-1\frac{1}{4}$ **23.** 14

Write the rational number represented by each letter.

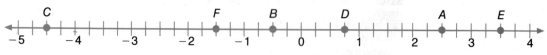

24. A **25.** B **26.** C **27.** D **28.** E **29.** F

30. Arrange the rational numbers in Exercises 24–29 in order from least to greatest.

Solve and check each equation.

31. $6d = -24$ **32.** $-3s = 36$ **33.** $4x = 48$ **34.** $-5y = -100$

35. $\frac{y}{-4} = -6$ **36.** $\frac{x}{-2} = -4$ **37.** $\frac{x}{5} = -13$ **38.** $\frac{n}{-6} = 5$

39. $2b - 3 = 7$ **40.** $-4a + 3 = -13$ **41.** $6c - 4 = -40$ **42.** $-8s - 5 = -29$

For Exercises 43–44 use positive and negative integers to represent each situation by a multiplication problem. Then find the product.

43. Five times a debt of $25

44. Twice a temperature drop of 4

Write an equation for each problem. Then solve the equation.

45. Eight hours times the hourly wage is $40.00. What is the hourly wage?

46. Two minutes more than Pat's time is 8 minutes. What is her time?

Choose the correct answer. Choose a, b, c, or d.

1. Larry earns $8 per hour and works 38 hours per week. Deductions amount to $64. Find the net pay.

   a. $304   **b.** $474
   c. $368   **d.** $240

2. Which shows the numbers listed in order from least to greatest?

   a. $\frac{1}{5}$, 0.6, -1, -10

   b. -10, -1, $\frac{1}{5}$, 0.6

   c. -1, -10, $\frac{1}{5}$, 0.6

   d. 0.6, $\frac{1}{5}$, -1, -10

3. Divide: $15 \div (-3)$

   a. 5   **b.** -5 **c.** 45   **d.** -45

4. Ten hours times the hourly wage is $120.00. What is the hourly wage?

   a. $120   **b.** $1200
   c. $12   **d.** $10

5. What is the volume in cubic inches of this bread box?

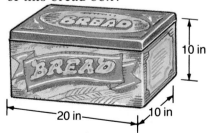

10 in

20 in   10 in

   a. 40   **b.** 200
   c. 2000   **d.** 100

6. The rate of sales tax in one state is 7%. What is the amount of tax on a car that costs $6500?

   a. $455   **b.** $445
   c. $45.50   **d.** $44.50

7. Multiply 0.56 by 0.05.

   a. 0.28   **b.** 0.028
   c. 0.0028   **d.** 28

8. Subtract.
$$8\tfrac{1}{3} - 2\tfrac{2}{3}$$

   a. $5\frac{2}{3}$ **b.** $6\frac{1}{3}$

   c. $5\frac{1}{3}$ **d.** $6\frac{2}{3}$

9. Which number of degrees represents an obtuse angle?

   a. 36°   **b.** 90°
   c. 91°   **d.** 85°

10. Divide.
$$-36 \div 12$$

   a. 3   **b.** -3
   c. $\frac{1}{3}$ **d.** $-\frac{1}{3}$

11. Add.
$$-15 + (-27)$$

   a. 12   **b.** -12
   c. -42 **d.** 42

12. A box contains 4 purple cards and 10 yellow cards. Without looking, Ramone picked a card. What is the probability that he picked a purple card?

a. $\frac{2}{5}$ **b.** $\frac{2}{7}$

c. $\frac{5}{7}$ **d.** 1

13. Multiply.
$$-5 \times 20$$

a. -100 **b.** 100
c. -25 **d.** -15

14. Solve for n.
$$\frac{n}{-6} = -12$$

a. 2   **b.** 72
c. -2 **d.** -72

15. Multiply.
$$17 \times (-8)$$

a. 136   **b.** 126
c. -136 **d.** -126

16. Use $c^2 = a^2 + b^2$ to find the length of the hypotenuse of a right triangle whose legs measure 6 meters and 8 meters.

a. 28   **b.** 14
c. 10   **d.** 48

17. Choose the best estimate.
$$705 \times 18$$

a. 20,000   **b.** 18,600
c. 16,000   **d.** 14,000

18. The fraction $\frac{1}{5}$ is equal to:

a. 50%   **b.** 20%
c. 0.02   **d.** 2.0

19. Triangle ABC is similar to triangle DEF. Which proportion would you use to find side d?

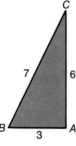 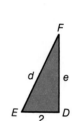

a. $\frac{3}{7} = \frac{d}{2}$ **b.** $\frac{3}{7} = \frac{2}{d}$

c. $\frac{6}{7} = \frac{2}{d}$ **d.** $\frac{3}{2} = \frac{d}{7}$

20. The number of home runs hit by six people are given below. Find the mode.

10   12   11   14   20   12

a. 14   **b.** 13
c. 12   **d.** 15

21. Subtract.
$$-32 - 17$$

a. 49   **b.** -49
c. -15 **d.** 15

22. Phil has 3 pennies, 4 nickels and 2 quarters. One coin is drawn. What is the probability that it is a penny?

a. $\frac{3}{4}$ **b.** $\frac{1}{3}$

c. $\frac{1}{6}$ **d.** $\frac{3}{5}$

APPENDIX A
SKILLS BANK

Estimating with Large Numbers (Pages 2–3)

Estimate. Find the actual sum or difference. Compare.

1. $4213 + 2876$

2. $6870 - 2105$

3. $718 - 395$

4. $43,785 + 67,933$

5. $81,003 - 32,881$

6. $7863 + 6401$

This table shows the wheat production for several countries. Refer to this table for Exercises 7–9.

7. Estimate the difference between the number of metric tons of wheat produced by China and the United States?

8. Which country produced about 27,000,000 metric tons of wheat?

9. Estimate the total metric tons of wheat produced by Argentina and Canada.

Wheat Production For One Year	
Country	**Metric Tons**
Argentina	11,256,000
Australia	8,600,000
Canada	26,866,000
China	68,420,000
United Kingdom	10,258,000
United States	76,443,000

Order of Operations (Pages 8–9)

Find each answer.

1. $(8 + 2) - (4 + 1)$

2. $11 - (4 - 3)$

3. $6 + 9 \div 3$

4. $4 \times 8 - 2 \times 6$

5. $23 - 9 + 2$

6. $(30 - 8) + 3 \times 5$

7. $28 - \frac{6 + 12}{3}$

8. $(9 + 12) \times 4 \div 3$

9. $\frac{72}{3 \times 4} + 15$

10. Which is greater?
$(3 \times 2) + 7$ or $3 \times (2 + 7)$

11. Which is less?
$40 \div 5 - 3$ or $40 \div (5 - 3)$

Insert parentheses to make each statement true.

12. $4 + 3 \times 2 = 14$

13. $8 \times 4 - 2 + 6 = 22$

14. $7 + 8 \div 4 - 1 = 5$

15. $9 \div 1 + 2 \times 5 = 15$

Exponents (Pages 14–15)

Write the standard form of each number.

1. 5^4

2. 7^2

3. 2^4

4. 3^4

5. 4^3

6. 10^6

7. 16×10^4

8. 6×2^3

9. 4×5^3

Complete.

10. $\square^2 = 64$

11. $3 \times \square^7 = 3$

12. $7 \times 10^? = 70,000$

Decimals: ADDITION/SUBTRACTION

(Pages 28—29)

Add or subtract as indicated.

1.   5.280
   18.050
+ 9.075
	2. 4.036
	  11.050
	+ 7.352
		3. 83.76
		− 5.036
			4. 62.12
			−53.04
				5. 62.800
				   5.640
				+32.005

6.   186.9
− 57.32
	7. 56.040
	   8.752
	+127.160
		8. 18.24
		  3.56
		+27.85
			9. 2.7
			−0.93
				10. 14.20
				− 2.34

11.   14.4
  21.6
+ 7.3
	12. 0.916
	  0.32
	+24.7
		13. 0.64
		−0.07
			14. 88.44
			− 2.15
				15. 19.802
				−18.907

16. $34.5 - \underline{\ ?\ } = 17.8$

17. $843.2 + 95.03 + 13.5$

18. $1.53 + \underline{\ ?\ } + 12.64 = 20.42$

19. $4.75 + 11.09 + 21.78$

Multiplying Whole Numbers

(Pages 32—33)

Multiply.

1. 73 ×83	2. 28 ×32	3. 52 ×83	4. 67 ×38	5. 98 ×58
6. 136 × 55	7. 429 × 83	8. 321 × 84	9. 665 × 58	10. 964 × 37
11. 327 ×861	12. 914 ×632	13. 762 ×296	14. 539 ×291	15. 246 ×354

16. 624×92 17. 736×54 18. 482×107 19. 19×35 20. 714×206

Decimals: MULTIPLICATION

(Pages 41—42)

Multiply.

1. 607 ×0.09	2. 19.32 × 0.5	3. 276.5 × 9	4. 18.001 × 3.05	5. 16.2 ×0.13
6. 4.7 × 8	7. $8.31 × 9	8. 3.761 × 1.7	9. 5.9 × 16	10. $2.73 × 42

11. 32.15×2.03 12. 0.31×6.24 13. 9.4×3.19

14. 0.14×91 15. 3.4×8.12 16. 0.75×12.48

Division: WHOLE NUMBERS

(Pages 58–59)

Divide. Check each answer.

1. $38\overline{)2562}$ **2.** $42\overline{)629}$ **3.** $86\overline{)946}$ **4.** $24\overline{)7539}$ **5.** $43\overline{)4853}$

6. $41\overline{)1115}$ **7.** $75\overline{)1995}$ **8.** $63\overline{)3661}$ **9.** $74\overline{)2146}$ **10.** $39\overline{)2538}$

11. $29\overline{)6245}$ **12.** $73\overline{)5331}$ **13.** $36\overline{)1242}$ **14.** $52\overline{)4576}$ **15.** $46\overline{)5602}$

Zeros in the Quotient

(Pages 60–61)

Divide.

1. $42\overline{)13449}$ **2.** $37\overline{)7663}$ **3.** $28\overline{)11289}$ **4.** $35\overline{)21079}$ **5.** $63\overline{)33392}$

6. $81\overline{)16689}$ **7.** $32\overline{)9765}$ **8.** $54\overline{)5847}$ **9.** $67\overline{)54273}$ **10.** $96\overline{)27898}$

11. $31\overline{)6417}$ **12.** $18\overline{)5548}$ **13.** $50\overline{)2007}$ **14.** $70\overline{)4930}$ **15.** $58\overline{)8723}$

Dividing a Decimal by a Whole Number

(Pages 64–65)

Divide. Round each quotient to the nearest hundredth.

1. $21\overline{)27.4}$ **2.** $46\overline{)189.7}$ **3.** $70\overline{)8.638}$ **4.** $13\overline{)3.762}$ **5.** $15\overline{)16.02}$

6. $23\overline{)6.31}$ **7.** $73\overline{)18.05}$ **8.** $13\overline{)7.056}$ **9.** $19\overline{)97.85}$ **10.** $56\overline{)176.6}$

Divide. Round each quotient to the nearest tenth.

11. $16\overline{)7.7}$ **12.** $57\overline{)77.5}$ **13.** $13\overline{)16.6}$ **14.** $24\overline{)35.2}$ **15.** $86\overline{)67.6}$

Dividing by a Decimal

(Pages 68–69)

Divide. Round each quotient to the nearest tenth.

1. $0.37\overline{)18.4}$ **2.** $6.2\overline{)674}$ **3.** $3.5\overline{)6.25}$ **4.** $0.71\overline{)12.5}$

$2.7\overline{)1.32}$ **6.** $0.86\overline{)42.5}$ **7.** $8.9\overline{)1.3}$ **8.** $0.28\overline{)12}$

Divide. Round each quotient to the nearest hundredth.

9. $7.4\overline{)2.86}$ **10.** $6.8\overline{)7.2}$ **11.** $0.32\overline{)2.44}$ **12.** $0.57\overline{)12.81}$

Bar Graphs and Applications   (Pages 84–85)

1. Which animal can run the fastest?

2. Which is the slowest runner?

3. About how fast can an elk run?

4. Which animal(s) can run faster than 45 miles per hour?

5. List the animals in order from slowest to fastest runners.

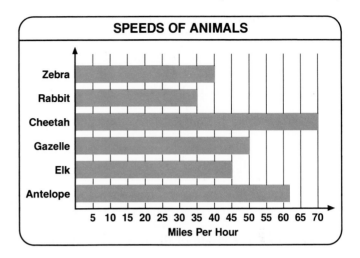

SPEEDS OF ANIMALS

6. Use the information in the table to construct a vertical bar graph for the areas of the Great Lakes.

Lake	Superior	Michigan	Huron	Erie	Ontario
Area (square miles)	81,000	67,900	73,700	32,630	30,740

Line Graphs and Applications   (Pages 86–87)

1. During which years were there more than 900 tornadoes?

2. During which year were there the most tornadoes?

3. During which year were there the fewest tornadoes?

4. Did the number of tornadoes increase or decrease from 1971 to 1972?

5. Did the number of tornadoes increase or decrease from 1978 to 1979?

6. Use graph paper to construct a line graph to show the number of cars crossing a bridge.

TORNADOES IN THE UNITED STATES

Hour	6 A.M.	7 A.M.	8 A.M.	9 A.M.	10 A.M.	11 A.M.	Noon
Number of Cars	5	12	32	48	25	20	45

The Mean and the Mode

(Pages 105–106)

Find the mean, range, and the mode for each list of numbers.

1. 2, 5, 6, 7, 4, 6

2. 8, 5, 12, 3, 4, 5, 14, 5

3. 9, 12, 7, 5, 11, 16, 14, 11

4. 15, 8, 1, 8, 12, 9, 11, 2

5. 2, 4, 9, 8, 4, 7, 1, 10

6. 4, 17, 12, 6, 17, 9, 5, 7

7.

Scores of Eight Bowlers			
124	128	132	155
118	125	137	145

8.

Scores of Ten Golfers				
78	84	80	82	80
86	80	85	79	84

In the 100-yard dash, 5 runners had these times: 10.4 seconds, 11.2 seconds, 10.6 seconds, 12.1 seconds, and 11.2 seconds.

9. What was the mode of the running times?

10. What was the range of the running times?

11. What was the mean time?

The Median

(Pages 108–109)

For Exercises 1–8, find the median.

1. 85, 96, 83, 90, 96

2. 19, 26, 17, 28, 35, 13

3. 8.3, 5.6, 3.7, 7.4

4. 2.8, 2.7, 2.3, 2.2, 2.6, 2.4

5. 56, 72, 63, 87, 42

6. 28.3, 25.6, 30.7, 27.4

7.

Miles Driven to Work	
Employee	**Miles Driven**
T. Gonzales	12
A. McVean	8
R. Trividi	7
B. Benzak	15
L. Greco	6

8.

Enrollment Per Grade	
Grade	**Enrollment**
7	192
8	198
9	205
10	215
11	195
12	200

9. **a.** Find the median of these scores. 87, 83, 93, 89, 97

b. Replace 83 with 71. Find the median.

c. Are the medians in **a** and **b** the same? Explain why.

10. **a.** Find the median of these scores. 36, 28, 42, 39, 45, 31, 47

b. Omit the largest and smallest scores. Find the new median.

c. Are the medians in **a** and **b** the same? Explain why.

Addition and Subtraction: LIKE FRACTIONS   (Pages 134–135)

Add or subtract. Write each answer in lowest terms.

1. $\frac{1}{7} + \frac{5}{7}$ **2.** $\frac{5}{6} - \frac{1}{6}$ **3.** $\frac{3}{5} - \frac{1}{5}$ **4.** $\frac{5}{8} + \frac{7}{8}$

5. $\frac{6}{7} - \frac{3}{7}$ **6.** $\frac{2}{6} + \frac{5}{6}$ **7.** $\frac{5}{12} + \frac{3}{12}$ **8.** $\frac{9}{10} - \frac{5}{10}$

9. $2\frac{5}{12} + 3\frac{6}{12}$ **10.** $8\frac{3}{4} - 2\frac{2}{4}$ **11.** $6\frac{1}{5} + 2\frac{3}{5}$ **12.** $2\frac{11}{12} - 1\frac{1}{12}$

13. $9\frac{5}{8} - 2\frac{3}{8}$ **14.** $3\frac{5}{8} + 7\frac{2}{8}$ **15.** $9\frac{11}{12} - 4\frac{5}{12}$ **16.** $1\frac{5}{9} + 6\frac{2}{9}$

17. $\frac{7}{9} + \frac{5}{9}$ **18.** $12\frac{7}{9} - 4\frac{2}{9}$ **19.** $\frac{13}{14} - \frac{3}{14}$ **20.** $\frac{5}{12} + \frac{4}{12}$ $\frac{3}{4}$

21. $19\frac{4}{5} - 10\frac{2}{5}$ **22.** $7\frac{1}{8} + 6$ **23.** $9 + 14\frac{2}{3}$ **24.** $14\frac{5}{6} - 6\frac{4}{6}$

Addition and Subtraction: UNLIKE FRACTIONS   (Pages 140–141)

Add or subtract. Write each answer in lowest terms.

1. $\begin{array}{r}\frac{1}{3}\\+\frac{7}{8}\\\hline\end{array}$ **2.** $\begin{array}{r}\frac{3}{4}\\-\frac{1}{3}\\\hline\end{array}$ **3.** $\begin{array}{r}\frac{4}{5}\\-\frac{5}{8}\\\hline\end{array}$ **4.** $\begin{array}{r}\frac{3}{4}\\+\frac{2}{3}\\\hline\end{array}$ **5.** $\begin{array}{r}\frac{1}{2}\\-\frac{4}{11}\\\hline\end{array}$ **6.** $\begin{array}{r}\frac{2}{7}\\+\frac{13}{14}\\\hline\end{array}$

7. $\frac{3}{8} + \frac{9}{16}$ **8.** $\frac{3}{4} - \frac{2}{7}$ **9.** $\frac{7}{8} - \frac{2}{9}$ **10.** $\frac{4}{11} + \frac{6}{22}$

11. $\begin{array}{r}2\frac{1}{6}\\+3\frac{1}{4}\\\hline\end{array}$ **12.** $\begin{array}{r}\frac{11}{12}\\-\frac{1}{4}\\\hline\end{array}$ **13.** $\begin{array}{r}7\frac{4}{8}\\-7\frac{1}{2}\\\hline\end{array}$ **14.** $\begin{array}{r}9\frac{1}{2}\\+6\frac{8}{9}\\\hline\end{array}$ **15.** $\begin{array}{r}5\frac{7}{8}\\-1\frac{3}{4}\\\hline\end{array}$ **16.** $\begin{array}{r}2\frac{7}{9}\\+4\frac{1}{6}\\\hline\end{array}$

17. $2\frac{3}{5} + 6\frac{3}{10}$ **18.** $10\frac{11}{12} - 7\frac{3}{4}$ **19.** $30\frac{2}{3} - 17\frac{1}{4}$ **20.** $10\frac{5}{12} + 4\frac{1}{4}$

21. $\frac{7}{12} + \frac{1}{3}$ **22.** $2\frac{5}{8} + 4\frac{3}{4}$ **23.** $\frac{8}{9} - \frac{3}{4}$ **24.** $7\frac{3}{4} + 7\frac{3}{4}$

25. $\frac{7}{8} - \frac{1}{2}$ **26.** $7\frac{3}{4} - 4\frac{1}{3}$ **27.** $\frac{3}{4} + \frac{1}{3}$ **28.** $21\frac{4}{15} - \frac{1}{5}$

Subtraction: MIXED NUMBERS   (Pages 142–143)

Subtract. Write each answer in lowest terms.

1. $\begin{array}{r}6\frac{1}{3}\\-1\frac{2}{3}\\\hline\end{array}$ **2.** $\begin{array}{r}5\frac{1}{5}\\-2\frac{3}{5}\\\hline\end{array}$ **3.** $\begin{array}{r}3\frac{1}{6}\\-1\frac{3}{4}\\\hline\end{array}$ **4.** $\begin{array}{r}9\frac{7}{10}\\-5\frac{4}{5}\\\hline\end{array}$ **5.** $\begin{array}{r}10\frac{11}{12}\\-7\frac{3}{4}\\\hline\end{array}$ **6.** $\begin{array}{r}4\frac{3}{16}\\-1\frac{1}{4}\\\hline\end{array}$

7. $6\frac{1}{2} - 2\frac{4}{5}$ **8.** $5\frac{1}{3} - 2\frac{1}{2}$ **9.** $7\frac{3}{8} - 4\frac{3}{4}$ **10.** $4\frac{1}{4} - 3\frac{1}{3}$

11. $\begin{array}{r}7\\-3\frac{1}{3}\\\hline\end{array}$ **12.** $\begin{array}{r}10\\-3\frac{2}{9}\\\hline\end{array}$ **13.** $\begin{array}{r}2\\-1\frac{5}{6}\\\hline\end{array}$ **14.** $\begin{array}{r}7\\-5\frac{3}{10}\\\hline\end{array}$ **15.** $\begin{array}{r}8\\-4\frac{1}{6}\\\hline\end{array}$ **16.** $\begin{array}{r}3\\-2\frac{3}{10}\\\hline\end{array}$

17. $7 - 4\frac{1}{3}$ **18.** $10 - 1\frac{1}{2}$ **19.** $8 - 7\frac{2}{3}$ **20.** $10 - 4\frac{4}{7}$

21. $12 - 11\frac{3}{8}$ **22.** $17\frac{2}{3} - \frac{5}{40}$ **23.** $1 - \frac{7}{12}$ **24.** $1 - \frac{3}{7}$

Multiplication: FRACTIONS

(Pages 155–156)

Multiply. Use the method you prefer. Write answers in lowest terms.

1. $\frac{1}{8} \times \frac{1}{5}$
2. $\frac{11}{12} \times \frac{7}{22}$
3. $\frac{8}{15} \times \frac{5}{12}$
4. $\frac{5}{9} \times \frac{3}{8}$
5. $\frac{3}{14} \times \frac{7}{8}$
6. $\frac{2}{3} \times \frac{9}{10}$

7. $\frac{7}{8} \times \frac{3}{4}$
8. $\frac{5}{8} \times 16$
9. $\frac{7}{15} \times \frac{20}{21}$
10. $\frac{9}{16} \times \frac{4}{12}$
11. $\frac{2}{3} \times \frac{2}{9}$
12. $\frac{4}{25} \times \frac{5}{8}$

13. $\frac{5}{6} \times \frac{3}{10}$
14. $5 \times \frac{3}{15}$
15. $\frac{3}{8} \times \frac{16}{45}$
16. $\frac{5}{16} \times \frac{7}{15}$
17. $\frac{1}{12} \times 4$
18. $\frac{2}{9} \times \frac{5}{8}$

19. $\frac{7}{12} \times \frac{11}{14}$
20. $\frac{7}{8} \times \frac{4}{5}$
21. $\frac{3}{7} \times \frac{5}{6}$
22. $\frac{9}{10} \times \frac{5}{8}$
23. $\frac{11}{15} \times \frac{5}{9}$
24. $\frac{9}{10} \times \frac{20}{21}$

Multiplication: MIXED NUMBERS

(Pages 158–159)

First estimate the answer. Then find the exact answer. Compare.

1. $\frac{7}{9} \times 2\frac{1}{5}$
2. $\frac{8}{9} \times 9\frac{1}{2}$
3. $\frac{9}{11} \times 4$
4. $2\frac{2}{3} \times 7$
5. $5\frac{1}{3} \times 6\frac{1}{2}$

6. $4\frac{3}{5} \times 7\frac{3}{4}$
7. $3\frac{2}{3} \times 9\frac{3}{8}$
8. $3\frac{1}{3} \times 5\frac{1}{8}$
9. $5\frac{1}{4} \times 6\frac{2}{3}$
10. $1\frac{1}{3} \times 3\frac{3}{4}$

11. $5\frac{2}{3} \times \frac{1}{3}$
12. $2\frac{5}{6} \times 4\frac{1}{4}$
13. $2\frac{3}{8} \times 2\frac{4}{5}$
14. $7\frac{3}{4} \times 6\frac{4}{5}$
15. $3\frac{2}{3} \times 1\frac{4}{5}$

16. $5 \times 3\frac{1}{4}$
17. $5\frac{3}{4} \times 2\frac{2}{3}$
18. $1\frac{1}{3} \times 2\frac{2}{9}$
19. $1\frac{1}{2} \times 3\frac{2}{5}$
20. $3\frac{5}{6} \times 4\frac{2}{3}$

Division: FRACTIONS

(Pages 160–161)

Divide. Write each answer in lowest terms or as a mixed number.

1. $\frac{2}{3} \div \frac{8}{15}$
2. $\frac{23}{24} \div \frac{9}{15}$
3. $\frac{7}{8} \div \frac{7}{15}$
4. $\frac{1}{9} \div \frac{9}{10}$
5. $\frac{9}{20} \div \frac{5}{6}$

6. $\frac{5}{7} \div \frac{13}{14}$
7. $\frac{4}{9} \div \frac{4}{9}$
8. $\frac{9}{10} \div 9$
9. $\frac{2}{5} \div 18$
10. $\frac{1}{2} \div 24$

11. $\frac{6}{5} \div 3$
12. $\frac{8}{11} \div \frac{8}{11}$
13. $\frac{14}{15} \div 2$
14. $6 \div \frac{14}{15}$
15. $\frac{3}{8} \div \frac{6}{11}$

16. $1 \div \frac{2}{3}$
17. $1 \div \frac{3}{4}$
18. $\frac{3}{8} \div \frac{1}{2}$
19. $\frac{5}{6} \div 3$
20. $\frac{2}{9} \div \frac{2}{3}$

Division: MIXED NUMBERS

(Pages 162–163)

Divide. Write each answer in lowest terms or as a mixed number.

1. $\frac{2}{3} \div 7\frac{1}{3}$
2. $\frac{2}{3} \div 1\frac{1}{3}$
3. $6\frac{3}{5} \div \frac{5}{8}$
4. $7\frac{2}{3} \div 8$
5. $8\frac{2}{9} \div 1\frac{2}{3}$

6. $7\frac{3}{5} \div 6\frac{1}{2}$
7. $12 \div 2\frac{2}{3}$
8. $9 \div 3\frac{1}{3}$
9. $2\frac{1}{6} \div 4\frac{1}{5}$
10. $8\frac{1}{2} \div 4\frac{5}{9}$

11. $1\frac{1}{4} \div 3\frac{1}{3}$
12. $5\frac{1}{2} \div 7\frac{1}{3}$
13. $3\frac{1}{2} \div 2\frac{1}{6}$
14. $1\frac{1}{8} \div 1\frac{1}{3}$
15. $5 \div 8\frac{1}{3}$

16. $2\frac{3}{5} \div 2\frac{3}{5}$
17. $8\frac{1}{2} \div 3$
18. $2\frac{3}{4} \div 4$
19. $1\frac{4}{5} \div 5\frac{2}{5}$
20. $8\frac{1}{4} \div 5\frac{1}{2}$

452   APPENDIX A

Perimeter

(Pages 186–188)

Complete.

	Length	Width	Perimeter
1.	$4\frac{1}{2}$ in	5 in	?
2.	6.1 cm	5.3 cm	?
3.	7 yd	?	22 yd
4.	2 mi	6 mi	?

	Length	Width	Perimeter
5.	?	30 m	360 m
6.	17 ft	?	51 ft
7.	?	4.8 m	21.6 m
8.	20 cm	?	70 cm

Area: RECTANGLES AND SQUARES

(Pages 191–193)

Complete.

	Length	Width	Area
1.	3 km	4 km	? km^2
2.	5 mi	7.4 mi	? mi^2
3.	7 m	? m	49 m^2
4.	? yd	$8\frac{1}{4}$ yd	99 yd^2

	Length	Width	Area
5.	? in	$9\frac{1}{2}$ in	114 in^2
6.	17.6 cm	? cm	396 cm^2
7.	? m	$\frac{5}{6}$	$5\frac{1}{4}$ m^2
8.	5 ft	5 ft	? ft^2

Area: PARALLELOGRAMS/TRIANGLES

(Pages 196–198)

Complete.

	Figure	Base	Height	Area
1.	Parallelogram	7 cm	8 cm	?
2.	Triangle	10 in	6 in	?
3.	Triangle	9 m	6.4 m	?
4.	Parallelogram	?	4.4 yd	34.32 yd^2

	Figure	Base	Height	Area
5.	Parallelogram	6 ft	?	42 ft^2
6.	Triangle	5 yd	$6\frac{1}{2}$ yd	?
7.	Parallelogram	9.7 m	5.4 m	?
8.	Triangle	32 cm	41 cm	?

Area: TRAPEZOIDS

(Pages 200–201)

Complete.

	Upper Base	Lower Base	Height	Area
1.	5 in	17 in	6 in	?
2.	4 cm	6 cm	9 cm	?
3.	12 yd	8 yd	4 yd	?
4.	3 m	11 m	7 m	?

	Upper Base	Lower Base	Height	Area
5.	8.3 m	9.5 m	6.4 m	?
6.	$15\frac{1}{2}$ ft	$12\frac{1}{2}$ ft	13 ft	?
7.	7.8 cm	4.7 cm	6 cm	?
8.	$9\frac{1}{4}$ in	$11\frac{3}{4}$ in	4 in	?

Solving Equations: ADDITION

(Pages 210–211)

Solve and check each equation.

1. $n + 7 = 18$ **2.** $n + 5 = 16$ **3.** $n + 21 = 42$ **4.** $n + 2 = 81$

5. $n + 19 = 57$ **6.** $n + 37 = 42$ **7.** $n + 3.6 = 9.8$ **8.** $n + \frac{3}{10} = \frac{9}{10}$

Solving Equations: SUBTRACTION

(Pages 212–213)

Solve and check each equation.

1. $n - 17 = 5$ **2.** $n - 1 = 38$ **3.** $n - 43 = 18$ **4.** $n - 9 = 10$

5. $n - 12 = 75$ **6.** $n - 10 = 11$ **7.** $n - 2 = 8.3$ **8.** $n - \frac{1}{6} = \frac{1}{3}$

Solving Equations: MULTIPLICATION/DIVISION

(Pages 216–217)

Solve and check.

1. $7n = 343$ **2.** $8n = 128$ **3.** $12n = 96$ **4.** $5n = 225$

5. $\frac{n}{14} = 2$ **6.** $\frac{n}{16} = 9$ **7.** $1.9n = 5.7$ **8.** $\frac{n}{72} = 2.4$

More on Solving Equations

(Pages 221–222)

Solve and check.

1. $3n + 5 = 56$ **2.** $4n + 11 = 55$ **3.** $5n + 6 = 36$

4. $8n - 4 = 36$ **5.** $2n - 11 = 17$ **6.** $3n - 17 = 13$

7. $\frac{n}{6} - 3 = 8$ **8.** $\frac{n}{8} - 6 = 1$ **9.** $\frac{n}{7} - 3 = 7$

10. $\frac{n}{4} + 6 = 15$ **11.** $\frac{n}{3} + 7 = 10$ **12.** $\frac{n}{7} + 5 = 12$

Proportion

(Pages 228–229)

Solve each proportion for n.

1. $\frac{n}{8} = \frac{3}{6}$ **2.** $\frac{n}{20} = \frac{3}{12}$ **3.** $\frac{3}{11} = \frac{9}{n}$ **4.** $\frac{14}{6} = \frac{7}{n}$ **5.** $\frac{1}{6} = \frac{n}{30}$

6. $\frac{60}{n} = \frac{4}{3}$ **7.** $\frac{27}{n} = \frac{9}{7}$ **8.** $\frac{4}{n} = \frac{2}{5}$ **9.** $\frac{6}{35} = \frac{12}{n}$ **10.** $\frac{3}{4} = \frac{n}{36}$

Percents/Decimals/Fractions

(Pages 246—247)

Write a percent for each fraction.

1. $\frac{2}{5}$ 2. $\frac{37}{50}$ 3. $\frac{3}{4}$ 4. $\frac{19}{20}$ 5. $\frac{9}{20}$ 6. $\frac{8}{25}$

7. $\frac{7}{10}$ 8. $\frac{3}{5}$ 9. $\frac{1}{2}$ 10. $\frac{2}{25}$ 11. $\frac{1}{20}$ 12. $\frac{18}{50}$

13. $\frac{7}{40}$ 14. $\frac{5}{16}$ 15. $\frac{5}{9}$ 16. $\frac{3}{40}$ 17. $\frac{15}{16}$ 18. $\frac{11}{12}$

19. $\frac{2}{3}$ 20. $\frac{7}{12}$ 21. $\frac{5}{6}$ 22. $\frac{1}{15}$ 23. $\frac{3}{8}$ 24. $\frac{9}{16}$

Finding a Percent of a Number

(Pages 250—251)

Find each answer.

1. 12% of 250   2. 25% of 84   3. 9% of 800   4. 3% of 400

5. 1% of 650   6. 10% of 76   7. 20% of 90   8. 60% of 130

9. $87\frac{1}{2}$% of 104   10. $33\frac{1}{3}$% of 69   11. $37\frac{1}{2}$% of 80   12. $16\frac{2}{3}$% of 150

Discount

(Pages 252—253)

Solve.

1. A camera listed at $95 was sold at a discount of 15%. What was the selling price?

2. A toaster with a regular price of $28.70 was sold at a 40% discount. What was the selling price?

3. Dan Woods bought a shirt at a 20%-off sale. The original price was $11.50. How much did Dan pay?

4. A rug-cleaning company offered a 20% discount on all cleaning done in April. How much would it cost to clean a rug when the regular price for cleaning is $28?

Savings Account/Interest

(Pages 256—257)

Find the interest.

	Principal	Rate	Time		Principal	Rate	Time
1.	$800	6%	9 months	4.	$1000	9%	3 months
2.	$1200	8%	4 months	5.	$2400	5%	2 months
3.	$900	8%	6 months	6.	$3200	6%	6 months

Finding the Percent

(Pages 268—269)

Solve.

1. What percent of 70 is 7?

2. What percent of 64 is 16?

3. What percent of 45 is 30?

4. What percent of 4 is 16?

5. What percent of 8 is 10?

6. What percent of 80 is 16?

7. Twelve is what percent of 15?

8. Thirteen is what percent of 65?

9. Twenty-four is what percent of 64?

10. Fifteen is what percent of 60?

11. Eighty-four is what percent of 42?

12. Ninety-five is what percent of 76?

13. What percent of 80 is 25?

14. Nine is what percent of 60%

15. What percent of 30 is 5?

16. Twenty-eight is what percent of 8?

17. Six is what percent of 72?

18. What percent of 108 is 90?

19. Eight is what percent of 64?

20. What percent of 90 is 135?

Finding a Number Given the Percent

(Pages 272—273)

Find the number.

1. 13 is 25% of what number?

2. 8 is 20% of what number?

3. 68 is 85% of what number?

4. 84 is 30% of what number?

5. 24 is 15% of what number?

6. 74 is 37% of what number?

7. 21 is 60% of what number?

8. 46 is 20% of what number?

9. 64 is 40% of what number?

10. 18 is 24% of what number?

11. 35 is 50% of what number?

12. 6 is 12% of what number?

13. 17 is 1% of what number?

14. 5 is 2% of what number?

15. 38 is 10% of what number?

16. 28 is 40% of what number?

Surface Area

(Pages 300–302)

Find the surface area of each rectangular prism.

1. File Cabinet

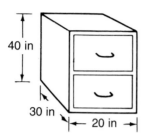

2. Box

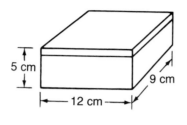

3. Refrigerator

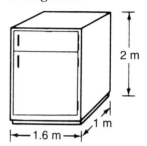

4. A box is 24 centimeters long, 18 centimeters wide, and 12 centimeters high. How many square centimeters of cardboard are needed to make 12 boxes?

5. A room is 5 meters long, 4 meters wide, and 2.8 meters high. How many square meters of wallpaper are needed to cover the four walls?

Volume: RECTANGULAR PRISMS

(Pages 304–305)

Find the volume.

1. Storage Bin

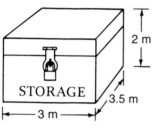

2. Aquarium

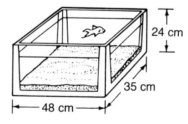

3. Suitcase

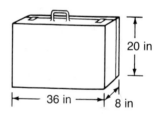

4. Box of Cereal

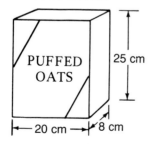

5. Desk Drawer

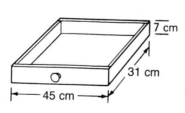

6. Wall Oven

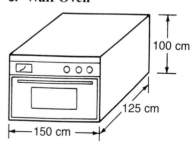

7. A swimming pool 33 meters long and 10 meters wide is filled to a depth of 1.4 meters. What is the volume of the water in the pool?

8. A freight car is 14 meters long, 3.2 meters wide and 3.2 meters high. Find the volume of the freight car.

Circumference and Applications

(Pages 319–320)

Find the circumference. Round each answer to the nearest whole number.

1. $r = 70$ yd

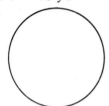

2. $d = 7$ ft

3. $r = 25.5$ cm

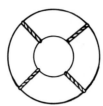

4. $d = 1.2$ m

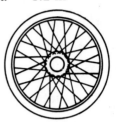

Circle Graphs

(Pages 323–324)

Draw a circle graph to show the data.

1.

Vehicles Sold by One Dealer

Type	Compact	Standard	Vans	Trucks
Percent	45%	25%	20%	10%

2.

Family Budget

Item	Shelter	Food	Savings	Other
Percent	25%	30%	5%	40%

Area and Applications

(Pages 330–331)

Find the area. Round each number to the nearest whole number.

1. $r = 6$ in

2. $r = 7\frac{1}{2}$ ft

3. $r = 15$ cm

4. $r = 10.4$ cm

Volume and Applications: CYLINDERS

(Pages 332–333)

Find the volume. Round each answer to the nearest whole number.

1. $r = 2$ ft;
  $h = 3$ ft

2. $r = 2$ in;
  $h = 4$ in

3. $r = 3.5$ cm;
  $h = 10$ cm

4. $r = 5.1$ m;
  $h = 3$ m

The Probability Ratio

(Pages 350–351)

The pointer on the spinner is spun once. Write each probability as a fraction in lowest terms and as a percent.

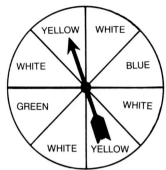

1. The pointer stops on white.

2. The pointer stops on green.

3. The pointer stops on yellow.

4. The pointer stops on blue.

5. The pointer doesn't stop on white.

6. The pointer doesn't stop on yellow.

Compound Events: AND

(Pages 354–355)

The pointers on two spinners are each spun once. The table lists all the possible ways in which the arrows can end up. Write each probability as a fraction in lowest terms.

	Second Spinner				
First Spinner	1	2	3	4	5
A	A1	A2	A3	A4	A5
B	B1	B2	B3	B4	B5
C	C1	C2	C3	C4	C5

1. Spinning B <u>and</u> 4

2. Spinning A <u>and</u> 5

3. Spinning C <u>and</u> an odd number

4. Spinning A <u>and</u> a number less than 4

5. Spinning B <u>and</u> a number less than 5

6. Spinning C <u>and</u> a number greater than 0

7. Spinning A <u>and</u> an even number

Compound Events: OR

(Pages 356–357)

In a game, each player spins the arrow twice and adds the score. The table shows the possible sums. Find each probability.

	Second Spin				
+	2	4	6	8	10
2	4	6	8	10	12
4	6	8	10	12	14
6	8	10	12	14	16
8	10	12	14	16	18
10	12	14	16	18	20

(First Spin)

1. Spinning a sum of 4 <u>or</u> a sum of 14

2. Spinning a sum of 8 <u>or</u> a sum of 10

3. Spinning a sum of 12 <u>or</u> a sum of 20

4. Spinning a sum of 6 <u>or</u> a sum of 18

Using a Table of Squares and Square Roots   (Pages 370–372)

Use the table on page 371 to find each square or square root.

1. 31^2 **2.** 131^2 **3.** 53^2 **4.** 143^2 **5.** 95^2 **6.** 19^2

7. 126^2 **8.** 35^2 **9.** 59^2 **10.** 113^2 **11.** 71^2 **12.** 27^2

13. $\sqrt{37}$ **14.** $\sqrt{83}$ **15.** $\sqrt{4761}$ **16.** $\sqrt{125}$ **17.** $\sqrt{576}$ **18.** $\sqrt{8649}$

19. $\sqrt{6561}$ **20.** $\sqrt{139}$ **21.** $\sqrt{1156}$ **22.** $\sqrt{78}$ **23.** $\sqrt{6084}$ **24.** $\sqrt{119}$

25. $\sqrt{8}$ **26.** $\sqrt{1936}$ **27.** $\sqrt{102}$ **28.** $\sqrt{14{,}161}$ **29.** $\sqrt{19}$ **30.** $\sqrt{1369}$

The Rule of Pythagoras   (Pages 373–374)

Use the Rule of Pythagoras to determine which triangles are right triangles. Write Yes or No.

1. 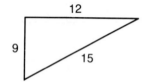   **2.** 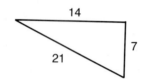   **3.**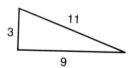

4. 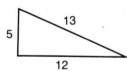   **5.** 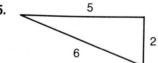   **6.**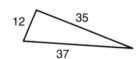

Tangent Ratios and Applications   (Pages 383–385)

For Exercises 1–12, use the Table of Tangents on page 383. Find each of the following.

1. $\tan 15°$ **2.** $\tan 70°$ **3.** $\tan 40°$ **4.** $\tan 75°$ **5.** $\tan 35°$

Solve. Round each answer to the nearest whole number.

6. $\tan 15° = \frac{n}{25}$ **7.** $\tan 75° = \frac{n}{12}$ **8.** $\tan 5° = \frac{n}{35}$ **9.** $\tan 50° = \frac{n}{50}$

Find the length of each side. Round the answers to the nearest whole number.

10. 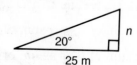   **11.** 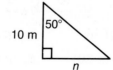   **12.**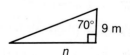

Adding Integers: LIKE SIGNS

(Pages 399-400)

Use a number line to add.

1. $2 + 4$ **2.** $5 + 5$ **3.** $9 + 3$ **4.** $5 + 0$

5. $10 + 4$ **6.** $5 + 3$ **7.** $0 + 12$ **8.** $8 + 4$

9. $-2 + (-2)$ **10.** $-4 + (-5)$ **11.** $-5 + (-4)$ **12.** $-4 + (-3)$

13. $-3 + (-3)$ **14.** $-2 + 0$ **15.** $0 + (-8)$ **16.** $-6 + (-3)$

17. $0 + (-5)$ **18.** $-8 + (-2)$ **19.** $-7 + (-4)$ **20.** $-2 + (-6)$

Adding Integers: UNLIKE SIGNS

(Pages 402-403)

Use a number line to add.

1. $-9 + 3$ **2.** $-10 + 5$ **3.** $-4 + 6$ **4.** $-3 + 9$

5. $-8 + 9$ **6.** $-12 + 7$ **7.** $-7 + 11$ **8.** $-5 + 10$

9. $6 + (-4)$ **10.** $7 + (-11)$ **11.** $3 + (-2)$ **12.** $7 + (-7)$

13. $8 + (-8)$ **14.** $8 + (-6)$ **15.** $1 + (-8)$ **16.** $5 + (-9)$

17. $-3 + 12$ **18.** $-10 + 13$ **19.** $13 + (-9)$ **20.** $13 + (-6)$

Subtracting Integers

(Pages 406-407)

Use a number line to subtract.

1. $4 - 7$ **2.** $5 - 7$ **3.** $-2 - (-3)$ **4.** $-3 - (-8)$

5. $-7 - 1$ **6.** $-6 - 7$ **7.** $12 - 7$ **8.** $9 - 11$

9. $-4 - 9$ **10.** $-3 - 4$ **11.** $-15 - (-7)$ **12.** $-9 - (-11)$

13. $0 - 3$ **14.** $0 - 6$ **15.** $0 - (-2)$ **16.** $0 - (-15)$

17. $-4 - (-6)$ **18.** $-3 - (-13)$ **19.** $5 - (-1)$ **20.** $7 - (-12)$

21. $14 - 9$ **22.** $12 - 9$ **23.** $10 - (-10)$ **24.** $9 - (-7)$

25. $-5 - 6$ **26.** $-9 - 8$ **27.** $-1 - (10)$ **28.** $-5 - (-1)$

29. $-7 - 0$ **30.** $-12 - 0$ **31.** $-8 - (-4)$ **32.** $-6 - (-8)$

Add or subtract. Use a number line.

33. $6 + 5$ **34.** $4 - 1$ **35.** $4 + (-7)$ **36.** $-5 - 12$

37. $-9 + 9$ **38.** $-3 + 0$ **39.** $2 - 12$ **40.** $-3 - 6$

41. $9 + (-5)$ **42.** $-6 + (-9)$ **43.** $6 - 11$ **44.** $7 - (-4)$

45. $6 - 3$ **46.** $11 - 5$ **47.** $4 + (-15)$ **48.** $-15 + (-4)$

Multiplication: PATTERNS AND UNLIKE SIGNS (Pages 422–423)

Multiply.

1. -5×3

2. -7×16

3. $4 \times (-19)$

4. $5 \times (-6)$

5. $12 \times (-5)$

6. $0 \times (-33)$

7. -100×8

8. $5 \times (-17)$

9. -35×2

10. $7 \times (-3)$

11. -12×9

12. $15 \times (-15)$

Use positive and negative integers to represent each situation by a multiplication problem. Then find the product.

13. Five times a drop in temperature of 6°.

14. Three times a debt of $42 each.

Multiplication: PATTERNS AND LIKE SIGNS   (Pages 424–425)

Multiply.

1. $-2 \times (-31)$

2. 6×9

3. $-32 \times (-56)$

4. 5×14

5. $-12 \times (-12)$

6. $-7 \times (-25)$

7. 14×23

8. $-19 \times (-100)$

9. $-16 \times (-50)$

10. $-105 \times (-6)$

11. $-34 \times (-200)$

12. $70 \times (-36)$

13. A river flows at a rate of 14 feet per minute. How many feet will it flow in 15 minutes? **210**

14. The temperature fell 4° an hour for 3 hours. How much did the temperature fall in all?

Dividing Integers   (Pages 426–427)

Divide.

1. $25 \div (-5)$

2. $32 \div 4$

3. $-60 \div 5$

4. $81 \div (-9)$

5. $-84 \div 7$

6. $130 \div (-26)$

7. $48 \div (-3)$

8. $-72 \div (-12)$

9. $-225 \div (-15)$

10. $-200 \div (-40)$

11. $558 \div (-84)$

12. $57 \div (-3)$

13. One day the low temperature is $-8°C$ and the high temperature is 12°C. What is the mean temperature?

14. Low temperatures for five days are $-5°C$, 0°C, $-3°C$, $-8°C$, and 1°C. What is the average temperature?

Using the Calculator

A calculator is a tool that you must learn to use properly. Just as there are different kinds of tools, there are different calculators with different keys and different modes of operation. Here are some ideas for working with an unfamiliar calculator.

> **a.** Carefully read the instruction booklet.
>
> **b.** Estimate each answer to be sure the calculator's result makes sense.
>
> **c.** Experiment with solving problems in different ways.

Don't be afraid to experiment with different sequences of keys.

If the calculator has these keys:	To clear the last number entered, press:	To clear the entire problem, press:
C and CE	CE	C
CE / C	CE / C	CE/C CE/C
ON/C	ON/C	ON/C ON/C

The following examples show how to correct entries.

A. Correcting a number that has been entered incorrectly

Add: $235 + 168 =$ ___?___

235 [+] 167 [CE] 168 [=] 403.

B. Clearing all the entries

Add: $235 + 168 =$ ___?___

233 [+] 168 [C] 235 [+] 168 [=] 403.

C. Correcting an operation that has been entered incorrectly

Add: $235 + 168 =$ ___?___

235 [×] [+] 168 [=] 403.

Addition/Subtraction: WHOLE NUMBERS

When adding on a calculator you may become careless and enter a wrong number. It is important to **estimate** the sum before you add.

EXAMPLE 1

$8,417 + 9,205 + 1,875 = \underline{\ ?\ }$

Round to the nearest thousand.

$$
\begin{array}{rcl}
8,417 & \longrightarrow & 8,000 \\
9,205 & \longrightarrow & 9,000 \\
+\,1,875 & \longrightarrow & 2,000 \\
\end{array}
$$

> *To estimate a sum, round each addend. Then add.*

Estimated sum: 19,000

Press: 8417 [+] 9205 [+] 1875 [=] $19497.$

> *Commas are not entered.*

The sum is **19,497.** The answer is close to the estimate.

When subtracting you should always **check** your answer.

EXAMPLE 2

In 1970 the population of Pine Falls was 79,084. By 1980 the population had grown to 82,321. What was the increase?

Press: 82321 [−] 79084 [=] $3237.$

To check your answer, add.

Check: 3237 [+] 79084 [=] $82321.$

The population increased by **3,237.**

TRY THESE

First estimate the answer. Then find the exact answer. Compare the answer with the estimate to see if it is reasonable.

1. One year 436,725 personal computers were sold. The next year 1,217,841 were sold. How many computers were sold?

2. Carla says she is 6,307,200 minutes old. Todd is only 6,044,400 minutes old. How many minutes older is Carla?

3. One year a band sold 1,420,760 records. The next year the band sold 884,912 records. How many records were sold in the two years?

An Addition/Subtraction Shortcut

You can add or subtract a number repeatedly by entering that number, called a **constant,** into the calculator only once.

Problem: 45 + 15 + 15 + 15 + 15 = __?__

45 ⊞ 15 ⊟ 60. ⊟ 75. ⊟ 90. ⊟ 105.

Problem: 360 − 30 − 30 − 30 − 30 = __?__

360 ⊟ 30 ⊟ 330. ⊟ 300. ⊟ 270. ⊟ 240.

TRY THESE

1. Marva Johnson is a research scientist. She recently received a grant to help her finish a project. The terms of the grant entitle her to an initial payment of $5,000 and additional payments of $2,500 a year for 5 years. How much money will she receive under this grant?

2. The Oceangate Library has a collection of 36,489 books. The town council voted to increase the library's budget. The library will now be able to buy 725 new books each year. How many books will be in the library's collection after 6 years?

3. A neighborhood group of volunteers opened a soup kitchen to help feed needy people in their section of the city. When the soup kitchen opened, there were 2,800 cans of stew in the storeroom. The cooks estimated that about 220 cans will be used each week. About how many cans of stew will be in the storeroom after 8 weeks?

4. Yoshi Keto is a computer programmer. His employer offered him a 5-year contract. According to the contract, his salary during the first year will be $25,000. During the next four years, his salary will increase by $2,750 a year. If Yoshi accepts the contract, what will his annual salary be during the fifth year of the contract?

5. The AQ Comp Manufacturing Company announced that it was accepting advance orders for its new line of minicomputers. The company expects to produce 15,000 computers within the next three months. During the first week after the announcement, the company received orders for 1,685 computers. If orders are received at the same rate each week, how many computers will the company have left to sell after 7 weeks?

CALCULATOR MANUAL

Using the Calculator Memory

Many calculators have a **memory.** The number in the memory is controlled by using either three or four keys, depending on the calculator.

If the calculator has this key:	the key:
M+ *or* M±	Adds the displayed number to the number in the memory.
M- *or* M_	Subtracts the displayed number from the number in the memory.
RM *or* MR	Recalls the number in the memory for display or use.
CM *or* MC	Clears (erases) the number in the memory.
RM/CM *or* RCM *or* RMC	When you press this key once, it recalls the number from the memory. When you press it twice, it clears the number in the memory.

When a number is stored in the memory, a small ^{M} is displayed to remind you. Before starting a new problem, always press [CM] and [C] to clear the calculator.

EXAMPLE

Mrs. Weiss bought 6 bars of soap at $0.39 each and 3 boxes of cereal at $1.19 each. How much money did she spend? How much change did she receive from a $20.00 bill?

[CM] [C]	Make sure the memory is clear.
6 [×] .39 [=] [M+] 2.34^m	Find the cost of the soap ($2.34) and save it in the memory.
3 [×] 1.19 [=] [M+] 3.57^m	Find the cost of the cereal ($3.57) and add it to the cost of the soap stored in the memory.
[RM] 5.91^m	Display the amount spent ($5.91).
[2] [0] [−] [RM] [=] 14.09^m [CM]	Subtract the amount spent from $20.00.

Mrs. Weiss spent **$5.91** and received **$14.09** in change.

TRY THESE *Solve.*

1. Alfredo purchased 3 pads at $0.79 each, and 5 blank cassettes at $2.29 each. What did he spend? How much change did he receive from a $20 bill?

2. A salesperson sold 2 typewriters at $189.95 each, 7 calculators at $9.99 each, and a camera for $139.49. What were the total sales?

Order of Operations/Powers

Different answers can be obtained for the same problem, especially when a calculator is used. Consider the problem $700 - 200 \times 3$.

On many calculators: 700 $\boxed{-}$ 200 $\boxed{\times}$ 3 $\boxed{=}$ 1500. **Which is correct?**

On some calculators: 700 $\boxed{-}$ 200 $\boxed{\times}$ 3 $\boxed{=}$ 100.

Mathematicians have agreed upon the **order of operations:**

> **a.** Perform operations within parentheses first.
> **b.** Multiply and divide, in order from left to right.
> **c.** Add and subtract, in order from left to right.

Therefore, $700 - 200 \times 3 = 700 - 600$, or **100.**

If you follow the rules for order of operations, you will get the correct answer no matter which type of calculator you use.

EXAMPLE 1 $25 \times 35 + (400 - 125) = \underline{\ ?\ }$

a. Within parentheses. 400 $\boxed{-}$ 125 $\boxed{=}$ $\boxed{M+}$ 275.

b. Multiply. Then add. 25 $\boxed{\times}$ 35 $\boxed{=}$ 875. $\boxed{+}$ $\boxed{RM}$ $\boxed{=}$ 1150.ᵐ

c. Press $\boxed{CM}$. The result is **1,150.**

EXAMPLE 2 $9^3 = \underline{\ ?\ }$

a. Method 1

9 $\boxed{\times}$ 9 $\boxed{\times}$ 9 $\boxed{=}$ 729.

b. Method 2 (Scientific Calculator)

9 $\boxed{y^x}$ 3 $\boxed{=}$ 729.

TRY THESE

Use your calculator. Remember to clear the memory after each problem.

1. $639 - 41 \times 9$
2. $123 + 18 \times 23$
3. $(947 + 787) \div (120 - 86)$
4. $(82 - 58) \times (96 + 47)$
5. 15^4
6. 6^5
7. 5^6
8. 12^3

Addition/Subtraction: DECIMALS

The calculator automatically places the decimal point in the answer. It is easy to make an error in entering or reading a decimal point. It is important that you estimate first to see that the answer displayed makes sense.

EXAMPLE 1

6.875 + 3.25 + 4.1875 = ?

Estimate: Round to the nearest whole number: 7 + 3 + 4 = **14**

6.875 $\boxed{+}$ 3.25 $\boxed{+}$ 4.1875 = 14.3125

The answer, **14.3125**, is close to the estimate of 14.

EXAMPLE 2

215.4 − 92.625 = ?

Estimate: Round to the nearest ten: 220 − 90 = **130**

215.4 $\boxed{-}$ 92.625 $\boxed{=}$ 122.775

The answer, **122.775**, is close to the estimate of 130.

When a problem involving decimals has too many digits for the calculator, the digits farthest to the right are dropped or rounded off. There is no error signal to show that digits have been lost.

EXAMPLE 3

50,000,000 + 215.375 = ?

Think: The sum is 50,000,215.375.

50000000 $\boxed{+}$ 215.375 $\boxed{=}$ $50000215.$

The calculator displays an answer that is only approximate.

TRY THESE

First estimate the answer. Then find the exact answer. Compare the answer with the estimate to see if it is reasonable.

1. 94.7 + 8.85 + 46.3

2. 113.7 − 49.85

3. 139.8 + 86.95 + 62.4

4. 72.3 − 38.675

5. 7.875 + 13.375 + 9.5

6. 400 − 129.63

7. $426,723.19 + $217,406.58

8. $600,000 − $139,416.27

Multiplication and Estimation:
WHOLE NUMBERS/DECIMALS

You can tell whether the answer displayed by the calculator makes sense by estimating the answer first.

EXAMPLE 1

$539 \times 78 = \underline{\quad ? \quad}$

Think: Round 539 to 500. Round 78 to 80. $500 \times 80 = \mathbf{40,000}$

$539 \boxed{\times} 78 \boxed{=} \; \text{42042.}$

The estimate of 40,000 suggests that the product **42,042** is correct.

EXAMPLE 2

$387 \times 6.2 = \underline{\quad ? \quad}$

Think: Round 387 to 400. Round 6.2 to 6. $400 \times 6 = \mathbf{2,400}$

$387 \boxed{\times} 6.2 \boxed{=} \; \text{2399.4}$

The estimate of 2,400 suggests that the product **2,399.4** is correct.

TRY THESE

First estimate the answer. Then find the exact answer.
Compare the answer with the estimate to see if it is reasonable.

	Estimate	Product		Estimate	Product
1. 91×87	?	?	**13.** 83×7.9	?	?
2. 51×52	?	?	**14.** 62×6.1	?	?
3. 28×33	?	?	**15.** 34×8.9	?	?
4. 77×61	?	?	**16.** 71×5.2	?	?
5. 68×84	?	?	**17.** 58×3.8	?	?
6. 83×219	?	?	**18.** 82×41.5	?	?
7. 58×643	?	?	**19.** 91×63.3	?	?
8. 72×307	?	?	**20.** 21×30.8	?	?
9. $66 \times 1,259$	?	?	**21.** $1,126 \times 6.7$	?	?
10. $68 \times 3,269$	?	?	**22.** $5,231 \times 5.7$	?	?
11. 228×684	?	?	**23.** 318×68.4	?	?
12. $26 \times 4,349$	?	?	**24.** 709×42.6	?	?

Multiplying Large Numbers: OVERFLOW

When the answer to a problem has too many digits to be displayed, the calculator **overflows.**

EXAMPLE 1

$52,645 \times 7,983 = \underline{\ ?\ }$

$$52645 \boxed{\times} 7983 \boxed{=} \ 4.2026503_E$$

The E indicates that the entire answer has not been displayed. The decimal point helps to tell how many digits are missing. There is one digit before the decimal. So one digit is missing.

Look at the original problem. The first factor ends in 5. The second factor ends in 3. So $5 \times 3 = 15$. Therefore, the final digit of the answer must be a 5.

$$52,645 \times 7,983 = \textbf{420,265,035}$$

Now press $\boxed{C}$ to clear the calculator.

EXAMPLE 2

$98,542 \times 81,086 = \underline{\ ?\ }$

$$98542 \boxed{\times} 81086 \boxed{=} \ 79.903766_E.$$

There are two digits before the decimal. So two digits are missing.

Think: The last two digits of the product depend on the last two digits of each factor.

$$\boxed{C} \ 42 \boxed{\times} 86 \boxed{=} \ 3612.$$

The last two digits of this product are 1 and 2.

$$98,542 \times 81,086 = \textbf{7,990,376,612}$$

TRY THESE

1. $13,407 \times 9,526$
2. $37,512 \times 3,589$
3. $61,039 \times 4,203$
4. $53,826 \times 7,816$
5. $43,609 \times 4,933$
6. $92,125 \times 6,515$
7. $51,439 \times 24,302$
8. $82,065 \times 46,127$
9. $29,842 \times 38,094$
10. $43,663 \times 63,693$
11. $58,391 \times 42,107$
12. $64,089 \times 29,312$
13. $111,111 \times 11,111$
14. $412,205 \times 16,519$
15. $653,417 \times 13,921$

Multiplication: DECIMALS

A calculator multiplies decimals as easily as it multiplies whole numbers. The result displayed may be approximate, or the calculator may **underflow** and display a zero, without any indication or error signal.

EXAMPLE 1 Find when the calculator:
   a. Displays an approximate product;
   b. Underflows and displays a zero.

.025 ⊠ .035 ⊟ **0.000875**

.0025 ⊠ .035 ⊟ **0.0000875**

.0025 ⊠ .0035 ⊟ **0.0000087** ◀ *When the product requires more than eight digits, an approximate answer is displayed.*

.00025 ⊠ .0035 ⊟ **0.0000008**

.00025 ⊠ .00035 ⊟ **0.** ◀ *When the product is less than 0.0000001, a nonscientific calculator underflows and displays 0.*

EXAMPLE 2 0.00025 × 0.00034

Think: The product will have 10 decimal places.

25 ⊠ 34 ⊟ **850.** ◀ *When the calculator underflows, this method will often give the product.*

Write: 0.0000000850 or 0.000000085

TRY THESE
First estimate the answer. Then find the exact answer. Compare the answer with the estimate to see if it is reasonable.

1. 46.4×36.25

2. 31.75×6.64

3. 78.35×11.8

4. 5.375×49.6

5. 631.4×916.3

6. 147.8×802.6

7. 925.73×412.9

8. 302.75×98.43

9. 692×723.85

10. 47.88×531.25

11. 0.09×0.36

12. 0.285×0.4

13. 0.0045×0.063

14. 0.175×0.0049

15. 2.0083×1.0092

16. 0.64175×0.55413

17. 0.00017×0.00039

18. 0.00023×0.00021

Division: WHOLE NUMBERS/DECIMALS

When you are using a calculator to do a division problem, remember that the dividend must always be entered first.

EXAMPLE 1 Janine drove 255 miles in 4.25 hours. Find the speed of the automobile in miles per hour.

Think: 255 ÷ 4.25   **Estimate:** 260 ÷ 4 = 65

255 $\boxed{÷}$ 4.25 $\boxed{=}$ **60.**   Janine drove at a rate of 60 miles per hour.

Multiplication and division are **inverse operations.**
Use multiplication to **check** the answer to a division problem.

Think: 60 miles per hour for 4.25 hours

60 $\boxed{×}$ 4.25 $\boxed{=}$ **255.**   The answer checks.

Solving problems on a calculator sometimes involves difficulties.

EXAMPLE 2 Fred has 81 books to put in boxes. He puts one dozen books into each box. How many boxes does he fill? How many books are left over?

81 $\boxed{÷}$ 12 $\boxed{=}$ **6.75**   **Think:** The display shows that Fred can fill 6 boxes. But the remainder has been expressed as a decimal. It does not tell how many books are left over.

Here are two ways to find the remainder.

12 $\boxed{×}$ .75 $\boxed{=}$ **9.**   1. Multiply the decimal part of the answer, 0.75, by the divisor, 12.

12 $\boxed{×}$ 6 $\boxed{=}$ **72.**   2. Multiply the whole-number part of the answer, 6, by the divisor, 12. Then
81 $\boxed{-}$ 72 $\boxed{=}$ **9.**   subtract the result from the dividend.

Both methods show that Fred has **9** books left over.

TRY THESE
First estimate the answer. Then find the exact answer.
Compare the answer with the estimate to see if it is reasonable.

1. 8.6) 51,944 ·   **2.** 24.68) 21,328,456   **3.** 24) 150   **4.** 7380 ÷ 75

Fractions and Decimals

On most calculators fractions and mixed numbers must be expressed in decimal form.

To indicate a **repeating decimal,** draw a bar over the digit(s) that repeat.

EXAMPLE 1

a. Find the decimal for $\frac{5}{12}$.

$$5 \boxed{\div} 12 \boxed{=} \; 0.4166666$$

$\frac{5}{12} = 0.4166666$ or $0.41\overline{6}$

A calculator that rounds may display 0.4166667.

b. Find the decimal for $9\frac{7}{16}$.

Think: $9\frac{7}{16} = \frac{7}{16} + 9$

$$7 \boxed{\div} 16 \boxed{+} 9 \boxed{=} \; 9.4375$$

$9\frac{7}{16} = 9.4375$

EXAMPLE 2

a. Write the decimal for $\frac{11}{15}$.

$$11 \boxed{\div} 15 \boxed{=} \; 0.7333333$$

$\frac{11}{15} \boxed{=} \; 0.7\overline{3}$

b. Write the decimal for $5\frac{3}{11}$.

$$3 \boxed{\div} 11 \boxed{+} 5 \boxed{=}$$

$$5.2727272$$

A calculator that rounds may display 5.2727273.

$5\frac{3}{11} = 5.\overline{27}$

TRY THESE   *Find the decimal for each fraction.*

1. $\frac{31}{80}$
2. $\frac{15}{16}$
3. $\frac{113}{160}$
4. $21\frac{35}{56}$
5. $\frac{23}{40}$
6. $57\frac{5}{32}$

7. $\frac{157}{200}$
8. $\frac{29}{116}$
9. $4\frac{307}{500}$
10. $48\frac{58}{87}$
11. $\frac{101}{128}$
12. $\frac{91}{143}$

13. $\frac{59}{64}$
14. $17\frac{56}{111}$
15. $\frac{14}{15}$
16. $31\frac{13}{64}$
17. $\frac{117}{143}$
18. $92\frac{113}{200}$

19. $\frac{19}{30}$
20. $11\frac{37}{40}$
21. $\frac{196}{252}$
22. $104\frac{13}{16}$
23. $\frac{119}{280}$
24. $58\frac{313}{400}$

25. $\frac{148}{407}$
26. $\frac{236}{649}$
27. $\frac{468}{1,287}$
28. $5\frac{35}{112}$
29. $5\frac{195}{624}$
30. $5\frac{1}{32}$

Proportions

Two **equivalent ratios** can be written to form a true **proportion.** When a proportion is true, its cross products are equal.

EXAMPLE 1

Does $\frac{39}{26} = \frac{51}{34}$?

$$26 \boxed{\times} 51 \boxed{=} 1326.$$

$$39 \boxed{\times} 34 \boxed{=} 1326.$$

The ratios are equivalent.

A missing term of a proportion can be found if the other three terms are known.

EXAMPLE 2

Solve for n. $\quad \frac{9}{n} = \frac{57}{152}$

Think: Find the cross products.

$$9 \times 152 = n \times 57$$

$$\frac{9 \times 152}{57} = \frac{n \times \overset{1}{\cancel{57}}}{\cancel{57}_{1}}$$

$$\frac{9 \times 152}{57} = n$$

$$9 \boxed{\times} 152 \boxed{\div} 57 \boxed{=} 24.$$

Therefore $n = $ **24.**

TRY THESE

Is it a true proportion? Write yes *or* no.

1. $\frac{28}{42} = \frac{38}{57}$ **2.** $\frac{51}{68} = \frac{213}{284}$ **3.** $\frac{52}{65} = \frac{92}{125}$ **4.** $\frac{46}{73.6} = \frac{37.5}{60}$ **5.** $\frac{22}{49.5} = \frac{26}{59.5}$

Solve each proportion for n. Use the cross products to check.

6. $\frac{52}{65} = \frac{92}{n}$ **7.** $\frac{68}{153} = \frac{n}{135}$ **8.** $\frac{n}{91} = \frac{152}{133}$ **9.** $\frac{161}{n} = \frac{98}{112}$ **10.** $\frac{84}{112} = \frac{n}{156}$

11. $\frac{69}{n} = \frac{138}{276}$ **12.** $\frac{57}{9} = \frac{n}{57}$ **13.** $\frac{12}{n} = \frac{0.25}{12}$ **14.** $\frac{n}{49.6} = \frac{42.5}{68}$ **15.** $\frac{92.5}{222} = \frac{48}{n}$

16. $\frac{208}{260} = \frac{n}{123}$ **17.** $\frac{12.6}{33.6} = \frac{6.375}{n}$ **18.** $\frac{48}{51.2} = \frac{n}{56}$ **19.** $\frac{n}{16} = \frac{14.1}{60}$ **20.** $\frac{35.6}{n} = \frac{88}{211.2}$

Percent

Many calculators have a percent key, but they do not all work the same way. It is important to estimate the answer and then check to see that the display makes sense. Percent problems can be solved by using decimals instead of the percent key. (See Method 4.)

EXAMPLE 1 How much is 45% of 780?

Estimate: 45% is close to one half. One half of 800 is 400. The answer is less than 400.

Method 1 780 [×] 45 [%] 351.

Method 2 780 [×] 45 [%] [=] 351.

Find which methods work on your calculator.

Method 3 45 [%] [×] 780 [=] 351.

Method 4 Remember: 45% = 0.45

.45 [×] 780 [=] 351. Therefore, 45% of 780 is **351.**

EXAMPLE 2 A family with weekly take-home pay of $250 budgets these amounts.

Rent: $100	Carfare: $10	Food: $80
Clothes: $20	Savings: $15	Other: $25

What percent is budgeted for each?

Ignore the [=] if the calculator you are using does not need it.

100 [÷] 250 [%] [=] 40. 10 [÷] 250 [%] [=] 4. 80 [÷] 250 [%] [=] 32.

20 [÷] 250 [%] [=] 8. 15 [÷] 250 [%] [=] 6. 25 [÷] 250 [%] [=] 10.

The family budgets **40% for rent, 4% for carfare, 32% for food, 8% for clothes, 6% for savings,** and **10% for other items.**

Check: 40% + 4% + 32% + 8% + 6% + 10% = 100%

TRY THESE

1. 17% of 1,400 **2.** 49% of 6,400 **3.** 9.75% of 132

Compound Interest

When interest on a bank account is compounded, the interest earned is added to the principal at certain intervals during the year.

EXAMPLE

Maria deposited $1,000 in a bank offering an annual interest rate of 15%, compounded monthly. What is the balance in her account at the end of the second month?

Step 1 Calculate the monthly rate.

.15 $\boxed{\div}$ 12 $\boxed{=}$ 0.0125

Step 2 Add 1.00 (or 100%) to represent the previous balance.

.0125 $\boxed{+}$ 1 $\boxed{=}$ 1.0125

Step 3 Use 1.0125 as the constant factor for each month.

1.0125 $\boxed{\times}$ 1000 $\boxed{=}$ 1012.5

At the end of 1 month, Maria has a balance of $1,012.50.

Step 4 Continue to press $\boxed{\times}$ 1.0125 $\boxed{=}$ for each of the remaining 11 months.

$\boxed{\times}$ 1.0125 $\boxed{=}$ 1025.1562

If your calculator rounds, the display may be 1025.1563

At the end of the second month, Maria has a balance of **$1,025.16.**

TRY THESE

Calculate the balance. The interest is compounded monthly.

	Deposit	Annual Rate	Time	Balance		Deposit	Annual Rate	Time	Balance
1.	$5,000	12%	1 yr	?	**2.**	$2,000	18%	1 yr	?
3.	$3,000	15%	2 yr	?	**4.**	$1,200	9%	$2\frac{1}{2}$ yr	?
5.	$ 4,500	12%	15 mo	?	**6.**	$4,500	15%	10 mo	?
7.	$10,500	6%	9 mo	?	**8.**	$12,500	9%	8 mo	?

Powers and Roots

The second **power** of 5 is 25.   5 is a **square root** of 25.

$$5^2 = 25 \qquad\qquad \sqrt{25} = 5$$

Which of these methods works on your calculator?

EXAMPLE 1

Find the value of 7^4.

a. $7\ \boxed{\times}\ 7\ \boxed{\times}\ 7\ \boxed{\times}\ 7\ \boxed{=}$ ᒲ2401.   **b.** $7\ \boxed{\times}\ 7\ \boxed{=}\ \boxed{=}\ \boxed{=}$ 2401.

On a scientific calculator, you find powers by using the $\boxed{y^x}$ key.

EXAMPLE 2

Find the value of 2^3.

$2\ \boxed{\times}\ 2\ \boxed{\times}\ 2\ \boxed{=}$ 8.   or $2\ \boxed{y^x}\ 3\ \boxed{=}$ 8.

If your calculator has a $\boxed{\checkmark}$ key, you can easily find the square root of a number.

EXAMPLE 3

a. Find the square root of 1,369.   **b.** $\sqrt{75} =$ _?_

$1369\ \boxed{\checkmark}$ 37. $75\ \boxed{\checkmark}$ 8.6602540

Check: $37 \times 37 = 1,369$ **Check:** 8.660254×8.660254

$= 74.999999$

The answer is approximate.

TRY THESE

Find the value of the following.

1. 15^3	**2.** 99^2	**3.** 8^6	**4.** 6^7	**5.** 114^3
6. 5^{11}	**7.** 347^3	**8.** 2^{24}	**9.** 37^5	**10.** 7.5^2
11. 3.25^3	**12.** 0.3^5	**13.** 8.35^2	**14.** 0.09^2	**15.** 9.8^3
16. $\sqrt{34{,}969}$	**17.** $\sqrt{50}$	**18.** $\sqrt{1.1}$	**19.** $\sqrt{79}$	**20.** $\sqrt{0.0144}$
21. $\sqrt{113}$	**22.** $\sqrt{271}$	**23.** $\sqrt{317}$	**24.** $\sqrt{805}$	**25.** $\sqrt{31}$
26. $\sqrt{250}$	**27.** $\sqrt{1{,}000}$	**28.** $\sqrt{12.7}$	**29.** $\sqrt{50.41}$	**30.** $\sqrt{61.83}$

Operations with Integers

Change-sign keys such as $\boxed{+/-}$ or $\boxed{\text{cs}}$ are usually used to enter negative numbers. When such a key is pressed, the sign of the displayed number changes.

EXAMPLE

a. $-24 + 31 = \underline{\ ?\ }$

$\quad 24\ \boxed{\text{cs}}\ \boxed{+}\ 31\ \boxed{=}\ \ 7.$

b. $-33 - (-19) = \underline{\ ?\ }$

$\quad 33\ \boxed{\text{cs}}\ \boxed{-}\ 19\ \boxed{\text{cs}}\ \boxed{=}\ \ -14.$

c. $-112 \div 7 = \underline{\ ?\ }$

$\quad 112\ \boxed{\text{cs}}\ \boxed{\div}\ 7\ \boxed{=}\ \ -16.$

d. $-36 \times -14 = \underline{\ ?\ }$

$\quad 36\ \boxed{\text{cs}}\ \boxed{\times}\ 14\ \boxed{\text{cs}}\ \boxed{=}\ \ 504.$

TRY THESE

Use your calculator.

1. $36 + (-57)$
2. $-56 + 83$
3. $-68 + (-21)$
4. $29 + (-14)$
5. $72 + 38$
6. $-19 + (-54)$
7. $-83 + 59$
8. $-51 + 98$
9. $92 + (-92)$
10. $45 + 45$
11. $-78 + 78$
12. $33 + (-79)$
13. $-34 - (-46)$
14. $28 - 59$
15. $52 - 35$
16. $-47 - (-29)$
17. $-85 - 51$
18. $47 - (-80)$
19. $91 - (-43)$
20. $-70 - 87$
21. $-63 - (-63)$
22. $19 - 19$
23. $29 - (-29)$
24. $-41 - 41$
25. $215 + (-86)$
26. $-113 + (-78)$
27. $59 - 131$
28. $-71 - 129$
29. $-10 \times (-16)$
30. $-100 \times (-21)$
31. $-100 \times (-46)$
32. $-10 \times (-92)$
33. $-21 \times (-21)$
34. $-36 \times (-12)$
35. $-46 \times (-25)$
36. $-72 \times (-81)$
37. -38×42
38. $24 \times (-27)$
39. $53 \times (-16)$
40. -87×29
41. $-345 \div 23$
42. $765 \div (-17)$
43. $957 \div (-11)$
44. $-352 \div (-16)$
45. $-510 \div (-15)$
46. $-672 \div 16$
47. $882 \div (-14)$
48. $-860 \div (-43)$
49. $-714 \div 51$
50. $-832 \div (-26)$
51. $-612 \div 18$
52. $483 \div (-21)$

GLOSSARY

The following definitions and statements reflect the usage of terms in this textbook.

Acute angle An angle whose measure is less than 90°. (Page 321)

Acute triangle A triangle with three acute angles. (Page 322)

Angle Two rays with the same endpoint. (Page 321)

Area The measure in square units of the amount of surface inside a closed, plane figure. (Page 191)

Average, or mean

$$average = \frac{sum\ of\ measures}{number\ of\ measures}$$ (Page 4)

Axis (Plural: axes) A horizontal or vertical number line used to locate points. (Page 84)

Balance Amount left in a bank account after a withdrawal is made or a check is written. (Page 43)

Bar graph A *bar graph* uses horizontal or vertical bars to show data. (Page 84)

Budget A plan for balancing income and expenses. (Page 325)

Celsius scale A scale used to measure temperature. On this scale, the freezing point of water is 0°C and the boiling point is 100°C. (Page 306)

Central Angle Each *angle* in a circle graph. (Page 323)

Circle A closed curve in a plane. (Page 318)

Circle graph A graph in the shape of a circle used to show data. The graph uses percents to show parts of a whole. (Page 323)

Circumference The distance around a circle. $C = 2 \times \pi \times r$ or $C = \pi \times d$ (Page 318)

Commission An amount, usually a percent of goods sold, given to a salesperson, real estate agent, and so on, for services. (Page 275)

Cone A space figure such as the one shown below. (Page 334)

Cone

Congruent triangles Triangles that have the same size and shape are *congruent triangles*. (Page 379)

Corresponding angles Two angles in *corresponding* positions in different triangles. (Page 380)

Corresponding sides Two lines in *corresponding* positions in different triangles. (Page 379)

Cubic centimeter The capacity of this container is 1 *cubic centimeter* (abbreviated: 1 cm³). (Page 304)

Customary measures The system of measurement commonly used in the United States. (Page 182)

Cylinder A space figure such as the one shown on page 332.

Decimal Numbers such as 7.8, 0.03, and 12.0 that are written using a decimal point and place value. (Page 16)

Degree A measure used to describe the size of an angle such as 30°, 90°, and so on. (Page 321)

Denominator In the fraction $\frac{3}{5}$, the *denominator* is 5. (Page 127)

Deposit Money put into a bank account. (Page 36)

Diameter A line segment through the center of a circle having its endpoints on the circle. (Page 318)

Discount   An amount subtracted from the regular (list) price to obtain the sale price or a percent of the regular price. (Page 252)

Divisible   One number is *divisible* by a second number if the second number divides exactly into the first with no remainder. (Page 56)

Equation   A mathematical statement that uses "$=$," such as $x + 6 = 9$. (Page 210)

Equivalent fractions   Equal fractions such as $\frac{2}{3}$ and $\frac{4}{6}$. (Page 129)

Equivalent ratios   Equal ratios such as $\frac{3}{4}$ and $\frac{12}{16}$. (Page 224)

Estimation   The process of calculating with rounded or compatible numbers. (Page 2)

Excise tax   An internal tax levied on the manufacture, sale, or consumption of a commodity within a country. (Page 255)

Exponent   A number that tells how many times a number is used as a factor. (Page 368)

Factor   In multiplication, the numbers that are multiplied, such as 8 in $8 \times 5 = 40$. (Page 32)

Factorial   The product of all the positive integers from one to a number. 5! (read: five factorial) is $5 \cdot 4 \cdot 3 \cdot 2 \cdot 1$, or 120. (Page 346)

Formula   A rule stated in words or in symbols that can be used in solving problems. (Page 190)

Fraction   The quotient of two whole numbers written in the form, $\frac{1}{3}$, $\frac{5}{6}$, $\frac{12}{7}$, and so on. The denominators cannot be zero. (Page 127)

Frequency table   A table that shows how many times items appear within given data. (Page 114)

Gram   A commonly used unit of mass in the metric system. *One gram* equals 0.001 kilogram. (Page 294)

Greater than   An inequality relation (symbol: $>$) between two numbers, such as $47 > 45$, $1.1 > 0$, $\frac{3}{4} > \frac{1}{2}$. (Page 16)

Histogram   A bar graph that lists data by intervals. (Page 116)

Hypotenuse   The side opposite the right angle in a right triangle. (Page 373)

Integer   The whole numbers and their opposites, such as -162, -51, 0, 36, 210, and so on. (Page 394)

Interest   An amount paid for the use of money. *Interest* is usually a percent of the amount invested, lent, or borrowed. (Page 256)

Kilogram   The base unit of mass in the metric system. (Page 296)

Least common denominator (LCD)   The smallest denominator exactly divisible by each of two or more denominators. For example, the LCD of $\frac{3}{5}$ and $\frac{5}{6}$ is 30. (Page 138)

Less than   An inequality relation (symbol: $<$) between two numbers, such as $6 < 10$, $7.06 < 7.1$; $\frac{7}{10} < \frac{5}{6}$. (Page 16)

Like fractions   Fractions having the same denominator, such as $\frac{1}{6}$ and $\frac{5}{6}$. (Page 134)

Liter   The base unit of capacity in the metric system. (Page 294)

Line graph   A graph that shows the amount of change over a period of time. (Page 86)

Lowest terms   A fraction is in *lowest terms* when its numerator and denominator have no common factors other than 1. (Page 130)

Mass   The amount of matter an object contains. (Page 294)

Mean   Another name for *average*. The *mean* of 2, 5, 6, and 7 is $(2 + 5 + 6 + 7) \div 4$, or 5. (Page 105)

Median   When a series of numbers are listed in order, the middle number is the *median*. The median of 1.6, 2.9, 3.4, 7.8 and 12.2 is 3.4. (Page 108)

Meter   The base unit of length in the metric system. (Page 184)

Milligram   A unit of mass in the metric system. *One milligram* equals 0.001 gram. (Page 294)

Milliliter   A unit of capacity in the metric system. *One milliliter* equals 0.001 liter. (Page 294)

Millimeter   A unit of length in the metric system. *One millimeter* equals 0.001 meter. (Page 296)

Mixed number   A number such as $4\frac{2}{3}$, $5\frac{3}{8}$, and so on. (Page 132)

Mode   In a series of numbers, the number that occurs most often. (Page 105)

Negative number   A number less than zero, such as -19, -100, -238, and so on. (Page 394)

Net price   The cost after the discount is subtracted from the regular price. (Page 165)

Numerator   In the fraction $\frac{9}{10}$, 9 is the *numerator*. (Page 127)

Obtuse angle   An angle whose measure is greater than 90°. (Page 321)

Obtuse triangle   A triangle with an obtuse angle. (Page 322)

Order of operations   When more than one operation $(+, -, \times, \div)$ is involved, the order in which the operations are performed. (Page 8)

Parallelogram   A four-sided polygon whose opposite sides are parallel. (Page 196)

Percent   *Percent* means per hundred or hundredths. $\frac{7}{100} = 0.07 = 7\%$ (Page 240)

Perfect Square   A number that has two equal factors. (Page 368)

Perimeter   The sum of the lengths of the sides of a polygon, such as a rectangle. (Page 186)

π (pi)   The ratio of the circumference of a circle to its diameter. The ratio is approximately equal to 3.14. (Page 318)

Pictograph   A graph that uses pictures or symbols to represent data. (Page 82)

Positive number   A number greater than zero, such as 19, 57, 12, 608, and so on. (Page 394)

Probability   A number from 0 to 1 which tells how likely it is that an event will happen. (Page 348)

Proportion   An equation which states that two ratios are equal. (Page 228)

Pyramid   A space figure with one base. The sides of a *pyramid* are triangles. (Page 308)

Quotient   In a division problem such as $102 \div 6 = 17$, the *quotient* is 17. (Page 58)

Radius   A line segment having one end at the center of the circle and the other on the circle. (Page 318)

Range   The difference between the lowest number and the highest number within a set of data. (Page 105)

Ratio   A comparison of two numbers expressed as 2 to 5, or $\frac{2}{5}$, or 2:5. (Page 224)

Rational number   A number which can be written as a fraction. (Page 172)

Reciprocal   Two numbers whose product is 1 are *reciprocals* of each other. Thus, $\frac{3}{4}$ and $\frac{4}{3}$ are *reciprocals*, because $\frac{3}{4} \times \frac{4}{3} = 1$. (Page 160)

Rectangle   A four-sided polygon whose opposite sides are equal and whose angles are right angles. (Page 186)

Rectangular prism   A space figure having two equal rectangles as bases. The bases are parallel. (Page 304)

Right angle   An angle whose measure is 90°. (Page 321)

Right triangle   A triangle having one right angle. (Page 322)

Rule of Pythagoras   In a right triangle, $(\text{hypotenuse})^2 = (\text{leg})^2 + (\text{leg})^2$, or $c^2 = a^2 + b^2$. (Page 373)

Similar triangles   Triangles that have the same shape. (Page 380)

Sphere   A round space figure shaped like a basketball. (Page 336)

Square   A rectangle with four equal sides. (Page 186)

Square of a number   The product of a number and itself. (Page 368)

Square root   One of two equal factors of a number. The symbol for square root is $\sqrt{\ }$. (Page 368)

Tangent ratio

Tangent of angle A $= \dfrac{\text{length of side opposite angle A}}{\text{length of side adjacent to angle A}}$ (Page 383)

Trapezoid   A four-sided polygon with one pair of parallel sides. (Page 200)

Triangle   A polygon with three sides. (Page 322)

Unit price   The cost per gram, per pound, per liter and so on. The *unit price* of a 16-ounce container of cottage cheese that sells for $0.96 is 6¢ per ounce. (Page 66)

Volume   The measure of the amount of space inside a space figure. (Page 304)

Whole number   A number such as 0, 1, 2, 3, 4, and so on (Page 10)

INDEX

Boldfaced numerals indicate the pages that contain formal or informal definitions.

ANSWERS TO ODD-NUMBERED EXERCISES

CHAPTER 1   USING WHOLE NUMBERS

Page 3   Exercises   1. 540; 500   **3.** 1450; 1500   **5.** 97,900; 97,900   **7.** 89,000; 90,000   **9.** 28,000; 30,000   For Ex. 11–15, the estimate is shown in brackets.   **11.** [13,000]; 12,561   **13.** [200]; 223   **15.** [40,000]; 38,129   **17.** 7757 square miles   **19.** 51,747 square miles   **21.** 400,000 square miles   **23.** Use exact numbers. Reasons will vary.

Page 5   Exercises   1. 413 pages   **3.** 43 passengers   **5.** 50   **7.** 50   **9.** No. Answers will vary.   **11.** 43 seconds   **13.** Four students   **15.** 95,000

Page 7   Math and Travel   1. 1   **3.** 3   **5.** 4:00 A.M. Mountain time   **7.** 11:00 A.M. Mountain time   **9.** 1:00 P.M. Eastern time   **11.** 12:10 P.M. Pacific time   **13.** 4 hours 45 minutes   **15.** 10:15 P.M. Pacific time

Page 9   Exercises   1. Divide.   **3.** Subtract.   **5.** Multiply.   **7.** 8   **9.** 8   **11.** 4   **13.** 8   **15.** 10   **17.** 90   **19.** 6   **21.** 22   **23.** 13   **25.** 27   **27.** 16   **29.** 56   **31.** 30   **33.** 64   **35.** 38   **37.** a   **39.** $3 + (17 - 8) \times 2 = 21$ **41.** $7 \times 8 - (3 + 2) = 51$ **43.** c

Page 11   Exercises   1. 63 and 36   **3.** 30   **5.** 120   **7.** Answers will vary.

Page 12   Mid–Chapter Review   1. [1400]; 1404   **3.** [11,000]; 10,724   **5.** [5000]; 5202   **7.** 86   **9.** 11   **11.** 55   **13.** 30   **15.** 58 and 85   **17.** b   **19.** About twice as high

Page 13   Math in Social Studies   1. Florida   **3.** Rhode Island   **5.** 2,099,000   **7.** 56,000 fewer

Page 15   Exercises   1. base; exponent   **3.** factor   **5.** 8   **7.** 64   **9.** 64   **11.** 100,000   **13.** 4   **15.** 64   **17.** 900   **19.** 17,000   **21.** 250,000   **23.** 3   **25.** 5   **27.** 3   **29.** 2   **31.** 4   **33.** 1   **35.** 1   **37.** 1   **39.** 3^1 **41.** 2^4 **43.** 3^2 **45.** 3^4 **47.** equal   **49.** 5^3 **51.** 9   **53.** 100   **55.** 441   **57.** $3^2, 6^2, 10^2, 15^2, 21^2, 28^2$; The difference between successive bases is 3, 4, 5, 6, and so on. The exponent is always 2.

Page 17   Exercises   1. thousandths   **3.** hundredths   **5.** hundred millions   **7.** tenths   **9.** ones   **11.** nine and six tenths   **13.** six hundred twenty–five thousandths   **15.** four hundred eighty–six and five tenths   **17.** fifty–one and twenty–nine hundredths   **19.** five thousand one hundred twenty–nine   **21.** 0.002   **23.** 0.2   **25.** 819,000,000   **27.** 0.00015   **29.** 600.05   **31.** <   **33.** <   **35.** <   **37.** <   **39.** <   **41.** >   **43.** 98.75

Pages 18–19   Exercises   1. Twenty–five and 12/100   **3.** Fourteen and 00/100   **5.** Three hundred twenty–five and 75/100   **7.** Nineteen and 08/100   **9.** One thousand seven hundred twenty–five and 00/100   **11.** The date is missing.   **13.** The check is not signed.
15.

Page 20   Patterns and Codes   1. ⌐ ∟ o ⊓   **3.** ⌡ ⌐ ⌐⌐   **5.** ⊓ ⌐ ⊏ ⌡   **7.** $3,200

Pages 22–23   Chapter Review   1. estimated   **3.** addition   **5.** exponent   **7.** value   **9.** 60; 100   **11.** 1660; 1700   **13.** 64,000; 64,000   **15.** 50,000; 50,000   **17.** 145,000; 150,000   **19.** [7000]; 7330

21. [3000]; 3402   **23.** [36,000]; 35,863   **25.** [60,000]; 61,962   **27.** [140,000]; 140,019   **29.** $4,872
31. 86 passengers   **33.** 27   **35.** 4   **37.** 34   **39.** 6   **41.** 6   **43.** 3   **45.** twelve and twenty–five
hundredths   **47.** fifty–seven and thirty–five hundredths   **49.** one hundred eighty–two and four tenths
51. 0.02   **53.** >   **55.** <   **57.** <   **59.** Oklahoma   **61.** 20   **63.** 98.6

Page 24   Chapter Test   1. 10   **3.** 1000   **5.** 6000   **7.** [4000]; 3651   **9.** [6000]; 6346   **11.** [230,000];
233,413   **13.** 19 points   **15.** 26   **17.** 22   **19.** 4   **21.** <   **23.** >   **25.** 240

CHAPTER 2   WHOLE NUMBERS AND DECIMALS

Pages 26–27   Exercises   1. nearest hundredth   **3.** nearest ten-thousandth   **5.** nearest tenth   **7.** 3
9. 10   **11.** 11   **13.** 3.7   **15.** 0.7   **17.** 0.98   **19.** 60.07   **21.** 0.03   **23.** $6.53   **25.** $9.85
27. 1.520   **29.** 0.004   **31.** 6.010   **33.** 6.0001   **35.** 9.0082   **37.** $20   **39.** $70   **41.** $130   **43.** $200
45. $3,700   **47.** 7.5 million   **49.** 8.3 million   **51.** 4.7 million

Page 29   Exercises   1. 11.36   **3.** 6.26   **5.** 8.018   **7.** 45.84   **9.** 1.765   **11.** 119.055   **13.** 0.0433
15. 25.456   **17.** 21.275   **19.** 10.238   **21.** 141.39   **23.** 407.275   **25.** 0.72   **27.** 1.872   **29.** $2.55
31. $2.40   **33.** $35.15

Page 31   Exercises   1. 1   **3.** 3   **5.** 1   **7.** 10   **9.** 100   **11.** 2700   **13.** 5,893,000   **15.** 100
17. 2,500,000   **19.** 2; 2; 30,000   **21.** 10,000; 6,250,000; 25,000,000   **23.** Six hundred fifty $20–bills
25. Seven thousand $10–bills   **27.** 6–$100; 1–$50   **29.** 2–$100; 1–$50; 1–$10   **31.** 7–$100; 1–$20; 1–$5
33. Exchange is $50 too little; $350 for $400.   **35.** Exchange is $10 too little; $380 for $390.

Page 33   Exercises   1. F   **3.** P   **5.** >   **7.** >   **9.** >   **11.** 6068   **13.** 6020   **15.** 323,748
17. 273,052   **19.** 3654   **21.** 8722   **23.** 81,885   **25** 196,460   **27.** 324,810   **29.** 362,986   **31.** 8099
33. 91,353   **35.** 187,860   **37.** 145,200   **39.** 742,552   **41.** 26,460   **43.** 6   **45.** 1; 2; 7; 9   **47.** 3; 4; 1;
4; 1; 5

Page 34   Mid–Chapter Review   1. 5   **3.** 50.06   **5.** 5.0135   **7.** 8.69   **9.** 32.15   **11.** 11.47
13. 7.915   **15.** 10   **17.** 100   **19.** 3367   **21.** 63,856   **23.** 2408   **25.** 32,340   **27.** 276,920
29. $10.55   **31.** >   **33.** =   **35.** <   **37.** >   **39.** Brooklyn Bridge

Page 35   Math and Making Change   1. c   **3.** c   **5.** $5.45; 0; 1; 0; 1; 2; 0; 0   **7.** $15.37; 1; 1; 0; 1; 1; 0; 2

Page 37   Exercises   1. Addition   **3.** Deposit   **5.** Subtraction   **7.** $449.38; $324.38   **9.** Subtotal & Net
Deposit: $448.69   **11.** Net Deposit: $461.62   **13.** $925.45

Page 39   Exercises   1. left   **3.** greater than   **5.** 27   **7.** 6.2   **9.** 5360   **11.** 1.8   **13.** 0.048
15. 0.000136   **17.** 9.76   **19.** 0.36   **21.** 0.01   **23.** 2.5   **25.** 1   **27.** 40   **29.** 3.42   **31.** 4.59
33. 100   **35.** 50

Page 40   Exercises   1. 72; 1; 1; 2; 2; 74.52   **3.** 62 × 2 = 124; 2; 2; 4; 4; 125.8803   **5.** 62 × 20 = 1240; 2;
1; 3; 3; 1258.803   **7. a.** whole   **b.** factors   **c.** sum

Page 42   Exercises   1. whole numbers   **3.** four   **5.** 2.88   **7.** 0.288   **9.** 0.0288   **11.** $57.00
13. 52.020, or 52.02   **15.** 8.944   **17.** 0.00003   **19.** 0.7865   **21.** 9.64566   **23.** 0.004136   **25.** 2.472309
27. $473.00   **29.** 0.001236   **31.** 0.03614   **33.** 2.8   **35.** $212.40   **37. a.** more   **b.** $59.10 more

Pages 43–44   Exercises   1. $162.43; $300.03; $147.13   **3.** $120.68; $76.88; $228.48; $203.31
5. $778.58; $478.58; $450.73   **7.** $130.10; $114.24   **9.** $774.11

Page 46   Exercises   1. 12; 18   **3.** 85; 75; 70   **5.** 24; 96   **7.** 39; 45   **9.** 625; 3125   **11.** 17; 30
13. 84   **15. a.** Answers may vary.   **b.** Answers may vary.   **17.** 43, 45, 47, 49, 51, 53, 55   **19.** No

Page 48   Planning a Vacation   1. $367 + $50 (meals); $508 (4 nights); $195; $15; $75; Total cost: $1,210
3. Answers will vary.

ANSWERS TO ODD-NUMBERED EXERCISES   **489**

Pages 50–51   Chapter Review   1. decimal points   **3.** product   **5.** right   **7.** decimal   **9.** 37   **11.** 1
13. 7   **15.** 2.9   **17.** 6.0   **19.** 83.5   **21.** 4.681   **23.** 12.359   **25.** 38.460   **27.** 61.49   **29.** 32.687
31. 35.506   **33.** 8.7   **35.** 1.326   **37.** 26.53   **39.** 2018.58   **41.** 1000   **43.** 810   **45.** 45,625
47. 125,972   **49.** 628   **51.** 238   **53.** 0.162   **55.** 0.00562   **57.** 55.2   **59.** 970.2   **61.** 0.0136
63. 1.6261   **65.** 0.273585   **67.** $197.54   **69.** $337.92   **71.** 162; 486

Page 52   Chapter Test   1. 8   **3.** 29.65   **5.** 184.93   **7.** 30.88   **9.** 280.04   **11.** 1504   **13.** 14,040
15. 164,424   **17.** 4.56   **19.** 0.0352   **21.** 620   **23.** 32,800   **25.** 0.014   **27.** 15,000   **29.** $13.53
31. 34; 22   **33.** 4

Pages 53–54   Cumulative Maintenance: Chapters 1–2   1. b   **3.** d   **5.** b   **7.** a   **9.** b   **11.** d   **13.** a
15. d   **17.** b   **19.** a   **21.** d

CHAPTER 3   DIVISION: WHOLE NUMBERS/DECIMALS

Page 57   Exercises   1. 11   **3.** 10   **5.** 10   **7.** 117, 507, 792, 834, 333   **9.** 920   **11.** 920   **13.** Yes;
Yes; Yes; Yes; No; Yes   **15.** Yes; No; Yes; No; No; No   **17.** Yes; No; Yes; Yes; No; Yes   **19.** No; No; No; No; No;
No   **21.** 465, 546, 564, 645, 654   **23.** No

Page 59   Exercises   1. 4   **3.** 30   **5.** 20   **7.** 30   **9.** 40   **11.** two   **13.** three   **15.** two   **17.** 3 r9
19. 3   **21.** 6   **23.** 6 r19   **25.** 7 r3   **27.** 119 r15   **29.** 68   **31.** 43 r56   **33.** 132 r25   **35.** 199 r12
37. 111 r59   **39.** 4 r79   **41.** 33 r10   **43.** 114 r19   **45.** 21 r4   **47.** $672 \div 12 = 56$

Page 61   Exercises   1. 20   **3.** 5   **5.** 40   **7.** 10; $960 \div 80 = 12$ **9.** 200; $6000 \div 30 = 200$ **11.** 50;
$4000 \div 80 = 50$ **13.** 205   **15.** 1608   **17.** 10 r39   **19.** 105 r45   **21.** 360 r18   **23.** 280 r11   **25.** 40 r9
27. 70 r5   **29.** 103   **31.** 407   **33.** 209 r60   **35.** 101 r12   **37.** 380   **39.** 506 r14   **41.** 30 r9
43. 104 r25   **45.** 150 r27   **47.** 1084 r33   **49.** $440

Page 62   Mid–Chapter Review   1. 2, 4   **3.** 5   **5.** 2, 5, 10   **7.** 2, 3, 5, 9, 10   **9.** 5   **11.** 2, 4   **13.** 3, 9
15. 2   **17.** 21 r14   **19.** 23 r21   **21.** 7 r54   **23.** 71   **25.** 30 r8   **27.** 105 r25   **29.** 40 r55   **31.** 360 r18
33. No   **35.** $1,709   **37.** 5.00, or 5   **39.** 134.55   **41.** 7.828   **43.** 14.16   **45.** $175

Page 63   Math and Finance Charges   1. $5,256; $438   **3.** $5,016; $209   **5.** $59   **7.** $1,188

Page 65   Exercises   1. decimal point; decimal point   **3.** thousandths   **5.** 1.74   **7.** 17.40   **9.** 0.174
11. 2; $6 \div 3 = 2$ **13.** 2; $180 \div 90 = 2$ **15.** 0.1; $3.7 \div 37 = 0.1$ **17.** 6.28   **19.** 0.053   **21.** 1.38
23. 0.36   **25.** 5.52   **27.** 0.08   **29.** 0.56   **31.** 0.01   **33.** 3.58   **35.** 6.00   **37.** 5.8   **39.** 0.9
41. 3.0   **43.** 0.1   **45.** 3.7   **47.** $0.96 \div 6 = 0.16$ **49.** $0.096 \div 6 = 0.016$ or $3.096 \div 6 = 0.516$

Page 67   Exercises   1. 7.6¢   **3.** 3.9¢   **5.** 8.1¢   **7.** 46.5¢   **9.** 50.5¢   **11.** 23.6¢   **13.** 11.3¢
15. 53.8¢   **17.** 3–kilogram bag   **19.** Buy at 23¢ per pound.   **21.** 10 ounces for 91¢   **23.** Answers will vary.

Page 69   Exercises   1. 100   **3.** 1000   **5.** 55; 275   **7.** 6; 360   **9.** 15; 720   **11.** 16; 165.44   **13.** 8.1
15. 47   **17.** 3200   **19.** 5800   **21.** 35   **23.** 400   **25.** 255   **27.** 40   **29.** 5.1   **31.** 9.2   **33.** 52.1
35. 5.2   **37.** 681.3   **39.** 16.7   **41.** 27.43   **43.** 386.84   **45.** 74.23   **47.** 0.67   **49.** 50.4 miles per hour

Pages 70–71   Exercises   1. 0.763   **3.** 0.763   **5.** 0.0763   **7.** 7.63   **9.** 0.00763   **11.** 0.7   **13.** 0.09
15. 1000   **17.** 88.92   **19.** 3.98   **21.** 0.13   **23.** 10   **25.** 11   **27.** 1000   **29.** 50   **31. a.** 30 **b.** 90; 30
c. same answer **d.** 30   **33.** 100; 4   **35.** 80   **37.** 120   **39.** 3   **41.** 300   **43.** 3   **45.** 400   **47.** 15
49. 300   **51.** 6   **53.** 120   **55.** 190   **57.** $60   **59.** $0.20

Page 73   Exercises   1. Exact; To check the paycheck amount   **3.** $77.70; $68.04; $56.70; $77.43; $93.10
5. Paper and pencil. Reasons may vary.   **7.** Exact. Reasons may vary.   **9.** 2.597; 3.082; 2.703; 2.682; 3.206
11. Answers may vary.

Page 74   Making a Model   1. $99.50 to $100.49   **3.** $24.50 to $25.49   **5.** $9,499.50 to $9,500.49
7. $67.50 to $72.49

Pages 76–77 Chapter Review 1. compatible **3.** zeros **5.** 5 **7.** 3 **9.** 2, 3, 5, 10 **11.** 3, 9
13. 3, 9 **15.** 2, 4 **17.** 2, 4, 5, 10 **19.** 3 **21.** 15 r28 **23.** 147 r4 **25.** 64 **27.** 32 r64 **29.** 30 r18
31. 206 r22 **33.** 130 r25 **35.** 105 r25 **37.** 40 r35 **39.** 0.42 **41.** 0.28 **43.** 0.02 **45.** 0.55
47. 5.69 **49.** 7.2 ÷ 6 = 1.2 **51.** 78.0 ÷ 6 = 13.0 or 81.0 ÷ 6 = 13.5 **53.** 84.245 ÷ 7 = 12.035
55. 4.9¢ **57.** 40.5¢ **59.** 0.62 **61.** 255 **63.** 7060 **65.** 0.55 **67.** 730 **69.** 10 **71.** 0.15
73. 13,900 **75.** 40 **77.** $609 **79.** $8,000

Page 78 Chapter Test 1. 2, 4 **3.** 2, 3, 4 **5.** 2, 3, 4, 9 **7.** 2, 3, 5, 9, 10 **9.** 56 r7 **11.** 302 r37
13. 0.57 **15.** 35.1 **17.** 15 **19.** 10 **21.** 15.2 **23.** The 32–ounce bottle

Pages 79–80 Cumulative Maintenance: Chapters 1–3 1. c **3.** b **5.** b **7.** c **9.** d **11.** c **13.** b
15. c **17.** a **19.** d **21.** a **23.** d

CHAPTER 4 GRAPHS AND APPLICATIONS

Pages 82–83 Exercises 1. Engineers **3.** $550 **5.** $350
7. $400 **9.** Architects **11.** 200 **13.** Allen, Glencove, Brooks
5. Justin High and Bradley **17. a.** See the graph at the right.
b. Answers may vary.

Pages 84–85 Exercises 1. Vertical **3.** August
5. Freshmen **7.** Juniors **9.** Juniors **11.** Australia
13. N. America **15. a.** See the graph at the right below.
b. Answers may vary.

Pages 86–87 Exercises 1. May and June
3. January, February, April, May, October, November,
December **5.** April, May, October **7.** January
9. July and August **11.** Answers may vary. **13.** See
the graph below. **15.** See the graph at the right below.

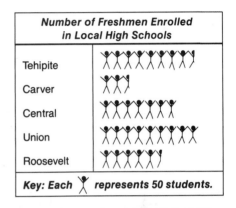

Number of Freshmen Enrolled
in Local High Schools

Key: Each ⅄ represents 50 students.

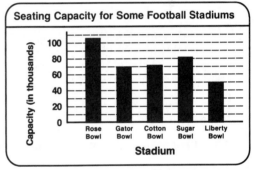

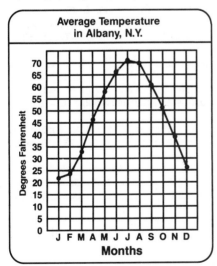

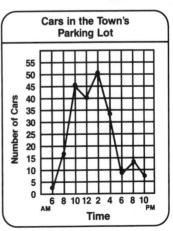

Page 88   Mid–Chapter Review   1. Superior   **3.** Michigan and Huron   **5.** April and May   **7.** July and August   **9.** 16.66   **11.** 66.83   **13.** 5.3   **15.** 122   **17.** $148   **19.** 18–ounce box

Page 89   Math and Scattergrams   1. 6 days   **3.** 11 days   **5.** 8 days   **7.** 9 days; 8.5 on the scale is the middle of the ninth day.

Pages 90–91   Exercises   1. Softball   **3.** Swimming   **5.** Basketball, Track   **7.** 15   **9.** 10   **11.** 150   **13.** Breed of dog   **15.** Cocker Spaniels   **17.** Cocker Spaniels   **19.** Cocker Spaniels   **21.** German Shepards and Beagles   **23.** May and September   **25.** December

Pages 92–93   Exercises   1. Answers may vary.   **3.** 2600   **5.** Answers may vary.   **7.** 2 times   **9.** 400 miles   **11.** Answers may vary.   **13.** 24   **15.** Subtract number for Jan.–Mar. from number for Jan.–June   **17.** Answers may vary.

Page 95   Using Graphs   1. B   **3.** B   **5.**

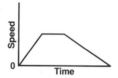

Page 96   Math in Art   1. 180   **3.** 20   **5.** $31,500,000

Pages 98–99   Chapter Review   1. numbers   **3.** length   **5.** horizontal   **7.** 20   **9.** San Antonio, TX   **11.** Norfolk, VA   **13.** Miles per hour   **15.** Chicken   **17.** 40 miles per hour   **19.** Cheetah   **21.** July   **23.** April and May

Page 100   Chapter Test   1. 100 million   **3.** 225 million   **5.** 300 million   **7.** Texas Commerce Tower   **9.** Texas Commerce Tower   **11.** July and August   **13.** 24°C

Pages 101–102   Cumulative Maintenance: Chapters 1–4   1. a   **3.** a   **5.** b   **7.** d   **9.** a   **11.** a   **13.** c   **15.** a   **17.** b   **19.** d

CHAPTER 5   STATISTICS

Page 104   Exercises   3. Less than; Some bars are shorter than the approximate average.

Page 106   Exercises   1. mean   **3.** 5   **5.** Mean: 9; Range: 8; No mode   **7.** Mean: 11; Range: 5; Mode: 10   **9.** Mean: 7.5; Range: 12; Mode: 12   **11.** Mean: $6.15; Range: $3.20; No mode   **13.** 30   **15.** 78   **17.** 91   **19.** 6   **21.** They differ by $2400.

Page 107   Math and Managing a Store   1. 6   **3.** 13   **5.** See diagram at right.   **7.** 12:00–1:00 P.M.   **9.** Transfer dots to make all columns the same height. Count the number of dots in a column to find the mean.

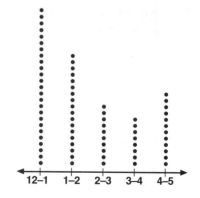

Page 109   Exercises   1. middle   **3.** middle   **5.** 64   **7.** 8   **9.** 26   **11. a.** 92 **b.** 92 **c.** Yes; 92 is still the middle score.   **13.** Insert 26 between 25 and 27.

Pages 110–111   Exercises   1. 13   **3.** Size 10   **5.** Sizes 8, 10, 12, 14, 16, 18; She sells more of these sizes.
7. Route A: 12; Route B: 28   **9.** $499.50   **11.** 7   **13.** $385   **15.** b; It is not practical to determine the capacity of the elevator by the weight of the passengers.

Page 112   Mid–Chapter Review   1. Mean: 23; Range: 20; Mode: 22   **3.** 80.5   **5.** $18,833.33   **7.** No mode   **9.** 4°C   **11.** May and June

Page 113   Math in Science   1. Oak   **3.** 75   **5.** 286.7 feet   **7.** Median; It is closer to more of the heights.

Page 115   Exercises   1.

Number of Letters	Tally	Frequency
2–3	||	2
4–5	⅂⅃⅂ |||	8
6–7	||||	4
8–9	|	1

3. 5; 22; 23; 9; 1
5. 10
7. 75–79
9. 17
11. Answers may vary.

Pages 116–117   Exercises   1. 1945–1954   **3.** No; Graph does not show 1968 alone.   **5.** 8   **7.** 50
9. 86–90   **11.** Highest   **13.** 34

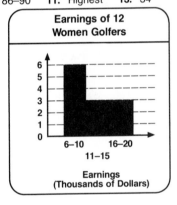

17.

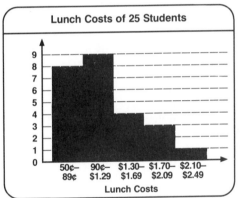

Page 118   Sampling   1.–7. Answers will vary.

Pages 120–121   Chapter Review   1. mean   **3.** mode
5. intervals   **7.** $2.48   **9.** $0.28   **11.** 180   **13.** 56
15. 388.25 mi   **17.** 515 mi   **19.** Tally: ⅂⅃⅂ \| ; ⅂⅃⅂ ; ⅂⅃⅂ \| ;
\|\|\| ; Frequency: 6; 5; 6; 3   **21.** See graph at right.

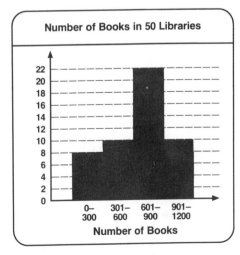

Page 122 Chapter Test 1. Mean: 8.5; Range: 14 **3.** 12.1 **5.** 15.5 **7.** Since they are close together, either the mean or the median. **9.** Tally: $|||$; $\cancel{||||}\,|$; $\cancel{||||}\,|$; $|$; Frequency: 3; 6; 6; 1

Pages 123–124 Cumulative Maintenance: Chapters 1–5 1. a **3.** c **5.** a **7.** a **9.** a **11.** b **13.** d **15.** d **17.** b **19.** d **21.** a

CHAPTER 6 FRACTIONS: ADDITION/SUBTRACTION

Page 126 Exercises 3. Yes; Answers may vary. **5.** ▯ **7.** △△△ △△△ / △△△ △△△

Page 128 Exercises 1. part **3.** numerator **5.** zero **7. a.** $\frac{3}{4}$ **b.** $\frac{1}{4}$ **9. a.** $\frac{6}{8}$ **b.** $\frac{2}{8}$ **11.** $\frac{1}{4}$
13. $\frac{5}{8}$ **15.** 1 **17.** 4 **19.** $\frac{7}{8}$ **21.** $\frac{9}{9}$ **23.** $\frac{2}{6}$ **25.** $\frac{13}{16}$

Page 129 Exercises 3. $\frac{2}{8}; \frac{4}{16}$ **5.** $\frac{6}{8}; \frac{12}{16}$

Page 131 Exercises 1. equivalent fractions **3.** is not **5.** 2 **7.** 3 **9.** 2 **11.** 2 **13.** 2 **15.** 2
17. 6 **19.** 3 **21.** 2 **23.** 7 **25.** 16 **27.** $\frac{1}{3}$ **29.** $\frac{1}{5}$ **31.** $\frac{13}{15}$ **33.** $\frac{2}{3}$ **35.** $\frac{7}{12}$ **37.** $\frac{1}{5}$ **39.** $\frac{16}{21}$
41. $\frac{1}{8}$ **43.** $\frac{7}{4}$ **45.** $\frac{7}{4}$ **47.** $\frac{9}{4}$ **49.** $\frac{3}{2}$ **51.** $\frac{3}{2}$ **53.** $\frac{7}{6}$ **55.** $\frac{9}{2}$ **57.** $\frac{9}{2}$ **59.** $\frac{11}{3}$ **61.** $\frac{10}{7}$ **63.** $\frac{27}{8}$
65. $\frac{14}{15}$ **67.** $\frac{6}{13}$ **69.** No; The numerator and denominator have a common factor, 2.

Page 133 Exercises 1. less than; less than **3.** equal to; equal to **5.** Less than one; $\frac{3}{9} = \frac{1}{3}$, which is less
than 1. **7.** > **9.** > **11.** = **13.** < **15.** = **17.** $1\frac{2}{7}$ **19.** $1\frac{1}{2}$ **21.** $2\frac{2}{3}$ **23.** $2\frac{1}{2}$ **25.** $2\frac{5}{6}$
27. $3\frac{1}{4}$ **29.** $3\frac{3}{5}$ **31.** 3 **33.** 4 **35.** 12 **37.** $1\frac{1}{2}$ **39.** $2\frac{6}{7}$ **41.** $3\frac{1}{3}$ **43.** $\frac{8}{2}, \frac{4}{4}, \frac{7}{8}, \frac{0}{9}$ **45.** $\frac{10}{2}, \frac{7}{4}, \frac{8}{8}, \frac{5}{6}$
47. $\frac{3}{1}, \frac{4}{2}, \frac{4}{3}, \frac{10}{10}$

Page 135 Exercises 1. Less than 1; $\frac{4}{5}$ is less than $\frac{5}{5}$. **3.** Closer to 1; $1\frac{1}{4}$ is closer to 1. **5.** $\frac{1}{5} + \frac{3}{5} = \frac{4}{5}$
7. $\frac{5}{12}$ **9.** $\frac{3}{12}$, or $\frac{1}{4}$ **11.** $\frac{2}{5}$ **13.** $\frac{1}{5}$ **15.** $\frac{1}{10}$ **17.** $\frac{1}{2}$ **19.** $5\frac{3}{5}$ **21.** $6\frac{3}{4}$ **23.** $3\frac{1}{3}$ **25.** $4\frac{1}{6}$ **27.** $1\frac{2}{3}$
29. $\frac{1}{2}$ **31.** $9\frac{1}{5}$ **33.** $18\frac{6}{7}$ **35.** $\frac{2}{5}$ **37.** $17\frac{1}{3}$ **39.** $\frac{4}{12}$ **41.** 1 **43.** $5\frac{6}{10}$ **45.** One answer is given.
$\frac{3}{10} + \frac{5}{10} = \frac{8}{10}$

Page 136 Mid–Chapter Review 1. $\frac{3}{5}$ **3.** $\frac{4}{6}$ **5.** $\frac{1}{4}$ **7.** $\frac{2}{9}$ **9.** $\frac{3}{2}$, or $1\frac{1}{2}$ **11.** $\frac{9}{4}$, or $2\frac{1}{4}$ **13.** $\frac{5}{3}$, or $1\frac{2}{3}$
15. $\frac{13}{14}$ **17.** < **19.** > **21.** < **23.** $2\frac{1}{3}$ **25.** $5\frac{4}{5}$ **27.** $3\frac{1}{4}$ **29.** $\frac{16}{25}$ **31.** 7.668 **33.** 84
35. $52.85 **37.** 15 miles per hour

Page 137 Exercises 1. 15 ; $165 **3.** 20; $260 **5.** $0.50 **7.** $10 **9.** $165, $8.25; $168, $8.40; $260, $13.00

Page 139 Exercises 1. 3, 6, 9, 12, 15 **3.** 13, 26, 39, 52, 65 **5.** 20, 40, 60, 80, 100 **7.** 100, 200, 300, 400, 500 **9.** $\frac{15}{24}$ **11.** $\frac{15}{24}$ and $\frac{4}{24}$ **13.** 6 **15.** 10 **17.** 6 **19.** 6 **21.** 12 **23.** $\frac{12}{30}$ and $\frac{5}{30}$ **25.** $\frac{3}{6}$ and $\frac{2}{6}$ **27.** $\frac{7}{12}$ and $\frac{9}{12}$ **29.** $\frac{5}{15}$ and $\frac{12}{15}$ **31.** $\frac{1}{8}$ and $\frac{4}{8}$ **33.** $\frac{35}{40}$ and $\frac{8}{40}$ **35.** $\frac{2}{20}$ and $\frac{15}{20}$ **37.** No. The numerator and denominator have a common factor, 3.

Page 141 Exercises 1. 20 **3.** $\frac{13}{20}$ **5.** Less than 1 because $\frac{7}{8}$ is less than $\frac{8}{8}$. **7.** Closer to 0 because $\frac{2}{10}$ is closer to $\frac{0}{10}$ than to $\frac{5}{10}$. **9.** $1\frac{7}{24}$ **11.** $\frac{11}{20}$ **13.** $\frac{5}{9}$ **15.** $\frac{13}{16}$ **17.** $\frac{1}{5}$ **19.** $7\frac{3}{4}$ **21.** 0 **23.** $1\frac{13}{24}$ **25.** $46\frac{17}{24}$
27. $1\frac{1}{4}$ **29.** $\frac{13}{15}$ **31.** $\frac{9}{20}$ **33.** $\frac{1}{3}$ **35.** $1\frac{1}{12}$ **37.** $\frac{1}{5}$ **39.** $\frac{1}{3}$ **41.** $\frac{1}{4}$ **43.** $5\frac{3}{4}$ **45.** $\frac{12}{3} - \frac{4}{6} = 3\frac{1}{3}$

Page 143 Exercises 1. $4\frac{2}{3} = 4 + \frac{2}{3} = 3 + \frac{3}{3} + \frac{2}{3} = 3 + \frac{5}{3} = 3\frac{5}{3}$ **3.** $5\frac{5}{4}$ **5.** $7\frac{13}{10}$ **7.** $5\frac{1}{3} - 2\frac{3}{4} =$
$5\frac{4}{12} - 2\frac{9}{12} = 4\frac{16}{12} - 2\frac{9}{12} = 2\frac{7}{12}$ **9.** Closer to 0 because $\frac{3}{4}$ is close to 1 and $1 - 1 = 0$. **11.** Greater than
because $\frac{5}{6}$ is greater than $\frac{1}{2}$. **13.** Closer to 6 because $5\frac{9}{10}$ is close to 6 and $\frac{1}{10}$ is close to 0. **15.** $1\frac{3}{4}$
17. $4\frac{7}{10}$ **19.** $5\frac{5}{12}$ **21.** $1\frac{11}{14}$ **23.** $\frac{17}{18}$ **25.** $\frac{1}{2}$ **27.** $2\frac{3}{4}$ **29.** $5\frac{1}{10}$ **31.** $\frac{4}{5}$ **33.** $\frac{1}{3}$ **35.** $16\frac{5}{6}$ **37.** $4\frac{13}{15}$
39. $5\frac{7}{8}$ **41.** $10\frac{1}{6}$ **43.** $7\frac{1}{4}$

Pages 144–145 Exercises 1. Last **3.** Cloud Trio **5.** Reapers and Plows **7.** 72 **9.** $\frac{1}{2}$

Page 146 Math in Music 1. half note **3.** 2 **5.** sixteenth note **7.** $\frac{4}{4}$

Pages 148–149 Chapter Review 1. terms **3.** greater **5.** unlike **7.** $\frac{1}{5}$ **9.** $\frac{7}{8}$ **11.** $\frac{9}{5}$ **13.** $\frac{2}{3}$
15. $\frac{8}{7}$, or $1\frac{1}{7}$ **17.** $\frac{1}{5}$ **19.** < **21.** > **23.** < **25.** > **27.** = **29.** $2\frac{1}{4}$ **31.** $5\frac{1}{2}$ **33.** $1\frac{3}{7}$
35. $4\frac{1}{4}$ **37.** $1\frac{1}{2}$ **39.** $\frac{3}{10}$ **41.** $2\frac{1}{2}$ **43.** $4\frac{1}{9}$ **45.** 13 **47.** 11 **49.** $5\frac{1}{3}$ **51.** 10 **53.** 10
55. $\frac{1}{4}$ and $\frac{2}{4}$ **57.** $\frac{30}{40}$ and $\frac{32}{40}$ **59.** $\frac{8}{24}$ and $\frac{9}{24}$ **61.** $\frac{5}{6}$ **63.** $2\frac{3}{8}$ **65.** $13\frac{4}{15}$ **67.** $1\frac{11}{24}$ **69.** $27\frac{3}{20}$ **71.** $24\frac{9}{56}$
73. $31\frac{23}{24}$ **75.** $3\frac{11}{40}$ **77.** $1\frac{3}{5}$ **79.** $4\frac{3}{4}$ **81.** $16\frac{3}{10}$ **83.** $2\frac{23}{40}$ **85.** $23\frac{19}{24}$ **87.** $\frac{11}{14}$ **89.** $\frac{19}{36}$ **91.** $9\frac{4}{5}$
93. a. 14 **b.** $210

Page 150 Chapter Test 1. $\frac{1}{2}$ **3.** $\frac{1}{3}$ **5.** $\frac{4}{3} = 1\frac{1}{3}$ **7.** $\frac{3}{4}$ **9.** $\frac{4}{2}, \frac{4}{4}, \frac{7}{9}, \frac{0}{8}$ **11.** $1\frac{1}{2}$ **13.** $5\frac{1}{4}$ **15.** $4\frac{1}{2}$
17. $\frac{3}{4}$ and $\frac{2}{4}$ **19.** $\frac{4}{24}$ and $\frac{9}{24}$ **21.** $\frac{49}{56}$ and $\frac{8}{56}$ **23.** 7 **25.** $\frac{1}{8}$ **27.** $21\frac{13}{20}$ **29.** $\frac{18}{35}$ **31.** $10\frac{5}{8}$ **33.** $2\frac{8}{15}$
35. Fourth

Pages 151–152 Cumulative Maintenance: Chapters 1–6 1. c **3.** b **5.** b **7.** b **9.** d **11.** b
13. a **15.** d **17.** d **19.** a **21.** a

CHAPTER 7   FRACTIONS: MULTIPLICATION/DIVISION

Page 154 Exercises 1. $\frac{3}{10}$ **3.** $\frac{1}{10}$ **5.** $\frac{2}{3}$

Page 156 Exercises 1. numerators; denominators **3.** multiply **5.** $\frac{6}{35}$ **7.** $\frac{8}{45}$ **9.** $\frac{1}{6}$ **11.** $\frac{1}{8}$
13. $\frac{3}{16}$ **15.** $\frac{4}{35}$ **17.** $\frac{5}{21}$ **19.** 8 **21.** $\frac{1}{6}$ **23.** $\frac{3}{5}$ **25.** $\frac{2}{5}$ **27.** $\frac{1}{6}$ **29.** $\frac{3}{40}$ **31.** $\frac{8}{45}$ **33.** 5 **35.** $\frac{20}{63}$
37. $\frac{5}{96}$ **39.** $\frac{1}{15}$ **41.** $\frac{3}{5}$ **43.** $\frac{2}{3}$ **45.** $\frac{2}{3}$ **47.** $\frac{1}{6}$ cup

Page 157 Math in Health 1. $\frac{3}{8}$ cup **3.** $\frac{3}{2}$, or $1\frac{1}{2}$ cups **5.** 2 cups

Page 159 Exercises 1. 1 **3.** 9 **5.** 1 **7.** 11 **9.** 8 **11.** 11 **13.** 5 **15.** 8 **17.** 8 **19.** 9
21. 36 **23.** 45 **25.** $\frac{5}{2}$ **27.** $\frac{31}{5}$ **29.** $\frac{15}{4}$ **31.** $\frac{93}{8}$ **33.** $\frac{8}{3}$ **35.** $\frac{57}{8}$ **37.** $\frac{193}{12}$ **39.** [3]; $2\frac{3}{4}$ **41.** [4];
$3\frac{3}{5}$ **43.** [12]; $12\frac{7}{15}$ **45.** [25]; $24\frac{1}{9}$ **47.** [56]; $60\frac{1}{2}$ **49.** [3]; $2\frac{2}{5}$ **51.** [8]; $7\frac{3}{16}$ **53.** [1]; $1\frac{1}{2}$ **55.** [8]; $9\frac{1}{3}$
57. [6]; $5\frac{19}{48}$ **59.** [36]; $38\frac{13}{40}$ **61.** [21]; $20\frac{1}{2}$ **63.** [20]; $20\frac{1}{8}$ **65.** [16]; $16\frac{1}{2}$ **67.** [24]; $25\frac{3}{5}$ **69.** Less;
factors for estimate were greater than factors for exact answer.

Page 161 Exercises 1. $\frac{7}{4}$ **3.** $\frac{1}{4}$ **5.** 3 **7.** $\frac{4}{21}$ **9.** 2 **11.** $\frac{3}{2}$ **13.** $\frac{4}{3}$ **15.** $\frac{5}{6}$ **17.** $\frac{48}{49}$ **19.** $\frac{1}{2}$
21. $\frac{7}{24}$ **23.** $1\frac{1}{3}$ **25.** $\frac{5}{6}$ **27.** 1 **29.** $\frac{1}{4}$ **31.** $\frac{1}{30}$ **33.** 21 **35.** 12 **37.** 28 **39.** $\frac{2}{27}$ **41.** $1\frac{1}{4}$
43. $1\frac{1}{2}$ **45.** $\frac{1}{12}$ **47.** $\frac{18}{25}$ **49.** $1\frac{7}{18}$ **51.** $1\frac{1}{3}$ **53.** $2\frac{2}{3}$ **55. a.** 12 **b.** 720

Page 163 **Exercises** **1.** 1 **3.** 2 **5.** 3 **7.** $\frac{1}{2}$ **9.** 8 **11.** 64 **13.** $\frac{5}{54}$ **15.** $17\frac{13}{16}$ **17.** 6 **19.** $4\frac{1}{2}$ **21.** $\frac{235}{256}$ **23.** $4\frac{3}{4}$ **25.** $2\frac{7}{9}$ **27.** $1\frac{3}{5}$ **29.** $2\frac{5}{6}$ **31.** $\frac{1}{4}$ **33.** 20 **35.** $\frac{1}{2}$ **37.** 1 **39.** $[\frac{1}{4}]$; $\frac{9}{32}$ **41.** [2]; $2\frac{2}{39}$ **43.** [3]; $2\frac{1}{2}$ **45.** $[\frac{1}{2}]$; $\frac{12}{25}$ **47.** [2]; $1\frac{23}{25}$ **49.** $1\frac{1}{3}$ minutes

Page 164 **Mid–Chapter Review** **1.** $\frac{11}{40}$ **3.** $\frac{4}{9}$ **5.** 1 **7.** $\frac{1}{24}$ **9.** $\frac{9}{28}$ **11.** [6]; $5\frac{9}{40}$ **13.** [2]; $1\frac{19}{25}$ **15.** [3]; $3\frac{2}{3}$ **17.** [12]; $12\frac{1}{2}$ **19.** [12]; $13\frac{7}{32}$ **21.** $[1\frac{1}{2}]$; $1\frac{4}{9}$ **23.** $[\frac{5}{6}]$; $\frac{4}{5}$ **25.** [9]; $8\frac{32}{41}$ **27.** $\frac{4}{5}$ **29.** 16 **31.** $5\frac{1}{4}$ **33.** $\frac{1}{18}$ **35.** 1 **37.** $3\frac{7}{11}$ minutes **39.** $\frac{2}{3}$ **41.** $\frac{5}{24}$ **43.** $4\frac{3}{5}$ **45.** $4\frac{1}{10}$ **47.** $3\frac{11}{20}$

Page 165 **Math in Selling** **1.** $33.00 **3.** $55.47 **5.** $\frac{1}{5}$ off; 17¢ is $\frac{1}{5} \times$ 85¢.

Page 167 **Exercises** **1.** < **3.** > **5.** > **7.** < **9.** > **11.** < **13.** > **15.** < **17.** < **19.** > **21.** < **23.** < **25.** < **27.** $\frac{7}{12}, \frac{3}{4}, \frac{5}{6}, \frac{8}{9}$ **29.** $\frac{3}{10}, \frac{1}{2}, \frac{3}{4}, \frac{4}{5}, \frac{7}{8}$ **31.** $\frac{1}{8}, \frac{3}{10}, \frac{1}{3}, \frac{2}{5}, \frac{3}{7}$ **33.** > **35.** < **37.** $\frac{1}{9}, \frac{1}{3}, \frac{3}{9}$ **39.** $\frac{1}{9}, \frac{1}{3}, \frac{3}{9}$ **41.** Ellen **43.** freestyle

Pages 168–169 **Exercises** **1.** $160.00 **3.** $204.44 **5.** $301.00 **7.** $1\frac{1}{2}$ **9.** $5\frac{3}{4}$ **11.** $228.00; $9.00; $40.50; $268.50 **13.** **a.** $280 **b.** $12 **c.** $60 **d.** $340 **15.** 4:45 P.M. **17.** Answers may vary. One answer is: ACME, because they pay more for 40 hours work.

Page 171 **Exercises** **1.** $86.31; $163.69 **3.** $43.67; $136.33 **5.** $286.09; $313.91 **7.** $448.03 **9.** $220 **11.** $26.46 **13.** $160.20 **15.** $18

Page 172 **Rational Numbers** **1.** $\frac{14}{1}$ **3.** $\frac{9}{2}$ **5.** $\frac{25}{3}$ **7.** $\frac{11}{24}$ **9.** $\frac{5}{12}$ **11.** 0.45 **13.** $1\frac{9}{20}$

Pages 174–175 **Chapter Review** **1.** numerators; denominators **3.** multiplying **5.** overtime pay **7.** $\frac{8}{27}$ **9.** $1\frac{1}{3}$ **11.** $\frac{5}{21}$ **13.** $1\frac{1}{3}$ **15.** $\frac{2}{5}$ **17.** $\frac{1}{2}$ **19.** $\frac{20}{3}$ **21.** $\frac{19}{5}$ **23.** $\frac{61}{15}$ **25.** $\frac{43}{6}$ **27.** 4 **29.** 7 **31.** 13 **33.** 2 **35.** 5 **37.** 5 **39.** 8 **41.** [4]; $3\frac{24}{25}$ **43.** [4]; $5\frac{1}{4}$ **45.** [8]; $7\frac{7}{8}$ **47.** [12]; $12\frac{3}{4}$ **49.** 6 **51.** $\frac{1}{5}$ **53.** $\frac{8}{5}$ **55.** $\frac{1}{3}$ **57.** $1\frac{2}{3}$ **59.** $\frac{3}{10}$ **61.** 10 **63.** $\frac{1}{4}$ **65.** $\frac{2}{3}$ **67.** 6 **69.** $\frac{1}{15}$ **71.** 6 **73.** $\frac{5}{12}$ **75.** $1\frac{5}{7}$ **77.** $\frac{4}{5}$ **79.** < **81.** < **83.** > **85.** < **87.** > **89.** Rosa **91.** $310.39

Page 176 **Chapter Test** **1.** $\frac{3}{10}$ **3.** $3\frac{3}{4}$ **5.** $\frac{2}{3}$ **7.** $\frac{1}{4}$ **9.** $\frac{2}{5}$ **11.** 8 **13.** 4 **15.** 10 **17.** 25 **19.** [30]; 30 **21.** [4]; $4\frac{3}{8}$ **23.** $1\frac{1}{2}$ **25.** $1\frac{1}{2}$ **27.** $\frac{2}{3}$ **29.** $\frac{64}{99}$ **31.** 5 **33.** $\frac{1}{6}, \frac{1}{4}, \frac{1}{3}, \frac{1}{2}$ **35.** $\frac{1}{2}, \frac{7}{12}, \frac{5}{6}, \frac{9}{10}$ **37.** $555.60

Pages 177–178 **Cumulative Maintenance: Chapters 1–7** **1.** b **3.** c **5.** d **7.** d **9.** c **11.** c **13.** a **15.** c **17.** a **19.** c

CHAPTER 8 **MEASUREMENT: PERIMETER/AREA**

Page 181 **Exercises** **1.** 2 **3.** $1\frac{3}{4}$ **5.** $1\frac{13}{16}$ **7.** $\frac{6}{8}$, or $\frac{3}{4}$ inch **9.** 1 in; $1\frac{1}{2}$ in; $1\frac{1}{4}$ in; $1\frac{1}{4}$ in **11.** 1 in; $\frac{1}{2}$ in; $\frac{5}{8}$ in; $\frac{5}{8}$ in

Pages 182–183 **Exercises** **1.** foot **3.** mile **5.** 80 in **7.** 2 mi **9.** 400 in **11.** 8 **13.** 29 **15.** 1; 1720 **17.** 77 **19.** 4 ft 12 in, or 5 ft **21.** 2 yd 2 ft **23.** 10 ft 8 in **25.** 1 yd 2 ft 8 in **27.** 5 **29.** 2 yd 2 ft **31.** 1 mi 33 ft **33.** 15,000 ft **35.** 30 yd 1 ft

Pages 184–185 **Exercises** **1.** cm **3.** mm or cm **5.** m **7.** 3 cm **9.** 18 mm **11.** 8 cm; 81 mm **13.** b **15.** b **17.** No; 1.8 × 14 is less than 27. **19.** The measurement is never exact because the measuring instruments are never exact.

Pages 187–188 **Exercises** **1.** 12.8 cm **3.** 52 ft **5.** 1441 ft **7.** 19 in **9.** 3 yd **11.** $5\frac{1}{2}$ ft **13.** 208 ft **15.** 20.8 m **17.** 6.2 m **19.** 693.6 m **21.** $117.75 **23.** The figures are: 1 unit by 11 units, 2 units by 10 units, 3 units by 9 units, 4 units by 8 units, 5 units by 7 units, and 6 units by 6 units.

Page 189 **Mid–Chapter Review** **1.** 11 ft 1 in **3.** 4 ft 9 in **5.** 8 cm; 80 mm **7.** 86 ft **9.** 302 m **11.** 68.12 **13.** 40 **15.** 6 **17.** $3.00 and $3.75

Page 190 **Math and Formulas** **1.** $140.50 **3.** $272.50 **5.** $159.80

Pages 192–193 **Exercises** **1.** square units **3.** $A = s \times s$ **5.** $4 \times s$ **7.** 1134 m^2 **9.** 1.21 m^2 **11.** 4518 mm^2 **13.** 12 **15.** 8 **17.** 4 **19.** 36.4 **21.** 1.5 **23.** l: 5; w: 3 **25.** 4 **27.** 486 m^2 **29.** 1440 in^2

Page 195 **Exercises** **1.** $2,250 **3.** $675 **5.** $1,250 **7.** $375 **9.** $1,250 **11.** $2,400 **13.** $4,800; $3,200; $1,600; $400 **25.** $\frac{1}{10,000}$

Pages 197–198 **Exercises** **1.** 2a + 2b; a × b **3.** 2a + 2b; a × h **5.** 63 ft^2 **7.** 5.7 m^2 **9.** 544 cm^2 **11.** 12.8 ft^2 **13.** 6.4 in^2 **15.** 6 in^2 **17.** 6 ft **19.** 36.34 m^2 **21.** 396 m^2 **23.** $4

Page 199 **Math in Languange Arts** **1.** 7.2 m **3.** 0.4 m **5.** Sixth

Pages 200–201 **Exercises** **1.** two **3.** two **5.** same **7.** Triangle I; Triangle II **9.** 1820 cm^2 **11.** 20 in^2 **13.** 15 yd^2 **15.** 190 ft^2 **17.** Answers may vary. **19.** 8 m^2 **21.** 246 m^2

Page 202 **Squares in Design** **1.** 49 cm^2 **3.** 40 in^2

Pages 204–205 **Chapter Review** **1.** length **3.** rectangle **5.** $P = 2 \times (l + w)$ **7.** square units **9.** $A = \frac{1}{2} \times b \times h$ **11.** 16 ft 12 in, or 17 ft **13.** 9 ft 8 in **15.** 3 yd 1 ft 9 in **17.** 4 cm; 35 mm **19.** b **21.** 34 in **23.** 1.56 m **25.** 364 ft^2 **27.** 32.49 cm^2 **29.** 116.25 cm^2 **31.** 121 in^2 **33.** 495 in^2 **35.** 630 cm **37.** $7,500; $4,500; $2,250; $750

Page 206 **Chapter Test** **1.** 14 yd 1 ft **3.** 3 yd 1 ft 9 in **5.** 6 cm; 64 mm **7.** 120 in **9.** 34 cm **11.** 49 ft^2 **13.** 62 mm^2 **15.** 22.96 cm^2 **17.** 6.09 m^2 **19.** $500

Pages 207–208 **Cumulative Maintenance: Chapters 1–8** **1.** c **3.** b **5.** d **7.** a **9.** a **11.** b **13.** b **15.** c **17.** b **19.** c **21.** a

CHAPTER 9 EQUATIONS: RATIO AND PROPORTION

Page 211 **Exercises** **1.** 2 **3.** 8 **5.** n = 3 **7.** n = 1 **9.** n = 5 **11.** n = 4 **13.** n = 8 **15.** n = 9 **17.** n = 2 **19.** n = 15 **21.** n = 59 **23.** n = 73 **25.** n = 24 **27.** n = 45 **29.** n = 8 **31.** n = $\frac{2}{3}$ **33.** n = 0.7 **35.** n = $\frac{1}{4}$ **37.** Let n = number of shells Denise had originally; n + 6 = 18; n = 12 **39.** Let n = amount of money Robert has now; n + 37 = 100; n = 63

Page 213 **Exercises** **1.** 5 **3.** 7 **5.** 9 **7.** 5 **9.** 5 **11.** 3 **13.** 2 **15.** 16 **17.** 8 **19.** 2.1 **21.** n = 19 **23.** n = 71 **25.** n = 57 **27.** n = 45 **29.** n = 44 **31.** n = 63 **33.** n = 20 **35.** n = 45 **37.** n = 11.2 **39.** n = 1.5 **41.** n = 1 **43.** n = $1\frac{1}{4}$ **45.** b; n = 12 **47.** a; n = 26 **49.** b; n = 12

Pages 214–215 **Exercises** **1.** a; 2.5 pounds **3.** c; 60 feet **5.** a; 231 miles per hour **7.** c; 212°F **9.** c; 152 miles per hour

Page 217 **Exercises** **1.** 6 **3.** 7 **5.** 0.3 **7.** n = 5 **9.** n = 9 **11.** n = 9 **13.** n = 15 **15.** n = 16 **17.** n = 13 **19.** n = 22 **21.** n = 5 **23.** n = 15 **25.** n = 6 **27.** 9 **29.** 13 **31.** n = 30 **33.** n = 180 **35.** n = 30 **37.** n = 72 **39.** n = 35 **41.** n = 21 **43.** n = 56

45. $n = 66$ **47.** $n = 28.7$ **49.** $n = 1.6$ **51.** $n = 5$ **53.** $n = 1.5$ **55.** $n = 7.2$ **57.** $n = 5$
59. $n = 126$ **61.** $\frac{n}{4} = 12; n = 48$ **63.** $4n = 12; n = 3$

Pages 219–220   **Exercises**   **1.** 60   **3.** 30.5   **5.** 45 kilometers per hour   **7.** 0.3 mile per minute   **9.** 47
seconds longer   **11.** 17 hours   **13.** 3200 hours   **15.** Number of miles   **17.** About 270 miles   **19.** About $3\frac{1}{3}$
hours   **21.** 60 miles per hour

Page 222   **Exercises**   **1.** Subtract 7 from each side.   **3.** Add 2 to each side.   **5.** Add 0.4 to each side.
7. Subtract $\frac{1}{3}$ from each side.   **9.** Subtract 2.1 from each side.   **11.** $n = 4$ **13.** $n = 6$ **15.** $n = 10$
17. $n = 7$ **19.** $n = 132$ **21.** $n = 9$ **23.** $n = 96$ **25.** $n = 90$ **27.** $n = 189$ **29.** $n = 10$
31. $n = 28$ **33.** $n = 7.3$ **35.** b; 10 years   **37.** b; 298 students

Page 223   **Mid–Chapter Review**   **1.** $n = 5$ **3.** $n = 15$ **5.** $n = 18$ **7.** $n = 6.2$ **9.** $n = 8$ **11.** $n = 32$
13. $n = 97$ **15.** $n = 3.7$ **17.** $n = 6$ **19.** $n = 7$ **21.** $n = 4$ **23.** $n = 72$ **25.** $n = 4$ **27.** $n = 6$
29. $n = 15$ **31. a.** $n + 5 = 20$ **b.** $n = 15$; 15 seconds   **33.** $\frac{5}{6}$ **35.** $\frac{1}{6}$ **37.** $2\frac{17}{20}$ **39.** $4\frac{1}{24}$ **41.** 52 cm²

Page 225   **Exercises**   **1.** 6 to 4; 6:4; $\frac{6}{4}$ **3.** 6 to 10; 6:10; $\frac{6}{10}$ **5.** $\frac{3}{7}$ **7.** $\frac{14}{42}$ **9.** $\frac{7}{21}$ **11.** $\frac{5}{25}$ **13.** $\frac{16}{64}$
15. $\frac{10}{100}$ **17.** No   **19.** No   **21.** No   **23.** No   **25.** No   **27.** Yes   **29.** Yes   **31.** No   **33.** Yes
35. Yes   **37.** Yes   **39.** Yes   **41.** No   **43.** No   **45.** 4 to 10   **47.** 2; 7

Page 226   **Math in Photography**   **1.** 1 to 2   **3.** 1 to 8   **5.** 1 to 4   **7.** 4   **9.** $\frac{1}{4}$

Page 227   **Math in Music**   **1.** $\frac{12}{8}$ **3.** $\frac{3}{4}$ **5.** 12

Page 228–229   **Exercises**   **1.** =   **3.** ≠   **5.** ≠   **7.** =   **9.** =   **11.** $n = 9$ **13.** $n = 20$
15. $n = 91$ **17.** $n = 6$ **19.** $n = 6$ **21.** $n = 21$ **23.** $n = 9$ **25.** $n = 7$ **27.** $n = 30$ **29.** $n = 16$
31. $n = 5$ **33.** $n = 8$ **35.** $n = 7$ **37.** $n = 12$ **39.** $n = 1$ **41.** $n = 19$ **43.** $n = 3$ **45.** $n = 14$
47. $n = 12$ **49.** $n = 50$ **51. a.** $\frac{3}{5} = \frac{n}{500}$ **b.** 300 voted for Gregg.   **53. a.** $\frac{5}{2} = \frac{10}{n}$ **b.** 4 pints   **55. a.** $\frac{1}{2} = \frac{5}{n}$
b. 10 gallons

Page 231   **Exercises**   **1.** 24 ft   **3.** 16 ft   **5. a.** More than one step **b.** $\frac{7}{12}$ **7.** Choosing a computation
method; **a.** An exact answer because the pay must be correct.   **b.** Calculator or paper and pencil; Reasons may
vary.   **c.** Plumber: $308.55; Carpenter: $197.40   **9.** Answers may vary.

Page 232   **Golden Ratio**   **1.** 0.6436   **3.** Answers may vary.   **5.** length: 10 mm

Pages 234–235   **Chapter Review**   **1.** subtract   **3.** divide   **5.** time   **7.** proportion   **9.** $n = 13$
11. $n = 9$ **13.** $n = 12$ **15.** $n = 47$ **17.** $n = 1.4$ **19.** $n = 5.4$ **21.** $n = 21$ **23.** $n = 28$
25. $n = 51$ **27.** $n = 60$ **29.** $n = 7.9$ **31.** $n = 12.2$ **33.** $n = 8$ **35.** $n = 8$ **37.** $n = 12$
39. $n = 14$ **41.** $n = 52$ **43.** $n = 182$ **45.** $n = 153$ **47.** $n = 468$ **49.** $n = 2$ **51.** $n = 4$
53. $n = 6$ **55.** $n = 189$ **57.** $\frac{2}{4}$ **59.** $\frac{36}{9}$ **61.** $\frac{18}{45}$ **63.** $\frac{60}{48}$ **65.** No   **67.** No   **69.** Yes   **71.** Yes
73. $n = 4$ **75.** $n = 6$ **77.** $n = 6$ **79.** $n = 2$ **81.** $n = 20$ **83.** $n = 24$ **85.** $n = 12$ **87.** $n = 4$
89. 1425 miles   **91.** 12 gallons

Page 236   **Chapter Test**   **1.** $n = 16$ **3.** $n = 14$ **5.** $n = 12$ **7.** $n = 75$ **9.** $n = 3$ **11.** $n = 28$
13. $\frac{5}{16}$ **15.** $\frac{14}{49}$ **17.** No   **19.** Yes   **21.** $n = 6$ **23.** $n = 7$ **25.** $n = 70$ **27.** $n = 7$ **29.** $n = 3$
31. c; $n = 68$ **33.** 13.5 feet

Pages 237–238   **Cumulative Maintenance: Chapters 1–9**   **1.** c   **3.** a   **5.** d   **7.** c   **9.** d   **11.** a
13. c   **15.** d   **17.** c

Pages 240–241   Exercises   1. 100   **3.** $\frac{36}{100}$ **5.** 4%   **7.** Living Room and Kitchen, or Living Room and either Bedroom   **9. a.** 83% **b.** 17%   **11. a.** 74% **b.** 26%   **13.** 40%   **15.** 21%   **17.** 11%   **19.** 7%   **21.** 3%   **23.** 1%   **25.** 83%   **27.** 77%   **29.** 99%   **31.** 35%   **33.** 83%   **35.** 6%   **37.** 7%   **39.** 30%   **41.** 50%   **43.** 0.11; 11%   **45.** 0.09; 9%   **47.** 0.80; 80%   **49.** 0.75; 75%   **51.** 0.83; 83%   **53.** 0.13; 13%

Page 243   Exercises   1. $\frac{7}{100}$; 0.07   **3.** $\frac{7.5}{100}$; 0.075   **5.** $\frac{700}{100}$; 7   **7.** $\frac{13}{100}$; 0.13   **9.** $\frac{3}{100}$; 0.03   **11.** $\frac{60}{100}$; 0.6   **13.** $\frac{112}{100}$; 1.12   **15.** $\frac{370}{100}$; 3.7   **17.** $\frac{208}{100}$; 2.08   **19.** $\frac{87.5}{100}$; 0.875   **21.** $\frac{52.9}{100}$; 0.529   **23.** $\frac{9}{100}$; 0.09   **25.** 21%   **27.** 4%   **29.** 80.5%   **31.** 57.5%   **33.** 175%   **35.** 30%   **37.** 90%   **39.** 0.8%   **41.** 390%   **43.** 0.3   **45.** 0.055   **47.** 0.6   **49.** 25%   **51.** 92%

Page 245   Exercises   1. $\frac{35}{100} = \frac{7}{20}$ **3.** $\frac{4}{100} = \frac{1}{25}$ **5.** $\frac{4\frac{1}{2}}{100} = \frac{9}{200}$ **7.** $\frac{2}{5}$ **9.** $\frac{1}{2}$ **11.** $\frac{3}{4}$ **13.** $\frac{7}{20}$ **15.** $\frac{11}{25}$ **17.** $\frac{49}{50}$ **19.** $\frac{43}{100}$ **21.** $\frac{18}{25}$ **23.** $\frac{22}{25}$ **25.** $\frac{3}{200}$ **27.** $\frac{3}{400}$ **29.** $\frac{1}{160}$ **31.** $\frac{5}{8}$ **33.** $\frac{7}{8}$ **35.** $\frac{5}{6}$ **37.** $\frac{1}{20}$ **39.** $\frac{1}{40}$ **41.** $\frac{1}{30}$ **43.** <   **45.** >   **47.** >   **49.** <   **51.** $\frac{3}{20}$ **53.** $\frac{7}{10}$

Page 247   Exercises   1. 50; 50%   **3.** 40; 40%   **5.** 35; 35%   **7.** 75%   **9.** 20%   **11.** 5%   **13.** 2%   **15.** 15%   **17.** 16%   **19.** 0.33$\frac{1}{3}$; 33$\frac{1}{3}$%   **21.** 0.625; 62.5%   **23.** 0.41$\frac{2}{3}$; 41$\frac{2}{3}$%   **25.** 0.225; 22.5%   **27.** 0.83$\frac{1}{3}$; 83$\frac{1}{3}$%   **29.** 0.22$\frac{2}{9}$; 22$\frac{2}{9}$%   **31.** $\frac{85}{100}$; 0.85; 85%   **33.** $\frac{13}{20}$; $\frac{65}{100}$; 65%   **35.** $\frac{60}{100}$; 0.60; 60%   **37.** $\frac{7}{8}$; $\frac{87.5}{100}$; 87.5%   **39.** $\frac{32}{100}$; 0.32; 32%   **41.** 83$\frac{1}{3}$%   **43.** $\frac{1}{5}$ **45.** $\frac{4}{5}$ **47.** $\frac{2}{3}$ or $\frac{7}{10}$ **49.** $\frac{3}{10}$ **51.** $\frac{4}{5}$ **53.** $\frac{1}{3}$

Page 248   Math in Social Studies   1. 40   **3.** 40%   **5.** 10%   **7.** 2.5%   **9.** 25%

11.

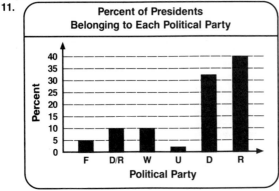

Page 249   Mid–Chapter Review   1. 5%   **3.** 40%   **5.** 25%   **7.** 0.18   **9.** 1.04   **11.** 0.09   **13.** 24%   **15.** 5%   **17.** 72.1%   **19.** $\frac{1}{2}$ **21.** $\frac{57}{100}$ **23.** $\frac{7}{800}$ **25.** 50%   **27.** 15%   **29.** 26%   **31.** 0.875; 87.5%   **33.** 0.44$\frac{4}{9}$; 44$\frac{4}{9}$%   **35.** 0.91$\frac{2}{3}$; 91$\frac{2}{3}$%   **37.** $\frac{16}{25}$ **39.** $\frac{1}{8}$ **41.** 3   **43.** 7$\frac{1}{4}$ **45.** English   **47.** Science

Page 251   Exercises   1. 30   **3.** 90   **5.** 108   **7.** 6.4   **9.** 21   **11.** 0.81   **13.** 178.6   **15.** 38.7   **17.** 32   **19.** 798   **21.** 4   **23.** 48   **25.** 51$\frac{3}{5}$ **27.** 195   **29.** 129$\frac{3}{5}$ **31.** 21   **33.** 40   **35.** 0.27   **37.** 24   **39.** 18   **41.** 70   **43.** 1.8   **45.** 228   **47.** 70   **49.** 40% of 32   **51.** 50% of 600

Page 253   Exercises   1. regular price   **3.** discount   **5.** discount   **7.** $9   **9.** $39.50   **11.** $34.17   **13.** $45   **15.** $40.50

Page 255   Exercises   1. $0.96   **3.** $0.73   **5.** $0.65   **7.** $0.70   **9.** $2.14   **11.** $0.32; $7.47   **13.** $0.68; $15.68   **15.** $0.96; $22.21   **17.** $17.01   **19.** $25.75   **21.** $0.23; $7.87   **23.** $8.49; $197.08   **25.** $36.16

Page 257   Exercises   1. withdrawal   **3.** added to   **5.** $67.50   **7.** $300.00   **9.** $466.67   **11.** $42.00
13. $920.34

Pages 258–259   Exercises   1. $10.50   **3.** $12.54   **5.** $25.20   **7.** $3.25   **9.** $4.60   **11.** $18
13. $4.80   **15.** $33   **17.** $14   **19.** $2.10   **21.** $11.25   **23.** $50   **25.** $42   **27.** $180   **29.** $60
31. $59.85   **33.** $10 off   **35.** $10 off

Page 260   Compound Interest   1. 1.5%; 2; $772.65   **3.** $80

Pages 262–263   Chapter Review   1. hundredths   **3.** right   **5.** discount   **7.** years   **9.** 4%   **11.** 35%
13. 18%   **15.** 0.54   **17.** 2.8   **19.** 0.179   **21.** 0.03   **23.** 0.079   **25.** 0.005   **27.** 31%   **29.** 46%
31. 6%   **33.** 79.2%   **35.** 7.1%   **37.** 0.3%   **39.** $\frac{6}{25}$ **41.** $\frac{1}{25}$ **43.** $\frac{1}{140}$ **45.** $\frac{1}{180}$ **47.** $\frac{23}{400}$ **49.** $\frac{31}{200}$
51. 70%   **53.** 95%   **55.** 78%   **57.** $77\frac{7}{9}$%   **59.** 87.5%   **61.** 43.75%   **63.** 46   **65.** 14.7   **67.** 0.82
69. 73.95   **71.** 7   **73.** 26   **75.** 21   **77.** 9   **79.** $1.28; $19.53   **81.** $3.85; $58.90   **83.** $8.33
85. $90.00   **87.** $8   **89.** $13.05   **91.** $3.30   **93.** 10   **95.** $180

Page 264   Chapter Test   1. 0.25   **3.** 1.81   **5.** 0.03   **7.** 32%   **9.** 58%   **11.** 7.3%   **13.** $\frac{23}{100}$
15. $\frac{1}{20}$ **17.** $\frac{1}{120}$ **19.** 80%   **21.** 35%   **23.** $88\frac{8}{9}$%   **25.** 4   **27.** 13.8   **29.** $9   **31.** $18.00
33. $296.09

Pages 265–266   Cumulative Maintenance: Chapters 1–10   1. a   **3.** b   **5.** b   **7.** d   **9.** b   **11.** c
13. b   **15.** d   **17.** b   **19.** a

CHAPTER 11   MORE ON PERCENT

Page 269   Exercises   1. less than 100%   **3.** less than 100%   **5.** greater than 100%   **7.** n × 20 = 15
9. n × 56 = 60   **11.** 30 = n × 66   **13.** 25%   **15.** $33\frac{1}{3}$%   **17.** 300%   **19.** 75%   **21.** 87.5%
23. 125%   **25.** $83\frac{1}{3}$%   **27.** $16\frac{2}{3}$%   **29.** 37.5%   **31.** $23\frac{1}{3}$%   **33.** 80%   **35.** $16\frac{2}{3}$%

Page 271   Exercises   1. $\frac{48}{320}$ **3.** $\frac{930}{4650}$, or $\frac{93}{465}$ **5.** $\frac{1.21}{24.20}$ **7.** 75% increase   **9.** 50% decrease   **11.** 200%
increase   **13.** 80% decrease   **15.** 60%   **17.** $33\frac{1}{3}$%   **19.** 50%   **21.** 1930–1970

Page 273   Exercises   1. 18 = 10% × n   **3.** 6 = 15% × n   **5.** 2 = 25% × n   **7.** 450   **9.** 20
11. 35   **13.** 20   **15.** 46   **17.** 20   **19.** 1500   **21.** 25 games   **23.** 20 free throws   **25.** 20 days

Page 274   Mid–Chapter Review   1. n × 70 = 14; 20%   **3.** 3 = n × 15; 20%   **5.** n × 30 = 25; $83\frac{1}{3}$%
7. $66\frac{2}{3}$% increase   **9.** $33\frac{1}{3}$% decrease   **11.** 80% decrease   **13.** 20% increase   **15.** 50 = 0.2 × n; 250
17. 18 = 0.75 × n; 24   **19.** 25 = 0.5 × n; 50   **21.** 75%   **23.** 4 to 5, or $\frac{4}{5}$ **25.** $520

Page 275   Math in Business   1. $1250   **3.** $675   **5.** $9660   **7.** $1042.80

Pages 277–278   Exercises   1. a. 30%; 60; n   **b.** $\frac{30}{100} = \frac{n}{60}$ **3. a.** n; 75; 12   **b.** $\frac{n}{100} = \frac{12}{75}$ **5. a.** n; 32; 8
b. $\frac{n}{100} = \frac{8}{32}$ **7. a.** 63%; n; 63   **b.** $\frac{63}{100} = \frac{63}{n}$ **9.** 3   **11.** 9   **13.** 50%   **15.** 60%   **17.** 200   **19.** 100
21. 12   **23.** 20%   **25.** 30 students   **27.** 75%   **29.** 450 students   **31.** 300 cans   **33.** 4 days

Page 280   Exercises   1. b   **3.** b   **5.** c   **7.** a   **9.** f   **11.** 60%   **13.** 20   **15.** 100   **17.** $83\frac{1}{3}$%

Pages 282–283   Exercises   1. c   **3.** b   **5.** b   **7.** c   **9.** $448   **11.** 15%   **13.** $2.40   **15.** 14%
17. $62\frac{1}{2}$%   **19.** $9.03

Page 285 **Exercises** **1.** About $36 **3.** About $54 **5.** About $63 **7.** About $40 **9.** About $1.35 **11.** About $1.50 **13.** About $8.40 **15. a.** Making a model **b.** Jon's **17. a.** More than one step **b.** 30% discount on a price of $90

Page 286 **Averages and Percents** **1.** 58% **3.** 82%

Pages 288–289 **Chapter Review** **1.** increase **3.** commission **5.** n × 8 = 2; 25% **7.** n × 25 = 5; 20% **9.** 9 = n × 15; 60% **11.** 12 = n × 30; 40% **13.** 25% increase **15.** $16\frac{2}{3}$% decrease **17.** $33\frac{1}{3}$% decrease **19.** 19 = 0.5 × n; 38 **21.** 8 = 0.1 × n; 80 **23.** 25 = 0.2 × n; 125 **25.** 72 = 0.3 × n; 240 **27.** 5 **29.** 80 **31.** 780 **33.** b **35.** b **37.** n = 0.05 × 60; 3 **39.** 25 = $\frac{1}{3}$ × n; 75 **41.** n × 200 = 180; 90% **43.** About $48 **45.** About $48 **47.** 5% **49.** $24

Page 290 **Chapter Test** **1.** a **3.** a **5.** n = 0.6 × 400; 240 **7.** 16 = n × 64; 25% **9.** 75 = 0.3 × n; 250 **11.** 10% **13.** 10 **15.** 90 **17.** $40 **19.** About $80

Pages 291–292 **Cumulative Maintenance: Chapters 1–11** **1.** d **3.** a **5.** b **7.** c **9.** b **11.** b **13.** b **15.** c **17.** a **19.** c **21.** b

CHAPTER 12   MEASUREMENT: SURFACE AREA/VOLUME

Pages 294–295 **Exercises** **Across: 1.** milliliter **2.** kilogram **Down: 1.** milligram **3.** liter **4.** gram **5.** b **7.** a **9.** a **11.** b **13.** a **15.** $54.28

Page 297 **Exercises** **1.** Divide **3.** Multiply **5.** Multiply **7.** Divide **9.** 4000 **11.** 2 **13.** 500 **15.** 5.4 **17.** 5.76 **19.** 3 **21.** 5300 **23.** 58 **25.** 0.01 **27.** 10.021 kilograms **29.** $36.50

Pages 298–299 **Exercises** **1.** 4 gal 2 qt **3.** 3 lb 10 oz **5.** 3 ft 9 in **7.** 2 mi 4440 ft **9.** $3\frac{5}{8}$ **11.** $3\frac{1}{2}$ **13.** $1\frac{1}{4}$ **17.** 19 ft 2 in **19.** 10 yd 2 ft **21.** 5 gal 3 qt **23.** 3 lb 14 oz **25.** 11 lb 4 oz **27.** 1 pt **29.** 1800 lb **31.** 10 qt **33.** 10 pt **35.** 5 yd 1 ft **37.** About 6 ounces

Pages 301–302 **Exercises** **1.** Back: 5 cm × 3 cm, 15 cm²; Top: 5 cm × 4 cm, 20 cm²; Bottom: 5 cm × 4 cm, 20 cm²; Left Side: 4 cm × 3 cm, 12 cm²; Right Side: 4 cm × 3 cm, 12 cm²; Surface Area: 94 cm² **3.** bottom **5.** 2; 2 **7.** 4456 cm² **9.** 37,080 cm² **11.** 4600 in² **13.** 174 ft² **15.** 270 in² **17.** 504 ft² **19.** 61.28 m² **21.** 4 in² **23.** 6 **25.** 122 m² **27.** 6 yd

Page 303 **Mid–Chapter Review** **1.** a **3.** a **5.** 2400 **7.** 5 **9.** 6.25 **11.** 4 lb 14 oz **13.** 6 qt 1 pt **15.** 1470 cm² **17.** 57 ft² **19.** 16 m² **21.** 300 students **23.** 30 cm²

Pages 304–305 **Exercises** **1.** cubic **3.** a **5.** 23 **7.** 21 **9.** 120 in³ **11.** 168 m³ **13.** 120 ft³ **15.** 1000 in³ **17.** 0.009 cm³ **19.** 280 ft³ **21.** 3 meters

Page 307 **Exercises** **1.** About 32°C **2.** About 20°F **5.** About 44°C

Pages 308–309 **Exercises** **1.** 3 **3.** 37 **5.** 8 m³ **7.** 8 in³ **9.** 10.5 yd³ **11.** 448 ft³ **13.** 12.5 in³ **15.** 8.64 m³ **17. a.** The Pyramid of the Sun **b.** 832,598 yd³

Page 310 **Math in Wallpapering** **1.** 7 **3.** About $54

Pages 312–313 **Chapter Review** **1.** capacity **3.** 10 **5.** rectangular prism **7.** 0 **9.** a **11.** c **13.** 13,000 **15.** 0.4 **17.** 5000 **19.** 1.5 **21.** 875,000 **23.** 3.721 **25.** 10 gal **27.** 1400 lb **29.** 2088 cm² **31.** $43\frac{3}{4}$ ft² **33.** 20.32 m² **35.** 978 yd² **37.** 64 in³ **39.** 17.92 m³ **41.** 18 yd³ **43.** 102 ft³ **45.** 27°F

Page 314 **Chapter Test** **1.** b **3.** b **5.** 3200 **7.** 0.43 **9.** 12 qt **11.** 1700 lb **13.** 88 cm² **15.** $307\frac{1}{3}$ ft² **17.** 12 in³ **19.** 72 m³ **21.** 15 m³ **23.** $24\frac{2}{3}$ ft³ **25.** 52.2°F

CHAPTER 13 CIRCLES AND APPLICATIONS

Page 318 Exercises 1. Answers will vary. **3.** 6.28 cm

Pages 319–320 Exercises 1. 2 **3.** π **5.** 3.14; $\frac{22}{7}$ **7.** 9 in **9.** 7071 mi **11.** 69 cm **13.** 80 cm **15.** About 24,913 mi **17.** About 31 m **19. a.** 88 in **b.** 720 times

Page 322 Exercises 1. 40°; acute **3.** 90°; right **5.** 20°; acute **7.** angle 1: 50°; angle 2: 70°; angle 3: 60°; 180°; Yes **9.** angle 1: 125°; angle 2: 25°; angle 3: 30°; 180°; Yes **11.** 70°; Acute **13.** 90°; Right

Page 324 Exercises 1. 100% **3.** 20% **5.** $168 **7.** See circle graph below. **9. a.** Football: $\frac{250}{1000} = \frac{1}{4}$; 25%; 90°; Baseball: $\frac{150}{1000} = \frac{3}{20}$; 15%; 54°; Soccer: $\frac{50}{1000} = \frac{1}{20}$; 5%; 18°; Track: $\frac{100}{1000} = \frac{1}{10}$; 10%; 36°; Other: $\frac{50}{1000} = \frac{1}{20}$; 5%; 18° **b.** See circle graph below.

7. Immigration to the U.S. Since 1820

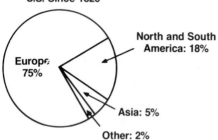

9b. Favorite Sports at Weldon High

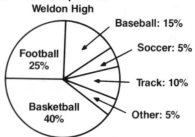

Page 325–326 Exercises 1. $10.40 **3.** $20.00 **5.** $\frac{12.00}{80} = 15\%$ **7.** $16.00; $\frac{16.00}{80} = 20\%$ **9.** $12.00; $\frac{12.00}{80} = 15\%$ **11.** 100% **13.** $18.00 **15.** $18.00 **17.** See circle graph at right. **19.** $16.50 **21.** $11.00 **23.** $8.80 **25.** $11.00 **27.** Some expense items occur once per month.

Maria's Budget

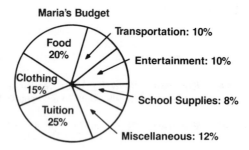

Page 327 Mid–Chapter Review 1. 20 cm **3.** 28 in **5.** acute **7.** obtuse **9.** acute **11.** See circle graph at right. **13.** $39 **15.** $1\frac{1}{10}$ **17.** $5\frac{9}{20}$ **19.** 1260 square miles

Uses of Paper

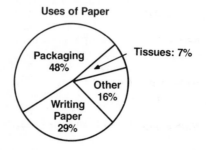

Pages 328–329 Exercises 1. b **3.** a **5.** ◇◇ / ◇◇ **7.** ■ **9.** △▽ **11. a.** Making a model **b.** $1\frac{1}{4}$ mi

Page 331 Exercises 1. π **3.** 6.28 cm **5.** 314 ft^2 **7.** 154 in^2 **9.** 79 m^2 **11.** 1661 cm^2 **13.** 62 cm^2 **15.** The area of the square is greater; $7\frac{5}{7}$ in^2

Page 333 Exercises 1. inches **3.** square inches **5.** inches **7.** 28,286 ft^3 **9.** 226 in^3 **11.** 100 cm^3 **13.** 663 cm^3 **15.** 3077 m^3

Page 335 Exercises 1. 452 cm^3 **3.** cylinder; 301 cm^3 more **5.** 42 ft^3 **7.** 212 in^3 **9.** 2093 cm^3 **11.** 170 mm^3

Page 336 Volume of Spheres 1. 195,511 ft^3 **3.** 310 in^3

Pages 338–339 Chapter Review 1. diameter **3.** budget **5.** cubic **7.** 38 m **9.** 41 ft **11.** 90°; right **13.** 112°; obtuse **15.** See circle graph at right. **17.** 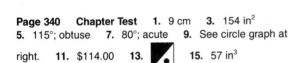 **19.** 113 cm^2 **21.** 151 yd^3 **23.** $16.80 **25.** $117.60 **27.** 2,718,571 mi

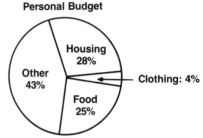

Page 340 Chapter Test 1. 9 cm **3.** 154 in^2 **5.** 115°; obtuse **7.** 80°; acute **9.** See circle graph at right. **11.** $114.00 **13.** 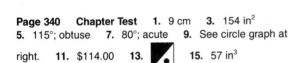 **15.** 57 in^3

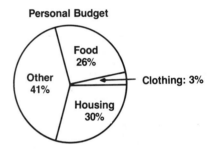

Pages 341–342 Cumulative Maintenance: Chapters 1–13 1. d **3.** d **5.** b **7.** d **9.** c **11.** d **13.** b **15.** b **17.** a **19.** b **21.** b

CHAPTER 14 PROBABILITY

Page 345 Exercises 1. White: 12 oz: regular, thin; 16 oz: regular, thin; 24 oz: regular, thin; Pumpernickel: 12 oz: regular, thin; 16 oz: regular, thin; 24 oz: regular, thin **3.** 2 **5.** 6 **7.** 80

Page 347 Exercises 1. 6 **3.** 24 **5.** 20 **7.** 3 **9.** 1 **11.** 6 **13.** 4 **15.** 2

Page 349 Exercises 1. P **3.** 1 **5.** $\frac{1}{3}$; $33\frac{1}{3}$% **7.** 1 **9.** $\frac{1}{3}$; $33\frac{1}{3}$% **11.** $\frac{1}{4}$; 25% **13.** $\frac{3}{4}$; 75% **15.** Drawing a black card **17.** $\frac{1}{10}$; 10% **19.** $\frac{1}{2}$; 50%

Page 351 Exercises 1. $\frac{1}{12}$, or $8\frac{1}{3}$% **3.** 7 **5.** $\frac{1}{3}$, or $33\frac{1}{3}$% **7.** $\frac{2}{9}$, or $22\frac{2}{9}$% **9.** E; there are more E's than any other letter. **11.** N and S; there are two of each.

Page 352 Mid–Chapter Review 1. 12 **3.** 6 **5.** 120 **7.** $\frac{1}{2}$, or 50% **9.** $\frac{1}{2}$, or 50% **11.** $\frac{11}{28}$ **13.** $\frac{1}{2}$ **15.** $\frac{1}{6}$ **17.** $\frac{1}{36}$ **19.** $\frac{1}{2}, \frac{4}{7}, \frac{2}{3}, \frac{5}{6}$

Page 353 Math in Advertising 1. No. All voters are not friends. **3.** No. The sample is too small.

Page 355 Exercises 1. $\frac{3}{16}$ **3.** $\frac{1}{16}$ **5.** Green shirt and red sweater **7.** $\frac{1}{4}$ **9.** $\frac{1}{16}$ **11.** $\frac{1}{8}$ **13.** $\frac{1}{4}$

Page 357 Exercises 1. $\frac{1}{9}$ **3.** $\frac{1}{6}$ **5.** $\frac{1}{36}$ **7.** $\frac{7}{36}$ **9.** 16 **11.** $\frac{1}{8}$ **13.** $\frac{13}{16}$

Page 359   Exercises   1. $\frac{1}{50}$; 2%   **3.** $\frac{1}{10}$; 10%   **5.** $\frac{49}{50}$; 98%   **7.** 360   **9.** 1176   **11.** 480   **13.** 900
15. 100

Page 360   What are the Odds   1. 4 to 3   **3.** 1 to 2

Pages 362–363   Chapter Review   1. tree diagram   **3.** probability   **5.** 1   **7.** 6   **9.** 3,628,800   **11.** $\frac{1}{7}$
13. $\frac{1}{7}$ **15.** $\frac{2}{7}$ **17.** $\frac{1}{5}$ **19.** $\frac{1}{6}$ **21.** $\frac{1}{15}$ **23.** $\frac{1}{2}$ **25.** $\frac{4}{5}$ **27.** $\frac{7}{10}$ **29.** 60   **31.** 30

Page 364   Chapter Test   1. 720   **3.** $\frac{1}{12}$ **5.** $\frac{1}{4}$ **7.** $\frac{1}{2}$; 50%   **9.** $\frac{1}{4}$; 25%   **11.** $\frac{1}{5}$; 20%   **13.** $\frac{1}{10}$; 10%
15. $\frac{1}{5}$ **17.** $\frac{2}{5}$ **19.** $\frac{3}{5}$

Pages 365–366   Cumulative Maintenance: Chapters 1–14   1. d   **3.** c   **5.** d   **7.** a   **9.** d   **11.** c
13. b   **15.** a   **17.** c   **19.** d

CHAPTER 15   SQUARES AND SQUARE ROOTS

Pages 368–369   Exercises   1. perfect square   **3.** exponent   **5.** 2   **7.** 5   **9.** 49   **11.** 64   **13.** 121
15. 100   **17.** 900   **19.** 441   **21.** 361   **23.** 3600   **25.** 40,000   **27.** 1   **29.** 9   **31.** 10   **33.** 30
35. 25   **37.** 14   **39.** 60   **41.** 13 m   **43.** 729   **45.** $\frac{1}{4}$ **47.** 50   **49.** 1.69   **51.** 0.2   **53.** $\frac{1}{5}$
55. $\frac{5}{9}$ **57.** $\frac{1}{6}$ **59.** 400 ft   **61.** 1 ft   **63.** 4 sec

Page 372   Exercises   1. 7569   **3.** 9.274   **5.** 89   **7.** 2025   **9.** 6400   **11.** 14,641   **13.** 15,625
15. 21,904   **17.** 4356   **19.** 8.367   **21.** 11.874   **23.** 10.247   **25.** 7.937   **27.** 11.225   **29.** 9.487
31. 1.732   **33.** 7.000   **35.** 9.165   **37.** 18   **39.** 26   **41.** 40   **43.** 22   **45.** 99   **47.** 141   **49.** 29
51. 41   **53.** 117   **55.** 66 mi   **57.** 173 mi

Page 374   Exercises   1. 90°   **3.** hypotenuse   **5.** 30   **7.** Yes   **9.** No   **11.** Yes   **13.** Yes   **15.** No

Pages 376–377   Exercises   1. 10   **3.** 11.314   **5.** 7.810 ft   **7.** 9.487 m   **9.** 7.2 m   **11.** 16
13. 35   **15.** 12.0 m   **17.** 5.4 m   **19.** 34 yd

Page 378   Mid–Chapter Review   1. 1   **3.** 289   **5.** $\frac{1}{16}$ **7.** 1   **9.** 11   **11.** 0.6   **13.** 5041
15. 19,321   **17.** 11.958   **19.** No   **21.** Yes   **23.** 26 m   **25.** 36 ft   **27.** 3.3 m   **29.** 72   **31.** 12 in

Page 379   Exercises   1. ∠A and ∠E; ∠B and ∠F   **3.** DE; CA

Pages 381–382   Exercises   1. ∠P; ∠Q; ∠R   **3.** 9   **5.** 15   **7.** $\frac{1.6}{2} = \frac{h}{70}$; h = 56 m   **9.** $\frac{1.5}{2.4} = \frac{h}{8}$; h = 5 m

Pages 384–385   Exercises   1. 0.40   **3.** 0.67   **5.** 0.75   **7.** 2.22   **9.** 0.176   **11.** 0.466   **13.** 1.00
15. 6   **17.** 57   **19.** 62 m   **21.** 24 m   **23.** 76 m

Page 386   Finding Reaction Time   1. Answers will vary.

Pages 388–389   Chapter Review   1. square   **3.** right   **5.** similar   **7.** 36   **9.** 225   **11.** 169   **13.** 3
15. 9   **17.** 15   **19.** 729   **21.** 7921   **23.** 12,544   **25.** 324   **27.** 5476   **29.** 17,956   **31.** 8.660
33. 10.677   **35.** 9.747   **37.** 59   **39.** 95   **41.** 106   **43.** No   **45.** Yes   **47.** 5.8   **49.** 8   **51.** 7; n = 14
53. 6.3 m   **55.** 56 m

Page 390   Chapter Test   1. 9   **3.** 4   **5.** 12   **7.** 1369   **9.** 10.198   **11.** No   **13.** 10   **15.** 36
17. 2   **19.** 25 m

Pages 391–392   Cumulative Maintenance: Chapters 1–15   1. a   **3.** a   **5.** b   **7.** b   **9.** a   **11.** d
13. b   **15.** d   **17.** d   **19.** d

Pages 394–395   Exercises   1. 120   **3.** -5 **5.** -500 **7.** 10   **9.** -3 **11.** 6   **13.** -4 **15.** 80
17. -45 **19.** 40   **21.** -23 **23.** 10   **25.** -40 **27.** 70   **29.** 10,200   **31.** -282 **33.** 20,320

Pages 396–397   Exercises   1. 30; left   **3.** same   **5.** opposites   **7.** 10   **9.** 0   **11.** -87 **13.** 32
15. -311 **17.** 5000   **19.** -9125 **21.** 20 miles north   **23.** 3 hours after   **25.** Gain 5 pounds.   **27.** $>$
29. $>$ **31.** $>$ **33.** $>$ **35.** $<$ **37.** $>$ **39.** $-8, -7, -3, -1, 7$ **41.** $-25, -8, 0, 5, 16$ **43.** $>$

Page 398   Exercises   1. -6 **3.** -5 **5.** -7 **7.** right   **9.** 8

Pages 399–400   Exercises   1. c   **3.** a   **5.** 8   **7.** 4   **9.** 6   **11.** 10   **13.** -7 **15.** -5
17. -1 **19.** -7 **21.** -8 **23.** -6 **25.** -9 **27.** -8 **29.** -11 **31.** 10   **33.** -1 **35.** -10
37. 12   **39.** 7   **41.** 9   **43.** -12 **45.** negative   **47.** 5   **49.** 9   **51.** -1 **53.** 0   **55.** -15
57. $0 + (-3) = -3$ **59.** $-4 + (-7) = -11$

Page 401   Exercises   1. -2 **3.** 0   **5.** -5 **7.** left   **9.** 2   **11.** opposites   **13.** The sum of opposites
is zero.

Page 403   Exercises   1. c   **3.** d   **5.** -4 **7.** 2   **9.** 1   **11.** 4   **13.** 2   **15.** 1   **17.** 0   **19.** -5
21. 8   **23.** 2   **25.** 2   **27.** 0   **29.** -5 **31.** 6   **33.** -13 **35.** 9   **37.** 6   **39.** 6   **41.** 5
43. $0 + (-3) = -3$ **45.** $-4 + 4 = 0$

Page 404   Mid–Chapter Review   1. -175 **3.** 2000   **5.** 6   **7.** -15 **9.** 0   **11.** -78 **13.** $<$
15. $>$ **17.** $<$ **19.** $>$ **21.** 10   **23.** 8   **25.** -7 **27.** -7 **29.** 1   **31.** 1   **33.** 3   **35.** 4
37. $0 + (-4) = -4$ **39.** 10–ounce can   **41.** $\frac{1}{3}$

Page 405   Math and Automobile Maintenance   1. 20W–40   **3.** 5W–30   **5.** 100°F

Page 407   Exercises   1. $-2; (-2)$ **3.** $-1; (-1)$ **5.** 6; 6   **7.** -1 **9.** 3   **11.** -19 **13.** -4
15. -10 **17.** 6   **19.** -7 **21.** 12   **23.** 9   **25.** 18   **27.** 2   **29.** 13   **31.** -14 **33.** -3 **35.** -9
37. 2   **39.** 1   **41.** -14 **43.** -7 **45.** -7 **47.** -15 **49.** 13   **51.** 7   **53.** -13 **55.** $9 + (-8) = 1$
57. $-4 + 2 = -2$ **59.** $-3 + 9 = 6$ **61.** $-11 + 9 = -2$ **63.** $-6 + 5 = -1$ **65.** $-16 + (-10) = -26$

Page 409   Exercises   1. $-22°F$ **3.** $-13°F$ **5.** 35° colder   **7.** 21° colder   **9.** 30 mph   **11.** Using a
formula; 849.2°F

Page 411   Exercises   1. ■   **3.** ■ ■   **5.** (-2) **7.** (-8) **9.** (-4) **11.** (-6) **13.** -5
15. -8 **17.** -10 **19.** -10 **21.** -2 **23.** 2   **25.** 4   **27.** -6 **29.** -19 **31.** 5   **33.** -6
35. -40 **37.** 18   **39.** -19 **41.** 12   **43.** -63

Page 413   Exercises   1. $x + (-3) = 1$ **3.** $x + (-1) = -6$ **5.** $x + 2 = 4$ **7.** $x + 7 = -4$ **9.** 2
11. 9   **13.** -5 **15.** -8 **17.** -1 **19.** -4 **21.** 8   **23.** 4   **25.** -1 **27.** 11   **29.** -14 **31.** 1
33. 0   **35.** 16   **37.** 28   **39.** -26 **41.** -12 **43.** 40   **45.** -27 **47.** 15   **49.** -9 **51.** -1
53. 60

Page 414   The Slippery Cricket   1. 43 days; Methods will vary.

Pages 416–417   Chapter Review   1. positive   **3.** opposite   **5.** 3   **7.** 9   **9.** -6 **11.** -200
13. 81   **15.** 10   **17.** $>$ **19.** $<$ **21.** $<$ **23.** $>$ **25.** 14   **27.** -4 **29.** 25   **31.** -13 **33.** -4
35. 6   **37.** -4 **39.** -2 **41.** -1 **43.** 8   **45.** -9 **47.** 7   **49.** -17 **51.** -17 **53.** -5
55. -7 **57.** 8   **59.** 28   **61.** -32 **63.** 9   **65.** 9   **67.** -6 **69.** 29   **71.** $>$ **73.** $7 + 3 = 10$
75. $-3 + (-2) = -5$

Page 418   Chapter Test   1. -3 **3.** -9 **5.** -1 **7.** $<$ **9.** $<$ **11.** 11   **13.** -3 **15.** -3
17. -4 **19.** -4 **21.** -12 **23.** 8   **25.** -2 **27.** -12 **29.** 21   **31.** -7 **33.** -22 **35.** -38
37. 3   **39.** -7 **41.** -23 **43.** $3 + 4 = 7$ **45.** 15° colder

Cumulative Maintenance: Chapters 1–16 **1.** a **3.** c **5.** b **7.** a **9.** d **11.** b
13. a **15.** a **17.** a

CHAPTER 17   INTEGERS: MULTIPLICATION/DIVISION

Pages 422–423   **Exercises**   **1.** -16 **3.** -18 **5.** $-11; -22$ **7.** $-20; -25$ **9.** $-7; -14$
11. -6 **13.** -100 **15.** 0   **17.** -63 **19.** -99 **21.** -120 **23.** -156 **25.** 0   **27.** -84
29. -160 **31.** -90 **33.** -5000 **35.** -119 **37.** -581 **39.** -5000 **41.** -60 **43.** -18
45. -21 **47.** -24 **49.** -6

Pages 424–425   **Exercises**   **1.** 10   **3.** 6   **5.** 8; 16   **7.** 6; 12   **9.** 12; 24   **11.** 25   **13.** 12
15. 80   **17.** 42   **19.** 128   **21.** 119   **23.** 60   **25.** 42   **27.** 30   **29.** 128   **31.** 42   **33.** 48
35. 1258   **37.** 1701   **39.** 3200   **41.** 990   **43.** 154   **45.** 640   **47.** 2400   **49.** 13,736   **51.** -1785
53. 410   **55.** 777   **57.** 1264   **59.** $-8°$ **61.** $-3; -3$

Page 427   **Exercises**   **1.** $-9; -6$ **3.** $12; -11$ **5.** 12; 16   **7.** 5   **9.** 3   **11.** 9   **13.** 3   **15.** -9
17. -7 **19.** -3 **21.** -23 **23.** -20 **25.** 93   **27.** 4   **29.** 27   **31.** -6 **33.** -7 **35.** 43
37. -36 **39.** -5 **41.** -4 **43.** -48 **45.** -29 **47.** 3880   **49.** -39 **51.** $-25°C$ **53.** $-3°C$

Pages 428–429   **Exercises**   **1.** $\frac{5}{1}$ **3.** $-\frac{10}{1}$ **5.** $\frac{9}{4}$ **7.** $-\frac{13}{4}$ **9.** $-\frac{7}{1}$ **11.** $-\frac{10}{3}$ **13.** $\frac{17}{1}$ **15.** $\frac{0}{1}$
17. $-\frac{9}{5}$ **19.** $-\frac{6}{1}$ **21.** $\frac{51}{10}$ **23.** $-\frac{15}{2}$ **25.** $-\frac{1}{2}$ **27.** $-\frac{4}{1}$ **29.** $-\frac{17}{2}$ **31.** $-\frac{17}{2}, -\frac{11}{2}, -\frac{4}{1}, -\frac{1}{2}, 0, \frac{3}{2}$
33. $-\frac{1}{4}$ **35.** $\frac{1}{4}$ **37.** $-\frac{1}{2}$ **39.** $=$ **41.** $=$ **43.** $<$ **45.** $=$ **47.** $>$ **49.** $=$ **51.** $-5\frac{1}{5}; -2\frac{1}{5}; -\frac{1}{4};$
$\frac{1}{4}; 1; 3\frac{1}{3}$ **53.** $-1\frac{1}{2}; -\frac{3}{4}; -\frac{2}{3}; \frac{4}{5}; 1; 2$ **55.** $-1\frac{7}{8}; -1\frac{5}{6}; -\frac{11}{16}; 0; 1\frac{1}{2}; 2\frac{1}{4}$ **57.** $-3; -2\frac{7}{8}; -2\frac{1}{2}; -1\frac{1}{8}; -1; -\frac{1}{4}$
59. $3 = \frac{3}{1}$ **61.** $0 = \frac{0}{1}$

Page 430   **Mid–Chapter Review**   **1.** -12 **3.** -9 **5.** -168 **7.** -1000 **9.** 28   **11.** 6   **13.** 30
15. 1148   **17.** 9   **19.** -9 **21.** 8   **23.** 9   **25.** $\frac{7}{1}$ **27.** $-\frac{8}{1}$ **29.** $-\frac{21}{5}$ **31.** $-\frac{10}{1}$ **33.** $\frac{17}{4}$
35. $-\frac{17}{1}$ **37.** $\frac{11}{4}$ **39.** $-\frac{3}{4}$ **41.** $\frac{3}{2}$ **43.** $-2; -\frac{5}{4}; -\frac{3}{4}; \frac{1}{4}; \frac{3}{2}; \frac{11}{4}$ **45.** 4.5 L   **47.** $\frac{1}{2}$

Page 431   **Math and Saving Energy**   **1.** -4068 Btu's   **3.** -1548 Btu's   **5.** Brick, 8 inches thick

Page 433   **Exercises**   **1.** 2   **3.** -2 **5.** -10 **7.** 6   **9.** 2   **11.** 2   **13.** 1   **15.** -1 **17.** -6
19. -4 **21.** -8 **23.** -3 **25.** -9 **27.** -3 **29.** -5 **31.** 3   **33.** 16   **35.** 8   **37.** -11
39. 40   **41.** -5 **43.** -5 **45.** -19 **47.** 12   **49.** -28 **51.** 12   **53.** 101   **55.** -11 **57.** -13
59. -39 **61.** 0   **63.** -73

Page 434   **Exercises**   **1.** -5 **3.** 4   **5.** 3   **7.** -8 **9.** 6   **11.** -5 **13.** 100   **15.** -72 **17.** 0
19. -180 **21.** -600 **23.** 63   **25.** 48   **27.** -90

Page 435   **Exercises**   **1.** $3k + 2 + (-2) = 5 + (-2)$ **3.** $7n + 20 + (-20) = 13 + (-20)$
5. $\frac{r}{-4} - 13 + 13 = 23 + 13$ **7.** $\frac{n}{-9} + 14 + (-14) = 10 + (-14)$ **9.** 8   **11.** -3 **13.** 2   **15.** 9
17. 64   **19.** -2

Page 437   **Exercises**   **1.** $2r = 82$ **3.** $d - 40 = 98$ **5.** $\frac{p}{10} = 12$ **7.** $l - 3 = 12$ **9.** $\frac{c}{24} = 30$
11. $s - 45 = 50$ **13.** 83   **15.** 750   **17.** 7   **19.** 24   **21.** 192   **23.** 12

Page 438   **Equations and Patterns**   **1.** $y = 2x - 3$ **3.** $y = 3x + 7$ **5.** $y = 4x + 3$ **7.** $y = \frac{x}{2}$

11. -200 **13.** -136 **15.** 45 **17.** 35 **19.** 20 **21.** 18 **23.** -4 **25.** -32 **27.** -5
29. -52 **31.** $-\frac{12}{1}$ **33.** $-\frac{29}{4}$ **35.** $\frac{25}{6}$ **37.** $\frac{7}{4}$ **39.** 1 **41.** $-\frac{13}{4}$ **43.** $-3\frac{1}{4}$; $-2\frac{1}{2}$; 0; 1; $3\frac{1}{5}$; $6\frac{1}{2}$
45. $-4\frac{1}{2}$; -3; -1; 0; 2; $4\frac{1}{4}$ **47.** 6 **49.** 5 **51.** 9 **53.** -16 **55.** 42 **57.** 18 **59.** 40 **61.** -120
63. 3 **65.** 2 **67.** -2 **69.** 4 **71.** -6 **73.** -24 **75.** -21 **77.** $h - 1 = 4$ **79.** $\frac{b}{5} = 12$
81. $h - 3 = 4$; 7 ft **83.** $\frac{b}{3} = 30$; $90

Page 442 **Chapter Test** **1.** 18 **3.** -24 **5.** 84 **7.** -273 **9.** -336 **11.** 8 **13.** -5
15. -49 **17.** -17 **19.** $-\frac{3}{1}$ **21.** $\frac{11}{2}$ **23.** $\frac{14}{1}$ **25.** $-\frac{1}{2}$ **27.** $\frac{3}{4}$ **29.** $-\frac{3}{2}$ **31.** -4 **33.** 12
35. 24 **37.** -65 **39.** 5 **41.** -6 **43.** $5(-25)$; $-$125 **45.** $8w = 40.00$; $5.00

Pages 443–444 **Cumulative Maintenance: Chapters 1–17** **1.** d **3.** b **5.** c **7.** b **9.** c **11.** c
13. a **15.** c **17.** d **19.** b **21.** b

APPENDIX A **SKILLS BANK**

Page 446 **Estimating Large Numbers** **1.** [7000]; 7089 **3.** [300]; 323 **5.** [50,000]; 48,122
7. 10,000,000 **9.** 38,000,000

Page 446 **Order of Operations** **1.** 5 **3.** 9 **5.** 16 **7.** 22 **9.** 21 **11.** $40 \div 5 - 3$
13. $8 \times (4 - 2) + 6 = 22$ **15.** $9 \div (1 + 2) \times 5 = 15$

Page 446 **Exponents** **1.** 625 **3.** 16 **5.** 64 **7.** 160,000 **9.** 500 **11.** 1

Page 447 **Decimals: ADDITION/SUBTRACTION** **1.** 32.405 **3.** 78.724 **5.** 100.445 **7.** 191.952
9. 1.77 **11.** 43.3 **13.** 0.57 **15.** 0.895 **17.** 951.73 **19.** 37.62

Page 447 **Multiplying Whole Numbers** **1.** 6059 **3.** 4316 **5.** 5684 **7.** 35,607 **9.** 38,570
11. 281,547 **13.** 225,552 **15.** 87,084 **17.** 39,744 **19.** 665

Page 447 **Decimals: MULTIPLICATION** **1.** 54.63 **3.** 2488.5 **5.** 2.106 **7.** $74.79 **9.** 94.4
11. 65.2645 **13.** 29.986 **15.** 27.608

Page 448 **Division: WHOLE NUMBERS** **1.** 67 r16 **3.** 11 **5.** 112 r37 **7.** 26 r45 **9.** 29 **11.** 215 r10
13. 34 r18 **15.** 121 r36

Page 448 **Zeros in the Quotient** **1.** 320 r9 **3.** 403 r5 **5.** 530 r2 **7.** 305 r5 **9.** 810 r3 **11.** 207
13. 40 r7 **15.** 150 r23

Page 448 **Dividing a Decimal by a Whole Number** **1.** 1.30 **3.** 0.12 **5.** 1.07 **7.** 0.25 **9.** 5.15
11. 0.5 **13.** 1.3 **15.** 0.8

Page 448 **Dividing by a Decimal** **1.** 49.7 **3.** 1.8 **5.** 0.5 **7.** 0.1 **9.** 0.39 **11.** 7.63

Page 449 **Bar Graphs and Applications** **1.** Cheetah **3.** 45 miles per hour **5.** Rabbit, Zebra, Elk, Gazelle,
Antelope, Cheetah

Page 449 **Line Graphs and Applications** **1.** 1973, 1974, 1975 **3.** 1970 **5.** Increase

Page 450 **The Mean and the Mode** **1.** Mean: 5; Range: 5; Mode: 6 **3.** Mean: 10.625; Range: 11;
Mode: 11 **5.** Mean: 5.625; Range: 9; Mode: 4 **7.** Mean: 133; Range: 37; No mode **9.** 11.2 seconds
11. 11.1 seconds

Page 450   **The Median**   **1.** 90   **3.** 6.5   **5.** 63   **7.** 8   **9. a.** 89 **b.** 89 **c.** Yes; 89 is still the middle score.

Page 451   **Addition and Subtraction: LIKE FRACTIONS**   **1.** $\frac{6}{7}$ **3.** $\frac{2}{5}$ **5.** $\frac{3}{7}$ **7.** $\frac{2}{3}$ **9.** $5\frac{11}{12}$ **11.** $8\frac{4}{5}$ **13.** $7\frac{1}{4}$ **15.** $5\frac{1}{2}$ **17.** $1\frac{1}{3}$ **19.** $\frac{5}{7}$ **21.** $9\frac{2}{5}$ **23.** $23\frac{2}{3}$

Page 451   **Addition and Subtraction: UNLIKE FRACTIONS**   **1.** $1\frac{5}{24}$ **3.** $\frac{7}{40}$ **5.** $\frac{3}{22}$ **7.** $\frac{15}{16}$ **9.** $\frac{47}{72}$ **11.** $5\frac{5}{12}$ **13.** 0   **15.** $4\frac{1}{8}$ **17.** $8\frac{9}{10}$ **19.** $13\frac{5}{12}$ **21.** $\frac{11}{12}$ **23.** $\frac{5}{36}$ **25.** $\frac{3}{8}$ **27.** $1\frac{1}{12}$

Page 451   **Subtraction: MIXED NUMBERS**   **1.** $4\frac{2}{3}$ **3.** $1\frac{5}{12}$ **5.** $3\frac{1}{6}$ **7.** $3\frac{7}{10}$ **9.** $2\frac{5}{8}$ **11.** $3\frac{2}{3}$ **13.** $\frac{1}{6}$ **15.** $3\frac{5}{6}$ **17.** $2\frac{2}{3}$ **19.** $\frac{1}{3}$ **21.** $\frac{5}{8}$ **23.** $\frac{5}{12}$

Page 452   **Multiplication: FRACTIONS**   **1.** $\frac{1}{40}$ **3.** $\frac{2}{9}$ **5.** $\frac{3}{16}$ **7.** $\frac{21}{32}$ **9.** $\frac{4}{9}$ **11.** $\frac{4}{27}$ **13.** $\frac{1}{4}$ **15.** $\frac{2}{15}$ **17.** $\frac{1}{3}$ **19.** $\frac{11}{24}$ **21.** $\frac{5}{14}$ **23.** $\frac{11}{27}$

Page 452   **Multiplication: MIXED NUMBERS**   **1.** [2]; $1\frac{32}{45}$ **3.** [4]; $3\frac{3}{11}$ **5.** [35]; $34\frac{2}{3}$ **7.** [36]; $34\frac{3}{8}$ **9.** [35]; 35   **11.** [2]; $1\frac{8}{9}$ **13.** [6]; $6\frac{13}{20}$ **15.** [8]; $6\frac{3}{5}$ **17.** [18]; $15\frac{1}{3}$ **19.** [6]; $5\frac{1}{10}$

Page 452   **Division: FRACTIONS**   **1.** $1\frac{1}{4}$ **3.** $1\frac{7}{8}$ **5.** $\frac{27}{50}$ **7.** 1   **9.** $\frac{1}{45}$ **11.** $\frac{2}{5}$ **13.** $\frac{7}{15}$ **15.** $\frac{11}{16}$ **17.** $1\frac{1}{3}$ **19.** $\frac{5}{18}$

Page 452   **Division: MIXED NUMBERS**   **1.** $\frac{1}{11}$ **3.** $10\frac{14}{25}$ **5.** $4\frac{14}{15}$ **7.** $4\frac{1}{2}$ **9.** $\frac{65}{126}$ **11.** $\frac{3}{8}$ **13.** $1\frac{8}{13}$ **15.** $\frac{3}{5}$ **17.** $2\frac{5}{6}$ **19.** $\frac{1}{3}$

Page 453   **Perimeter**   **1.** 19 in   **3.** 4 yd   **5.** 150 m   **7.** 6 m

Page 453   **Area: RECTANGLES AND SQUARES**   **1.** 12 km²   **3.** 7 m   **5.** 12 in   **7.** $6\frac{3}{10}$ m

Page 453   **Area: PARALLELOGRAMS/TRIANGLES**   **1.** 56 cm²   **3.** 28.8 m²   **5.** 7 ft   **7.** 52.38 m²

Page 453   **Area: TRAPEZOIDS**   **1.** 66 in²   **3.** 40 yd²   **5.** 56.96 m²   **7.** 37.5 cm²

Page 454   **Solving Equations: ADDITION**   **1.** $n = 11$ **3.** $n = 21$ **5.** $n = 38$ **7.** $n = 6.2$

Page 454   **Solving Equations: SUBTRACTION**   **1.** $n = 22$ **3.** $n = 61$ **5.** $n = 87$ **7.** $n = 10.3$

Page 454   **Solving Equations: MULTIPLICATION/DIVISION**   **1.** $n = 49$ **3.** $n = 8$ **5.** $n = 28$ **7.** $n = 3$

Page 454   **More on Solving Equations**   **1.** $n = 17$ **3.** $n = 6$ **5.** $n = 14$ **7.** $n = 66$ **9.** $n = 70$ **11.** $n = 9$

Page 454   **Proportion**   **1.** $n = 4$ **3.** $n = 33$ **5.** $n = 5$ **7.** $n = 21$ **9.** $n = 70$

Page 455   **Percents/Decimals/Fractions**   **1.** 40%   **3.** 75%   **5.** 45%   **7.** 70%   **9.** 50%   **11.** 5%   **13.** $17\frac{1}{2}$%   **15.** $55\frac{5}{9}$%   **17.** $93\frac{3}{4}$%   **19.** $66\frac{2}{3}$%   **21.** $83\frac{1}{3}$%   **23.** $37\frac{1}{2}$%

Page 455   **Finding a Percent of a Number**   **1.** 30   **3.** 72   **5.** 6.5   **7.** 18   **9.** 91   **11.** 30

Page 455   **Discount**   **1.** $80.75   **3.** $9.20

Page 455   **Savings Account/Interest**   **1.** $36.00   **3.** $36.00   **5.** $20.00

Page 456   **Finding the Percent**   **1.** 10%   **3.** $66\frac{2}{3}$%   **5.** 125%   **7.** 80%   **9.** 37.5%   **11.** 200%   **13.** 31.25%   **15.** $16\frac{2}{3}$%   **17.** $8\frac{1}{3}$%   **19.** 12.5%

508

Page 456   **Finding a Number Given the Percent**   **1.** 52   **3.** 80   **5.** 160   **7.** 35   **9.** 160   **11.** 70
13. 1700   **15.** 380

Page 457   **Surface Area**   **1.** 5200 in^2 **3.** 13.6 m^2 **5.** 50.4 m^2

Page 457   **Volume: RECTANGULAR PRISMS**   **1.** 21 m^3 **3.** 5760 in^3 **5.** 9765 cm^3 **7.** 462 m^3

Page 458   **Circumference and Applications**   **1.** 440 yd   **3.** 160 cm

Page 458   **Circle Graphs**   **1.**

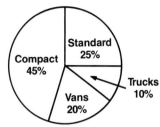

Vehicles Sold

Page 458   **Area and Applications**   **1.** 113 in^2 **3.** 707 cm^2

Page 458   **Volume and Applications: CYLINDER**   **1.** 38 ft^3 **3.** 385 cm^3

Page 459   **The Probability Ratio**   **1.** $\frac{1}{2}$, or 50%   **3.** $\frac{1}{4}$, or 25%   **5.** $\frac{1}{2}$, or 50%

Page 459   **Compound Events: AND**   **1.** $\frac{1}{15}$ **3.** $\frac{1}{5}$ **5.** $\frac{4}{15}$ **7.** $\frac{2}{15}$

Page 459   **Compound Events: OR**   **1.** $\frac{1}{5}$ **3.** $\frac{6}{25}$

Page 460   **Using a Table of Squares and Square Roots**   **1.** 961   **3.** 2809   **5.** 9025   **7.** 15,876
9. 3481   **11.** 5041   **13.** 6.083   **15.** 69   **17.** 24   **19.** 81   **21.** 34   **23.** 78   **25.** 2.828
27. 10.100   **29.** 4.359

Page 460   **The Rule of Pythagoras**   **1.** Yes   **3.** No   **5.** No

Page 460   **Tangent Ratios and Applications**   **1.** 0.268   **3.** 0.839   **5.** 0.700   **7.** 45   **9.** 60   **11.** 12 m

Page 461   **Adding Integers: LIKE SIGNS**   **1.** 6   **3.** 12   **5.** 14   **7.** 12   **9.** -4 **11.** -9 **13.** -6
15. -8 **17.** -5 **19.** -11

Page 461   **Adding Integers: UNLIKE SIGNS**   **1.** -6 **3.** 2   **5.** 1   **7.** 4   **9.** 2   **11.** 1   **13.** 0
15. -7 **17.** 9   **19.** 4

Page 461   **Subtracting Integers**   **1.** -3 **3.** 1   **5.** -8 **7.** 5   **9.** -13 **11.** -8 **13.** -3
15. 2   **17.** 2   **19.** 6   **21.** 5   **23.** 20   **25.** -11 **27.** 9   **29.** -7 **31.** -4 **33.** 11   **35.** -3
37. 0   **39.** -10 **41.** 4   **43.** -5 **45.** 3   **47.** -11

Page 462   **Multiplication: PATTERNS AND UNLIKE SIGNS**   **1.** -15 **3.** -76 **5.** -60 **7.** -800
9. -70 **11.** -108 **13.** -30

Page 462   **Multiplication: PATTERNS AND LIKE SIGNS**   **1.** 62   **3.** 1792   **5.** 144   **7.** 322   **9.** 800
11. 6800   **13.** 210

Page 463   **Dividing Integers**   **1.** -5 **3.** -12 **5.** -12 **7.** -16 **9.** 15   **11.** -7 **13.** 2°C

Page 465   Try These   1. 1,654,566   **3.** 2,305,672

Page 466   Try These   1. $17,500   **3.** 1040   **5.** 3205

Page 467   Try These   1. $13.82; $6.18

Page 468   Try These   1. 270   **3.** 51   **5.** 50,625   **7.** 15,625

Page 469   Try These   1. 149.85   **3.** 289.15   **5.** 30.75   **7.** $644,129.77

Page 470   Try These   1. 8100; 7917   **3.** 900; 924   **5.** 5600; 5712   **7.** 36,000; 37,294   **9.** 91,000; 83,094   **11.** 140,000; 155,952   **13.** 640; 655.7   **15.** 270; 302.6   **17.** 240; 220.4   **19.** 5400; 5760.3   **21.** 7700; 7544.2   **23.** 21,000; 21,751.2

Page 471   Try These   1. 127,715,082   **3.** 256,546,917   **5.** 215,123,197   **7.** 1,250,070,578   **9.** 1,136,801,148   **11.** 2,458,669,837   **13.** 1,234,554,321   **15.** 9,096,218,057

Page 472   Try These   1. 1682   **3.** 924.53   **5.** 578,551.82   **7.** 382,233.917   **9.** 500,904.2   **11.** 0.0324   **13.** 0.0002835   **15.** 2.02677636   **17.** 0.0000000663

Page 473   Try These   1. 6040   **3.** 6 r6

Page 474   Try These   1. 0.3875   **3.** 0.70625   **5.** 0.575   **7.** 0.785   **9.** 4.614   **11.** 0.7890625   **13.** 0.921875   **15.** $0.9\overline{3}$ **17.** $0.\overline{81}$ **19.** $0.6\overline{3}$ **21.** $0.\overline{7}$ **23.** 0.425   **25.** $0.\overline{36}$ **27.** $0.\overline{36}$ **29.** 5.3125

Page 475   Try These   1. Yes   **3.** No   **5.** No   **7.** 60   **9.** 184   **11.** 138   **13.** 576   **15.** 115.2   **17.** 17   **19.** 3.76

Page 476   Try These   1. 238   **3.** 12.87

Page 477   Try These   1. $5,634.13   **3.** $4,042.05   **5.** $5,224.36   **7.** $10,982.06

Page 478   Try These   1. 3375   **3.** 262,144   **5.** 1,481,544   **7.** 41,781,923   **9.** 69,343,957   **11.** 34.328125   **13.** 69.7225   **15.** 941.192   **17.** 7.0710678   **19.** 8.8881944   **21.** 10.630146   **23.** 17.804494   **25.** 5.5677644   **27.** 31.622777   **29.** 7.1

Page 479   Try These   1. −21   **3.** −89   **5.** 110   **7.** −24   **9.** 0   **11.** 0   **13.** 12   **15.** 17   **17.** −136   **19.** 134   **21.** 0   **23.** 58   **25.** 129   **27.** −72   **29.** 160   **31.** 4600   **33.** 441   **35.** 1150   **37.** −1596   **39.** −848   **41.** −15   **43.** −87   **45.** 34   **47.** −63   **49.** 14   **51.** −34

Key: (t) top, (c) center, (b) bottom, (l) left, (r) right.

COVER PHOTOS: (tl, cl, c, cr, bl, bc, br), HBJ Photo; (tc) Michael Melford/The Image Bank; (tr), Jeffrey Sylvester/ FPG.